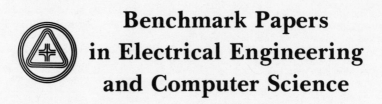

Benchmark Papers
in Electrical Engineering
and Computer Science

Series Editor: John B. Thomas
Princeton University

Published Volumes

Additional Volumes in Preparation

**Benchmark Papers
in Electrical Engineering
and Computer Science**

—— A *BENCHMARK*® Books Series ——

ENVIRONMENTAL
MODELING:
Analysis and Management

Edited by
DOUGLAS DAETZ and RICHARD H. PANTELL
Stanford University

Dowden, Hutchinson & Ross, Inc.
Stroudsburg, Pennsylvania

301.31
D13e
92273
Feb. 1975

Library of Congress Cataloging in Publication Data

Daetz, Douglas, 1941-   comp.
   Environmental modeling.

   (Benchmark papers in electrical engineering and
computer science, v. 6)
   Bibliography: p.
   1. Environmental engineering--Mathematical models--
Addresses, essays, lectures. 2. Electronic data
processing--Environmental engineering. I. Pantell,
Richard H., 1927-   joint comp. II. Title.
TD153.D33   301.31   73-22191
ISBN 0-87933-082-1

Acknowledgments
and Permissions

ACKNOWLEDGMENTS
IX International Grassland Congress—*Proceedings of the XI International Grassland Congress*
   "A Systems Approach to Grasslands"

National Academy of Sciences—*Energy Resources*
   "Introduction"

United States Geological Survey—*Circular 554*
   "Hydrology for Urban Land Planning—A Guidebook on the Hydrologic Effects of Urban Land Use"

PERMISSIONS
The following papers have been reprinted with the permission of the authors and copyright holders.

Academic Press, Inc.—*The Ecosystem Concept in Natural Resource Management*
   "A Study of an Ecosystem: The Arctic Tundra"

American Association for the Advancement of Science—*Science*
   "Systems Studies of DDT Transport"
   "Forest Fuel Accumulation—A Growing Problem"
   "The Nation's Rivers"

American Economic Association—*American Economic Review*
   "Production, Consumption, and Externalities"

American Geophysical Union—*Water Resources Research*
   "Watershed Management: A Systems Approach"

American Institute of Physics—*Physics Today*
   "Physics Looks at Waste Management"

Baywood Publishing Company, Inc.—*Journal of Environmental Systems*
   "A Systems Approach to an Analysis of the Terrestrial Nitrogen Cycle"
   "Cost-Effectiveness Analysis of Waste Management Systems"
   "Law, Operations Research, and the Environment"

Duke University Press for the Ecological Society of America—*Ecology*
   "Energy Storage and the Balance of Producers and Decomposers in Ecological Systems"

Institute of Electrical and Electronics Engineers, Inc.
   IEEE Transactions on Systems, Man, and Cybernetics
      "Principles of Ecosystem Design and Management"
      "Engineering for Ecological, Sociological, and Economic Compatibility"
   IEEE Transactions on Systems Science and Cybernetics
      "A Water Pricing and Production Model"

International Business Machines Corporation, Inc.—*IBM Journal of Research Development*
   "Air Quality Diffusion Model; Application to New York City"
   "Interactive Computer-based Game for Decision-making in Ecology"

Microforms International Marketing Corporation—*Socio-Economic Planning Sciences*
  "Computers in Urban Air Pollution Control Systems"
  "Determination of Optimal Air Pollution Control Strategies"

Operations Research Society of America—*Operations Research*
  "Scheduling Bioproduction Harvest"

President and Fellows of Harvard College—*Review of Economics and Statistics*
  "Environmental Repercussions and the Economic Structure: An Input–Output Approach"

Water Pollution Control Federation—*Journal of the Water Pollution Control Federation*
  "Systems Analysis for Optimal Water Quality Management"
  "Marine Waste Disposal—A Comprehensive Environmental Approach to Planning"

Donovan Young
  "Optimal Pollution Regulation—A Data-based Study"

Series Editor's Preface

This Benchmark Series in Electrical Engineering and Computer Science is aimed at sifting, organizing, and making readily accessible to the reader the vast literature that has accumulated. Although the series is not intended as a complete substitute for a study of this literature, it will serve at least three major critical purposes. In the first place, it provides a practical point of entry into a given area of research. Each volume offers an expert's selection of the critical papers on a given topic as well as his views on its structure, development, and present status. In the second place, the series provides a convenient and time-saving means for study in areas related to but not contiguous with one's principal interests. Last, but by no means least, the series allows the collection, in a particularly compact and convenient form, of the major works on which present research activities and interests are based.

Each volume in the series has been collected, organized, and edited by an authority in the area to which it pertains. In order to present a unified view of the area, the volume editor has prepared an introduction to the subject, has included his comments on each article, and has provided a subject index to facilitate access to the papers.

We believe that this series will provide a manageable working library of the most important technical articles in electrical engineering and computer science. We hope that it will be equally valuable to students, teachers, and researchers.

This volume, *Environmental Modeling: Analysis and Management,* has been edited by Douglas Daetz and Richard H. Pantell of Stanford University. It contains twenty-six papers chosen to present a unified view of the techniques developed for the analysis and management of environmental systems. We are particularly pleased that this volume illustrates the increasing interaction between systems analysts and those concerned with societal problems of an economic and ecological nature. It should prove valuable to a wide audience both in the social sciences and in the engineering disciplines.

John E. Thomas

Preface

The purpose of this book is to illustrate various techniques that have been or are being applied to the analysis and management of environmental systems. We do not restrict "environmental systems" solely to physical components (i.e., land, water, and air); we include economic and social considerations as well.

A great deal of concern has been expressed in the past few years regarding possible deterioration of some aspects of our society. It is suggested that we are running out of critical minerals and fossil fuels, that our air and water systems are being polluted, that food chains are being disrupted, that cities are congested, and so on. Responses to these problems range from an expressed confidence that technology will overcome every difficulty to a call for the cessation of economic and population growth. The course of action that is selected is usually determined by the magnitude of the pressures that are exerted. When an environmental problem (e.g., the smog over a metropolitan area) grows in annoyance and visibility, public protest may become sufficiently vigorous to persuade officials to take some action. But what action is to be taken? In many environmental systems it is difficult to determine cause and effect relationships. The papers included in this book all have one element in common: an attempt to identify and/or utilize cause and effect dependencies to improve some environmental situation.

The articles vary in terms of the specialization and mathematical background required in order to understand the material. Many papers are introductory in nature and, therefore, will be useful to the individual interested in a general statement and basic understanding of the problem. These papers may be of particular importance to a person who wishes to broaden his perspective on environmental problems, or who may, perhaps, want to redirect his career.

Some papers, such as those that illustrate a particular optimization technique, a method for handling uncertainty, or a procedure for selecting between alternative policy options, are intended primarily for the working professional. Almost all the papers concern the solution of some practical problem, as opposed to a strictly theoretical analysis.

The topics discussed include air and water pollution, solid waste management, energy balance, and ecosystems. Some papers focus on a problem that is contained primarily within a single discipline, whereas others are interdisciplinary. Discussions of economic strategies for controlling effluence and of legal constraints imposed on environmental decision making fall into the latter category.

Some systems analysts believe that the function of an analysis should be to answer specific questions related to a well-defined problem. The overview, the interdisciplinary perspective, should be derived from the judgments of the planner or appropriate public representative. Alternatively, others believe that environmental problems are inherently interdisciplinary and that a narrow focus may lead to an undesirable decision. Both types of approaches are presented in the text; the reader may draw his own conclusions.

This book is divided into six major parts. Part I contains papers that are concerned with ecosystem analysis, as opposed to ecosystem management, which is discussed in Part II. A paper that describes energy flows and estimates of the amount of different types of energy sources and uses is included in the analysis section. Specific chemical paths are followed in two papers, one dealing with nitrogen and the other with DDT. One of the remaining papers is a study of an isolated region, the Arctic tundra, and the other is an attempt to define general energy and mass flows in an ecosystem.

The primary purpose of ecosystem analysis is to define cause and effect relationships so that ecosystems may be managed toward some particular goals. In Part II, papers that consider methods for achieving objectives in ecological systems are presented. Grassland management, forest fire control, and watershed design and management are discussed.

Parts III and IV deal with specific environmental problem areas, air quality and water management. Articles on air pollution discuss meteorological models of the atmosphere and strategies for achieving satisfactory air quality. Water management papers are concerned with the identification of the impact of man's activities on our water systems and methods for ameliorating the problems that arise.

Waste Management is the subject of the papers in Part V. Man's living habits have resulted in a clustering of people in densely populated regions, with the attendant difficulties of waste disposal. The papers in this section suggest various approaches to improve effluence-handling procedures.

Each of the first five parts of the book deals with some specific aspect of our environment. Social, economic, legal, and natural factors, however, cannot be managed independently. We have, therefore, included a final section that provides a more general overview than the previous material. The first paper describes legal factors in environmental management; the second attempts to link ecological, sociological, and economic factors; and the final three papers emphasize the economic ties to the environment.

The matrix array presented on the facing page indicates some of the salient features of each article: its level of complexity, its consideration of analysis or management, its subject matter, the academic disciplines represented in it, the analysis tools employed, its applicability to a specific problem, and its mathematical level. Almost every paper has considerable nonmathematical (qualitative) content that the reader should find valuable regardless of his background in quantitative methods. This matrix may be used to obtain, in summary fashion, an idea of which articles will be of greatest interest to the reader.

The editors wish to thank Janet Dinkey for her secretarial contributions to this volume and Iris Riley for her assistance in preparing the subject index.

Paper Number	Audience	Orientation	Appreciably Mathematical	Specific Application	Energy	Ecosystems	Natural Cycles	Systems Analysis	Air Quality	Water Management	Waste Management	ENGINEERING (including Operations Research)	APPLIED MATHEMATICS (Mathematical Modeling)	PHYSICS	CHEMISTRY	BIOLOGY/ECOLOGY/BOTANY/ZOOLOGY	FORESTRY/RANGE MANAGEMENT/SOIL SCIENCE	STATISTICS (including Probability)	LAW	ECONOMICS	Conservation of Energy	Conservation of Mass/Material Balance	Flow Diagrams	Computer Simulation	Time Series	Cost–Benefit Analysis	Rate Equations	Graph Theory	Mathematical Programming	Game Theory	Scheduling Techniques (PERT, CPM)	Input–Output Analysis	Control Theory/Feedback Analysis
1	G	A			X	X								X	X	X					X		X										
2	G	A		X		X		X						X	X	X					X	X		X								X
3	M	A					X	X				X	X	X	X	X	X	X			X		X									
4	M	A	X		X	X	X							X	X		X	X					X			X						
5	M	A	X			X						X	X	X	X						X		X									
6	M/R	M	X		X	X		X				X	X		X				X	X	X	X	X				X					
7	G	M	X															X															
8	M	M	X	X	X	X		X					X	X		X			X	X	X	X			X							
9	M/R	M	X			X		X					X				X											X		X		
10	M/R	M	X	X		X		X				X				X											X		X		
11	G	M					X	X				X										X	X									X
12	M	A	X	X				X				X				X						X										
13	R	A/M	X					X				X	X							X								X			X
14	G	M	X						X	X		X																				
15	G	A/M			X	X		X				X			X		X						V X								
16	M	M	X	X			X		X		X	X				X		X			X		X		X	X	X			
17	M	M	X	X			X		X		X	X				X		X	X	X										
18	G	M		X				X		X		X							X	X										
19	G	M	X	X			X		X	X	X											X									
20	M	A/M	X				X		X	X	X									X					X	X				
21	M	A/M		X		X		X	X	X	X			X		X								X						
22	G	A/M				X			X	X			X				X	X			X		X					X	X	X	X
23	M	A/M				X		X					X		X	X	X	X					X						
24	M	M	X				X		X		X									X		X							
25	M	A/M	X			X		X					X							X									X
26	M/R	A	X				X	X	X	X				X				X		X	X	X						X

Contents

I. ECOSYSTEM ANALYSIS

II. MANAGEMENT OF ECOLOGICAL SYSTEMS

III. AIR QUALITY

Contents by Author

Introduction

In recent years systems techniques have been applied increasingly to environmental problems. According to E. S. Quade,* systems analysis involves

> a systematic approach to helping a decision maker choose a course of action by investigating his full problem, searching out objectives and alternatives, and comparing them in the light of their consequences, using an appropriate framework—insofar as possible analytic—to bring expert judgment and intuition to bear on the problem.

Some advantages to this approach, as opposed to a primary reliance upon intuitive or visceral responses, are that a study of cause-and-effect relationships may lead to better understanding of the problem, and that an explicit indication of the method for reaching a decision allows for modification, updating, and improvement. However, it is generally less expensive to rely upon the intuition and judgment of a decision maker; and it may be better, in the sense that it is often difficult, if not impossible, to put a complex strategy situation in model form. We have not yet designed a computer that can beat champion chess players. Perhaps the role of the systems analyst is best confined to the contribution of information to the decision process rather than to involvement in the normative aspects of the problem.

Difficulties Associated with Systems Analysis

To those who had envisioned environmental systems analysis as a panacea for societal problems, the results have been meager and disappointing. The reasons for this, which are numerous, must be studied if we are to cure the present deficiencies.

One difficulty has been that systems analysts often define their problem in a sufficiently simple manner so that a known programming procedure can be applied to provide a solution, but in so doing they may have divorced their analysis from reality by neglecting to consider political, legal, economic, and social constraints.

---

*E. S. Quade and W. I. Boucher, eds., *Systems Analysis and Policy Planning: Applications in Defense,* American Elsevier, New York, 1968.

1

Another difficulty is that data are frequently inaccurate or unavailable, and information on such matters as future technological development and population growth does not, of course, exist. The existence of uncertainty permits the analyst to speculate and by so doing to reach completely subjective conclusions.

Because their vocabularies and methods are different, communication problems exist between the systems analyst and the decision maker. The analyst is often slow to reach a conclusion, even with many computer memory banks of stored information at his disposal; a decision maker will often arrive at a decision rapidly and with minimal data.

The assumptions in a study may be difficult to justify and difficult to prove. A regression analysis can provide an accurate rendition of the past, but it may have little relevance to a future situation. The time constants for many environmental systems are so long that it is almost hopeless to check a hypothesis, even if the relevant variables could be manipulated and measured.

In the past, analysts have tended to oversell their product by promising solutions to all problems. Some decision makers have been eager to grasp at these promises and have been disappointed when the results fell short of expectations. It has not been uncommon for those who perform the analysis to operate in isolation from those who want the information, with the result that very little understanding of cause-and-effect relationships was communicated.

Decisions involve both subjective and objective information. Therefore, one person's perceived benefit may be another person's perceived cost, and no amount of additional information can alter this conflict. In such a case, the systems analyst may not serve a particularly useful function.

What does all this mean in terms of the future of environmental systems analysis? After the initial rapid development of the field, the pace has slackened because of realization of the aforementioned difficulties. Attempts are being made to ameliorate as many of these shortcomings as possible. Models of the environment are being checked more carefully, analysts are working in closer cooperation with potential users to define a realistic system and to improve communication, additional data are being collected, and various methods for dealing with environmental uncertainties are being explored.

Perhaps some of the skepticism concerning systems modeling will diminish as the analyst improves his procedures and becomes more realistic in evaluation of his capabilities. We have attempted, in gathering the papers in this volume, to seek out those that consider the pragmatic aspects of environmental analysis and management.

Procedures for Analysis

There is no single best method for performing an environmental systems analysis and management study, but most comprehensive approaches have a number of features in common. The first step is identification and formulation of the problem, the second is modeling or analysis, and the third is evaluation of management alternatives.

Formulation includes specification of the goals of management and criteria for evaluation of the extent to which these goals are met. Variables must be identified as *exogenous* (input variables over which management has no control), *endogenous* (variables that change as the system is altered), or *controllable* (variables that may be altered to influence performance). Alternatives are enumerated, and each alternative is analyzed to select an optimum course of action.

The purpose of the modeling or analysis step is to identify relationships among variables. Relationships may be based upon conservation conditions, accounting studies, or intuitive judgments; derived from regression analysis; obtained from an experiment; or constructed from the opinions of experts. In general, a model is developed from a combination of approaches. Whenever possible, these relationships should be checked against available data, assuming, of course, that the data used for checking are independent of the data used for forming the model.

Evaluation of management alternatives is often a very difficult part of the systems study. In general, the evaluation criteria consist of two or more incommensurable variables (e.g., costs and environmental quality), so a vector description of each alternative is required. However, the designation of one alternative as "best" is based on a scalar description, and it is therefore necessary to find some method of collapsing a vector into a scalar. This step is always necessary, whether the decision is based on intuitive judgment or on a complex computer simulation. One purpose of systems analysis is to specify the procedure used to obtain the scalar so that the process is open for review and improvement.

A scalar representation can be obtained by several methods. One approach is to specify constraints on all variables but one and to optimize that variable. In terms of cost and environmental quality, this might be expressed in terms of maximum allowable pollutant concentrations with the selection of the least-cost alternative, or as a maximum-cost specification with the selection of the lowest-pollutant alternative.

Another approach is to identify an equivalence relationship among variables. What cost increase would one tolerate to accomplish a given pollutant reduction? A linear equivalence is often used, in which an objective function is defined as a linear, weighted summation of variables. In this case the preferred alternative is the one that optimizes the objective function.

The Challenge

The increasing interest in environmental systems analysis is a response to society's recognition that some aspects of our interaction with the environment are harmful. More careful planning is necessary; we can no longer consider our air, water, land, and ecosystems as inexhaustible resources. The role of systems analysis has yet to be defined, and it will depend upon the amount of creative thought that is applied to this problem. New ideas are needed with regard to ways to resolve differences in objectives among conflicting groups, to improve communication with decision makers, to reduce uncertainties, to check a model, and so on. The papers included in this volume should stimulate you to look for new answers to these perplexing problems.

I
Ecosystems Analysis

Editors' Comments on Papers 1 Through 5

1 **Hubbert:** Introduction to *Energy Resources*
   Publ. 1000–D, National Research Council, Washington, D.C., 1962, pp. 1–8

2 **Schultz:** A Study of an Ecosystem: The Arctic Tundra
   The Ecosystem Concept in Natural Resource Management, G. M. Van Dyne, ed., Academic Press,
   Inc., New York, 1969, pp. 77–93

3 **Endelman, Northup, Keeney, Boyle, and Hughes:** A Systems Approach to an Analysis of
   the Terrestrial Nitrogen Cycle
   J. Environ. Sys., 2, 3–19 (Mar. 1972)

4 **Olson:** Energy Storage and the Balance of Producers and Decomposers in Ecological
   Systems
   Ecology, **44,** 322–331 (Spring 1963)

5 **Harrison, Loucks, Mitchell, Parkhurst, Tracy, Watts, and Yannacone:** Systems Studies of
   DDT Transport
   Science, **170,** 503–508 (Oct. 30, 1970)

The five papers in Part I are intended to provide background for environmental, ecological, and energetic ideas and systems analysis concepts. Part I involves broad ecological systems rather than such specific environmental topics as air quality, water management, and waste management, which are discussed in Parts III, IV, and V. Papers in this part are oriented toward the *analysis* of environmental systems, whereas Part II contains papers that deal with the *management* of ecological systems. Although no paper by Kenneth E. F. Watt, an entomologist-turned-systems-analyst at the Davis campus of the University of California, is included in Part I or II, his books *Systems Analysis in Ecology* and *Ecology and Resource Management* should be considered useful resources (see the Bibliography).

The opening paper, by M. King Hubbert, provides a very readable introduction to the earth's energy situation, a discussion of the energy available from each major energy source, and a figure that shows major energy flows and estimated magnitudes. The role of solar radiation as the dominant energy supplier for a "world-girdling" heat engine is described.

The U.S. population growth rate has dropped significantly in the eleven years since this paper was published. Including immigration, the doubling time for the U.S. population has increased from about 40 years (the figure mentioned in the paper) to about 80 years (1972 population growth rate). Another figure that should be updated is the installed generating capacity of electric utilities in the United States. The current figure is approximately 40×10^{10} W (computed on the basis of a 7 percent annual increase) compared to the 15.7×10^{10} W cited in the paper. Correspondingly, the ratio of solar radiation power to U.S. electric utilities power has shrunk from 1 million:1 to 400,000:1.

Arnold M. Schultz, of the School of Forestry at the University of California, Berkeley, has written an excellent, captivating piece that introduces both the systems approach and the general topic of ecological relationships. A system is defined in terms of elements, states, and relationships. Two main considerations in the modeling of systems are presented: *homomorphism* (in connection with the level of model detail, or degree of subdivision) and *isomorphism* (with regard to the type of relationships defined among system elements). Schultz makes the point that "the observer is always a part of the system." He also describes the use of inputs by an observer to stress a system in ways that may validate or invalidate hypotheses stemming from initial observations. The notion of a homeostatic system, one with self-regulating feedback, is introduced. Using an interesting case study involving the Arctic tundra, Schultz expresses skillfully the "everything is interrelated" principle of ecology.

One of the most difficult aspects of systems analysis work is that of choosing appropriate system boundaries. With regard to the location of a boundary, Schultz states:

> The system has within it all the elements the observer is interested in. . . . These elements have a certain density or concentration. Outside, the elements are different either in kind or in concentration. The observer is not interested in these to the same degree. If they had been of the same kind of elements or had the same density as inside, he would have included them in the system. Thus the environment (e) (outside) is different from the system (s) itself.

In seeming contradiction, on a later page Schultz remarks:

> In most ecosystems there is no actual membrane separating system from environment. The boundary is imaginary and is located at the convenience of the observer. One choice is to set the boundary in a zone where there is no gradient (where concentration$_e$ − concentration$_s$ = 0). Now any crucial variations in density are trapped within the system and must be measured as state transformations.

Perhaps a way to make sense of these coexisting statements is to assume that the first statement describes one kind of situation and the second one another kind. With the first situation—the easy case—one system can be clearly differentiated on the basis of observable "edges" of the systems. In the second case—both a frequent and a difficult one—the edges of systems may not be perceptible or may not even exist in the region of interest. Therefore, the strategy becomes one of defining tractable or convenient system boundaries instead of simply using clearly apparent natural boundaries. For example, in the soil–plant–herbivore–predator system in the vast expanse of arctic tundra, there may not be, for hundreds of miles in every direction, a natural boundary in terms of concentration differences of elements. A person who desires to study the ecological interrelationships of the tundra would simply have to choose a convenient demarcation between the area he observes (the system) and the area he does not observe (the outside environment). However, there may be one region where the nests of the local animal population (e.g., lemmings) are concentrated. The nest density may drop off continuously with distance from the central nesting area; and after a certain distance, nests may occur with an approximately constant spatial frequency. Schultz suggests, with merit, that an appropriate choice is to draw the system boundary somewhere in the constant-density region.

It is probably helpful to clarify the meaning of a few terms. Schultz states that samples from compartments of the system "can be digested for nutrients and bombed for energy." What he refers to here are measurement procedures, the first involving a vessel called a *digester* and a process for freeing nutrients from the sample material, and the second utilizing a bomb calorimeter to determine the amount of heat energy stored in the sample. Finally, for those for whom the word "jaeger" conjures up a jaguar-like predator, a pomarine jaeger is a large, dark-colored bird of prey.

Another characterization of the systems approach is presented and applied to the nitrogen cycle in the paper by Fred J. Endelman et al. The paper makes a case for interdisciplinary efforts to improve prospects for the understanding of ecological systems. In the words of the authors: "The model development combines the theory and analysis of the statistician and the chemical engineer with the experimental methods and results of the soil physicist and soil chemist." The authors' goal is the construction of a mathematical model of the nitrogen cycle that can be employed to evaluate the causative factors in nitrate pollution and to predict the effects of the use of nitrogen fertilizers and changing patterns of land use.

The law of the conservation of mass is used to derive a continuity equation for nitrogen in the soil as a function of time and depth. The paper contains a discussion of the theoretical equations of soil physics and the means used to determine the coefficients in these equations. Additionally, the work of Endelman et al. provides an example of a key modeling decision: the level of aggregation or disaggregation to use in defining variables. (Recall the discussion of homomorphic models in Paper 2.) Their explanation of Table 5 provides the rationale for their choice of the particular six species of nitrogen used in the model. Unfortunately, the paper does not report on the results of the planned application of the nitrogen-cycle model to two extreme systems, undisturbed deciduous forest and highly manipulated farmland.

The authors of Paper 3 appear to use the term "mechanistic model" where other investigators use the term "causal model" [i.e., to refer to models that relate dependent variable(s) to independent variable(s) through a functional relation that is deducible from the mechanism or circumstances of the interaction itself]. For example, the equation $s = \frac{1}{2}gt^2$ is a mechanistic, or causal, model which relates the distance fallen, s, to the amount of time, t, since the start of fall for a free-falling body released with zero velocity in a gravitational field subjecting the body to a constant acceleration, g.

The word "hysteretic" relates to the physical property of hysteresis, and a "lysimeter" is a device used to measure the percolation of water through soils and to determine the soluble constituents removed in the drainage. The v in the cv terms of Figure 4 should be ν (nu) to correspond to the symbol used in the text and in Equation (3).

Paper 4, by botanist Jerry S. Olson, complements Paper 3. In particular, nonbiologists should find it easier to understand Olson's paper if they first read the article by Endelman et al. In Paper 4, carbon, rather than nitrogen, is the principal element monitored in the producer–decomposer system, and litterfall (mostly leaves) is used as an index of ecosystem productivity.

The section on "Models and Methods" begins with a straightforward application of the conservation of matter and a clear development of a basic differential equation—the rate equation. The reader is shown how data on litter production and accumulation can be used to estimate the decomposition rate of organic carbon. The rate equation for carbon is applied to different cases—for example, to continuous litterfall and discrete annual litterfall. The author discusses properties of the exponential form of solution—in particular, "half-time" and "95 percent time" values for environmental accumulation or decay. A clever analogy between the "time constant" for an electrical circuit and for an ecosystem circuit is described. In addition, theoretical curves are presented to show the relationships among decay parameter value, litter productivity parameter, accumulation and decay half-time, and asymptotic (final) value of forest floor accumulation.

Following the mathematical analysis, Olson moves into a qualitative discussion of life at the soil level. He considers energy flow and, interestingly, introduces the notion of "the energy budget of the ecosystem as a whole." Aspects of natural (biological) communities, including their stability, disturbability, and developmental succession are described.

A final section of the paper discusses modifications to the basic model that would give increased realism to the analysis of ecosystems. Olson, citing his own and other investigators' work, mentions briefly the possibilities of using analog computers to solve the equations of more comprehensive and complicated models.

There are some typographical errors and points of possible confusion that should be noted. The first part of Equation (10) should read $k' = \Delta X / X_0$ and, contrary to normal expectations, ΔX must be understood as $\Delta X = X_0 - X$. The minus sign on the right-hand side of Equation (12) should be deleted. Finally, in Figure 7, the asymptotic value of the "very high" curve should be 25,600, not 256,000.

The last paper of Part I, entitled "Systems Studies of DDT Transport," was written to show the empirical and analytical evidence supporting public efforts to have DDT declared a pollutant in Wisconsin. In this case, environmental modeling was used as a lobbying tool.

An interesting description of the ways that DDT circulates in an ecosystem and its known effects on certain species draws the reader effortlessly toward the mathematical model—a mechanistic model based on the trophic-level conceptualization of an ecosystem and on the law of the conservation of mass. While the multidisciplinary group of authors describe their work as modeling the movement of DDT in a Wisconsin regional ecosystem, their mathematical model is completely general with respect to the transport of any substance in a terrestrial ecosystem. Because of the treatment of a chain of trophic levels, the model in Paper 5 represents an extension in complexity over the model for carbon concentration elucidated in Paper 4.

Through both the qualitative and quantitative presentations, Harrison et al. drive home the important ecological concept of "biological concentration" (i.e., the increased concentration of a substance in biological organisms at successively higher trophic levels). The authors conclude from their model that time lags in the propagation of DDT from lowest to highest trophic levels are such that the DDT "concentration in certain species at or near the top of the trophic structure could continue to rise for some years" even if no more DDT were added to the biosphere. This result reinforces the conclusion of Olson (Paper 4) regarding the long time for full effects of input changes to be felt in an ecosystem. Harrison et al. conclude with a discussion of another secondary effect of the use of DDT to control predators—that is, the temporary increase in population of the organisms that are a food source for the predator species.

Two technical clarifications may be helpful to the reader. First, the leftmost term in Equation (9) should read

$$\left(\frac{m_i}{\dot{m}_{d,i}}\right)\frac{dc_i}{dt}$$

(i.e., there should not be a dot indicating a time rate of change over m_i). Second, in Equations (19), the numbering for the constants K_i and W_i again begins at 1. Hence the K_1, W_1, K_2, and W_2 of Equation (18) become K_2, W_2, K_3, and W_3, respectively, in Equations (19).

1

Reprinted from *Energy Resources*, Publ. 1000–D, National Research Council, Washington, D.C., 1962, pp. 1–8

M. K. HUBBERT

INTRODUCTION

If we are to appreciate the significance of energy resources in the evolution of our contemporary society it will be necessary not only for us to understand the principal physical aspects of the conversion of energy in the complex of activities transpiring on the earth, but also to view these activities in a somewhat longer historical perspective than is customary. For those of us who live in the more industrialized areas of the world—particularly in the United States—it is difficult to appreciate the unique character of the industrial and social evolution in which we are participating. During our own lifetimes, and during the immediately preceding period of history with which we are most familiar, the pattern of activity we have observed most consistently has been one of continuous change, usually continuous growth or increase. We have seen a population begun by a small number of European immigrants to North America expand within a few centuries to over 200 million, while still maintaining such a growth-rate, even now, as to double within the next 40 years. We have seen villages grow into large cities. We have seen primeval forests and prairies transformed into widespread agricultural developments. We have seen a transition from a handicraft and agrarian culture to one of complex industrialization. Within a few generations we have witnessed the transition from human and animal power to continent-wide electrical power supernetworks; from the horse and buggy to the airplane.

Out of this experience it is not surprising that we have come to regard continual growth and increase as being the normal order of things.

However, if we are to appraise more accurately what our present position is in our social and industrial evolution, and what limitations may be placed upon our future, it is necessary that we consider, not only for the present but in historical perspective, certain fundamental relationships which underlie all our activities. Of these the most general are the properties of matter and those of energy.

- 1 -

From such a viewpoint the earth may be regarded as a material system whose gain or loss of matter over the period of our interest is negligible. Into and out of this system, however, there occurs a continuous flux of energy in consequence of which the material constituents of the outer part of the earth undergo continuous or intermittent circulation. The material constituents of the earth comprise the familiar chemical elements. These, with the exception of a small number of radioactive elements, may be regarded as being nontransmutable and constant in amount in processes occurring naturally on the earth.

For the present discussion our attention will be directed primarily to the flux and degradation of a supply of energy, and secondarily to the corresponding circulation of the earth's material components.

Flux of Energy on the Earth

The overall flux of energy on the earth is shown qualitatively and diagrammatically in the flow-sheet of Figure 1.

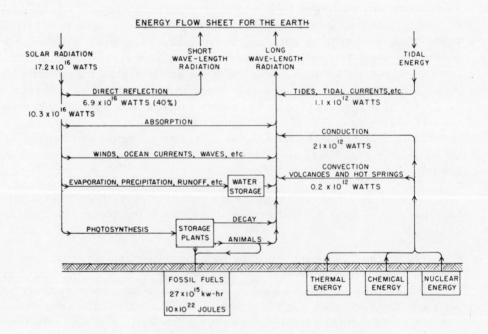

Figure 1. Energy Flow Sheet for the Earth

- 2 -

12

The energy inputs into the earth's surface environment are principally from three sources: (1) the energy derived from the sun by means of solar radiation, (2) the energy derived from the mechanical kinetic and potential energy of the earth-sun-moon system which is manifested principally in the oceanic tides and tidal currents, and (3) the energy derived from the interior of the earth itself in the form of outward heat conduction, and heat convected to the surface by volcanos and hot springs. Secondary sources of energy of much smaller magnitude than those cited are the energy received by radiation from the stars, the planets, and the moon, and the energy released from the interior of the earth in the process of erecting and eroding mountain ranges.

No definite quantity can be assigned to the energy from any of the foregoing sources because we are confronted not with a fixed quantity of energy but a continuous flux of energy from the various sources, at nearly constant rates. The rate of energy flux is measurable in terms of power, defined by

$$\text{power} = \frac{\text{energy}}{\text{time}},$$

and if the energy is measured in terms of the work unit, the joule, and the time in seconds, the power is then in joules per second, or watts.

Energy from Solar Radiation

The rate of energy flux from the sun, or the solar power, intercepted by the earth is readily obtainable from the solar constant, and the area of the earth's diametral plane. The solar constant is the quantity of energy which crosses unit area normal to the sun's rays in unit time in free space outside the earth's atmosphere, at a distance from the sun equal to the mean distance to the earth. It is, accordingly, the power transmitted by the sun's rays per unit cross-sectional area at the mean distance of the earth.

In heat units, the value of the solar constant, I, has been found to be 1.94 calories per minute per square centimeter (Landsberg, 1945, p. 929). This can be converted explicitly to power units by noting that 1 calorie of heat is equal to 4.19 joules of work, and 1 minute is 60 seconds. The solar constant in watts/cm^2 is, accordingly, given by

- 3 -

13

$$I = \frac{1.94 \times 4.19 \text{ joules/cm}^2}{60 \text{ seconds}}$$

$$= 0.135 \text{ watts/cm}^2.$$

The total solar power intercepted by the earth is then

$$P = IA = I\pi r^2,$$

where $\underline{A}$ is the diametral area of the earth and $\underline{r}$, equal to 6.37 $\times 10^8$ cm, is the mean radius of the earth. Supplying the numerical values of $\underline{I}$ and $\underline{r}$, we then obtain for the total solar power incident upon the earth

$$P = 17.2 \times 10^{16} \text{ watts.}$$

For comparison, the installed generating capacity of all the electric utilities in the United States in 1959 amounted to 15.7 $\times 10^{10}$ watts (Dept. of Commerce, 1961, p. 525). Hence, the power of the solar radiation intercepted by the earth is about a million times the power capacity of all the electric utilities in the United States in 1959.

Energy From the Earth's Interior
_____

The second largest input of energy into the earth's surface environment is that which escapes from the interior of the earth, which is estimated to be at a rate of about 21×10^{12} watts. Of this, about 99 per cent is by thermal conduction, and only about 1 per cent by convection in volcanos and hot springs.

Tidal Energy
_____

The tidal energy is derived from the combined potential and kinetic energy of the earth-moon-sun system. The total rate of dissipation of this energy, as indicated by the rates of change of the earth's period of rotation and the moon's period of revolution, is estimated by Harold Jeffreys (1952, p. 227, 231) to be about 1.4×10^{19} ergs/sec, or 1.4×10^{12} watts. Of this, about 1.1 x 10^{12} watts, or about 80 per cent, is estimated to be accounted for by oceanic tidal friction in bays and estuaries around the world.

Thus, tidal power is about an order of magnitude smaller than that of the heat escaping from the earth's interior, and both

- 4 -

14

together are less than one-thousandth of the power impinging upon the earth from solar radiation.

Energy Flow-Sheet

In view of its predominance, our principal concern is in tracing the flow of the 17.2×10^{16} watts of solar power that is being shed continuously on the earth. About 40 per cent of this, or 6.9×10^{16} watts (Landsberg, 1945, p. 933), known as the albedo, is directly reflected back into space. This leaves about 10.3×10^{16} watts which are effective in propelling the various material circulations occurring on the earth.

No further quantitative breakdown will be attempted. However, a part of the remaining solar power is absorbed directly by the atmosphere, the oceans, and the lithosphere, and is converted into heat. A large part of this heat is immediately reradiated back into space as long-wavelength thermal radiation. Another part, however, sets up differences of temperature in the atmosphere and the oceans, in such a manner that convective currents of both water and air are generated, producing the winds, ocean currents, and waves. The oceans and the atmosphere serve in this manner as the working fluids of a world-girdling heat engine whereby a fraction of the thermal energy from sunshine is converted into mechanical energy. The mechanical energy of the wind, waves, and currents is again dissipated by friction into heat at the lowest temperature of the surroundings.

Still another part of the solar energy follows the evaporation, precipitation, and surface run-off channel of the hydrologic cycle. Heat energy is absorbed during the evaporation of water, but it is again released when the water is precipitated. However, the water vapor, being a part of the atmosphere, is convected to high elevations by means of the convective energy already discussed; and, when precipitation occurs at these elevations, the water possesses potential energy, which again is dissipated back to low-temperature heat on the descent to sea level. It is this energy, however, that is responsible for all precipitation on the land, and for the potential and kinetic energy of surface lakes and streams.

A final fraction of incident solar radiation is that which is captured by the leaves of plants by the process of photosynthesis. Although enormously complex in detail, this is the driving

- 5 -

15

mechanism for the synthesis of common inorganic chemicals, such as H_2O, and CO_2, into the chemical compounds of living plants. Schematically this process is represented by the reaction

$$\text{Energy} + CO_2 + H_2O \rightarrow \text{Carbohydrates} + O_2,$$

during which solar energy becomes captured and stored as chemical energy. By the reverse reaction, as in the burning of wood,

$$O_2 + \text{Carbohydrates} \rightarrow CO_2 + H_2O + \text{Heat},$$

and the stored energy is released as thermal energy.

The energy-flow channel whose first step is photosynthesis is that which sustains the entire complex of organisms on the earth. We have the familiar food chain:

$$\text{Plants} \rightarrow \text{Herbivores} \rightarrow \text{Carnivores} \rightarrow \text{Parasites} \rightarrow \ldots.$$

in which the energy of each link is a small fraction of that of the preceding, the remainder being dissipated by heat. The end-product of this chain is the complete degradation of the photosynthetic energy to heat at the ambient temperature, and the conversion of the material constituents back to their initial inorganic state.

The Fossil Fuels

If the energy stored in plants by photosynthesis could be systematically retained, as for example in the form of firewood, it is clear that the aggregate amount would increase without limit, and could, in a few decades or centuries, become very large indeed. Actually, in the natural state, the rate of decay of organic compounds and the release of their stored energy as low-temperature heat is very nearly equal to the contemporary rate of photosynthesis. However, in a few favored places such as swamps and peat bogs, vegetable material becomes submerged in a reducing environment so that the rate of decay is greatly retarded and a storage of a small fraction of the photosynthesized energy becomes possible.

This, in principle, is what has been happening during the last 500 million years of geologic history. During that time a minute fraction of the existing organisms have become buried in

- 6 -

16

sedimentary muds under conditions preventing their complete decay. These accumulated organic remains comprise our present stores of the fossil fuels: coal, petroleum and natural gas, and related products, the energy content of these fuels being derived from the solar energy of this 500 million-year period which was stored chemically by contemporary photosynthesis.

Summary

The energy flow-diagram, which we have just reviewed, represents, in broad outline, all the major channels of energy flux into and out of the earth's surface environment. By the First Law of Thermodynamics, the quantity of energy in any particular channel, although repeatedly transformed in transit, remains constant in amount. It follows, therefore, that, with the exception of an insignificant amount of energy storage, the energy which leaves the earth by long-wavelength thermal radiation into space must be equal to the combined energy inputs from solar and stellar radiation, from tidal forces, and from the earth's interior.

By the Second Law of Thermodynamics, however, this flux of energy is unidirectional and irreversible. It arrives as short-wavelength electromagnetic radiation, corresponding to the temperature of the sun; or as mechanical energy of the tides; or as thermal energy from a temperature higher than that of the earth's surface environment. By a series of irreversible degradations it ultimately is reduced to thermal energy at the lowest temperature of its environment, after which it is radiated from the earth in the form of spent, long-wavelength, low-temperature radiation.

During this energy flux and degradation the material constituents of the earth's surface, while remaining essentially constant in amount, are circulated. The wind blows; oceanic currents, tides, and waves are formed; rain falls and rivers flow; volcanos erupt and geysers spew; and plants grow and animals eat, move about, procreate, and die.

But for this energy flux none of these things would or could happen and the matter of the earth's surface would be as dead or inactive as that of the moon.

Biologically, the human species is simply a member of the energy-consuming chain which begins with the energy capture and storage of plants by photosynthesis. Man is both an herbivore

- 7 -

17

and a carnivore, and, as such, is merely another member of the
biological complex, depending for his essential energy supply—
his food—upon other members of the complex, and ultimately on
the energy from the sun captured and stored in plants by photosynthesis.

In addition, however, man has been able to do what no other
animal has ever achieved; he has learned to tap other channels of
the energy flow-sheet, and he has managed to divert the energy
flow from its customary path into other channels appropriate to
his own uses.

An understanding of these processes is essential if we are
to appreciate the significance of energy resources in determining
what is possible and what is impossible in human affairs.

References

Commerce, Department of, 1961, Statistical Abstract of the
United States, 1961: U. S. Govt. Printing Office, 948 p.

Jeffreys, Harold, 1952, The Earth, Its Origin, History and
Physical Constitution: Cambridge Univ. Press, 376 p.

Landsberg, H., 1945, Climatology, p. 928-997 in Handbook of
Meteorology: New York, McGraw-Hill Book Co., 1056 p.

- 8 -

2

Reprinted from *The Ecosystem Concept in Natural Resource Management*, G. M. Van Dyne, ed., Academic Press, Inc., New York, 1969, pp. 77–93

Chapter V   A Study of an Ecosystem: The Arctic Tundra

ARNOLD M. SCHULTZ

I. INTRODUCTION

A few years ago some ecologists predicted that the ecosystem bubble would soon burst, after which investigators would "go back to tried and true methods" for probing nature. Others have said there is nothing new in the ecosystem concept at all except a fancy name and another language for students to learn. Like any other science, ecology has had its share of fads, some of which turned out to be nothing but old ideas dressed in new semantics.

The concept of system is indeed very old but not so the area we have come to call loosely systems analysis. The way in which modern systems

77

19

analysis has been applied to the ecosystem—just since 1962—is truly fantastic. It comes closer to a breakthrough than any other event in the history of ecology.

We now have a conceptual tool which allows us to look at big chunks of nature as integrated systems (Schultz, 1967). Also we now have the technical tools to handle the information obtained in this framework. Ecologists no longer fear complexity. These innovations have come along not any too soon. The alarm of "Silent Spring" is still ringing in our ears. Finally we realize that nature is not as piecemeal as science is.

Where does one begin, to study an ecosystem? The first part of this chapter explains some concepts fundamental to ecosystem study. In a discussion of models, it tells how the author decided to look at the tundra. The second part gives our first-hand experience in studying a tundra ecosystem in northern Alaska. In a discussion of results, it tells what we decided to look for.

II. THE REAL SYSTEM AND THE MODEL

A. The Language of Systems

The concept of system dates back to the very dawn of thought although its language is quite modern. From the beginning man has perceived only wholes; his penchant for taking them apart is rather a recent development but his ability to put the parts together again has scarcely developed at all. Our language of systems derives from these three ways of looking at things: taking things apart (analysis), assembling parts into wholes (synthesis), and seeing things only as wholes.

Therefore a system is a whole thing; it has three kinds of components (Fig. 1).

The elements of a system are the physical objects, often thought to be the "real" parts. In an ecosystem the elements are space-time units in that they occupy some volume in space for a certain length of time. Rain-

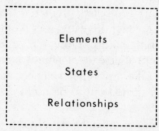

Elements

States

Relationships

FIG. 1. The components of a system including boundary.

drops, sand grains, and mosquito larvae are examples of elements in a system. Each element has a set of properties or states, e.g., number, size, temperature, color, age, or value. Between two or more elements or between two or more states there are relationships which can be expressed as mathematical functions or less formally, with plain English verbs.

A system can now be defined as a set of elements together with relations among the elements and among their states (Hall and Fagen, 1956).

The term "set" implies that the components can be bounded. In the diagram the boundary is permeable to indicate an open system into which elements can enter from outside.

We might think that the elements, being physical entities of some kind such as nitrogen ions or living organisms, are the real and important components while the others are mere abstractions. In system thinking, however, we put more emphasis on the state. In a thermostatic control system, for example, it is not the air in the room but its temperature that is important; nor is it the hardware of the furnace but its state of being off or on, low or high that we consider. The elements of most interest to us in a system are capable of taking on two or more alternative states. In other words, the element is a variable and over time its difference in state is what we observe, measure, and record. We cannot record the element itself.

There would be no need to invoke the systems concept were it not for the crucial component relationship. A system becomes a whole thing only because its elements and states are connected together in some way. Thus, by understanding the linkages we see how the whole system works.

We like to use this very simple model of a system (Fig. 1) as a reminder of what is real, what is abstraction, and why we have the systems concept in the first place.

How different is this approach from the one used in science before systems analysis? Scientists have always studied systems but they have shied away from complex ones. Early physicists learned that the mathematics needed to describe the attractions of more than two bodies at a time were beyond their powers of calculation or too time-consuming to carry out. Ecology is the science of relationships; yet ecologists, though awed by the complexity of nature, have long used methods which treat one factor at a time. How can one unravel the many interrelationships in a species-rich temperate forest, for example, or for that matter, in the arctic tundra?

B. Homomorphic Models

It is improbable that any ecosystem will ever be studied in its full complexity. In some systems there may be thousands of kinds of organisms and billions of interacting individuals. If we had spectroscopic X-ray

Arnold M. Schultz

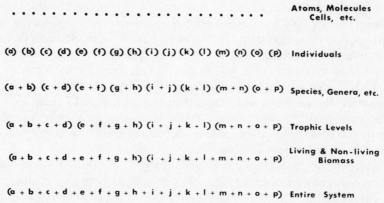

FIG. 2. Homomorphic models showing six levels of discrimination of system parts.

eyes we could see yet finer subdivisions: cells, molecules, atoms, perhaps even electrons. Obviously, to study the many states and relations in such a complex system would be too much for our best computers. The system must be simplified.

In the past ecologists would select a certain few parts of a larger system for study, for example, a species population or a relationship such as plant succession or competition. This kind of simplification falls short of studying the whole system; the other parts and the other relations that occur are completely neglected.

A complex system can be simplified by making a homomorphic model of it. Here the system remains intact. Its parts are discriminated at some level that can be handled conveniently. Figure 2 shows the ABC's of homomorphic models.

In the row second from the top are sixteen distinct individuals. Not everyone can see them, however, so my statement is only a point of view. Suppose the letters *a* to *p* represent grass plants in a dense sward. You would not be able to distinguish individuals at all. The next row down shows the units combined by twos into superunits. We can think of these as populations of taxonomic groups (species, genera, families) or of physiognomic ones (herbs, shrubs, trees). All of the finer units are still present but undiscriminated at the next higher level and so on through the hierarchy—trophic levels, then living and nonliving biomass, until the entire system emerges as a Gestalt and there is no discrimination of parts whatsoever.

Homomorphic models are not designed just for lazy taxonomists. There is a very practical reason for looking at systems this way. Let us look at the top and bottom lines of Fig. 2. At the top, every possible state has been distinguished but the sheer bulk of information is so overwhelming that we can make little use of it. At the other end, all the states are fused into one grand but platitudinous expression and all you can say for it is "There it is!" In between are a number of handy simplifications. On the fine end some realism is retained. On the coarse end we gain generality. We now have a set of models which allows us to coordinate all the discoveries made by specialists on the separate parts of the system. Any part can be handled as a black box coupled into the system at the appropriate level of organization.

C. The Tundra as a Simple Ecosystem

The history of scientific investigation of arctic tundra follows a pattern different from that of other regions of the world. Because of the sparse indigenous human population, pressure for agricultural research, as we know it in lower latitudes, has not occurred. Basic studies from diverse disciplines and with broad objectives were initiated; they did not assume the single-minded purpose of maximizing crop yields.

Many ecologists were lured to the tundra because it was supposed to be simple. Here in the arctic could be found a paucity of growth forms and species, shallow soil, a short growing season, an extreme climate, and essentially no disturbance of the landscape by humans. If ever the total processes of nature could be put in order, it would be done for the tundra. Let us see how simple the tundra really is.

I can cite numerous detailed descriptions of coastal tundra in the vicinity of Point Barrow, Alaska. These include my own investigations and those of my students which began in 1958 (Schultz, 1964; Pieper, 1964; Van Cleve, 1967) and the excellent work on microtines and their predators by F. A. Pitelka and others begun in 1952 (Pitelka *et al.*, 1955; Pitelka, 1958; Maher, 1960). Intensive studies on soils, meteorological phenomena, bacteriology, and other aspects of the tundra have been going on at Point Barrow since 1950.

It is not my intent here to redescribe the tundra. Rather I shall point out the proximity between the real ecosystem and its simplified model.

It would be possible to print on two pages of this book a list of all the species of plants, animals, and microorganisms known to occur in the massive ecosystem under study. Perhaps it would go on one page. The list could be reduced to about ten species, and still include 90% or more or the biomass in each major group. It would include sphagnum moss among the lower plants, several grasses and sedges among the higher plants, the brown lemming and pomarine jaeger among herbivores and carnivores, respectively. For quantitative studies of energy and major nutrients, an analysis of samples from these few predominant groups would give essentially the same results as would a total ecosystem study. To put it another way, the properties or states of the trophic levels are at least 90% predictable from just one or two of their component parts.

We have already seen how the complexity of a system is determined by the number of distinguishable parts. At the species population level of discrimination, the tundra is fairly simple compared to other ecosystems of the world. But there is another determinant of complexity: the number of recognizable states which the parts can assume. Some examples are depth of active soil layer, exchangeable calcium in the soil, phosphorus level of forage plants, population density of lemmings, jaegers, and owls, and decomposition rate of organic matter. The number of states depends entirely on the yardstick and stopwatch used to measure. The investigator can make it as simple or complex as he wishes.

We must come to the conclusion that the easiest way to analyze the tundra or any other ecosystem is to lump all populations of organisms into trophic levels and the resources (atmosphere and soil) into convenient compartments. For each compartment a sample, in exactly the proportion in which the various populations occur, can be digested for nutrients and bombed for energy. Egler (1964) calls this the "meat-grinder" approach. A wealth of information inside each box is conveniently ignored. Our primary interest is directed to relationships between boxes. How much phosphorus flows from one tropic level to another? In this framework, one system is as simple as another—tundra or temperate zone grassland.

D. Isomorphic Models

We can now return to the third component of systems, relationships. The homomorphic model does not help us here; it is concerned only with the power of resolution used for the elements and their states. We need an isomorphic model for studying relationships.

An isomorphic model is a map. A road map of New Mexico is a model of the real geographical area of the state. It shows among other things the distance and direction between Albuquerque and Santa Fe. By inference from the kind of highway, it also shows the rate of traffic flow between the two cities. So a map can be a flow chart.

Figure 3 is a map of a homomorphic model of an ecosystem—two kinds of models combined. This one has been proposed by the subcommittee on Terrestrial Productivity for the International Biological Program as a tentative model for studying all terrestrial ecosystems. The arrows indicate paths of nutrient or energy flow from one box to another. Research on ecosystems can be standardized and routinized by using

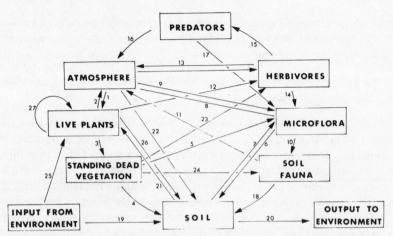

FIG. 3. Isomorphic model mapping the relations between ecosystem compartments.

such a scheme. One can use the numbers on the arrows as a checklist to
see which relations of Fig. 3 one has forgotten to measure.

Let us consider what kinds of relationships there are and how they
are measured.

Arrow No. 12 represents a process with a familiar name: grazing. The
model suggests that something travels physically along the path from
live plants to *herbivores*. We measure this something in units of chemical compounds, biomass, or energy. It is measured as a rate—so many
pounds of dry matter per unit time. Rarely is the transfer measured
directly. The *live plants* box is weighed at time one (T_1) and again at time
two (T_2). The difference in state between T_1 and T_2 is read as a gain or
loss. A series of exclosures and enclosures will indicate what proportion
has been gained or lost via pathways No. 1, 25, 2, 3, 12, or 21. The same
is done for *herbivores* and every other box in the model. We should note
here that enclosures and exclosures are experimental tools not to be preempted by range managers. The bacteriologists' agar plates and the radio-
biologists' isotope tracers are based on the same principle as a fenced plot.

Isomorphic models can be constructed for many kinds of relationships.
A correlation matrix is such a model; the coefficients show spatial or
phenetic distances. Regression coefficients and ratios may be used to
express causation, conjunction, or succession of events. Finally, a model
can be set up with vernacular expressions like "boy meets girl," or more
appropriately, as in André Voisin's definition of grazing, "cow meets
grass." You will recognize pathway 15 as owl eats lemming, 5 as bac-
terium rots straw, and 20 as ocean takes soil. I am not being trite. If we

could clearly establish verbally expressible relationships between all of the parts we would indeed understand ecosystems very well and could develop valuable theorems about them. The numbers would then be superfluous.

E. System Boundary and Environment

Earlier in this chapter an ecosystem was described as a space-time unit—a volume that exists in time. But you cannot see an ecosystem in the same way that you can see, for instance, a lemming. The skin clearly marks the boundary of the animal. It is simply a matter of perception and all observers would agree as to what is lemming and what is not lemming.

Not all ecosystems have a skin you can touch. Defining the boundaries of his ecosystem is one of the biggest problems an ecologist has to face. Some insight into the problem can be obtained by considering the nature of boundary for a generalized system (Fig. 4).

The system has within it all the elements the observer is interested in (the set, by definition). These elements have a certain density or concentration. Outside, the elements are different either in kind or in concentration. The observer is not interested in these to the same degree. If they had been of the same kind of elements or had the same density as inside, he would have included them in the system. Thus, the environment (e) (outside) is different from the system (s) itself. It is not entirely true that the observer is disinterested in the environment; he cares about what effect it has on the system. There is pressure for some of the elements to cross over the boundary.

Think of the boundary as a membrane with a certain thickness Δx. It has a texture which determines how easily any of the elements can flow across it. This can be thought of as a permeability factor (m). Since there are different densities of elements inside and outside, there must be a concentration gradient across the boundary. If Δx is narrow, the gradient will

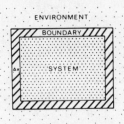

FIG. 4. The boundary between system and environment.

be steep. The flow of elements across the membrane is governed by the following three factors (Jenny, 1961):

$$\text{Flow} = -\left(\frac{\text{concentration}_e - \text{concentration}_s}{\Delta x}\right) m$$

This can be illustrated with an exclosure in a pasture. The fence, of course, represents the boundary. The exclosed plot has no animals, but outside are 10 steers per acre. Some of these are pressing against the fence. The gradient is sharp—from 10 to 0 over a distance the diameter of the wire. If the fence is strong ($m = 0$), the flow will be zero no matter what the steer pressure is.

In most ecosystems there is no actual membrane separating system from environment. The boundary is imaginary and is located at the convenience of the observer. One choice is to set the boundary in a zone where there is no gradient (where concentration$_e$ − concentration$_s$ = 0). Now any crucial variations in density are trapped within the system and must be measured as state transformations. The other approach is to use a natural boundary (where $m \to 0$) such as the shore line of an island. Here the problem is that the permeability coefficient may be low for only one kind of element, and as we have already seen, a complex may have a thousand kinds.

F. The Role of the Observer

One of the tenets of systems research is that the observer is always a part of the system. Refer back to the four diagrams and picture the role of the observer. He is an element of the system, with definite properties and unique relations to the other elements and states (Fig. 1). The observer decides the level of discrimination to be used in the study (Fig. 2). He selects from the large number of possible relations just those he wants to measure (Fig. 3). He fixes the boundary of his system according to his resources and his interests (Fig. 4). Apart from the observer there can be no unique ecosystem. No one can go to Point Barrow and see the same system that I see. It follows that I cannot possibly describe to you the real or absolute ecosystem, only my model of it.

Within the system he has circumscribed, the observer looks for time-invariant relationships. During the investigation he records the activity of the system. This includes haphazard events which may happen only once, activities which occur frequently, and those which occur every time. These represent, respectively, the temporary, the hypothetical, and the permanent or real behavior of the system. In graphical form, the three kinds of activity can be shown as scatter diagram, prediction curve, and equation for an absolute law.

It is within this framework that I present some of the results of my research.

III. THE TUNDRA AS A HOMEOSTATIC SYSTEM

A. Cyclic Phenomena

Many of the activities of the tundra ecosystem under study are cyclic, with a periodicity of three or four years, but with varying amplitudes between cycles. We can think of a cycle as a series of transformations of state (Ashby, 1963). Thus, if a subsystem (compartment) has four clearly recognizable alternative states, a, b, c, and d, and the transformation always goes a → b → c → d → a → b, etc., then the sequence of states is a cycle. This can be shown kinematically:

$$
\begin{array}{ccc}
a & \rightarrow & b \\
\uparrow & & \downarrow \\
d & \leftarrow & c
\end{array}
$$

or, when put on a time scale, as a sine wave:

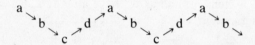

and so on.

At Point Barrow we have good records of yearly lemming population densities, starting from 1946 (Fig. 5). Lemming peaks occurred in 1946, 1949, 1953, 1956, 1960, and 1965. Neither the amplitude nor the wavelength of these cycles is always the same. Yet there are some striking similarities. Using the systems language given above with a 1-year time interval, generally we can recognize four states: a, high density; b, very low density; c, low density; and d, medium density. The states can be named by reading the histogram, without any knowledge of lemming population dynamics or life histories. Sometimes c is missing and once c and d are transposed. Always the high year was immediately followed by a very low year. But even during the low years, there were found local "pockets" supporting a denser population; for example, on the outskirts of the Eskimo village of Barrow, the fluctuations in lemming numbers have never been as pronounced as on the open tundra. Also of significance, some areas are out of phase with Point Barrow. At a point 100 miles east of Barrow, the population peaked in 1957, a year after the Point Barrow high. By 1960, it was in phase again.

What can be said about lemming population cycles? By Ashby's criterion, we have definitely observed cycles; but an engineer would say,

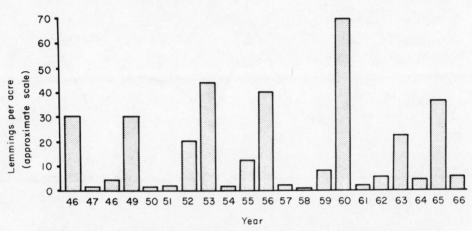

Fig. 5. Lemming population cycles at Point Barrow, Alaska, from 1946 to 1966.

if he saw the waves on an oscilloscope, that there was a lot of noise on the channel.

We have data on standing crop of plants (tops only), starting from 1958. Clippings were made in the vicinity of traplines used for the lemming census. Ninety percent of the dry matter was contributed by three species (*Dupontia fischeri, Eriophorum angustifolium,* and *Carex aquatilis*); these same species constitute the bulk of the lemming diet. When a graph is constructed for standing crop each year at phenologically equivalent dates, a cyclic pattern appears. Moreover, the pattern is in synchrony with that of the lemming population cycle, noise and all. A short lag develops between the two curves as the high point approaches. I do not want to explain the facts at this time, but I should say, in passing, that the high correlation between forage yield and lemming stocking rate would not-come as a surprise to a range management audience.

Samples of the material clipped for production records were analyzed for total nitrogen, phosphorus, calcium, and several other elements. Concentrations of nutrients in the herbage, at any given phenological stage (i.e., date) increased through the year corresponding to the peak lemming year, then dropped to low values the year after, only to increase again. Figure 6 shows the activity for phosphorus superimposed on the histogram of lemming population density. Within a season, nitrogen, phosphorus, and potassium decrease percentagewise as grasses mature, while calcium increases. The line in Fig. 6 should not be construed as a continuous increase in phosphorus level from 1957 through early 1960.

Calcium, potassium, and nitrogen show the same trends as phosphorus. Due to greater plot-to-plot variability, the nitrogen data are not as sig-

29

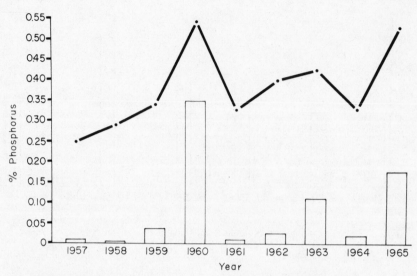

FIG. 6. Phosphorus levels in forage at Point Barrow, Alaska. Bars represent relative lemming numbers.

nificant as are those of the other three elements, but the trend is nevertheless the same. Magnesium and sodium show no relationship to lemming numbers at all, nor were the data cyclic.

Still another activity studied was decomposition of organic matter on the soil surface. This, too, turned out to be a cyclic phenomenon, and correlation with the activities already mentioned is high.

I have given a rough sketch of the behavior of the tundra, as discovered by survey techniques. The observations seem to fit closely a hypothesis of synchronous cycles. But at this stage, the results could be spurious. The close fit might result from artifacts of sampling. The transformation sequence a → b → c and/or d → a might occur frequently just by chance.

The next step is experimental: to introduce a disturbance at any one of the compartments and watch for reverberations throughout the system.

B. Experiments in Stressing the System

If the fluctuations in herbage production and nutrient level are related to immediate grazing history, then the cyclic aspect should disappear when grazing is eliminated. A simple exclosure, in effect, removes the herbivore from the system.

In 1950, a series of exclosed plots was established, alongside paired plots open to normal grazing (Thompson, 1955). Records kept for 13 years show cyclic variation on the outside paired plots, while the fenced plots show a constant decline. Since 1958, percentages of phosphorus, calcium, potassium, and nitrogen in the herbage from the grazed plots show the same marked cycles that occur elsewhere on the tundra (see Fig. 6). By comparison, year-to-year fluctuations inside the exclosures are slight and not cyclic.

With regard to decomposition of litter, the outside plots responded as did the tundra on the whole; inside the exclosures, decomposition rates were low and constantly decreasing.

An unexpected bonus came from the exclosure experiment. It gave an opportunity to assess the effect of lemming activity on the depth of thaw. By comparing, at the time of maximum thaw, soil depths inside and outside exclosures, I could separate the lemming-caused (within-system) effects from the summer temperature (environmental) effects. The results were most interesting. During a peak lemming year, the thickness of the active soil layer was maximum and it gradually diminished to the shallowest point the year before the next peak.

A second experiment was to stabilize artificially the fluctuating nutrient levels in the soil. This was done by fertilizing annually 6 acres of tundra with nitrogen, phosphorus, potassium, and calcium. Heavy applications were made to make sure that the variations in native soil nutrients were completely masked. What effect would this kind of disturbance have on primary and secondary production?

Net primary production, for the 4 years studied, was stabilized at a level 3–4 times that of the control plot. Annual variation was obliterated. Herbage quality was also stabilized. Protein levels, for example, were 4–5 times those of the vegetation of the control plot. Percent of calcium and phosphorus in the green tissue at equivalent dates remained high and constant in the four years.

The first fertilization was applied in 1961. No animals were seen either on fertilized or on control plots in 1961 or 1962. In 1963, animals were abundant all over the tundra (see Fig. 5 or 6), while in 1964, they were generally sparse. Immediately after the snowmelt in 1964, 30 winter nests per acre were counted on the fertilized area, none on the control plot, and less than 1 per 10 acres on the tundra in general. However, jaegers had found this 6-acre pantry and picked it clean. The few survivors observed at the time of the winter nest survey were large and fat. In 1965, a year of high lemming density all over the tundra, lemmings were abundant both on the fertilized and unfertilized control plot.

C. Hypotheses of Ecosystem Cycling

Only a fraction of the information collected so far can be presented in this paper. For the sake of brevity, I have shorn away all evidence of

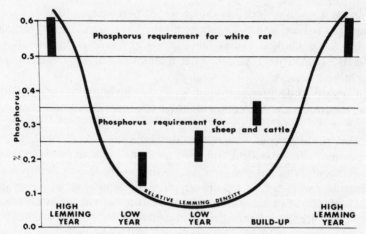

FIG. 7. Hypothesis relating lemming populations and nutritional quality of forage. Lemming density curve is generalized, as are the black bars showing phosphorus in forage.

"noise" and shown, so to speak, only the slick regression lines. These represent the hypothetical behavior of the system compartments.

With the evidence at hand, let us develop a more general hypothesis to explain the synchronous cycles apparent in our ecosystem. Tentatively it might be called the nutritional threshold hypothesis (Fig. 7) but such a name places undue emphasis on just one part of the system.

Let us review a generalized 4-year cycle.

1. Early in summer of the high lemming year, the forage is calcium- and phosphorus-rich. Because of high production and consumption, much of the available calcium and phosphorus is tied up in organic matter. The soil has thawed down deep into the mineral layer because grazing, burrowing, and nest-building has altered the albedo and insulation of the surface.

2. The next year, not only is forage production low, but also the percent of calcium and phosphorus in the diet is below that which would be required for lactation by sheep or cattle. Nutrients in organic matter have not yet been released by way of decomposition.

3. The following year production is up, the plants are recovering from the severe grazing two years earlier, and dead grass from the previous year insulates the soil surface. At the same time, decomposition of that dead material is speeding up. Forage quality is still quite low. Whether there is enough calcium and phosphorus in the diet to support lemming reproduction and lactation depends on how closely the species resembles domestic livestock, on the one hand, or the laboratory rat, on the other, in mineral requirements.

4. In the fourth year plants have fully recovered from grazing. Forage species accumulate minerals in their stem bases. Freezing action concentrates solutes in upper soil layers. Dead grass from several years has accumulated and plant cover is high; soil surface is well insulated and the thawed layer is very shallow. Decomposition rate is high. Calcium and phosphorus (and also potassium and nitrogen) content of forage is satisfactory for reproduction. There is enough food to support a large population of herbivores.

Next, the sequence is repeated.

Not until a nutritional threshold has been reached can a large lemming population build up. But the population does not keep getting bigger and bigger. This would be disastrous to the vegetation. So a deferred-rotation grazing scheme is built into the system. No grazing at all would also be disastrous to the vegetation and to the soil as well. Predators play a role at the time of herbivore decline. Indeed, all parts of the system play a role. It is a homeostatically controlled system.

This is only a hypothesis. It can be tested in the framework of the ecosystem concept: First, by showing that all parts bear some relationship to all others; second, by experimentally stressing the system to see how it adapts to disturbance; third, by opening the black box and studying its physiology—that is, explanation of a phenomenon at a lower level of organization.

D. Contemporary Hypotheses on Cycling

Needless to say, the nutritional threshold hypothesis is at variance with several prominent hypotheses that have been advanced in recent years. The hypotheses of Christian and Chitty minimize the role played by energy and nutrition in controlling animal populations. The stress hypothesis of Christian (1950) associates population declines with shock disease and changes in adrenal–pituitary functions. The increase in adrenal activity at high population densities lowers reproduction and raises mortality. The hypothesis involving genetic behavior (Chitty, 1960) suggests that when animal numbers fluctuate, the populations change in quality. This is brought about through selection resulting from mutual antagonisms at high breeding densities.

All hypotheses concerning animal population cycles have in common the notion of feedback. There are two kinds of feedback, negative and positive. The kind generally involved in control mechanisms is negative or deviation-counteracting while the "vicious circle" kind is positive or deviation-amplifying (Maruyama, 1963). Most ecosystems have both kinds. We can think of loops running through a series of compartments

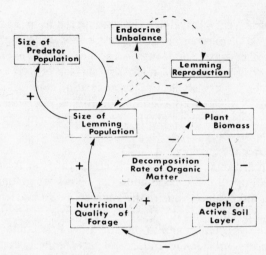

FIG. 8. Feedback-loop model showing homeostatic controls in the arctic tundra ecosystem.

so that the state of each compartment either counteracts (−) or amplifies (+) the change of state of the next (Fig. 8).

It is probable, in fact common, that any given element will be stationed on several loops. It may be checked via one loop and amplified via another. Consider the herbivore compartment of any ecosystem. Amount of forage, its quality, and availability of space are all positive; predators and pathogens are negatively related. In some cases, a density control mechanism operates within the compartment itself—as described by the stress theory or the genetic selection theory. This is simply an additional loop in the system. There is no reason why a control system should have but one governor.

The idea of one cause–one effect is left over from the nineteenth century when physics dominated science. The whole notion of causality is under question in the ecosystem framework. Does it make sense to say that high primary production causes a rich organic soil and a rich organic soil causes high production? This kind of reasoning leads up a blind alley. We are dealing with the different dependent properties of the same system. Only things outside the system can cause something to happen inside. For the same reason, we cannot say that the lemmings are the driving force, any more than the vegetation, the soil, or the microflora, in making the ecosystem tick.

ACKNOWLEDGMENTS

Work on the coastal tundra near Point Barrow, Alaska, was done under grants from the Arctic Institute of North America, the National Science Foundation, and the Office of

Naval Research. R. D. Pieper, who participated in some of these studies, is thanked for presenting this paper at the symposium while the author was in Great Britain.

REFERENCES

Ashby, W. R. 1963. "An Introduction to Cybernetics." Wiley, New York. 295 pp.

Chitty, D. 1960. Population processes in the vole and their relevance to general theory. *Can. J. Zool.* **38**, 99–113.

Christian, J. J. 1950. The adreno-pituitary system and population cycles in mammals. *J. Mammal.* **31**, 247–260.

Egler, F. E. 1964. Pesticides—in our ecosystem. *Am. Scientist* **52**, 110–136.

Hall, A. D., and R. E. Fagen. 1956. Definition of system. *Gen. Systems Yearbook* **1**, 18–28.

Jenny, H. 1961. Derivation of state factor equations of soils and ecosystems. *Soil Sci. Soc. Am. Proc.* **25**, 385–388.

Maher, W. J. 1960. The relationship of the nesting density and breeding success of the pomarine jaeger to the population level of the brown lemming at Barrow, Alaska. *Alaskan Sci. Conf., Proc.* **11**, 24–25.

Maruyama, M. 1963. The second cybernetics: Deviation-amplifying mutual causal processes. *Am. Scientist* **51**, 164–179.

Pieper, R. D. 1964. Production and chemical composition of arctic tundra vegetation and their relation to the lemming cycle. Ph.D. thesis, University of California, Berkeley, California.

Pitelka, F. A. 1958. Some characteristics of microtine cycles in the arctic. *Ann. Biol. Colloq.* **18**, 73–78.

Pitelka, F. A., P. Q. Tomich, and G. W. Treichel. 1955. Ecological relations of jaegers and owls as lemming predators near Barrow, Alaska. *Ecol. Monographs* **25**, 85–117.

Schultz, A. M. 1964. The nutrient-recovery hypothesis for arctic microtine cycles. II. Ecosystem variables in relation to arctic microtine cycles. *In* "Grazing in Terrestrial and Marine Environments" (D. J. Crisp, ed.), pp. 57–68. Blackwell, Oxford.

Schultz, A. M. 1967. The ecosystem as a conceptual tool in the management of natural resources. *In* "Natural Resources: Quality and Quantity" (S. V. Ciriancy-Wantrup and J. J. Parsons, eds.), pp. 139–161. Univ. of California Press, Berkeley, California.

Thompson, D. Q. 1955. The role of food and cover in population fluctuations of the brown lemming at Point Barrow, Alaska. *Trans. 20th N. Am. Wildlife Conf.*, pp. 166–175.

Van Cleve, K. 1967. Nutrient loss from organic matter placed in soil in different geographic regions. Ph.D. thesis, University of California, Berkeley, California.

3

Copyright © 1972 by the Baywood Publishing Company

Reprinted from *J. Environ. Sys.* **2**, 3–19 (Mar. 1972)

A Systems Approach
to an Analysis of the
Terrestrial Nitrogen Cycle

FRED J. ENDELMAN
Department of Chemical Engineering

MELVIN L. NORTHUP
Department of Soil Science

DENNIS R. KEENEY
Department of Soil Science

JAMES R. BOYLE
Departments of Soil Science and Forestry

RICHARD R. HUGHES
Department of Chemical Engineering

University of Wisconsin
Madison, Wisc.

ABSTRACT

In an interdisciplinary study, a new methodology for analysis of environmental systems is being applied to the terrestrial nitrogen cycle. It has already resulted in an informative, qualitative description of this complicated system and its components. With this system definition, we plan to develop a mathematical simulation of the dynamics of the nitrogen cycle in both forested and agricultural ecosystems. Initial results from on-going experimental and modeling studies in soil chemistry and soil physics have been encouraging. A critical evaluation of the final simulation model, using field data obtained from an extensive system monitoring program, will be a rigorous test of the feasibility and merit of the proposed systems method.

Introduction

Development and evaluation of measures for environmental protection require a quantitative knowledge both of the individual processes which

3

make up a particular ecological system, and of the interrelationships between these processes. This knowledge is essential to elucidate the cause and effect pathways and to identify the most suitable means of attacking the problem. The detailed investigation of fundamental biological processes is the responsibility of the biological scientist, but the analysis of an entire ecological system requires the cooperation of physical scientists, statisticians, and engineers as well.[1] Integration of their diverse contributions is achieved through systems analysis—a systematic mathematical approach to large, complex problems.

This paper is a progress report on an interdisciplinary study of this type. A diverse team of scientists and engineers is building a mathematical model for simulating the movement and biochemical transformation of inorganic and organic nitrogen species in selected soil systems. With such a model, we can analyze and evaluate the role played by soil drainage in the nitrate pollution of groundwater supplies and the eutrophication of our natural water systems, as well as predict, quantitatively, the effects of nitrogenous fertilizers and of changing land-use patterns. The model development combines the theory and analysis of the statistician and the chemical engineer with the experimental methods and results of the soil physicist and soil chemist. This systems approach is general enough to be useful in analyzing many environmental systems.

First we describe the methods used in our systems analysis, and define the terrestrial nitrogen cycle which is the basis of the ecosystems to be studied. Some details follow on the two particular systems to be investigated—deciduous forest and sandy soil. Next we describe the framework and basis of the mathematical model to be used. Our discussion closes with some details of the modeling and experimental studies of the soil physics and soil chemistry involved.

Systems Analysis

The numerous texts and articles on systems analysis contain many definitions of the subject. Most of these are conceptual and do not provide the layman or uninformed scientist with a fundamental understanding of the systems approach to problem solving. The following functional or operating definition of systems analysis is probably more useful. According to Nadler,[2] a system is a collection of real life phenomenon that possesses the seven elements defined in Table 1. In this context, systems analysis is the logical and comprehensive specification and study of these seven elements, both individually and collectively.

The systems approach can be clarified still further by consideration of a specific method for implementing a logical and comprehensive study of a

Table 1. The Seven Elements of a System*

1. *Function*	Objective which the system attempts to achieve.
2. *Inputs*	Materials and energy which are processed by the system to arrive at the outputs.
3. *Outputs*	Desired and undesired material and energy products of the system.
4. *Sequence*	Interrelationships among the component processes of the system, and the order in which they operate, such that the inputs are converted to the outputs.
5. *Environment*	Physical surroundings within which the system operates.
6. *Physical agents*	Material (non-human) resources which operate in at least one step of the system sequence, but which do not become a part of the output.
7. *Human agents*	Human resources which operate in the system sequence.

*Adapted from Nadler.[2]

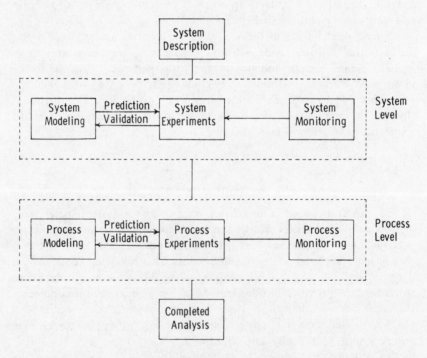

Figure 1. A schematic representation of a method for systems analysis.

system. Figure 1 is a schematic representation of such a method suggested in part by Loucks et al.[3] The first required step is a complete description of the seven system elements described in Table 1. This is followed by investigations at the system and process levels. While the end result of a systems analysis is usually an integrated system model, it must be emphasized that it can be no better than its component process models.

The iterative, feedback nature of systems analysis (Figure 1) is portrayed in more detail in Figure 2, which has been adapted from Van Dyne.[4] An investigator's definition and model of a system and its component processes must always be subject to alteration. As more is learned about the system during the analysis, the specification and models of the system can be improved by appropriate modification based on experimentally obtained evidence.

Another concept to be clarified is that of models and modeling. Nadler[2] defines a model as an abstraction of a real life phenomenon which is used as a means of representing some part of a defined system. This model need not be mathematical in form. In fact, during the definition phase of a systems analysis, it is usually in a schematic form.

A mathematical model is the usual formal objective of a modeling study. It consists of one or more mathematical relationships, which predict the values of a set of dependent variables Y, as a function of a set of independent variables X and parameters k. Thus, the general form of a mathematical model is the following:

$$Y = f(X,k) \tag{1}$$

At the initiation of a modeling study the set of the dependent variables of interest has already been specified. A prior specification of f, X, or k, however, is not required. Indeed, determination of any combination of these three factors is the goal of the study itself. Hunter[5] has defined the four general classes of modeling studies presented in Table 2. The actual techniques required in these studies vary from well-developed theories such as linear regression,[6] to comparatively new methods, such as model discrimination,[7] for which the theory is still being developed.

System Definition of
the Terrestrial Nitrogen Cycle

The first element of the system to be specified, as indicated in Table 1, is the function of the terrestrial nitrogen cycle.[8] This can be defined as maintenance of the balance between atmospheric and lithospheric nitrogen by the circulation, transformation, and accumulation of the various forms of inorganic and organic nitrogen through the biotic and abiotic segments of the terrestrial environment.

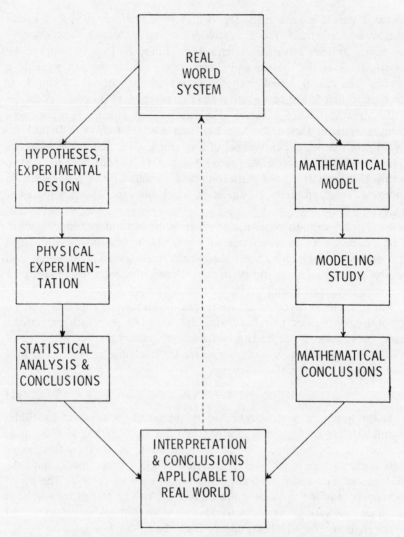

Figure 2. The interdependence of experimentation and modeling
in systems analysis adapted from an illustration in Van Dyne.[4]

A discussion of the inputs and outputs to this system can be facilitated
by introducing the carrier concept. A carrier is a material that can
transport or store one or more nitrogen compounds. The concentration of
a given nitrogen compound in a carrier is defined such that, when we
multiply the concentration by an appropriate measure of the bulk amount
of the carrier, we obtain the total amount of the given nitrogen compound
in the carrier. The different forms of the five carrier media in the terrestrial

Table 2. Classification of Studies of the Model $Y = f(X, k)$*

Phase	Unknown	Objective
1. Screening study	f,X,k	Determine the subset **X** of important independent variables from the complete set of all potentially important independent variables.
2. Empirical model building	f,k	Determine a suitable empirical representation of **f** along with associated values of **k**.
3. Mechanistic model building	f,k	Determine the true functional form of **f**.
4. Mechanistic model fitting	k	Determine the true value of **k** in the true form **f**.

*Taken from Hunter.[5]

nitrogen cycle are listed in Table 3, along with a specification of their roles as influents, effluents, or internal carriers.

These media can also be thought of as the physical agents of the system. The presence of human agents in the system depends on the type of region under consideration. For example, an undisturbed wilderness forest would have no human agents, while in an agricultural system, human agents would serve as both consumers and manipulators.

The sequence element of the terrestrial nitrogen system consists of three types of processes—transportation by a carrier, exchanges between carriers, and transformations between different forms of nitrogen in a given carrier. A detailed qualitative study of the nitrogen cycle to identify these component processes has resulted in Table 4.

A proposed characterization of the environment for the terrestrial nitrogen cycle is presented in Figure 3. The upper boundary of the system is the top of the layer of animal and vegetative litter which rests on the surface of the soil. The lower boundary is the level of the groundwater table. The vertical direction is the only spatial dimension which will be considered at this time. It is assumed that lateral variations can be accounted for sufficiently by dividing a region to be modeled into homogeneous subregions within which areal variations are negligible.

While we propose to predict the concentration of nitrogen as a continuous function of depth, we realize that various parameters and variables of soils have not often been studied in this manner. Many earth

Table 3. Functions of the Carrier Media in the Terrestrial Nitrogen Cycle

Carrier		Input	Internal	Output
Water	1. Precipitation	X		
	2. Run-on	X		
	3. Soil water		X	
	4. Ground water	X		X
	5. Water vapor		X	X
	6. Runoff			X
Fertilizer	7. Organic fertilizer	X		
	8. Inorganic fertilizer	X		
Organic	9. Living organisms	X	X	X
matter	10. Dead organic matter*	X		
Air	11. Atmosphere	X		X
	12. Soil air		X	
Soil	13. Soil matrix		X	

*Includes entire or parts of entire organisms and their metabolic wastes.

Table 4. Transportation, Exchange, and Transformation Processes
in the Terrestrial Nitrogen Cycle

Transportations
1. Gravity
2. Surface runoff and erosion
3. Infiltration
4. Redistribution
5. Diffusion and dispersion
6. Capillary rise
7. Faunalpedoturbation*
8. Translocation

Exchanges
1. Evaporation
2. Dissolution
3. Ion exchange and adsorption
4. Ammonium fixation

Transformations
1. Biological nitrogen fixation
2. Ammonification
3. Immobilization
4. Nitrification
5. Nitrate reduction
6. Denitrification (chemical and biological)
7. Plant uptake
8. Excretion
9. Death

*Movement of earthworms and other soil fauna as defined by Hole.[18]

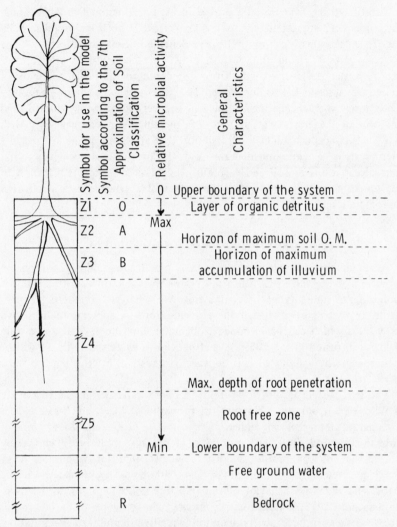

Figure 3. A proposed characterization of the environment for the terrestrial nitrogen system.

scientists have studied only selected regions of a soil system, e.g., topsoil, subsoil, root zone, or groundwater table. Therefore, to aid in obtaining data on the overall system from various researchers and from the literature, the soil system is divided conceptually into commonly-used horizontal layers for which some data are available. The criteria used to perform this division are spatial trends in the properties of a soil as viewed by a soil physicist, a soil chemist, and a pedologist. The latter scientist examines and classifies soils and soil profiles as they occur in the natural environment.

The first three zones of the profile in Figure 3, denoted as Z1, Z2, and Z3, correspond approximately to the O, A, and B horizons established by the 7th Approximation, Comprehensive Scheme of Soil Classification.[9] The Z4 zone extends from the bottom of Z3 to the depth to which roots penetrate; the Z5 layer extends from this depth to the lower boundary of the entire system—the water table. The thickness of each of these five zones varies with the nature of the soil and vegetation being studied, and for some of the zones, e.g., Z1 and Z5, the thickness may also vary with time. This concept includes the possibility that at a given point in time in a given system, a zone might have zero thickness. It is recognized that unusual variations of soil and vegetation might necessitate modification of this zonation scheme, but we believe that this scheme is a good first approximation.

Ecosystems Under Investigation

Assuming that our systems approach is practicable, and that the system definition of the terrestrial nitrogen cycle is workable, then both should be applicable to most types of ecosystems. We are now conducting experiments to assess the validity of these assumptions by applying them to two ecosystems which can be imagined as being at opposite ends of a spectrum of man's intrusion in the affairs of nature. If this work succeeds, we will be reassured of the validity of this approach to systems intermediate to these two.

The first system is the Noe Woods, a micro-watershed in the University of Wisconsin Arboretum. This woods, primarily of black and white oak, is intended to represent an undisturbed, forested ecosystem. In accordance with the method illustrated in Figure 1, the system is being intensively monitored. The soil profile and litter layer of the Noe Woods are being sampled extensively at nine areas. Data are being accumulated on the spatial and temporal behavior of soil moisture contents, moisture tensions, temperatures, pH's, and nitrogen contents. The level of the groundwater table, surface runoff, air temperature, relative humidity, precipitation, evaporation, dust fall, and wind velocity are also being monitored in the immediate vicinity of the Noe Woods. These data will aid in inferring which processes listed in Table 4 are of primary importance, and in evaluating the validity of the system model at the end of the modeling study.

To provide the required sharp contrast to the undisturbed Noe Woods, the second ecosystem chosen for study is a set of field plots at the University of Wisconsin Agricultural Experiment Station at Hancock, Wisconsin. This system is highly manipulated to produce high yields of agronomic and truck crops by the use of supplemental irrigation, fertilizers, and pesticides. The suction drainage and hydraulic weighing systems of the two field lysimeters installed at this site provide for convenient system monitoring.[10] For several years, University scientists have acquired data on the physical properties and behavior of the sandy soil at this station, and of the nutrient uptake characteristics of the variety of irrigated crops grown in this region.[11]

In some respects the dynamics of the Hancock system are simpler than those of the Noe Woods. The complex processes of litter fall and its subsequent decomposition are extremely important in the nitrogen dynamics of a forest soil, while in an agricultural system, litter fall and its incorporation by various cultivation practices can be controlled to alter its relative significance. The Z1 zone (Figure 3) of the Hancock model, therefore, will be smaller than that of the Noe Woods model. On the other hand, the Z5 of the Hancock model will be relatively larger than that in the Noe Woods model. This will be due to the shallowness of the roots, relative to the forested system, of most of the plants grown at the Hancock site, and to the irrigation and fertilization of the sandier soil.

System Model

The Law of Conservation of Mass provides a theoretical basis for developing a system mathematical model of the terrestrial nitrogen cycle. Imagine that at some depth z in the system there exists a volume element of soil. Nitrogen is traveling into and out of this volume element, as well as reacting within it. If mass is to be conserved within this element, the following relationship must be satisfied:

$$\begin{Bmatrix} \text{Rate of Mass} \\ \text{Accumulation} \end{Bmatrix} = \begin{Bmatrix} \text{Rate of} \\ \text{Mass In} \end{Bmatrix} - \begin{Bmatrix} \text{Rate of} \\ \text{Mass Out} \end{Bmatrix} + \begin{Bmatrix} \text{Net Rate of Appearance} \\ \text{of Mass by Reaction} \end{Bmatrix} \quad (2)$$

Figure 4 portrays a volume element of unit cross-sectional area and length $\triangle z$. A fraction (ϵ) of this volume element is occupied by a carrier. The (cv) expression represents the transport in the z direction of a nitrogen compound at concentration c via the flow of a carrier medium with volumetric flux v, a type of process often referred to as convection or bulk flow. The infiltration and redistribution phenomena of the hydrologic cycle referred to in Table 4 are examples of convective processes. Since (cv) is a function of depth, its value at any z is denoted by $(cv)_z$. Nitrogen may also be transported by dispersive mechanisms, denoted in Figure 4 by the product of a dispersivity coefficient D and a concentration gradient $(\partial c/\partial z)$. Finally, the R_i term refers to the rate of a process in which a nitrogen species is being consumed, produced, or exchanged.

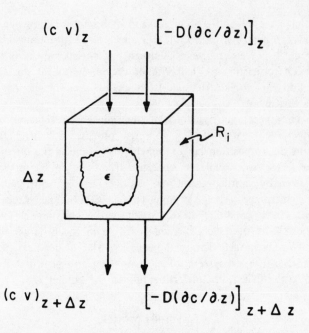

Figure 4. A theoretical volume element at depth z in the soil.

Equation 2 can now be written for the volume element of Figure 4 in a precise mathematical formulation known as the continuity equation (12):

$$\partial(\epsilon c)/\partial t = -\partial(cv)/\partial z + \partial\ [\epsilon D(\partial c/\partial z)]/\partial z + \Sigma R_i \qquad (3)$$

where: c = concentration of a nitrogen compound in a carrier (mass/unit volume of carrier)

ϵ = volume of a carrier per unit volume of soil

v = volumetric flux of a carrier (volume of carrier/unit cross-sectional area of soil/time)

D = dispersivity of a nitrogen compound in a carrier (cross-sectional area of carrier/time)

R_i = a rate of reaction involving the nitrogen compound (mass/unit volume of soil/time)

t = time

z = depth

Using the previously described terminology of equation (1) and Table 2, c is a dependent variable, z and t are independent variables, and D is a parameter. The terms ϵ, v, and R_i are dependent variables of the component processes of the system. Before equation (3) can be integrated to yield the response of c as a function of z and t, models which also express ϵ, v, and R_i as functions of z, t, and various parameters must be obtained. This requires process level studies in soil physics and soil chemistry, in a manner consistent with the scheme portrayed in Figure 1.

Dutt et al[13] followed essentially the above plan to develop a mathematical model for predicting the nitrate content of agricultural drain water from an arid soil. The result was an excellent first approximation for a system model, of the type proposed in equation (3), of the terrestrial nitrogen cycle. We believe, however, that the use of the more sophisticated modeling and experimental techniques in the process studies proposed in the following sections can yield a still more refined simulation than that achieved by Dutt et al.

Process Studies in Soil Physics

The movement of water through soil can be described by Darcy's law, a one-dimensional form of which is:

$$v = -K(h)\ \partial(h - z)/\partial z \tag{4}$$

where: v = volumetric flux in positive z direction (cm^3 of water/cm^2 of soil/day)

$K(h)$ = hydraulic conductivity (cm/day)

h = soil moisture pressure head (cm of water)

z = depth measured positive downward (cm)

When Darcy's law is substituted into the continuity equation for water,[14] the following dynamic equation results:

$$C(h)\ \partial h/\partial t = \partial\ [K(h)(\partial h/\partial z - 1)]/\partial z \tag{5}$$

$C(h)$, known as the specific moisture capacity (cm^{-1}), is actually the slope of the soil moisture characteristic curve, $\theta(h)$, which relates the soil moisture content θ(cm^3 water/cm^3 soil) to the pressure head h. For the soil water regime, θ is analagous to the ϵ of equation (3).

A numerical integration of equation (5), which requires prior knowledge of the way in which C and K vary with h, i.e., $C(h)$ and $K(h)$, results in a dynamic profile of h vs z. This profile can be used in equation (4) to obtain a velocity profile for inclusion in equation (3). It can also be converted into a profile of θ if the moisture characteristic curve is known. It is the common practice of scientists investigating this problem to represent the functional relationships for $\theta(h)$, $K(h)$, and $C(h)$ in tabular form.[14] Various interpolation schemes for these tables are then devised to produce a value of θ, K, or C for any given value of h. Our current process level studies are also directed toward developing models for producing moisture content and velocity profiles to substitute into equation (3). We plan to accomplish this, however, by independent development of analytical expressions for $\theta(h)$ and $K(h)$.

Because the $\theta(h)$ relationship is nonlinear and often hysteretic, its determination is a complex problem. Von Rooyen[15] has provided field data on θ vs. h in each of the four horizons of the sand in the Hancock lysimeters. Using a nonlinear least squares algorithm, as explained in detail by Draper and Smith[6], these data have been used to fit several nonlinear models in a search for the best equation for predicting the characteristic curve. Preliminary results from this modeling are very promising. Assuming that an adequate analytical representation of the moisture characteristic can be attained, the differentiation of this function will then yield the gradient $d\theta/dh$, i.e., the soil moisture capacity $C(h)$.

Inasmuch as hydraulic conductivity measurements in the field are difficult to accomplish, values of K vs. h have been obtained partially from theory. Green and Corey[16] recently reviewed the variety of modifications that have been made to a method for predicting the hydraulic conductivity of a porous medium based on pore-size distribution data. Since this distribution is actually manifested in the moisture characteristic, it is believed this method will dovetail conveniently with the efforts for modeling the moisture characteristic as described above. Preliminary results from the calculation and subsequent fitting of the generated K values to a nonlinear model in h are also encouraging.

Once analytical expressions for $\theta(h)$, $K(h)$, and $C(h)$ have been obtained, they will be included in an algorithm to numerically solve equations (5) and (4). The logic of this algorithm and the skeleton of a computer program for implementing it have already been devised. Once this model is built, it will be evaluated using data obtained from moisture flow experiments on the Hancock lysimeters. In accordance with the scheme for systems analysis illustrated in Figure 1, this initial testing of the overall model will lead to a modification of it and subsequent re-evaluation. Several iterations will be required before a final evaluation of the results and accuracy of the simulation can be made. If this work proves successful, a foundation will have been laid for achieving an analagous predictive capability on natural, more complex ecosystems, such as the Noe Woods, using similar experimental and modeling techniques.

Process Studies in Soil Chemistry

A list of the biochemical transformations in which soil nitrogen is involved was introduced in Table 4. Reaction equations for each of these are presented in Table 5 using convenient abbreviation symbols for the six

Table 5. Soil Nitrogen Transformations

Nitrogen species to be accounted for	Shorthand symbol	Symbolic reaction equations
1. NH_4^+ – Ammonium ion	NH4	1. Biological nitrogen fixation NGF→NLO
2. NO_2^- – Nitrite ion	NO2	2. Ammonification NDO→NH4
3. NO_3^- – Nitrate ion	NO3	3. Immobilization NH4→NLO; NO2→NLO; NO3→NLO
4. N bound in living organic matter	NLO	4. Nitrification NH4→NO2→NO_3
5. N bound in dead organic matter	NDO	5. Nitrate reduction NO3→NO2→NH4
6. N bound in gaseous form	NGF	6. Biological denitrification NO3→NO2→NGF
		7. Chemical denitrification NO2→NGF
		8. Plant uptake NH4→NLO; NO2→NLO; NO3→NLO
		9. Excretion NLO→NDO
		10. Death NLO→NDO

nitrogen species to be accounted for in the study. The first three of these species are common inorganic ions of nitrogen. The last three are not unique molecular compounds of nitrogen, but rather an aggregate of nitrogen containing substances with common characteristics. This characterization was thought to be a good compromise between the need to account for these obviously important forms of nitrogen, and the impracticability of distinguishing among the many unique compounds which constitute each of the three groups.

The objective of the proposed process studies in soil chemistry is the development of an analytical expression for the rate of each reaction in Table 5. These rates will be a function of independent variables such as temperature and moisture content, of parameters such as rate constants and metabolic efficiencies, and of dependent variables such as the concentrations of the nitrogen species themselves. While the modeling study of all these reactions will follow the pattern outlined in Table 2, in light of the differences in complexity between the transformations, not all the studies are expected to result in models of equal sophistication. The model of nitrogen immobilization, for example, may be no more than an empirical model based on linear regression of literature data, e.g.,

$$-d(NH4)/dt = a_0 + a_1(NH4) + a_2(NLO) + a_3 T + a_4 \theta \qquad (6)$$

where: ( ) = concentration of nitrogen species
a_i = regression coefficient
T = temperature

The kinetics of nitrification, on the other hand, may be studied theoretically and experimentally, resulting in nonlinear models such as:

$$-d(NH4)/dt = a_1 [\exp (-a_2/T)] (NH4)a_3 (NO3)a_4 \qquad (7)$$

McLaren[17] has proposed several Michaelis-Menten type equations which could also serve as potential mechanistic models for nitrification.

A preference for developing mechanistic models has motivated our initial process studies in soil chemistry. We are conducting a screening study (see Table 2), using a two-level factorial design, to determine the magnitude of the main and interaction effects of four variables on the overall nitrogen dynamics in undisturbed cores extracted from different horizons of the Noe Woods soil profile. The levels of these four variables—temperature, moisture content, concentration of ammonium ion, and of nitrate ion—will be evaluated periodically while the cores incubate for several weeks. As in our soil physics studies, nonlinear least squares theory will then be used to estimate the parameters in the proposed rate models. Based on analyses of the results of these experiments, further studies will be made of the kinetics of individual reactions, and of the mechanistic roles of these, and possibly other, independent variables.

ACKNOWLEDGEMENTS

Research supported by the College of Agricultural and Life Sciences, University of Wisconsin, Madison, and by grants from the National Science Foundation to the Institute for Environmental Studies and The Eastern Deciduous Forest Biome Project, International Biological Program, funded by the National Science Foundation (Subcontract 3351) under Interagency Agreement AG-199, 40-193-69 with The Atomic Energy Commission, Oak Ridge National Laboratory. The senior author also acknowledges the partial support provided by a graduate fellowship from the Chevron Research Company of the Standard Oil Company of California.

REFERENCES

1. National Research Council, Division of Biology and Agriculture, "Research Programs Constituting U.S. Participation in the International Biological Program," Report No. 4 of the U.S. National Committee for the International Biological Program, National Academy of Sciences, Washington, D.C., 1971.
2. Gerald Nadler, A collection of lecture and discussion notes from Industrial Engineering 816, "Planning of Large Scale Complex Systems," University of Wisconsin, Madison, 1970. Re: Gerald Nadler, *Work Design: A Systems Concept,* Richard D. Irwin, Homewood, Ill., 1970.
3. O. L. Loucks et al., "Integrated Studies of Land and Water Systems, Lake Wingra Basin," Research Proposal for the Lake Wingra Project, Institute of Environmental Studies, University of Wisconsin, Madison, 1969.
4. G. M. Van Dyne, "Ecosystems, Systems Ecology, and Systems Ecologists," ORNL - 3957, Oak Ridge National Laboratory, Oak Ridge, Tenn., 1966.
5. W. G. Hunter, A collection of lecture and discussion notes from Statistics 824, "Advanced Statistical Experimental Design for Engineers," University of Wisconsin, Madison, 1970. Re: G.E.P. Box, "Experimental Strategy," Technical Report No. 111, Dept. of Statistics, University of Wisconsin, Madison, 1967.
6. N. R. Draper and H. Smith, *Applied Regression Analysis,* John Wiley and Sons, Inc., New York, 1966.
7. G.E.P. Box and N.J. Hill, "Discrimination Among Mechanistic Models," *Technometrics, 9,* 57-71, 1967.
8. C. C. Delwiche, "The Nitrogen Cycle," *Scientific American, 223*(3), 136-147, 1970.
9. U. S. Department of Agriculture, Soil Conservation Service, Soil Survey Staff, "Soil Classification, a Comprehensive Scheme, 7th Approximation," Washington, D.C., 1960 (Supplement, 1967).
10. T. A. Black, G. W. Thurtell, and C. B. Tanner, "Hydraulic Load Cell Lysimeter, Construction, Calibration, and Tests," *Soil Sci. Soc. Amer. Proc., 32,* 623-629, 1968.
11. T. A. Black, W. R. Gardner, and G. W. Thrutell, "The Prediction of Evaporation, Drainage, and Soil Water Storage for a Bare Soil," *Soil Sci. Soc. Amer. Proc., 33,* 655-660, 1969.
12. R. B. Bird, W. E. Stewart, and E. N. Lightfoot, *Transport Phenomena,* John Wiley and Sons, Inc., New York, 1960.
13. G. R. Dutt et al., "Predicting the Nitrate Content of Agricultural Drain Water," Final Report on Contract 14-06-D-644 to U.S. Dept. Interior, Bureau of Reclamation, Department of Agriculture Chemistry and Soils, University of Arizona, Tucson, 1970.
14. R. Allan Freeze, "The Mechanism of Ground-water Recharge and Discharge 1. One-dimensional, Vertical, Unsteady, Unsaturated Flow above a Recharging or Discharging Ground-water Flow System," *Water Resources Research, 5,* 153-171, 1969.
15. Johan Van Rooyen, Unpublished data, Department of Soil Science, University of Wisconsin, Madison, 1971.
16. R. E. Green and J. C. Corey, "Calculation of Hydraulic Conductivity: A Further Evaluation of Some Predictive Methods," *Soil Sci. Soc. Amer. Proc., 35,* 3-8, 1971.

17. A. D. McLaren, "Kinetics of Nitrification in Soil: Growth of the Nitrifiers," *Soil Sci. Soc. Amer. Proc.,* **35**, 91-95, 1971.
18. F. D. Hole, "A Classification of Pedoturbations and Some Other Processes and Factors of Soil Formation in Relation to Isotropism and Antisotropism," *Soil Sci.,* **91**, 375-377, 1961.

4

Reprinted by permission of Duke University Press from *Ecology*, **44**, 322–331 (Spring 1963)

ENERGY STORAGE AND THE BALANCE OF PRODUCERS AND DECOMPOSERS IN ECOLOGICAL SYSTEMS[1]

JERRY S. OLSON

Radiation Ecology Section, Health Physics Division
Oak Ridge National Laboratory,[2] Oak Ridge, Tennessee

INTRODUCTION

The net rate of change in energy or material stored in an ecological system or its parts equals the rate of income minus the rate of loss. These rates may be expressed for various trophic levels (Lindeman 1942) or species, and also for the accumulated dead organic matter. In forests, both the living and the dead materials accumulate substantial reservoirs of energy, as shown by caloric measurements of Ovington (1961), Ovington and Heitkamp (1960), and others.

The rates of loss from all reservoirs can be expressed conveniently by a parameter k, which equals the fraction of the stored quantity that is lost per (short) unit time, without implying yet whether these fractions are approximately constant or not. Jenny, Gessel, and Bingham (1949) and recently Greenland and Nye (1959) have used such fractional loss rates, as constants, in characterizing the turnover and build-up of dead organic litter and soil humus. A confusing difference in approach and formulas which these papers use can be resolved in the following review of simple mathematical models for litter production and de-

---

[1] Parts of this paper were presented at the 9th International Botanical Congress (Olson 1959b) and the 1959 Symposium on "Energy Flow in Ecosystems" at Pennsylvania State University. Later references have been added to the present version for completeness.

[2] Operated by Union Carbide Corporation for the U. S. Atomic Energy Commission.

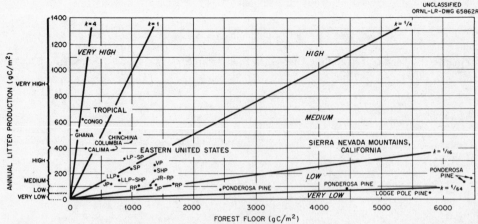

UNCLASSIFIED
ORNL-LR-DWG 65862R

Fig. 1. Estimates of decomposition rate factor k for carbon in evergreen forests, from the ratio of annual litter production L to (approximately) steady-state accumulation of forest floor X_{ss}. Tropical data for Ghana from Nye (1961), for Congo from Laudelot and Meyer (1954), and for Columbia from Jenny (1950) for mixed forests at 30 m above sea level (Calima) and 1,630 m (Chinchina). Southern pine forest data adapted from Heyward and Barnette (1936) for *Pinus palustris* (longleaf pine, LLP), *Pinus elliotii* (slash pine, SHP) and mixtures on southeastern U. S. coastal plain; from Metz (1952, 1954) for *Pinus echinata* (shortleaf pine, SP) and its mixtures with *Pinus taeda* (Loblolly pine, LP) on the South Carolina piedmont; from McGinnis (1958) for *Pinus virginiana* (Virginia or scrub pine, VP) in the Appalachian Ridge and Valley province at Oak Ridge, Tennessee. Northern pine forest data from Minnesota adapted from Alway and Zon (1930) for *Pinus banksiana* (jack pine, JP), *Pinus resinosa* (red or Norway pine, RP) and mixtures. Sierra Nevada data from Jenny, Gessel, and Bingham (1949) and Jenny (1950) for *Pinus ponderosa* and *Pinus contorta* (lodgepole pine) at various elevations above sea level.

cay in idealized evergreen and deciduous forests. The wide range of decay rates estimated here from data on forests of very contrasting climates helps to account for the great differences in total accumulation of organic carbon on top of mineral soil (horizontal axis of Fig. 1), and in the promptness in approaching their maximum storage capacity for dead organic matter.

Ovington's examples (1961) confirm that a substantial fraction (often one-third to one-half or more) of the energy and carbon annually fixed in forests is contributed to the forest floor as litter fall (mostly leaves). Because of this, and because litter fall is generally related to the quantity of photosynthetic machinery in the system, it is an interesting index of ecosystem productivity. Fig. 1 shows a wide range in litter production, plotted along the vertical axis in terms of grams of carbon per square meter per year. Litter production is "very high," sometimes above 400 g/m² in tropical forests of America (Jenny et al. 1949) and Africa (Greenland and Nye 1959). It is "medium" or "high," 100-200 or 200-400 g/m², in the northern and southern pine forests of the eastern United States. California mountain forest data (Jenny et al. 1949), when expressed in terms of carbon (using C contents based on Jenny 1950) show medium to low production (50-200 g/m²).

"Very low" litter production, below 50 g/m² carbon, would normally indicate relatively nonproductive forest.

The scatter in any one portion of Fig. 1 indicates that the production and storage of dead organic carbon are not closely related. In fact, the diagram as a whole demonstrates an inverse relation. Low storage of carbon in the highly productive tropical forests contrasts with high levels of carbon and energy accumulation in the relatively unproductive cool temperate forests. A major reason for this inverse relation clearly involves rates at which dead organic matter is broken down or incorporated into the mineral soil by organisms. Chemical composition of coniferous litter, as well as low temperature, tends to retard biological activity in the northern or subalpine forests. Under the assumption that the forest floors in the stands here selected may approximate a steady state, one method of estimating the decay parameter k can be made from the ratio of the vertical and horizontal coordinates of each point on Fig. 1; other methods are also indicated below.

MODELS AND METHODS

Let X be measured either as ovendry weight, organic carbon, or energy in dead organic matter per square meter of ground surface, and let the income rate, and either the amount or the fraction lost per unit time, be expressed in comparable units. The opening sentence on net rates of change for a discrete interval of time (day or year) Δt can be restated as:

$$\frac{\Delta X}{\Delta t} = \text{income for interval} - \text{loss for interval.} \tag{1}$$

For the model of steady income, L, the instantaneous rate of change is the limit as Δt and ΔX approach zero

$$\frac{dX}{dt} = L - kX \tag{2}$$

The loss rate kX is considered as a product of the amount accumulated (X) and the instantaneous fractional loss rate, k, which will first be considered for the special case of a constant loss rate.

If and when accumulation reaches a steady-state level, X_{ss}, then (by definition of a steady state) the rate of change in equation (2) is 0, so income = loss.

$$L = kX_{\text{steady state}} = kX_{ss} \tag{3}$$

For this case, the rate parameter k can be estimated by the ratio of income to steady-state total, as the ratio of the vertical axis over the horizontal axis of Fig. 1 (slopes of the diagonal lines).

$$k = L/X_{ss} \tag{4}$$

While equation (4) represents one method of estimating loss rates from harvest of litter and forest floor materials in the field, experimental approaches are also being used for more direct estimates of loss rates (Shanks and Olson 1961, Olson and Crossley 1963). Where L and k can be estimated independently of X, then their ratio might be used to predict the steady-state level yet to be accumulated in an ecosystem which has not yet come to a balance of income and loss.

$$X_{ss} = L/k \tag{5}$$

Decay with no production

The special case in which $L = 0$ approximates our current experiments (Shanks and Olson 1961) in which litter is confined in mesh bags and remeasured after loss of differential increments of material, dX. Equation (2) can be rearranged to express these losses as a fraction of the residue X currently remaining.

$$\frac{dX}{X} = - k \, dt \tag{6}$$

The model for constant fractional weight loss implies (by integration) a constant negative slope,

$-k$, on a semi-log graph of the amount remaining from an initial quantity X_o at $t = 0$.

$$\text{natural logarithm } (X) = - kt + \text{natural logarithm } (X_o) \tag{7}$$

The fraction remaining is

$$\text{natural log } \left(\frac{X}{X_o}\right) = - kt \tag{8}$$

Statistical estimation of this slope provides a second method for estimating the parameter k.

Taking antilogarithms of both sides of equation (8) gives the fraction remaining as a negative exponential function, like that shown on Fig. 2a.

$$\frac{X}{X_o} = e^{-kt} \tag{9}$$

For example, after 1 year of biological decay and physical breakdown, the fraction remaining would ideally be $X/X_o = e^{-k}$. A third method for estimating the instantaneous rate of breakdown (really a special case of the second method) makes allowance for the change in weight loss, kX, which is due to changes in X (assuming k constant). Jenny expressed this loss as a fraction of the original total and called it k':

$$k' = \Delta X/X = 1 - X/X_o = 1 - e^{-kt} \tag{10}$$

From equations (8) to (10), the relation of k' to the instantaneous decay rate k is:

$$k = - \text{natural log } (X/X_o) = - 1n(1 - k') \tag{11}$$

Where the time interval is short (e.g., expressed in loss per day), or where the decay is slow even for time units as long as a year, the change in X during the interval is small, k and k' are both small fractions of 1, and there is little numerical difference between them. However, for time spans as long as a year, the k' for litter decay is frequently a large fraction of 1, and k may be equal to 1 (as in Fig. 2) or may exceed 1 (see below). For $k = 1$, $k' = 1 - e^{-1} = 0.632$.

Accumulation with continuous litter fall (case 1)

For the case in which litter is almost steadily falling, at a rate L which the model assumes constant, equation (2) can be rewritten like equation (6) after dividing all terms by k:

$$\frac{dX}{(L/k - X)} = - k \, dt \tag{12}$$

This has an integral like equation (7).

$$1n(L/k - X) = - kt - \text{constant} \tag{13}$$

For an initial condition with no forest floor (e.g., burned off by ground fire), $X = 0$ at $t = 0$, and

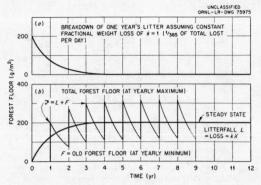

FIG. 2. a. Negative exponential curve for idealized litter decay, assuming weight loss proportional to amount remaining at any one time. b. Gradually rising exponential curve for accumulation under conditions of steady income and loss, compared with step-wise curve for additions and losses of litter in idealized deciduous ecosystem.

the constant in equation (13) is $- \ln(L/k)$. The antilog of equation (13) gives the solution—a rising curve like that shown in Fig. 2b.

$$X = (L/k) \, (1 - e^{-kt}) \qquad (14)$$

This curve is the mirror image of the curve for decay, shown in Fig. 2a, for the case $k = 1$ in yearly units, which is equivalent to a daily loss of 1/365 of the total weight remaining at any one time. As k increases or decreases, the steady-state level $X_{ss} = L/k$ decreases or increases accordingly. But there is also a speeding up or delay in the approach to this steady state, which is illustrated in Fig. 3 by the dashed lines (corresponding to solid lines for decay curves with the same value of k).

Accumulation with discrete annual litter fall (case 2)

Important differences between case 1, for steady fall, and case 2, for fall at the end of the growing season ("idealized deciduous forest") are illustrated by the jagged curve of Fig. 2b, for the same values of $L = 200 \, g/m^2$ produced per year, and $k = 1$ (or 1/365 of total lost per day). There is no longer a steady replacement of the litter decomposing between pulses of litter fall, and the remainder of the 200 g after 1 year of decay ($e^{-1} = 0.368$) is less than the amount which had accumulated after either 1 or 2 years of steady fall and decay in case 1. This deficit below the theoretical level for steady accumulation is then made up by the second sudden "autumn" litter fall; but the more rapid loss, kX, again lowers X past the curve given for case 1. Because kX is higher in the first half of each annual decay cycle than in the second half, the steady state of the smooth

curve lies a little below the half-way level between the peaks and troughs for the discontinuous case. (Considering intermediate cases, e.g., with quarterly or weekly "installments" of litter fall, would still give a jagged curve with peaks and valleys straddling the rising curve of Fig. 2b, but with less amplitude between the extremes.)

An equation for the annual peak values $\mathcal{J}_n$ which occur right after the nth year's annual fall of litter differs from equation (14) only by the constant $L/k' = T$ for the theoretical limiting value (Olson 1959a, 1959b):

$$\mathcal{J}_n = (L/k') (1 - e^{-kn}) \qquad (15)$$

This is illustrated in Fig. 4, for the value of $k' = 0.25$ (so that $k = 0.288$ from equation (11)). Still lower values for k or k' of course would show slower decay and slower approach to the limiting value for accumulation. The values given are equivalent to the geometric series of Jenny et al. (1949) which developed directly from the annual increments of L (which they called A).

Between annual events of litter fall in the ideal deciduous model, decay is governed by equations (9) and (10). The value $\mathcal{J}_n$ is reduced to a value $F_n = (1 - k') \mathcal{J}_n$.

$$F_n = (L(1 - k')/k') (1 - e^{-kn}) \qquad (16)$$

When $\mathcal{J}$ approaches an upper limit of $T = 800$ g/m^2 in Fig. 4, F approaches an upper limit of $0.75T$ or $T - L$, namely 600 g/m^2.

After the limiting value T is approached, it is possible to estimate

$$k' = L/T \qquad (17)$$

which equals $A/(F + A)$ in Jenny's terminology, by analogy with equation (4). The ratio of equation (17) may thus be useful in characterizing decay where the deciduous model is a good approximation, and where autumn-peak values can be measured directly, or summed by taking late summer values (F) and adding the litter which falls in the autumn. Even in this case, it is necessary to use equation (11) if it is desired to find actual decay rates k, which can be converted to short time units simply by a change in time scale. In the case of a forest with little or no seasonal alternation, equation (4) can be used, as Greenland and Nye proposed (1959, p. 287) for a direct measure of effective litter decay. It should be noted, however, that the term F_E in their paper refers to the steadily rising curve in Fig. 2b (including a *fraction* of the current year's litter, and some older material), not strictly material which had been subject to one or more years of decay (*cf.* Jenny et al. 1949).

For both deciduous and coniferous forests, of

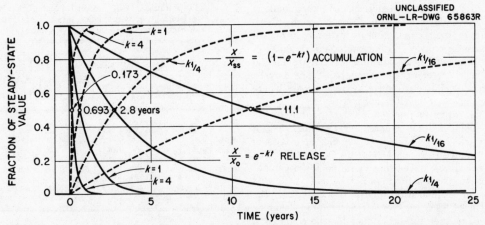

FIG. 3. Exponential equations for carbon or energy stored in dead organic matter in a model ecosystem, for four values of $k = 4$, 1, 1/4, and 1/16 for very high, high, medium, and low decay rates, respectively. Solid lines for decrease from steady state, assuming production drops from L units per year to 0 at time $t = 0$. Dashed lines for accumulation toward steady state, where production is continuous at L units per year after $t = 0$.

course, litter fall is actually spread over a period of time. Both the peaks and valleys of the stair-step curve should be rounded off to something like the form shown by the short dotted lines of Fig. 4. Periodic litter inputs that are not discrete are suggested elsewhere as a better approximation compromising between two extremes of case 1 and case 2 (Neel and Olson 1962), but the extreme cases suffice for discussion of the illustrative data in the present paper. Other important limitations on the assumption of year-to-year uniformity of L and k are acknowledged as noted below, and can be overcome by elaboration of the mathematics or by the aid of computer techniques (Neel and Olson 1962, Olson 1963). However, these objections do not invalidate the basic ideas of the models

or the kinds of conclusions drawn by Jenny et al. (1949) and by Greenland and Nye (1959).

RESULTS ON ESTIMATING DECOMPOSITION PARAMETERS

The estimates of k (from equation (4)) range from high values near $k = 4$ for the African forests, down to about 1 for two forests from Columbia. Pine forests of southeastern United States have values scattering around the line for $k = 0.25$ (1/4), while Minnesota pine forests range down toward the line for $k = 0.0625$ (1/16). Jenny et al. (1949) emphasized the high accumulation of litter in the California profiles and very slow decay parameters, down to 0.009 for lodge-pole pine at 3,000 m altitude. Because such a small fraction of any one year's production is spent in decomposition during early stages of forest floor accumulation, storage of organic matter and of energy must continue until the total becomes so large that the product kX gradually approaches the income L and approximates the balance in equation (3). The time required for such an adjustment is considered next.

DURATIONS AND LEVELS OF ACCUMULATION

A convenient virtue of the simple exponential model is that the time required to reach halfway to the asymptotic level is the same time as that required for decomposition of half of the accumulated organic matter (Fig. 3). For either equation (9), (14), (15), or (16), this time is given by the solution of $0.5 = e^{-kt}$ for t, which is $-\ln(0.5)/k = 0.693/k$. We have an analogy with radioactive half-life, or half-time for accumu-

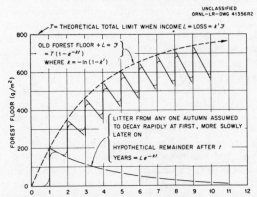

FIG. 4. Exponential decay and accumulation curves for idealized deciduous forest, with sudden annual litterfall of $L = 200$ g/m², and annual decomposition $k' = 1 - e^{-k} = 0.25$ of total present at any one time, so $k = 0.288$.

lation of radioactive materials. Another analogy is with "biological half-time" for either the elimination or the accumulation of materials in organisms or organs. The parameter $0.693/k = T_{0.5}$ may be viewed as a "half-time" for environmental accumulation or decay. Such half-times may be useful descriptive parameters, even in cases where an exponential model is not strictly applicable.

If an exponential model is valid, the time period $3/k$ should be that required for attaining 95% of the final level, while $5/k$ should approximate time needed to reach 99% of the final level. The reciprocal $1/k$ is the time required for decomposition to the fraction $1/e = 0.368$ of the initial level, or accumulation to $1 - 1/e = 0.632$ of the final level (see Fig. 3). This reciprocal can be viewed as the "time constant" for the component of an ecosystem circuit, analogous with that for discharging or charging of a condenser in a simple electrical circuit or in the integrator of an analog computer.

Such numbers are readily obtainable from tables of exponentials or logarithms, and a few have been selected in Tables I and II to represent the range

TABLE I. Parameters for exponential accumulation of organic matter or energy in ecosystems with steady litterfall rate (L)

DECAY PARAMETERS		"Half-time"	95% time	STEADY-STATE LEVEL FOR PRODUCTION OF L UNITS PER YEAR L/K			
$k = \dfrac{L}{X_{ss}}$	$\dfrac{1}{k} = \dfrac{X_{ss}}{L}$	$\dfrac{0.6931}{k}$	$\dfrac{3}{k}$	$L=50$	$L=100$	$L=200$	$L=400$
4	0.25	0.173	0.75	12.5	25	50	100
2	0.5	0.346	1.50	25	50	100	200
1	1.0	0.693	3.0	50	100	200	400
0.693	1.442	1.000	4.33	72	144	289	577
0.5	2	1.386	6	100	200	400	800
0.25	4	2.772	12	200	400	800	1,600
0.125	8	5.544	24	400	800	1,600	3,200
0.0625	16	11.09	48	800	1,600	3,200	6,400
0.0312	32	22.21	96	1,600	3,200	6,400	12,800
0.0156	64	44.42	192	3,200	6,400	12,800	25,600
0.01	100	69.31	300	5,000	10,000	20,000	40,000
0.003	333	232.3	1,000	15,000	30,000	60,000	120,000
0.001	1,000	693.1	3,000	50,000	100,000	200,000	400,000

of values particularly important for organic matter decay for cases 1 and 2 discussed above.

Figs. 5-7 provide illustrations of the great differences in the levels of accumulation and the promptness of equilibration to be expected for various combinations of productivity and decay parameters. Values of k plotted here are for the idealized maximum accumulation immediately following litter fall in a deciduous forest case from Table II. Comparison with Table I suggests that numerically the differences between deciduous and evergreen forests would not be great, especially for values of k or k' below about 0.06 or 0.016 which are shown in Figs. 6 and 7. For values

TABLE II. Parameters for seasonal accumulation and decay of organic litter, with sudden litterfall of L units, once each year (idealized deciduous forest)

DECAY PARAMETERS		TIME PARAMETERS		SEASONAL MAXIMA AND SEASONAL MINIMA			
				For $L=200$		For $L=400$	
k	$k' = \dfrac{1}{1-e^{-k}}$	$\dfrac{1}{k} \dfrac{L+F_{ss}}{L}$	$\dfrac{0.6931}{k}$	$\dfrac{3}{k}$	F_{ss}	F_{ss}	F_{ss}	F_{ss}
4	0.9717	1.010	0.173	0.75	204	4	407	7
2	0.865	1.156	0.346	1.50	231	31	462	62
1	0.632	1.582	0.693	3.00	316	116	632	232
0.693	0.50	2	1.00	4.33	400	200	800	400
0.288	0.25	4	2.41	10.4	800	600	1,600	1,200
0.136	0.125	8	5.19	22.5	1,600	1,400	3,200	2,800
0.0645	0.0625	16	10.7	46.5	3,200	3,000	6,400	6,000
0.0317	0.0312	32	21.8	94.5	6,400	6,200	12,800	12,800
0.0157	0.0156	64	44.0	190.5	12,800	12,600	25,600	25,200
0.0100	0.0100	100	69.3	300	20,000	19,800	40,000	39,600

higher than 0.1, numerical differences between k and k' become greater. The sawtooth character of the accumulation curve whose peak values are shown in Fig. 5 ($k' = 0.25$, $k = 0.288$) were already brought out on a larger scale, in Fig. 4.

DISCUSSION

The rate parameter k in part measures the effectiveness of decomposer organisms like fungi, bacteria, and certain animals in breaking down organic materials. Some of this breakdown of litter, accumulated on top of mineral soil, involves leaching and physical transport of materials into the mineral soil, providing income of carbon and energy for soil organic matter. But a large fraction presumably represents losses of energy from the ecosystem due to respiration of the decomposing organisms. These must be considered in the energy budget of the ecosystem as a whole.

At the 1959 Symposium on "Energy Flow in Ecosystems" at Pennsylvania State University, where this paper was presented, differences between aquatic and terrestrial systems and between the approaches of the workers investigating them were emphasized by several papers. The first two papers by Beyers (1962) and by Wilson illustrated progress and difficulties in the use of several methods aiming at direct measurement of rates of oxygen and CO_2 exchange and C^{14} uptake from which limnologists and oceanographers infer rates of energy flow. In planktonic microcosms decomposition rates and energy turnover are presumably high. But there is nevertheless deposition and accumulation of resistant organic materials in bottom sediments in many environments. This results in a "litter decay" situation and storage of sedimentary carbon somewhat analogous with that considered here for terrestrial systems.

The third paper (see Ovington 1957, 1961, Ovington and Heitkamp 1960) illustrated the

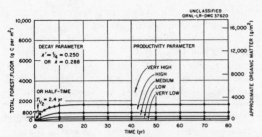

FIG. 5. Increase in annual "autumn maximum" of dead organic matter on top of mineral soil for several levels of litter productivity parameter L, for idealized deciduous forest. Medium decay rates, k and k'.

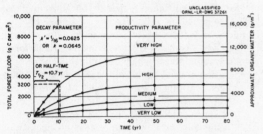

FIG. 6. Increase in anual "autumn maximum" of dead organic matter on top of mineral soil for several levels of litter productivity parameter L, for idealized deciduous forest. Low decay rates.

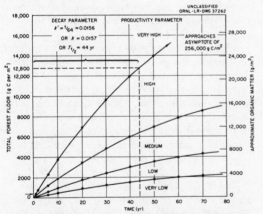

FIG. 7. Increase in annual "autumn maximum" of dead organic matter on top of mineral soil for several levels of litter productivity parameter L, for idealized deciduous forest. Very low decay rates. Note great differences in both the asymptotic accumulation level and the promptness of attaining this level; also the similarity of k and k' in Figs. 6 and 7.

contrasting approach of terrestrial harvest studies, typically measuring the accumulated net production and calculating average rates of energy flow over increments of time. Ovington estimated that over half of the net production during the development of a 55-year-old pine forest had been re-

leased by decomposition. The relative importance of the decomposer pathways of energy flow was even greater in younger forests and herbaceous ecosystems.

Implications for net production and succession

Present illustrations show the wide variation in the period during which the ecosystem as a whole may show a positive net storage (production minus loss) of energy in the form of dead organic matter, unincorporated in mineral soil. Thick humus layers in many northern regions require many decades to develop. Even longer durations will normally be required for equilibration of production and loss of organic matter or humus within the mineral soil. For example, rates of accumulation of soil nitrogen and humus in sand dune soils of approximately known ages indicated values of k near 0.003, so that $3/k$ or about 1,000 years would be needed to attain 95% of the steady-state level (Olson 1958). Depending on what assumptions are made about contributions of organic matter to the mineral soil from the forest floor and roots (cf. Jenny 1950, vs. Greenland and Nye 1959), decomposition rates for soil organic matter in many other soils may show similar lags in accumulation. Even maximum estimates involve only a few per cent decay per year, so $3/k$ is of the order of centuries.

Considering the total development of an ecosystem, a "climax" condition in the sense of a steady state, or zero net *community* storage of matter or energy, may not be attained until long after composition and average biomass of many living species has become nearly constant, or begun to oscillate around some average value.

One aspect of natural community development involves the readjustment of all components of an ecosystem toward an asymptotic condition. The condition approached is influenced by a given set of values for the productivity and decay parameters. Some biological developments may be specifically delayed until litter and soil humus have approached near their steady-state values.

A second aspect of developmental succession involves gradual or sudden changes in levels of production or rates of decay, controlled by the presence of new species of plants or animals, and new physical and chemical characteristics of the materials they produce. A typical succession may thus involve alternating episodes of adjustment toward specific levels and fairly sudden shifts when the parameters governing these levels change, as illustrated by E. P. Odum (1960). Slow and sudden changes in soil, which are regulated by parameters for losses of materials from the soil, may both contribute to long-term trends in succession.

As noted earlier (Olson 1958), quantitative models for changes in community succession and soil development can thereby impart a mathematical significance to Cowles's classic statement (1899): that succession represents a "variable approaching a variable, rather than a constant." The first variable consists of the vector sum of properties describing the state of the ecosystem. The second variable includes the asymptotic condition which is being approached during a given phase of development, and the parameters like L and k which govern the level of this condition. While the first variable may change rapidly at first, then slowly for a while as it approaches a steady state (or oscillates around it), a change in the second variable and the parameters which govern it will bring the whole system into a new period of readjustment.

Modifications of models

The assumptions of constant production and constant decomposition parameters will accordingly have to be modified to treat many special cases. One change is the allowance for low production during early stages of population development and succession, and a positive feedback to favor higher productivity as growth and development proceed. The result is a sigmoid curve pattern for accumulation of living organic matter, litter, and incorporated humus like the analog computer graphs of Neel and Olson (1962). H. T. Odum's paper (1960) on electrical network analogs did not account for any condition other than the steady state. It could be extended to the accumulations through time by the addition of capacitors into his circuit.[3]

Some aquatic systems and bogs (Lindeman 1942) have prolonged accumulation of peat or other biogenic sediments, where decomposition may be even slower than in any of the terrestrial systems. Some important ecosystems may share characteristics of terrestrial and aquatic communities and are modified by import and export of energy and materials by means of water currents, as noted in the papers on salt marshes by Smalley (1960) and Kuenzler (1961) (see also Odum and Smalley 1959). Their data on invertebrate herbivores and Golley's (1960) on vertebrates also show that a measurable but fairly small fraction of energy flow passes through consumers as com-

[3] This modification which I suggested to Odum in March 1959 has been made in his later electrical analog network. It introduces a positive feedback by manually charging the production parameter.

pared with that through decomposers, except in systems that are heavily grazed.

Decomposition may be so slow as to require a modification in the model for aquatic systems because of the possible failure to level off at any constant asymptote. If organic matter is essentially removed from access to decomposers during burial in bottom sediments, a given vertical column through the ecosystem extending down through these sediments may show the continued storage of energy in the forms which ultimately contribute to our fossil fuels of peat, coal, petroleum, or oil shale.

A particularly interesting case of the changing parameters mentioned earlier involves the increase in decomposition rate after early stages in community development or succession have passed. This often results in the breakdown of thick forest floor accumulations which commonly occur in first-generation conifer forests in many ecological successions. While lower storage of energy and mineral nutrients may be found in later successional stages (often deciduous or mixed forests), the faster turnover of the nutrient elements may permit the primary productivity of the system to be higher than before. On the other hand, there may be developments of succession which actually represent a degeneration, if the losses of important nutrients or of the favorable physical characteristics of organic humus proceed so rapidly that they cannot be balanced by production.

The equations outlined above are given in terms of deterministic models and are analogous with the differential equations and solutions in physics. Even for physical models (as in radioactive decay, for example) there are chance fluctuations which make these equations only approximations to what happens in reality. We could similarly visualize many complicating chance variables which could make the actual state of any given ecosystem fluctuate around the hypothetical conditions which are here projected on the basis of a simple model. In addition to chance fluctuations, there are important oscillations (besides the abrupt seasonal stops indicated in Fig. 2) which might be superimposed on the simple trends given here, and some of these are covered elsewhere (Neel and Olson 1962).

These further developments can be handled by direct extension of the simple differential equations and exponential equations used here. They are facilitated by the use of analog computers which perform the integrations electrically or electronically (Olson 1963). Furthermore, the use of computer components aids the connection of many components into a whole electrical or electronic circuit to simulate the trophic structure of the ecosystem. The physical operations simulating integration sidestep many of the complications in the analytical representation of integrals arising when the outputs of some components of the system serve as inputs to the next components (Neel and Olson 1962, Appendix). While the individual components might follow a simple exponential development controlled by income and loss, as in Figs. 2-7 if their inputs were constant, the various lags in the build-up of inputs to different parts of the system can be allowed for in the behavior of the analog model.

Summary

While some fraction of the solar energy fixed by producing plants is released by respiration of these plants and of animals, much of it is stored in dead organic matter until released by decomposing organisms, at rates which vary greatly from place to place. The general differential equation for the rate of change in energy storage is illustrated by models for build-up and decomposition of organic matter, particularly for litter in deciduous or evergreen forests. Equations of Jenny et al. (1949) and Greenland and Nye (1959) each have a useful place in estimating decay parameters. For the case of steady production and decay, the ratio of annual litter production, L, to the amount accumulated on top of mineral soil in a steady state, X_{ss}, provides estimates of the decomposition parameter k. Estimates range from over 4 in certain tropical forests to less than 0.01 in subalpine forests. Decomposition rates for organic matter within mineral soils may range from near 0.01 to 0.001.

Since it takes a period of about $3/k$ years before storage has attained 95% of its steady-state level, many ecosystems continue to show a positive net community production for centuries—perhaps long after changes in numbers and biomass of some species are reduced to minor fluctuations around a "climax" composition. On the other hand, the slow change in soil conditions may in some cases facilitate the introduction of new species after some delay during succession. The change in productivity or decomposition parameters controlled by these species may lead in turn to a series of later readjustments in energy storage and release, which modify litter and soil conditions. Modified microenvironments in turn may further alter the succession and "climax."

Literature Cited

Alway, F. J., and R. Zon. 1930. Quantity and nutrient content of pine leaf litter. J. Forestry 28:715-727.

Beyers, R. 1962. The metabolism of twelve aquatic laboratory microecosystems. Ph.D. Thesis, University of Texas, Austin, Texas. 123 p.

Cowles, H. C. 1899. The ecological relations of the vegetation on the sand dunes of Lake Michigan. Botan. Gaz. 27:361-391.

Golley, F. B. 1960. Energy dynamics of a food chain of an old-field community. Ecol. Monographs 30: 187-206.

Greenland, D. J., and P. H. Nye. 1959. Increases in the carbon and nitrogen contents of tropical soils under natural fallows. J. Soil Sci. 10:284-299.

Heyward, Frank, and R. M. Barnette. 1936. Field characteristics and partial chemical analyses of the humus layer of longleaf pine forest soils. Florida Agr. Expt. Sta. Bull. 302, 27 p.

Jenny, Hans. 1950. Causes of the high nitrogen and organic matter content of certain tropical forest soils. Soil Sci. 69:63-69.

Jenny, Hans, S. P. Gessel, and F. T. Bingham. 1949. Comparative study of decomposition rates of organic matter in temperate and tropical regions. Soil Sci. 68:419-432.

Kuenzler, E. J. 1961. Structure and energy flow of a mussel population in a Georgia salt marsh. Limnol. Oceanog. 6:191-204.

Laudelot, H., and J. Meyer. 1954. Les cycles d'elements minearaux et de matiere organique en foret equatoriale congolaise. Trans. 5th Int. Congr. Soil Sci. 2:267-272.

Lindeman, R. L. 1942. The trophic-dynamic aspect of ecology. Ecology 23:399-419.

McGinnis, John. 1958. Forest litter and humus types of East Tennessee. M.S. Thesis, Unversity of Tennessee.

Metz, L. J. 1952. Weight and nitrogen and calcium content of the annual litter fall of forests in the South Carolina Piedmont. Soil Sci. Soc. Am. Proc. 16:38-41.

———. 1954. Forest floor in the Piedmont Region of South Carolina. Soil Sci. Soc. Am. Proc. 18:335-338.

Neel, R. B., and J. S. Olson. 1962. The use of analog computers for simulating the movement of isotopes in ecological systems. Oak Ridge National Laboratory ORNL-3172.

Nye, P. H. 1961. Organic matter and nutrient cycles under moist tropical forest. Plant and Soil 13:333-346.

Odum, E. P. 1960. Organic production and turnover in old field succession. Ecology 41:34-49.

Odum, E. P., and A. E. Smalley. 1959. Comparison of population energy flow of a herbivorous and a deposit-feeding invertebrate in a salt marsh ecosystem. Proc. Natl. Acad. Sci. 45:617-622.

Odum, H. T. 1960. Ecological potential and analog circuits for the ecosystem. Am. Scientist 48:1-8.

Olson, J. S. 1958. Rates of succession and soil changes on southern Lake Michigan sand dunes. Botan. Gaz. 119:125-170.

———. 1959a. Forest studies, p. 41-45. *In* Health Physics Division Annual Progress Report period ending July 31, 1959, Oak Ridge Nat. Lab., ORNL-2806.

———. 1959b. Exponential equations relating productivity, decay, and accumulation of forest litter. IX Internat. Bot. Congr. Proc. 2:287.

———. 1963. Analog computer models for movement of radionuclides through ecosystems. In Radioecology: Proc. of First National Symposium. Reinhold Publ. Co. (in press). New York.

Olson, J. S., and D. A. Crossley, Jr. 1963. Tracer studies of the breakdown of forest litter. In Radioecology: Proc. of First National Symposium. Reinhold Publ. Co. (in press). New York.

Ovington, J. D. 1957. Dry-matter production by *Pinus sylvestris* L. Ann. Bot. NS 21:287-314.

———. 1961. Some aspects of energy flow in plantations of *Pinus sylvestris*. Ann. Bot. 25:12-20.

Ovington, J. D., and D. Heitkamp. 1960. Accumulation of energy in forest plantations in Britain. J. Ecology 48:639-646.

Shanks, R. E., and J. S. Olson. 1961. First-year breakdown of leaf litter in Southern Appalachian forests. Science 134:194-195.

Smalley, A. E. 1960. Energy flow of a salt marsh grasshopper population. Ecology 41:672-677.

Reprinted from *Science*, **170**, 503–508 (Oct. 30, 1970)

Systems Studies of DDT Transport

A systems analysis provides new insights for
predicting long-term impacts of DDT in ecosystems.

H. L. Harrison, O. L. Loucks, J. W. Mitchell,
D. F. Parkhurst, C. R. Tracy, D. G. Watts,
V. J. Yannacone, Jr.

During the recent hearings conducted by the State Department of Natural Resources considering a petition to ban the use of DDT (*1*) in Wisconsin (*2*, *3*), both the petitioners and the defendants produced witnesses who gave testimony from a wide range of scientific disciplines. The evidence presented included a description of the transport mechanisms of DDT, its chemical properties, its physiological effects on individual organisms, and its effects on whole populations. In this article we describe a model for DDT transport which was presented as testimony to integrate the range of evidence. The resulting systems analysis yields testable hypotheses, demonstrates gaps in our understanding of the impact of DDT, and indicates possible future consequences of its use.

First, a mechanistic model is developed to describe the movement of DDT and its breakdown product DDE (*1*) in an inland ecosystem. The analysis is based upon the trophic-level concept (*4*) which allows a simplified quantitative examination of complex "food web" processes in ecosystems. Trophic levels in their basic form may be represented by a pyramid of energy, or a pyramid of biomass, such that there is more energy or biomass in green plants than in herbivores, and more in herbivores than in their predators. These concepts allow use of mathematical formulations for the flows and storages of DDT in the ecosystem. The mechanisms leading to selective concentration of DDT in specific living organisms are then discussed and a mathematical model is formulated to describe how DDT or DDE concentration varies with time and with trophic level. Finally, a mathematical model is derived to indicate the dependence of population size in any trophic level

upon the populations in adjacent levels, and to provide a basis for predicting population changes attributable to DDT.

The results suggest, first, that even if no more DDT is ever added to the biosphere, its concentration in certain species at or near the top of the trophic structure could continue to rise for some years. In the light of the known broad range of DDT and DDE toxicity, additional species may decline or disappear. The population analysis indicates that secondary changes in prey populations would occur in response to direct effects on predator numbers. The methods described in this article should be viewed as an example of the approach that will have to be followed in many future studies of persistent contaminants in the environment if we are to make satisfactory estimates of long-term effects.

Descriptive Models for the Transport of DDT

The inputs, outputs, and storages of DDT in a Wisconsin regional ecosystem are shown schematically in Fig. 1. The ecosystem is divided into three levels—atmosphere, terrestrial biomass, and substrate water with its associated aquatic organisms. The major sources of DDT entering Wisconsin are listed as *Inputs*. DDT may be introduced into any one of the three levels. Examples are transport into the Wisconsin atmosphere by means of DDT attached to particulate matter carried by wind, commercial application of DDT to the land surface, and DDT brought into the Wisconsin water system by lake and river currents. DDT and its breakdown products leave the Wisconsin regional ecosystem by similar trans-

port processes. There do not appear to be any mechanisms that will completely degrade these toxic breakdown products at rates comparable to the present commercial application of DDT.

All available evidence indicates that the total output of DDT and its breakdown products is considerably less than the total input, the remainder being stored (*3*, *5*). First, there is buildup of DDT in the lipid portions of living organisms and this DDT is subsequently retained in dead tissue for varying periods of time. Second, there is long-term accumulation in the soil, in deep bodies of water, and in deep organic deposits in marshes and lake borders. A complete determination of DDT inputs and their redistribution requires a detailed examination of the mechanisms controlling redistribution of DDT in the natural environment. Research at the University of Wisconsin on the movement of nutrients through lake and stream systems has provided a means of examining in detail these inputs, storages, transformations, and losses of DDT from both terrestrial and aquatic environments (*6*).

A complete listing of the DDT inputs, transports, and outputs in air, in water, and in living organisms, as well as the potential transformation to breakdown products in the air and in the organisms, was prepared as a foundation to the generalized ecosystem trophic structure shown in Fig. 2. The three basic carrier systems—atmosphere, water, and living biomass—provide the basis for a series of differential equations which, taken together, would permit mathematical simulation of the flow of any transported material such as DDT or its breakdown products in an ecosystem (*6*). As represented in Fig. 2, all of the exchanges of DDT from one carrier variable to another within the system are included,

Professors Harrison and Mitchell are in the Mechanical Engineering Department, University of Wisconsin; Professor Loucks and Dr. Parkhurst are in the Botany Department; Mr. Tracy is in the Zoology Department; Professor Watts is in the Statistics Department; and Mr. Yannacone is a trial attorney and cochairman of the Environmental Law Section, American Trial Lawyers Association. This paper is a synthesis of material drawn from several sources and assembled as background to the testimony of Dr. Loucks, the last witness for the Environmental Defense Fund, Inc. of New York, the Citizens Natural Resources Association of Wisconsin, Inc., and the Izaac Walton League of Wisconsin, Inc., in hearings on their application to have DDT declared a pollutant under the laws of the State of Wisconsin.

and also the sites at which transformations of DDT to degradation products take place. All information available to date indicates that a degradation product such as DDE also will be transported and redistributed by the same mechanisms. The general analysis of transports showed that many of the exchanges of DDT and DDE occurred in the subsurface physical environment where degradation cannot take place. In addition, it showed the sites in both the physical environment and in living and dead biomass where significant storage of DDT and DDE can be expected because of either the slow exchange of the host material or a slow breakdown process.

Woodwell and others (7, 8) have described the role of the trophic structure in transporting and concentrating DDT in the ecosystem. The listing of the transport variables and flows of DDT in the environment through each successive level of the trophic structure can be summarized schematically as in Fig. 2, and more precisely in the equations that follow. Storage of DDT in each trophic level, transport from one trophic level to another, and transformation of DDT into DDE or DDD (1) by metabolism are examined in view of the processes involved within each trophic level and between each pair of trophic levels (6). Thus, all available information on the transport, accumulation, and transformation of DDT in natural biological systems was drawn together as a foundation for the mathematical analyses that follow.

DDT-Concentrating Mechanisms in Natural Systems

DDT has been described (3) as a chemical that "combines in a single molecule the properties of broad biological activity, chemical stability, mobility, and solubility characteristics that cause it to be accumulated by living organisms. . . ." DDT has a solubility in water of only 1.2 parts per billion (9) and a low vapor pressure (10). However, the vast amounts of air and water moving in the atmosphere and oceans transport significant quantities of DDT in relatively short periods of time. The result is that DDT has become ubiquitous.

In contrast to its near insolubility in water, the solubility of DDT in lipids and other organic materials is very high (5). These properties account for DDT accumulation in the lipids of plants and animals. DDT can be taken up actively with water and nutrients, or simply absorbed when an organism is exposed to water or air containing the pesticide. Uptake by exposure can occur, for example, through the gills of fish or the skin of terrestrial animals, or directly into the cells of aquatic plants.

The chlorinated hydrocarbons are chemically stable; DDT stored in the lipids of the organisms in a given trophic level often undergoes little degradation (5). The combination of high solubility and high stability allows "magnification" of DDT concentrations from lower to higher trophic levels within an ecosystem. Individuals in each trophic level feed on those in the levels below, and the proportion of food that an individual in a particular level converts into biomass of its own species is usually much less than 50 percent (7). The rest is excreted after the organism removes and respires much of the energy stored in it. Furthermore, organisms that grow to full size early in their life spans and others that give relatively little biomass to the egg or fetal stages of their young will tend to respire an even greater percentage of their food intake. Therefore, a substance like DDT, which is stored in the lipids and breaks down slowly, can accumulate to high concentrations in a trophic level.

In Table 1 are summarized available data on DDT concentration in a Lake Michigan ecosystem. Because of the concentrating mechanism described above, the concentration of DDT in the herring gull is some 7000 times that in the bottom muds. This increase in concentration may be even higher in other ecosystems; Woodwell et al. (8) found that concentrations in individuals from the top trophic level were some 10^6 times higher than those in the environment.

Data on the toxicity of DDT for the species in the Lake Michigan ecosystem are limited, but other data indicate that the species vary considerably in sensitivity to DDT. Some organisms in low trophic levels show adverse effects from concentrations much lower than those that are known to affect animals higher in the structure (3, 5). American kestrels fed DDT and dieldrin (1) in food (6 to 18 parts per million) underwent reproductive failure (11). Field observations of falcons and eagles show similar effects at relatively high concentrations in food and body tissues (12). The brain tissue of robins killed in an elm-spraying program showed high concentrations (50 parts per million) of DDT (13). Lower in the trophic structure, 39 percent of adult brine shrimp died within 3 weeks when placed in a solution of 1 part DDT in 10^{12} parts water, and all died within 5 days in concentrations of 1 part in 10^{10} (14). Finally, only a few parts DDT in 10^9 parts water are necessary to reduce photosynthesis in a number of species of phytoplankton (15).

Taken together, the chemical and physical properties of DDT allow it to "flow" to and concentrate in living tissue, especially in organisms in high

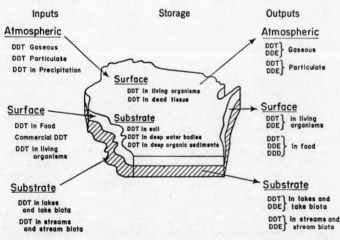

Fig. 1. Transport of DDT in the Wisconsin regional ecosystem. The major pathways by which DDT and its by-products flow throughout the ecosystem are shown.

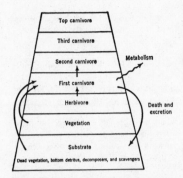

Fig. 2. Schematic representation of the flows of DDT in the ecosystem. The flows of DDT are shown for the first carnivore level only. Pathways similar to those indicated by the arrows exist between any other level and all other levels but are not shown for simplicity. Storage of DDT also occurs in each level.

trophic levels. However, low concentrations in individuals in low trophic levels may be equally alarming owing to the apparent greater sensitivity of some organisms in those levels.

Mathematical Analysis of DDT Concentration

A mathematical model of the movement of DDT from one trophic level to another in an ecosystem based on established information on the transporting and concentrating processes of DDT has been developed. The objectives of this model are: (i) to state quantitative relationships that show how DDT and its metabolites are concentrated in the various trophic levels, and (ii) to indicate the dynamic nature of the transport and concentrating processes. The transport of DDT is considered, but the analysis also applies to DDE and other products of similar chemical and biological properties.

The transports and storage of DDT identified in Fig. 1 are depicted schematically in trophic level form in Fig. 2. The possible pathways by which DDT may enter or leave are shown for the first carnivore level only.

DDT is carried by the flows of matter. The flows of mass into and out of a given trophic level i are related by the conservation of mass principle, where $\dot{m}$ is used to denote a mass flow rate (for example, in kilograms per day) entering or leaving a level.

$$\dot{m}_{in,i} = \dot{m}_{out,i} \quad (1)$$

The total mass of each trophic level is assumed to be constant. We recognize that population fluctuations exist, and in this analysis the time-average mass of the level is used. The inflow of mass is the sum of the rates of exposure to environment and of the ingestion of organisms in the lower levels, $\dot{m}_{j,i}$

$$\dot{m}_{in,i} = \sum_{j=1}^{i-1} \dot{m}_{j,i} \quad (2)$$

The outflow of mass from the level is due to death, $\dot{m}_{d,i}$, and excretion, $\dot{m}_{ex,i}$

$$\dot{m}_{out,i} = \dot{m}_{d,i} + \dot{m}_{ex,i} \quad (3)$$

Equations 2 and 3 can be combined with Eq. 1 to yield an expression for the flow of biomass through any trophic level

$$\sum_{j=1}^{i-1} \dot{m}_{j,i} = \dot{m}_{d,i} + \dot{m}_{ex,i} \quad (4)$$

For a given trophic level i, DDT is carried into and out of the level with these mass flows, destroyed by metabolic processes, and stored in the lipid biomass of the individuals in the trophic level. The mass flows are related by the conservation of mass principle, where $\dot{p}$ denotes the flow rate of the pesticide (DDT) (for example, in kilograms per day) and $\dot{p}_{met,i}$ is the rate of metabolism of DDT in level i

$$\dot{p}_{in,i} = \dot{p}_{out,i} + \dot{p}_{met,i} + \left(\frac{dp}{dt}\right)_i \quad (5)$$

where dp/dt is the rate of storage of DDT in the trophic level.

The flow of the pesticide into the i level is expressed as the product of the rate of ingestion or exposure to the species in lower levels $\dot{m}_{j,i}$ and the instantaneous average concentration c_j of DDT in that lower level. The inflow is the sum of these terms for all such lower levels

$$\dot{p}_{in,i} = \sum_{j=1}^{i-1} c_j \dot{m}_{j,i} \quad (6)$$

For application to DDE, or similar byproducts of DDT metabolism, the inflow term would also include creation of the substance from DDT.

The flow of DDT from the level by death, predation, or other causes is the product of the instantaneous average concentration c_i in the level and the death rate $\dot{m}_{d,i}$. In addition, DDT leaves by excretion in amount $c_{ex}\dot{m}_{ex,i}$ where c_{ex} is the DDT concentration.

$$\dot{p}_{out,i} = c_i \dot{m}_{d,i} + c_{ex}\dot{m}_{ex,i} \quad (7)$$

Table 1. Concentration of DDT in a Lake Michigan ecosystem (27).

Trophic level	Concentration (parts per 10^9)
Bottom muds	14
Amphipods	410
Fish	3,000–6,000
Herring gulls	99,000

The rate of storage of DDT in the i level is the product of the total mass in the level m_i and the rate of change of the average concentration with respect to time

$$\frac{dp}{dt} = m_i \frac{dc_i}{dt} \quad (8)$$

Combining Eqs. 6, 7, and 8 with Eq. 5 and rearranging yields an expression for the DDT concentration in a given level as a function of time

$$\left(\frac{\dot{m}_i}{\dot{m}_{d,i}}\right)\frac{dc_i}{dt} + c_i =$$
$$\frac{1}{\dot{m}_{d,i}}\left[\sum_{j=1}^{i-1} \dot{m}_{j,i}c_j - \dot{p}_{met,i} - c_{ex}\dot{m}_{ex,i}\right] \quad (9)$$

The expression for the substrate, level 1 of Fig. 2, has additional inflow terms due to man, $\dot{p}_{man}$, excretion from the upper levels, and the fraction f_i of those individuals in the upper levels that die naturally. DDT may also leave the substrate in amounts $\dot{p}_{out}$ through the mechanisms described in Fig. 1. Therefore

$$\frac{dc_1}{dt} + \left(\frac{\sum_{i=2}^{I} \dot{m}_{1,i}}{m_1}\right)c_1 =$$
$$\frac{1}{m_1}\left[\dot{p}_{man} - \dot{p}_{out} + \sum_{i=2}^{I}(c_{ex}\dot{m}_{ex,i} + c_i f_i \dot{m}_{d,i})\right] \quad (10)$$

where I is the total number of trophic levels.

Equations 9 and 10 are general equations that describe the rate of change of DDT or any other pesticide in the various levels of the ecosystem. At present, there are insufficient data to allow an evaluation of all the terms. Nevertheless, a qualitative study of these equations can provide valuable insight into pesticide transport through an ecosystem.

The term $(m_i/\dot{m}_{d,i})$ in Eq. 9 is related both to the time it takes the trophic level to respond and to the equilibrium concentrations of DDT in

65

the level. In order to show this more clearly, the average values for the mass and life span of the individuals in a trophic level will be used. The total mass in a level is the product of the number of individuals N_i and the average mass of each member M_i, or

$$m_i = N_i M_i$$

The death rate is the product of the average mass of the individuals and the number dying per unit time $N_{d,i}$, or

$$\dot{m}_{d,i} = M_i N_{d,i}$$

The average life span of the members is denoted T_i. The number dying per unit time is then $(1/T_i)$ times the total number of individuals, or

$$N_{d,i} = N_i / T_i$$

Combining these relations yields

$$\frac{m_i}{\dot{m}_{d,i}} = \frac{N_i M_i}{(M_i N_i / T_i)} = T_i \qquad (11)$$

It is now possible to deduce the equilibrium levels of DDT once the ecosystem reaches a steady state. With the simplifying assumption that there is no further addition of DDT to the system, and that the term representing the metabolism of DDT is small relative to the other terms (16), the equilibrium concentration C_i from Eq. 9 is

$$C_i = \frac{T_i}{m_i}\left[\sum_{j=1}^{i-1}\dot{m}_{j,i}C_j - C_{ex}\dot{m}_{ex,i}\right]$$
$$(12)$$

Equation 12 states that the equilibrium concentration of DDT in a level is: (i) directly proportional to the average life span T_i of its members; (ii) inversely proportional to the total mass m_i of the level; and (iii) proportional to the net retention of DDT in the level. The net retention in a level depends on the concentration of DDT in the lower levels. The rate at which organisms of the lower levels are ingested, and the amount of DDT excreted. Equation 12 is based on the assumption that the metabolism of DDT in the trophic structure is negligible; metabolism would serve to reduce the net retention in a level. Metabolism would also prevent the attainment of a true equilibrium condition. However, Eq. 12, modified to account for metabolism, would still provide an estimate of the concentration in a level under these conditions. The effects of life span and mass would be unchanged. In general, as one moves up the ecosystem (Fig. 2), the average life spans increase and the mass in the trophic level decreases. Equation 12 provides an ex-

planation for the observed increased concentration of DDT in the higher levels of the ecosystem.

The dynamic nature of pesticide flows in an ecosystem are determined from the solution of Eqs. 9 and 10. However, there are virtually no quantitative data on the flow of DDT into and out of natural populations and on the movement of biomass through predator-prey interactions. To indicate the dynamic nature of the concentrating process, we consider a simplified, but representative, situation. For this approximation, we assume that organisms in all consumer levels feed only on the organisms in the level immediately below, retain all DDT ingested, and neither metabolize nor excrete DDT. Equation 9, when combined with Eqs. 4 and 11, can then be simplified to

$$T_i\frac{dc_i}{dt} + c_i = \left(\frac{\dot{m}_{i-1}}{\dot{m}_{i-1} - \dot{m}_{ex,i}}\right)c_{i-1} \quad (13)$$

The equilibrium concentration is then

$$C_i = \left(\frac{\dot{m}_{i-1}}{\dot{m}_{i-1} - \dot{m}_{ex,i}}\right)C_{i-1} \qquad (14)$$

The coefficient of C_{i-1} in Eq. 14 is the ratio of the rate of mass ingested to the difference between the rates of ingestion and excretion. For an individual member, the difference between ingestion and excretion over the life span is the body weight. Thus, the coefficient of C_{i-1} can be written as the ratio of mass ingested over a lifetime to body weight. This coefficient is always greater than unity and probably ranges between 10 and 10,000. Thus, the equilibrium concentrations increase as one moves from lower to higher levels in the ecosystem.

The coefficient T_i in Eq. 13 is the time constant (17) of the level, and indicates the ecosystem response times. For example, a sudden sustained increase in concentration in the $i-1$ level yields an exponential increase in concentration c_i

$$c_i = \left(\frac{\dot{m}_{i-1}}{\dot{m}_{i-1} - \dot{m}_{ex,i}}\right)(1 - e^{-t/T_i})\, c_{i-1} \quad (15)$$

Calculations made with Eq. 15 show that the concentration in the i level reaches 98.2 percent of its equilibrium value in a time equal to $4\,T_i$. Each trophic level requires about four average life spans to reach equilibrium in response to changes in DDT concentration in the level below it.

This conclusion regarding response times is based on the simple system described by Eq. 13. Feedback loops, represented by the additional terms on the right side of Eq. 9, provide a more

accurate model of an ecosystem and these increase the time required to respond (17). Thus the prediction that the ecosystem cannot reach equilibrium until about four times the longest average life span is conservative.

The mathematical development is based on populations containing individuals at all ages, with deaths occurring in equal proportions in all age groups. This assumption is an oversimplification for natural populations when all ages are included, although adult birds tend to have a linearly decreasing expectation of further life as a function of time (18).

An alternate model for the age distribution in a population is that all deaths occur at an age corresponding to the average life span. With this model it can be shown that pesticide concentration in an ecosystem described by Eq. 13 cannot reach equilibrium until a time equal to the sum of the average life spans in all trophic levels. Concentrations in a more complex ecosystem described by Eqs. 9 and 10 would take longer to reach equilibrium because of the complex feedback loops.

Our estimates for the length of time necessary to reach equilibrium after the introduction of DDT, or any similar pesticide, depend upon the life spans in the trophic structure and the age distributions. We estimate that this time lies between four times the average life span of the longest-lived species and the sum of the life spans for all trophic levels.

We are not certain what value of average life span most closely approximates what occurs in nature, but it is most likely between the maximum attainable life and the expected life at birth (mean death age) for any species. In ecosystems with long-lived members such as the herring gull with a life span between 2.8 years (19) and 40 to 50 years (20), and the osprey, eagle, and falcon with life spans of as long as 60 to 100 years (20), it is quite apparent that the full effects of today's use of DDT will not be completely felt for many years to come. Furthermore, it is easily possible that ecosystems with such long-lived constituents have not yet felt the full impact of the original use of DDT in the late 1940's.

The model result, Eq. 15, describes the redistribution of an initial step input of DDT to the ecosystem. All inputs can be synthesized as a series of steps, and thus the time constant represents the time response of the system to all inputs. Therefore, the concentra-

66

tion of DDT or DDE in any species or trophic level at the present time reflects the addition of the responses to all of the step inputs of DDT to the present. In view of the continuing worldwide inputs of DDT, it is readily apparent that the Wisconsin ecosystem is not yet in equilibrium.

Population Response to Declining Predator Control

In the previous section, mathematical expressions relating the dynamic nature of pesticide flows were developed. In an earlier section, the impact of these DDT levels on major carnivores such as falcons, eagles, and ospreys was discussed. In this section we shall consider the population responses in the lower trophic levels (prey species) to variations in predator populations. The objective is to explore the overall impact on the system resulting from elimination or decline of certain species as a result of DDT.

The use of mathematical models to predict population responses is well established (18, 21). Lotka (22) and Volterra (23) formulated the first mathematical population model which assumes that population growth rate is proportional to population size. Predator-prey interactions, in which predator eats prey, are modeled by the product of the predator and prey populations (23, 24). The solutions for these models show that the prey populations react unstably to fluctuations in predator levels in a manner similar to some natural population fluctuations (25). It is the unstable behavior of these models, which does not correspond to the present behavior of the Wisconsin regional ecosystem, that has led to the development of the present population model.

To explore the possible effects of predator fluctuations in stable systems (17, 18) we considered a three-level model with a prey population n feeding on a food population f and being preyed upon by a predator population g. The system is assumed to be in equilibrium initially with population values N, F, and G. Deviations from equilibrium δf, δn, and δg, are assumed which are positive when a population exceeds its equilibrium value and negative when the population is less than that value. This approach is commonly used in control systems analysis (17).

For the food-prey interaction, the impact on population n of a change in food supply is a function of the differ-

ence between the food deviation and an equivalent population deviation, and is assumed to be proportional to the difference. Thus, in some period of time Δt, the change in population Δn can be expressed as

$$\Delta n = K_1 (\delta f - W_1 \delta n) \qquad (16)$$

where K_1 is a proportionality factor and W_1 is an equivalence factor.

Equation 16 states, for example, that if the food supply is great ($\delta f > 0$) at a time when the population is at its equilibrium value ($\delta n = 0$), the population will increase. Further, if decreased food ($\delta f < 0$) coincides with increased population ($\delta n > 0$), there would be a marked reduction in population.

The predator-prey interaction is developed similarly:

$$\Delta n = - K_2 (\delta g + W_2 \delta n) \qquad (17)$$

Equations 16 and 17 are combined to yield a defining equation for population n:

$d (\delta n)/dt =$
$\quad - (K_1 W_1 + K_2 W_2)\, \delta n + K_1 \delta f - K_2 \delta g \quad (18)$

Equation 18 is inherently stable as a result of the negative coefficient of the δn term. The population will not "blow up" because of the increased possibility of its members either being eaten or starving when the population increases ($\delta n > 0$).

The three-level population model is

$$d (\delta f)/dt = - K_1 W_1 \delta f - K_1 \delta n$$
$d (\delta n)/dt =$
$\quad K_2 \delta f - (K_2 W_2 + K_3 W_3)\, \delta n - K_3 \delta g$
$$d (\delta g)/dt = K_4 \delta n - K_4 W_4 \delta g \qquad (19)$$

Equation 19 produces oscillatory responses that qualitatively represent actual population fluctuations (18). For example, an abrupt reduction of predator population leads to an increase of population n followed by a decrease of food population. Ultimately, the values return to their equilibrium values. The continual oscillations observed in natural population numbers may simply be the result of random changes in external factors such as weather.

The increase in prey population due to a decrease in predator numbers (for example, due to DDT) has the potential of becoming a public nuisance. In addition, the resulting decrease in prey food supply is potentially damaging in that a particular food species may be eliminated entirely. Alternate food sources (for example, crops) may be sought by the prey in order to nourish the increased population.

The explosive nature of populations deprived of their natural predators is well documented (25). The classical example is the removal of wolves, coyotes, and mountain lions from the Kaibab Plateau in Arizona, which resulted in an explosive increase in the mule deer populations until the deer decimated their own food supply. Starvation and disease resulted with a tremendous mule deer population crash (26).

We recognize that the population models considered above represent a somewhat oversimplified view of actual population response. Nevertheless, these models are sufficiently descriptive to yield the general nature of population response, and to make possible the prediction that a significant variation of a predator population would cause upsets throughout the entire system, some of which might be of sufficient magnitude to create "out-of-control" conditions.

Discussion and Conclusion

Some predictions of the consequence of adding DDT to the environment based on the DDT transport, accumulation, and concentration mechanisms and on the evidence of the impacts of DDT in ecosystems are now possible. DDT concentrates in the higher trophic levels, and, depending on unknown rates of metabolic breakdown, the concentrations in long-lived species in higher levels can be expected to continue to increase long after the addition of DDT to the environment has ceased.

The presence of DDT in any trophic level can have three major consequences:

1) The concentration may be high enough to kill the members in that level. If this occurs, the entire ecosystem will move toward a new equilibrium no longer influenced by the removed trophic level.

2) The concentration may not be lethal, but may cause adverse sublethal effects such as reproductive failure (12). The affected trophic level will disappear just as if the dosage were lethal.

3) The concentration may have no apparent effect on the trophic level. The DDT in this level will then pass on to the next higher trophic level, which will then be subject to the same three consequences of DDT concentration.

The analysis of DDT diffusion through the trophic structure of the ecosystem indicates that the equilibrium concentrations will be a function of the "life spans" of the organisms in the system. Thus, the top carnivores, which play an important role in stabilizing the system, may take a long time to respond to the input of DDT. Since DDT is reducing predator numbers in present ecosystems, new population explosions may result.

The ecosystems making up the world biosphere might restabilize after the loss of a species, but with different population levels of the remaining species. Radical changes in population levels could have serious economic and public nuisance consequences. Further, the DDT once present in the obliterated populations will then be concentrated into fewer remaining species. Whether or not this process could be repeated in a series of systematic obliterations of the species in upper trophic levels with a consequent concentration of DDT into remaining species cannot be predicted at this time. However, with the models presented here, it can be predicted that the consequence of the present worldwide inputs of DDT in the environment will not become apparent for many years.

References and Notes

1. DDT, 1,1,1-trichloro-2,2-bis(p-chlorophenyl)ethane; DDE, 1,1-dichloro-2,2-bis(p-chlorophenyl)ethylene; DDD, 1,1-dichloro-2,2-bis-(p-chlorophenyl)ethane; dieldrin, 1,2,3,4,10,10-hexachloro-6,7-epoxy-1,4,4a,5,6,7,8,8a-octahydro-endo-exo-1, 4 : 5, 8-dimethanonaphthalene.
2. L. J. Carter, Science 163, 548 (1969). On 22 May 1970, Chief Examiner Maurice Van Sustern of the Wisconsin Department of Natural Resources released his findings from the Wisconsin hearings. Van Sustern summarized the substantial evidence in a 25-page report and issued a one-page ruling which states in part: "DDT, including one or more of its metabolites in any concentration or in combination with other chemicals at any level, within any tolerances, or in any amounts, is harmful to humans and found to be of public health significance. . . . DDT and its analogs are therefore environmental pollutants within the definitions of Sections 144.01 (11) and 144.30 (9), Wisconsin Statutes, by contaminating and rendering unclean and impure the air, land and waters of the state and making the same injurious to public health and deleterious to fish, bird and animal life."
3. C. F. Wurster, BioScience 19, 809 (1969).
4. E. P. Odum, Fundamentals of Ecology (Saunders, Philadelphia, ed. 2, 1959); Science 164, 262 (1969).
5. C. F. Wurster, Biol. Conserv. 1, 123 (1969).
6. D. D. Huff and P. Kruger, Advan. Chem. Ser. No. 93 (1970); D. G. Watts and O. L. Loucks, Models for Describing Exchanges within Ecosystems (Institute for Environmental Studies Paper, Univ. of Wisconsin, Madison, 1969).
7. G. M. Woodwell, Sci. Amer. 216, 24 (March 1967).
8. ————, C. F. Wurster, Jr., P. A. Isaacson, Science 156, 821 (1967).
9. M. C. Bowman, F. Acree, M. K. Corbett, J. Agr. Food Chem. 8, 406 (1960).
10. C. A. Edwards, Residue Rev. 13, 83 (1966).
11. R. D. Porter and S. N. Wiemeyer, Science 165, 199 (1969).
12. J. J. Hickey, Ed., Peregrine Falcon Populations, (Univ. of Wisconsin Press, Madison, 1969).
13. C. F. Wurster, Jr., D. H. Wurster, W. N. Strickland, Science 148, 90 (1965).
14. D. S. Grosch, ibid. 155, 592 (1967).
15. C. F. Wurster, Jr., ibid. 159, 1474 (1968).
16. W. W. Walley, thesis, Mississippi State University (1965); J. D. Judah, Brit. J. Pharmacol. 4, 120 (1949); Scientific Aspects of Pest Control (National Academy of Sciences Symposium, Washington, D.C., 1966).
17. H. L. Harrison and J. G. Bollinger, Introduction to Automatic Controls (International Textbook, Scranton, Pa., ed. 2, 1969).
18. D. L. Lack, Population Studies of Birds (Clarendon Press, Oxford, 1966).
19. R. A. Paynter, Auk 72, 79 (1955).
20. A. Comfort, Sci. Amer. 205, 108 (Feb. 1961); P. L. Altmann and D. S. Dittmer, Eds., Biology Data Book (Federation of American Societies for Experimental Biology, Washington, D.C., 1964).
21. K. E. F. Watt, Ecology and Resource Management (McGraw-Hill, New York, 1968); R. H. MacArthur and J. H. Connell, The Biology of Populations (Wiley, New York, 1966).
22. A. J. Lotka, Elements of Physical Biology (Williams & Wilkins, Baltimore, 1925).
23. V. Volterra, Mem. Accad. Lincei. Roma 2, 31 (1926).
24. E. L. Coultlee and R. I. Jennrich, Amer. Natur. 102, 6 (1968).
25. H. G. Andrewartha and L. C. Birch, Distribution and Abundance of Animals (Univ. of Chicago Press, Chicago, 1954).
26. C. B. Knight, Basic Concepts of Ecology (Macmillan, New York, 1965).
27. J. J. Hickey, J. A. Keith, F. B. Coon, J. Appl. Ecol. 3 (Suppl.) 141 (1966).
28. We thank C. F. Wurster for permission to paraphrase portions of references (3) and (5). Supported by the Fishing Tackle Manufacturers Association and the Citizens Natural Resources Association of Wisconsin.

II
Management of Ecological Systems

Editors' Comments on Papers 6 Through 10

6 **Koenig and Tummala:** Principles of Ecosystem Design and Management
IEEE Trans. Sys., Man Cybern., **SMC-2**, 449–459 (Sept. 1972)

7 **Dodge:** Forest Fuel Accumulation—A Growing Problem
Science, **177**, 139–142 (July 14, 1972)

8 **Van Dyne:** A Systems Approach to Grasslands
Proc. XI Intern. Grassland Congr., A131–A143 (1970)

9 **Eisel:** Watershed Management: A Systems Approach
Water Resources Res., **8**, 326–338 (Apr. 1972)

10 **Liang, Huang, and Wang:** Scheduling Bioproduction Harvest
Operations Res., **19**, 1968–1707 (Nov.–Dec. 1971)

The papers of Part I were chosen in an attempt to deepen ecological understanding and to sharpen the tools of analysis. The five papers of Part II deal with man's intervention in natural and mixed natural/man-made systems, in the following order: principles of ecosystem management, forest management, grassland management, watershed management, and management of a marketable crop.

The first paper in this section, Paper 6, is coauthored by Herman E. Koenig, Chairman of the Department of Electrical Engineering and Systems Science at Michigan State University. Koenig, one of the most seasoned of the "general systems" modelers, in 1965, publicly marked his leap out of pure electrical engineering systems into general systems with "Mathematical Models of Socio-Economic Systems: An Example," a paper which appeared in the first issue of the *IEEE Transactions on Systems, Science and Cybernetics* (**SSC-1**, No. 1, 41–45). Since that time Koenig has collaborated with various researchers to apply systems theory to educational institutions, the agricultural industry, and ecosystems.

"Principles of Ecosystem Design and Management," by Koenig and R. L. Tummala, originally appeared in September, 1972, in the *IEEE Transactions on Systems, Man and Cybernetics*, the successor to the journal cited above. This paper presents an engineering approach to the design and management of ecological systems. The authors have sought to integrate the physical world of *mass–energy* and the economic world of *mass–monetary cost* using a rigorous, generalized state-space framework. Pricing mechanisms of the economy are singled out for consideration as instruments of management. The analysis is based on viewing "each component of the ecosystem . . . as performing one or a combination of three basic functions: material transformation, transportation, or storage."

The first four sections of the paper are easily accessible to the layman and present a number of important ideas. For example, the authors suggest the use of pricing mechanisms "to 'hedge' against certain classes of uncertainties," and they correctly emphasize that "the physical structure of the ecosystem has a direct bearing on human values and social stresses." Koenig and Tummala affirm the need for analysis that improves "the capability to assess the mass-energy processing capacities of natural landscape components" (as, for example, in Papers 1 through 5), for without knowledge and information gained through analysis and experimental work, it is impossible to apply the design and management concepts they propose.

Starting with Section V of the paper, a background in engineering or applied mathematics and some experience in graph theory (network graphs) are almost essential. The key to understanding the equations of Section V is to consider the center of all the circles (the objects) of Figure 5 as a single, common datum point. Also, in trying to understand the meaning of a negative energy cost in the equations, the reader should think of the release of potential energy that was gained earlier as a result of a positive energy cost. In Section VI of the paper, the unit matrix U is the same as the identity matrix I.

Should progress in Paper 6 become difficult, a turn to Paper 23, a nonmathematical companion piece coauthored by Koenig, might well be helpful. In addition, the reader should note that, coincidentally, Papers 25 and 26 correspond to References (7) and (5), respectively, of Paper 6.

There are a number of typographical errors in the paper. Worth noting, perhaps, is that in Figure 2, the inward-pointing ray labeled m_i should read m, i. Reference (7) should give the title of Leontief's paper as "Environmental Repercussions and the Economic Structure: An Input–Output Approach." The title of the book in Reference (12) should read *The Limits to Growth*.

"Forest Fuel Accumulation—A Growing Problem" (Paper 7), has been included because it indicates how first notions for the management of an ecosystem may turn out to have very damaging effects after the natural system works out its own response to man's attempts at management. In the particular case of western forests, man's attempts to prevent forest fires over the past half-century have led to an accumulation of dead matter and smaller second-growth trees that exacerbate those fires that inevitably do occur.

Marvin Dodge reviews the state of knowledge concerning forest fires, pointing out that little is known quantitatively about the relationship between the type and amount of forest fuels and fire intensity. He notes that few studies to determine the annual accumulation of litter in a forest have been made. One of the studies Dodge does cite is Paper 4 by Olson. Forest managers currently make decisions with vastly improved qualitative understanding of forest dynamics but with severely limited quantitative models. Dodge considers the present management alternatives for preserving healthy forests and preventing disastrous forest fires. Perhaps the most promising alternative is to set controllable fires!

George M. Van Dyne is one of the few biologists to be lauded for his use of systems analysis and mathematical modeling in the study of ecological systems. "A Systems Approach to Grasslands" (Paper 8) provides a good, brief definition on the systems approach and its elements in relation to grasslands. Van Dyne conceptualizes a biological system as a set of interrelated compartments. A set of first-order differential equations (rate equations) is used to describe the relationships among compartments mathematically. Empirically fitted mathematical models are presented, with results for two cases involving the vegetation level of a grassland ecosystem. Van

Dyne then lays out (Figure 3) a nine-compartment multi-trophic-level model that he and his students at Colorado State University have developed into a FORTRAN-language computer program used for simulation of a grassland ecosystem; interesting computer results are shown in Figure 4. In addition, Van Dyne proposes a generalized grasslands model that allows for the incorporation of nonconstant coefficients into the model.

Van Dyne concludes with an excellent section, "Implications of Systems Approaches." He notes the need for more data and for more realism in models and argues that, since ecosystems are too complex for a single person to understand, interdisciplinary teamwork is mandatory. A strong case for the use of models in both understanding and planning is presented.

The only typographical error to be noted occurs in the first equation of the paper. There should be a dot (indicating differentiation with respect to time) over the v_i between the two equal signs. In addition to all the Latin names for plants in Figure 1a, the word "edaphic" may be unfamiliar; it means "relating to the soil."

The paper by L. M. Eisel presents a systems approach to the management of wildland areas. A mathematical model that decision makers can use to assist them in making land-use choices is developed. Risk and uncertainty are explicitly recognized as elements in the decision-making environment. The "true" relationships between inputs and outputs of a wildland ecosystem are generally stochastic (probabilistic) or unknown. There is uncertainty concerning future demand and future technology, and no objectively valid measure for aesthetic and recreational benefits exists. Eisel addresses himself to the goal of incorporating these realities into the model-building component of the decision-making process. He describes a chance-constrained mathematical programming model, one of the kinds of models that could be made applicable to land use management. (Recall the differential equation model for grassland management in Paper 8.) "Chance-constrained" indicates that some of the constraints on the values of the variables are stated in terms of probability relationships (e.g., Prob $(X \leq 10) \geq 0.95$).

The mathematical model is exercised with respect to a hypothetical watershed. Eisel mentions the difficulty of obtaining data for even a hypothetical case. He cites congressional directives that specify a "multiple use" policy and a "greatest good for the greatest number in the long run" criterion for wildland management. The translation of these directives and other considerations into a mathematical objective function and a constraint set is clearly explained. The resulting optimal management strategies and net benefits are presented for four cases that differ in assumed demand, benefits, and the discount rate applied to future recreational benefits. The sensitivity of the results to the level of probability associated with the chance constraints is shown graphically. Eisel refers to this sensitivity analysis as "parametric programming," that is, examination of a model as parameters are systematically varied over a range.

The management of a domesticated ecosystem in order to maximize economic profit is the subject of Paper 10, "Scheduling Bioproduction Harvest." The manager must select the optimal times to harvest a marketable crop. A formal mathematical model is developed. The first element modeled is the ripening of fruit, that is, "the growth process by which unmarketable fruits grow into various degrees of marketability and, finally, into overripe fruits without value." The fact that not all the fruit ripens simultaneously is taken into account by the authors, but the model does not deal with the biological aspects of the ripening process. Harvesting costs and market price enter into the economic-benefit model. The objective function to be maximized by the selection of the harvesting interval is the expected net return per day. Overripe fruit is considered a direct-loss item, perhaps because the authors had in mind the loss of goodwill (i.e., the cost of incurring a poor reputation as a supplier of market items). The implications of the distance to market and of the post-harvest handling system to the degree of ripeness at harvest are discussed. In particular, the model was applied to a papaya crop in Hawaii; the optimal solution of the model was found by direct enumeration using a computer. Several figures showing the results of sensitivity analysis are presented.

A few comments on details of the presentation in the paper may prove helpful. In the definition of X_k there should be an indication that this variable is the number of fruit arrival *per tree* on the kth day after the previous harvest. The left-hand variable of Equation (2) should be TF_k, of Equation (3) TF, and of Equation (6) $TF_{k,i}$, to match the notation of Equations (7) and (8). Equations (4) and (5) indicate that the authors do not count the day of harvest as a ripening day. Depending upon how one conceptualizes the day-to-day ripening interval and the hour of harvest, the sets of Equations (4) and (5) could be written with all the occurrences of $k + 1$ replaced by k. Equation (4c) needs a closing right parenthesis before the symbol $\cap$. In the first line of Equation (7), the superscript on the M in the final term should be L. Finally, Equation (14) contains an obvious typographical error; the expression $L - k = 1$ should be $L - k + 1$.

6

Reprinted from *IEEE Trans. Sys., Man Cybern.*, **SMC-2**, 449–459 (Sept. 1972)

Principles of Ecosystem Design and Management

HERMAN E. KOENIG, SENIOR MEMBER, IEEE, AND RAMAMOHAN L. TUMMALA, MEMBER, IEEE

Abstract—Basic principles of engineering, ecology, and economics are synthesized into logically consistent theoretical and computational procedures for a coordinated multilevel analysis of the tradeoffs in the static mass–energy and economic characteristics of alternate ecosystems (life support systems) and subsystems. Ecologically consistent pricing mechanisms are discussed as a means for regulating physical and technological succession.

I. INTRODUCTION

THE ECOSYSTEM of an industrialized society has been conceptualized by Koenig *et al.* [1] as a system of natural environmental components and man-made material transformation, transportation, and storage processes driven by solar, human, and physical forms of energy. Each component of the natural environment—viz., lake, stream, airshed, terrestrial region, etc.—is considered to have a limited capacity for processing restricted classes of man-made materials and energy, depending upon the "quality" of the environmental component to be maintained. From an engineering design perspective, these limited capacities represent ecological constraints against which the technological and spatial features of man-made processes in agriculture and industry and human habitats must be designed. From an economic point of view they represent potential constraints on regional economic developments; and from an ecological point of view the mass–energy features of the production–consumption processes of the economy must be in dynamic equilibrium with a heterogeneous pattern of biological communities as a closed ecosystem.

This paper synthesizes basic principles of engineering, ecology, and economics into logically consistent theoretical and computational procedures for a coordinated analysis of the tradeoffs in the mass–energy and economic characteristics of alternate ecosystems (life-support systems) and subsystems. The theory includes operational procedures for moving systematically from one level of spatial and/or functional organization to another, and the effects of the scale of production on energy and labor requirements and monetary values are included explicitly.

The capability to assess the mass–energy processing capacities of natural landscape components—lakes, streams, coastal regions, airsheds, terrestrial regions, etc.—is an absolute prerequisite to the practical application of the design and management concepts presented. Unfortunately the scientific base from which these capacities can be

evaluated in relation to the recreational, aesthetic, and health features of environmental components is very limited. The research required to provide reliable answers to these questions is sure to be extremely expensive, and a relatively long gestation period will be required before complete understanding is achieved. In the meantime, ecologically sound planning and regulation decisions must be made. It is shown that in the face of incomplete information on environmental capacities, it is possible to "hedge" against certain classes of uncertainties through the use of pricing mechanisms.

II. BASIC APPROACH

Policy decisions relating to the physical aspects of human life support are concerned primarily with the spatial, technological and economic features of the material processing substructure of our economy [1]. Cultural activities and service industries of the economy are assumed to have negligible interactions with the environment and, in the context of this paper they are considered to be self-adaptive to the basic life-support processes. To this extent the physical structure of the ecosystem has a direct bearing on human values and social stresses, stresses that as yet are not well understood. Consequently, it must be emphasized that all ecologically and economically feasible life-support systems are not necessarily politically and socially feasible. It is clear, however, that all economically and politically feasible life-support systems must be within the class of ecologically feasible structures. As one progresses from the identification of ecologically and economically feasible alternatives toward the identification of socially and culturally preferred structures, one moves from science and rational thought towards art forms, subjective judgment, and individual human values. The specific objective of ecosystem *design* is to articulate the ecological and economic options available at the various levels of technological, spatial, and social organization so that these normative decisions reflect well-informed input. It is in this context that we use the concept of ecosystem design. Ecosystem *management* is used to refer to the procedures through which society selects, implements and maintains the socio-ecological options implicit in the design process on preferential, humanitarian, and other normative scales. Pricing mechanisms of the economy represent one important class of social instruments of management, the only one considered in this paper. The design and management processes are the intimately related and enormously complex dynamic iterative processes through which man-made developments of the landscape take place. The objective is *not* to simulate these dynamic developments, even if such a simulation were feasible; it is rather to identify regulatory policies

Manuscript received April 26, 1972; revised May 5, 1972. This work was supported by the Office of Interdisciplinary Research, NSF, under Grant GI-20.

H. E. Koenig is with the Department of Electrical Engineering and Systems Science, Michigan State University, East Lansing, Mich. 48823.
R. L. Tummala is with the Division of Engineering Research, Michigan State University, East Lansing, Mich. 48823.

and pricing mechanisms (economic incentives) that will manage or direct these developments toward equilibrium states that are ecologically feasible. We are concerned first and foremost with the static or steady-state mass–energy and economics characteristics of alternate ecosystem structures as "targets" of regional development. The dynamics of the transitions from the existing structure to any given target structure is of central concern to the temporal implementation of the economic pricing mechanism and other regulatory policies, but it is not considered here.

A regional ecosystem is characterized as a set of interacting objects, such as agricultural and industrial production units, human settlements, and natural environmental components. The interactions, viewed as taking place through the exchange of materials and energy, unify the identified objects into a functional system. At a second level of analysis, this regional ecosystem can be viewed as a single entity interacting with other regional ecosystems through the exchange of materials and energy as though each regional ecosystem were a single object or a functional subsystem. By repeating this process, it is possible (at least conceptually) to move systematically from a micro to a macro level of analysis, from a lower level of aggregation to a higher level. The theoretical and computational procedures developed here are applicable at all these levels (macro and micro) to systematically evaluate the mass–energy characteristics, economic characteristics, and exchange rates with the environment as a function of spatial and technological features and size characteristics of the system processes.

In this perspective each component of the ecosystem (industrial, agricultural, urban or natural) is viewed as performing one or a combination of the three basic functions: material transformation, transportation, or storage. Each of these functions is carried out at a *cost* to society in human energy x^1 (labor); solar energy x^2 (land); and physical energy x^3. The energy cost vector $x = (x^1, x^2, x^3)$ essentially represents the cost in *nonrenewable resources* used to alter the molecular, chemical, physical, and technological forms and spatial distributions of *renewable* resources, that is, renewable in the sense that they can be recycled.

Conceptually each component of energy cost x^1, x^2, x^3 is handled in much the same way as the monetary cost in classical economic input–output analyses, except that labor is considered as an energy cost rather than a flow of services. Indeed, the monetary cost x^4 of the ith renewable resource is among other things a scalar function of the energy cost x_i^1, x_i^2, and x_i^3, thus

$$x_i^4 = f_i(x_i^1, x_i^2, x_i^3, \cdots, \text{other factors}). \qquad (1)$$

In principles the weight placed on x_i^1, x_i^2, x_i^3 by the function f_i depends upon the relative availability of the three forms of energy and the preference order of society for the ith renewable resource in relationship to other renewable resources.

Although monetary value is a prevalent measure of economic value, it is an inadequate metric for evaluating social cost-benefits. Its inadequacies in the context of ecologically based planning and policy decisions include:

1) The scalar function f_i in (1) can only be evaluated for items that enter the market place of bidding and trade. The waste assimilation capacities of the landscape components (lakes, streams, airsheds, etc.) do not enter the market place. Hence, monetary cost cannot be imputed to the materials exchanged with these landscape components when they are used as objects of the ecosystem.

2) The factors ultimately limiting the volume return of material goods to society by the ecosystem are scarce resources (land, labor, and physical energy) not dollars.

3) The monetary value of material goods determined by the market place at any time represents the weightings and preference orders according to the choices that are available within the context of prevailing conditions and existing market place items. It may not represent an unbiased basis for evaluating the relative merits of "alternatives yet to come."

This emphasizes the need for a coordinated analysis of the tradeoffs between the mass–energy and the economic (mass–monetary cost) characteristics of the ecosystem. The economic characteristics, as we shall see, depend explicitly upon the monetary value placed on the mass–energy exchange rates associated with the natural environmental components. It is through the pricing mechanisms associated with these mass–energy exchange rates that the economic equilibrium state is made to coincide with an ecologically feasible mass–energy equilibrium state. The ecologically feasible mass–energy equilibrium states, on the other hand, depend upon the mass–energy exchange rates assigned to the natural environmental components. In the context of this paper it is assumed that these exchange rates are assigned on the basis of auxiliary uses to which the environmental component is allocated and the "quality" to be maintained for these auxiliary uses.

III. ENVIRONMENTAL QUALITY

Environmental quality, particularly as it relates to aesthetic and recreational uses is recognized as a normative judgment. However, these normative decisions are based to a great extent upon the physical and ecological states of the component, states that can be related scientifically to the mass and energy exchange rates. To this end we define the *ecological* state S_e for each natural landscape component (lake, stream, airshed, etc.) as a vector of time functions of order n from which future behaviors of the component as a system object can be expressed as a function of current mass–energy stimuli. If we let $Y_e(t)$ be a vector of order m representing the *net* input rates of m material types (stimuli), then the rate of change in the ecological state of the landscape component can be expressed as

$$\dot{S}_e = F(S_e, Y_e). \qquad (2)$$

The set of all states that can occur for the set of stimuli under consideration is called the *ecological state space* $\overline{S}$.

Within the ecological state space $\overline{S}$ one can in principle identify regions to characterize general levels of "quality" as illustrated in Fig. 1. With regard to the future ecological

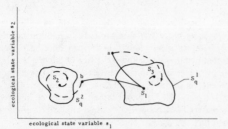

Fig. 1. Second-order quality states as regions of second-order of ecological state space.

state of the environmental component, three general levels of mass–energy exchange rates are distinguished:

1) *Nonpollution Levels:* If the exchange rates are such that the ecological state will remain within a given quality region S_q^1, for example, they are said to be of a non-polluting level. The lake or stream is said to have a corresponding material processing capacity.

2) *Reversible Pollution:* If the input rate is such that the ecological state may eventually move outside the quality regions S_q^1, for example, but will eventually return to quality region S_q^1 once the stimulus is removed, or is "neutralizable" by a counter stimulus, they are said to be at a reversible pollution level. The clean up of Lake Washington is well known [2], [3].

3) *Nonreversible Pollution:* If the input rates are such that the ecological states may eventually move outside quality regions S_q^1, for example, to the extent that the state cannot be returned to S_q^1 by reducing the stimuli, they are said to be at a nonreversible pollution level.

Note that all net input levels are specific to a quality region of the ecological state space *and* to a planning horizon. For example, if discharge of SO_2 into the regional airshed by industries is terminated, then air quality in the region usually is restored to its initial state by natural processes. On the other hand, phosphorous input levels to a lake may drive the ecological state from an equilibrium point S_1 in quality region S_q^1 to a point outside the region, such as point b in Fig. 1. When the stimulus is removed the ecological state may move to a new equilibrium state S_2 within a less desirable quality region S_q^2. Thus, if wastes are dumped into a lake for a sufficient length of time its use as a source of potable water and recreation may not be self-restoring.

The problem of identifying the function F in (2) which relates the change in ecological state S_e of natural ecosystems (lakes, streams, coastal regions, terrestrial regions, etc.) to the net input rate of materials Y_e is a scientific question that should be answerable through research. It requires major developments in the understanding of the functional structure of natural ecosystems and the laws governing the processes of the structure. Prior to the "environmental crisis" there was relatively little motivation for allocating research and development resources to a quantitative under-standing of natural ecosystems as material processing systems, and basic research in the area has not had such a specific objective or goal as a focal point. The problem of identifying the allowable mass–energy exchange rates with

natural, environmental components as a function of the ecological states to be maintained now stands as one of the most important and extensive areas of technical research and development in the support of sound land use and economic planning. But we shall not consider the subject further in this paper.

IV. COMPONENTS OF THE LIFE-SUPPORT SYSTEM

The components of the life-support substructure of our ecosystem fall into three general classes: material trans-formation processes, material transport processes, and material storage processes. The generic forms of the free-body models for components in each of these three classes are given in Figs. 2–4.

Transformation Processes

All production processes in industry and agriculture can be defined as a transformation of materials to achieve a well-defined change in their physical, chemical, tech-nological, or biological structure, through the application of energy in one or more of the three forms: solar, human, and/or physical. The transformation processes are organized in technical units called plants; a plant being a technically coordinated aggregate of fixed capital under common management. The term "process" is used in a semiabstract sense as distinct from the total productive activity of the plant. It specifically excludes the nonmaterial services, but explicitly includes technological treatment to which materials are subjected to bring them further toward their completed state as products useful to man. The concept of a process used here is related to the concept of *activity analysis* discussed by Koopmans [4].

Referring to Fig. 2 the material input and output rates $y_{ji}, j = 1,2,\cdots,n$ associated with a material transformation process P_i are identified with a set of directed line segments, as indicated. These line segments are used to establish a directional reference frame for the material input and output rates to the process. With each material flow rate y_{ji} we associate three energy costs $x_{ji}^1, x_{ji}^2, x_{ji}^3$, and a monetary cost x_{ji}^4. These costs represent, respectively, the energy (in the three forms) and the monetary costs *per unit* of y_{ji} required to provide the material j to process i at a given spatial location. If desired or necessary, capital equipment may also be identified explicitly as a cost variable. All costs are measured with respect to a datum point R_i associated with the process and are positive or negative depending upon the orientation of the flow reference frame with respect to this datum. The product

$$e_{ji}^l = x_{ji}^l y_{ji}, \qquad l = 1,2,3$$

can be viewed as an *energy* flow rate and y_{ji} as an energy flux rate, i.e., a carrier of energy bound in physical and technological form. Likewise, the product $e_{ji}^4 = x_{ji}^4 y_{ji}$ can be viewed as the monetary flow rate and y_{ji} as a monetary flux.

In general we shall assume that the material input and output rates with their associated energy and monetary costs are related by free-body models of the generic form given in Fig. 2. The coefficients k_{ji} are referred to as the

IEEE TRANSACTIONS ON SYSTEMS, MAN, AND CYBERNETICS, SEPTEMBER 1972

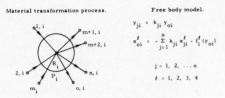

Material transformation process.   Free body model.

$$y_{ji} = k_{ji} y_{oi}$$

$$x_{oi}^{\ell} = -\sum_{j=1}^{n} k_{ji} x_{ji}^{\ell} - f_i^{\ell}(y_{oi})$$

$$j = 1, 2, \ldots n$$

$$\ell = 1, 2, 3, 4$$

Fig. 2. Material transformation process.

Material transport process.   Free body model.

$$x_o^{\ell} = -g^{\ell}(y_o)$$

$$\ell = 1, 2, 3, 4.$$

Fig. 3. Material transport process.

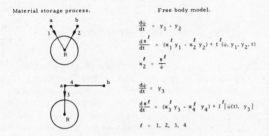

Material storage process.   Free body model.

$$\frac{d\psi}{dt} = y_1 - y_2$$

$$\frac{ds^{\ell}}{dt} = (x_1^{\ell} y_1 - x_2^{\ell} y_2) + t^{\ell}(\psi, y_1, y_2, t)$$

$$x_2^{\ell} = \frac{s^{\ell}}{\psi}$$

$$\frac{d\psi}{dt} = y_3$$

$$\frac{ds^{\ell}}{dt} = (x_3^{\ell} y_3 - x_4^{\ell} y_4) + t^{\ell}[\psi(t), y_3]$$

$$\ell = 1, 2, 3, 4$$

Fig. 4. Material storage process.

technical processing coefficients. The first equation of Fig. 2 represents the laws which govern how the input materials are combined to form the output materials, or in other words, characterizes the composition of the output materials in terms of the input materials. These laws of material combination are for the most part linear relations as indicated. A substitution of inputs or a change in their proportions to obtain essentially the same functional product is reflected in the model as a change in the coefficients k_{ji}.

The first term to the right of the equality sign in the cost equation in Fig. 2 represents the costs involved in making the inputs available to the process and to "remove" the outputs from the process. For $l = 1,2,3$ the last term represents the energy costs (per unit of output) required to carry out the material transformation process. For $l = 4$ it represents the monetary cost of production (per unit of output).

In the context of ecosystem design the output rate y_{oi} is defined as the *design capacity* of the processing unit. Note that the design capacity is specified in terms of one of the material input or output rates. The function $f^l(y_{oi})$ represents the variation in processing costs with design capacity. It accounts for the *scale economics* to energy and monetary cost associated with the processing unit.

The scale economics represented by the fundamentally nonlinear, largely monotonically decreasing function $f_i^l(y_{oi})$ are of central concern. As we shall see, to ignore this term

or to linearize it, is to throw out the very factor around which the tradeoffs must be made between ecological and social values of diversification and decentralization on one hand and improved economic and manpower efficiency made possible by large-scale mechanization and automation on the other. The effects of the externalities of production on the environment are well recognized in the economic literature [5]–[7]. The tradeoffs between scale economies and environmental quality, however, are virtually unformulated.

In principle all the technical coefficients k_{ji} and the scale economies $f_i^l(y_{oi})$ are known for industrial and agricultural processes, the processes were designed and made by man. The technical production coefficients for aggregated classes of industries are also available from the input–output table of national accounts. However, the information on scale economies to land, labor, physical energy, and monetary costs (including capital equipment) have not been systematically tabulated in any form. Subsequent developments in this paper will but amplify the central need for this information in long-range planning.

The generic form given in Fig. 2 is applicable to all types of material transformation processes associated with industrial, agricultural, and human habitats. In the case where the technical production coefficients are geographically specific as in the case of agriculture, each geographic region is appropriately subdivided into subregions over which these coefficients can be considered constant. The scale economies in agricultural production are directly related to the area of land holding, the scale of mechanization and the degree of specialization in products. Consequently, it is through this term that the improved scale economies to labor in the current trends away from diversified "family-type" agricultural operation toward large specialized operation, high-density feed lots, dairy barns, etc., are entered systematically into the model. Note, however, that an increase in specialization and in size of operating unit does *not* necessarily imply increased yields per acre. The increased solar energy conversion efficiency is due to the new genetic strains, improved soil technology, fertilizers, and pest control, and is largely independent of the scale of mechanization.

Transport Processes

Transport processes in the ecosystem can be viewed as a special class of material transformation processes wherein the process simply moves the material from one geographic location to another at a cost. The generic form of the model of the transportation process as a free-body object of the system is given in Fig. 3. In this model we identify y_0 as the material flow rate from region A_1 to region A_2 and x_0^l, $l = 1,2,3,4$, as the energy and monetary costs per unit of material required to carry out this translation. Note that x_0^l represents the *change* in costs—the economist's concept of marginal cost, the engineers concept of "potential" difference.

In general the transfer of material between regions A_1 and A_2 may involve a change in spatial concentration, as in collecting agricultural products and redistributing them in human settlements. These processes may be broken down conceptually by synthesizing a network of objects to represent the transport system. For example, the collection and redistribution can be represented simply as three transport processes in series: a process to characterize the collection, a process to represent translocation and another to represent redistribution. In all cases y_0 represents the *design capacity* of the transport facility and $x_0{}^l$ represents the scale economies to the three forms of energy and monetary costs as a function of design capacity. In the case where N materials y_i, $i = 1,2,\cdots,N$ of different types are carried over common transport facility, the energy costs are simply allocated according to the relation

$$x_0{}^l = -c_i f_0{}^l(y_0), \qquad i = 1,2,\cdots,N$$

where

$$y_0 = \sum_{i=1}^{N} c_i y_i$$

is the common carrier equivalent tonnage and c_i are weighting factors converting all materials to this equivalent base.

Material Storage Processes

Storage processes can also be viewed as a special class of material transformation processes, wherein the input and output materials are identical in form. Two generic forms of the model are shown in Fig. 4. In these models ψ and s^l are state variables representing, respectively, the accumulation of material and energy (or monetary value of inventory) within the process. The functions f^l in Fig. 4 represent the instantaneous energy input to the storage component, the maintenance energy. In general they depend upon the flow rates y_1 and y_2 and reflect any increases in energy and economic efficiencies with scale of process.

In reference to the operational procedures for developing ecosystem models discussed in subsequent sections, it must be pointed out that the flow rate and cost variables to the right of the equality signs in the free-body models in Figs. 2–4 are referred to as the *stimulus* or excitation variables and those to the left as the *response* variables. To the extent that it is possible to mathematically reorient the variables in these equations, the choice of stimulus and response variables is arbitrary. Any one of the rate variables in the first equation in Fig. 2, for example, can be selected as the stimulus variable.

V. STRUCTURAL CONSTRAINTS

A model of the physical and economic characteristics of any given or proposed ecosystem is obtained by constraining the free-body models of the system components according to the laws of material and energy balance imposed by the interconnections between the material transformation, transportation, and storage processes and the components of the natural environment. Operational procedures for generating these constraints are best illustrated by the means of the example structure shown in Fig. 5, an example which we will in fact use as a prototype

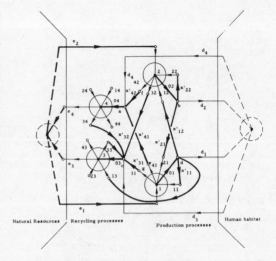

Fig. 5.   Topology of fourth-order life-support system.

in identifying some general principles of ecosystem design.

For discussion purposes, objects 1 and 2 can be regarded as an agricultural and an industrial process producing a "bundle" of food and a bundle of durable product at rates y_{01} and y_{02}, respectively. Objects 3 and 4 can be regarded as man-made or managed landscape components that process the joint products of the two basic industries. Object 3, for example, might be a lake allocated and managed to process the biodegradable wastes, and object 4 might represent a recycling process of nonbiodegradable materials. Many other interpretations are possible.

As the free-body reference frames indicate, object 1, for example, is assumed to have three joint output products y_{01}, y_{31}, and y_{41} from three inputs y_{11}, y_{21}, y_{e1} with corresponding costs $x_{01}{}^l$, $x_{31}{}^l$, $x_{41}{}^l$, $x_{11}{}^l$, $x_{21}{}^l$, and $x_{e1}{}^l$. The behavioral variables for all other objects are labeled in a similar manner.

The energy and monetary costs required to carry out the transport process are represented by the $x_{ij}{}'^l$. In the interest of preserving the visual tractability *not all edges corresponding to transport processes are shown explicitly* in the diagram.

The human habitat is assumed to receive the food and durable good packages at rates y_{d1} and y_{d2}, from which "waste" rates y_{d3} and y_{d4} are produced. The natural landscape delivers resources for the production of food and material products at rates y_{e1} and y_{e2}, and it receives "waste" materials of two classes at rates y_{e3} and y_{e4}. It must be emphasized that we do not attempt here to model the human habitat or the natural landscape as input–output objects in the same sense as the other objects. We rather take the view of providing a systematic procedure for evaluating the cost of providing various material returns to society, whatever they might be, *and* in disposing of whatever they wish, each as independent variables. If they choose to spend all the nonrenewable resources (energy) on inputs and allow the materials to accumulate under their feet after they have processed them; so be it! Likewise, the natural landscape is viewed essentially as a recipient and source of materials. Its limitations are characterized in terms of allowable exchange rates; the allowable exchange rates being related to the "quality" of the natural environment to be preserved as discussed in Section III.

The points in the diagram of Fig. 5 at which the reference frames of the free-body models of the objects are united are called *vertices*—in contrast to the datum node associated with the reference frame of each object. The diagram obtained by operationally taking all datum nodes as a common point, is called the system graph. It is the system "reference frame," from which a suitable and tractable set of constraint equations are identified.

The requirement that the materials in the system are *conserved* is expressed mathematically by the statement that the algebraic sum of the flow rates at vertices vanish identically. A generalization of the same constraint law states that the algebraic sum of the flow rates associated with a *cut-set* of edges of the graph vanishes identically; a cut-set being defined as a subset of edges which when removed leaves the graph in exactly two subparts, a vertex being accepted as one part [8]–[10].

The unit energy and monetary costs must be *compatible* throughout the system. These constraints are expressed mathematically by the statement that the algebraic sum of the cost vector associated with a circuit of edges vanishes. We have, for example, in Fig. 5

$$x_{31}{}^l + x_{31}{}'^l + x_{03}{}^l = 0, \qquad l = 1,2,3,4.$$

The compatibility and continuity constraints together imply a balanced *energy budget* for the systems. That is, the summation of the product $x_{ij}y_{ij}$, taken over all edges of the system graph, vanishes identically. This property, stated informally here and without proof, is a fundamental property of the orthogonality of the continuity and compatibility constraints and it is *independent of the object models*. The property is established in virtually every book on electrical network theory where the continuity and compatibility constraints are known as Kirchhoff's current and voltage laws. It is worth emphasizing that this is *not* to imply that ecological or economic systems have electrical analogs. To the contrary, there are *no* analogs of material transformation objects in electrical networks. Electrical network theory, by definition, is concerned only with the transmission of one type of energy flux (electrical currents) and therefore is concerned only with transmission objects, the majority of which have only two terminals like the transport objects as in Fig. 3. Thus it can be said that electrical network theory is a specialized technology built on the same fundamental principles described here and in Caswell *et al.* [11], and not the other way around. Attempts to develop electrical analogs of economic and ecological systems in this respect will continue to be frustrated.

There are many circuits and many cut-sets in the system graph. An operational procedure for selecting independent

circuits and cut-sets *and* for writing the corresponding constraint equations in a form that is mathematically compatible with the free-body models of the objects is provided in the simple concept of a *tree* in the system graph. A tree is defined as a subset of edges (a subgraph) that interconnect all the vertices and the datum points but form no circuits [9]–[11]. A tree in the system graph of Fig. 5 consists of heavyweight edges *plus* all the edges corresponding to the transport processes that are not shown explicitly. The tree is thus used to subdivide the variables of the system into two mutually exclusive, but all inclusive subsets.

1) The stimulus subset consisting of all material flow rate variables y_{ij} corresponding to the edges of the tree T and all costs $x_{km}{}^l$ corresponding to the complement of the tree, called the cotree CT.

2) The *response* subset consisting of all flow rates y_{km} corresponding to the cotree CT and all costs $x_{ij}{}^l$ corresponding to the tree T.

We require only that the tree be selected in such a way that the classification of the stimulus and response variables in the free-body models of the objects, is (or can be made to be) *consistent* with the stimulus and response subsets identified by the tree. Thus for the indicated tree T, free-body models of *material processing* objects in the graph of Fig. 5 are taken in the form already given by Fig. 2 with $j = 1,2,3,4,e$, and $i = 1,2,3,4$. The free-body models for the transport objects are taken as

$$x_{ij}{}'^l = -g_{ij}{}^l(y_{ij}), \qquad i,j = 1,2,3,4. \tag{3}$$

The tree in Fig. 5 identifies a unique set of independent continuity and compatibility constraints when we require that 1) all cut-sets include *exactly* one edge of the tree, and 2) all circuits include *exactly* one edge from the cotree.

The nontrivial continuity constraints defined by T are

$$y_{0j} = y_{j1} + y_{j2} + y_{j3} + y_{j4} + y_{dj}, \qquad j = 1,2,3,4. \tag{4}$$

The compatibility constraints defined by the cotree CT (the complement of T) are

$$x_{ji}{}^l = x_{ji}{}'^l - x_{0j}{}^l, \qquad i,j = 1,2,3,4 \tag{5}$$

$$x_{dj}{}^l = -x_{0j}{}^l, \qquad j = 1,2,3,4. \tag{6}$$

The free-body models of the objects of the system along with a compatible and consistent set of constraint equations represent the *mathematical structure* of the system. The graph or reference frame of the system with which these constraints are associated represents the *topological structure* of the system. The topological and mathematical properties of system structures are well understood for the class of objects encountered in electrical network design and synthesis. Counterpart developments for the class of objects encountered in ecosystem design and analysis remain as an essentially uncharted area of theoretical research for the network theorists interested in relating to this. In reference to Fig. 5 it is easy to see that this structure can be easily generalized to an arbitrary number of "production" and "recycling" processes. What are the general properties of such a structure and its implications to the physical and economic characteristics of our ecosystem? We see, for example, that the tree used to classify the behavioral variables of the system into stimulus and response subsets includes: 1) exactly one edge for each material processing object, and 2) it includes all transport links. As we shall see, this structural feature makes it possible to preserve all nonlinearities as additive terms in reduced models of subsystems. The authors conjecture that this very desirable structural feature is preserved for an arbitrary extention of the prototype structure if we view: 1) the "recycling" processes as input dependent and the "production processes" as output dependent, and 2) the human habitat material exchange rates and the natural landscape costs as stimulus variables of the ecosystem. Ayres and Kneese [6] and Leontief [7] deal with these structural features conceptually by considering waste processing industries as providing a pollution service to the production process. But in the authors' view, some of the important and essential structural features of the system are obscured by this procedure.

VI. MULTILEVEL ANALYSIS

Response of the ecosystem to changes in system structure (both component behavior and interconnection constraints) can, of course, be investigated by programming the object models and constraints for simultaneous solution on computing machines. Any logically consistent and valid computer simulation program directed at a study of the mass–energy and economic characteristics of ecosystems must, in fact, contain information equivalent to what has been defined here as the mathematical structure of the system. However, since the constraint equations are linear, the mathematical structure can always be reduced by eliminating either the stimulus or the response variables or a combination thereof. For the prototype system in Fig. 5 we reduce the mathematical structure to a minimum number of equation by:

1) substituting the constraints equations in (4) into the first free-body equations in Fig. 2 to eliminate the response flows internal to the system; and
2) substituting the second free-body equations in Fig. 2 and Fig. 3 into the constraint equations in (5) and (6) to eliminate the stimulus costs internal to the system.

To display these results let the material exchange rates with the human habitat and the natural landscape be represented, respectively, by the fourth-order vectors

$$Y_e = (y_{e1}, y_{e2}, \cdots)$$

$$Y_d = (y_{d1}, y_{d2}, \cdots)$$

and let each of the corresponding cost variables associated with these exchange rates be represented by

$$X_e^l = (x_{e1}^l, x_{e2}^l, \cdots)$$

$$X_{d1}^l = (x_{d1}^l, x_{d2}^l, \cdots), \qquad l = 1,2,3,4.$$

We also identify the rate capacities (inputs or outputs) of the four material transformation processes and their associated

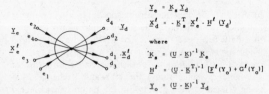

$$Y_e = K_s Y_d$$
$$X_d^l = -K_s^T X_e^l - H^l(Y_d)$$

where

$$K_s = (U - K)^{-1} K_e$$
$$H^l = (U - K^T)^{-1} [F^l(Y_o) + G^l(Y_o)]$$
$$Y_o = (U - K)^{-1} Y_d$$

Fig. 6. Input–output model of system in Fig. 5 as single object.

costs by the fourth-order vectors

$$Y_0 = (y_{01}, y_{02}, \cdots)$$

$$X_0^l = (x_{01}^l, x_{02}^l, \cdots).$$

A reduced form of the mathematical structure for the system in Fig. 5 can now be written as

$$(U - K)Y_0 = Y_d \qquad (7)$$

$$Y_e = K_e Y_0 \qquad (8)$$

and

$$(U - K^T)X_0^l = -K_e X_e^l - F^l(Y_0) - G^l(Y_0) \qquad (9)$$

$$X_d^l = -X_0^l \qquad (10)$$

where U is the fourth-order unit matrix and K is the fourth-order technical coefficient matrix

$$K = \begin{bmatrix} k_{11} & k_{21} & k_{31} & k_{41} \\ k_{12} & k_{22} & k_{32} & k_{42} \\ k_{13} & k_{23} & k_{33} & k_{43} \\ k_{14} & k_{24} & k_{34} & k_{44} \end{bmatrix} \qquad (11)$$

and K_e is a fourth-order diagonal matrix with diagonal entries K_{ei}, $i = 1,2,3,4$. The processing and transport costs as a function of design capacity Y_0 are represented, respectively, by the fourth-order vectors

$$F^l(Y_0) = [f_1^l(y_{01}), f_2^l(y_{02}) \cdots], \qquad l = 1,2,3,4 \qquad (12)$$

$$G^l(Y_0) = \sum_{i=1}^{4} \begin{bmatrix} k_{i1} g_{i1}^l & (k_{i1} y_{01}) \\ k_{i2} g_{i2}^l & (k_{i2} y_{02}) \\ k_{i3} g_{i3}^l & (k_{i3} y_{03}) \\ k_{i4} g_{i4}^l & (k_{i4} y_{04}) \end{bmatrix}, \qquad l = 1,2,3,4. \qquad (13)$$

Note in (7) through (10) that the vectors Y_0 and X_0^l represent a subset of behavioral variables internal to the system. If $(U - K)$ is nonsingular these internal behavior variables can also be eliminated. The result is a reduced mathematical structure in which the detail on all behavior internal to the system has been suppressed—a "black-box" representation, if you wish. It is appropriate therefore to refer to this reduced model as a free-body model of the ecosystem as a single object. The result is given in Fig. 6 along with the reduced form of the topological structure, a topological representation for which all details on internal structure has been eliminated. This reduced structure represents a free-body model of the ecosystem as a single object and is used at a second level of analysis, following the same principles used to develop it from the free-body models of

IEEE TRANSACTIONS ON SYSTEMS, MAN, AND CYBERNETICS, SEPTEMBER 1972

the objects in Figs. 2–4. The coefficient matrix K_s and the vector of functions $H\,(Y_d)$ represent, respectively, the production coefficients and the processing costs for the ecosystem as an object. They are directly computable from the production functions of the objects in the system, their processing costs and the transport costs interconnecting the objects of the ecosystem.

Since all nonlinearities remain additive, no time consuming iterative procedures are required to establish the reduced structure. A complete analysis of the internal behavior, of course, is available as an integral part of the reduction. The principles demonstrated therefore provide a systematic and effective procedure for progressing systematically from one level of analysis to another, using models of systems at one level as objects at the next higher level, a multilevel analysis procedure. These multilevel procedures also apply to dynamic systems but reduced structures in closed analytical form are possible for nonlinear systems only under special structural conditions: the additive nonlinearities and topological features encountered here being one such special condition.

VII. SOME PRINCIPLES OF ECOSYSTEM DESIGN AND ECONOMIC REGULATION

Many of the questions of contemporary national and international concern relate to economic growth and the ecological and sociological stresses generated by large scale mechanization, specialization and centralization of industry and agricultural processes [1], [13]. We shall attempt to bracket some of these questions and their relationship to the physical and technological features of the life-support system by looking at the ecological and economic equilibrium characteristics of two extreme designs, a zero material recycling life-support system and a complete (100 percent) recycling system.

The topological structure in Fig. 5 and the corresponding mathematical structures can be extended to include an arbitrary number N of distinct classes of production processes and M distinct classes of recycling processes. Let the matrix equations in (7) through (9) be partitioned so as to identify these N production and M recycling processes and associated material rates and energy costs explicitly. Thus let the material exchange rates with the environment and human habitat be represented in partitioned form

$$Y_e = \begin{bmatrix} Y_{e1} \\ Y_{e2} \end{bmatrix} \quad \begin{matrix} \text{outputs from environment} \\ \text{inputs to environment} \end{matrix}$$

$$Y_d = \begin{bmatrix} Y_{d1} \\ Y_{d2} \end{bmatrix} \quad \begin{matrix} \text{inputs to human habitat} \\ \text{outputs from human habitat.} \end{matrix}$$

Let the rate capacities of the production and recycling processes be represented in partitioned form

$$Y_0 = \begin{bmatrix} Y_{01} \\ Y_{02} \end{bmatrix} \quad \begin{matrix} \text{production processes} \\ \text{recycling processes} \end{matrix}$$

and write (7) and (8) as

$$\begin{bmatrix} (U - K_{11}) & -K_{12} \\ -K_{21} & (U - K_{22}) \end{bmatrix} \begin{bmatrix} Y_{01} \\ Y_{02} \end{bmatrix} = \begin{bmatrix} Y_{d1} \\ Y_{d2} \end{bmatrix} \quad (7a)$$

and

$$\begin{bmatrix} Y_{e1} \\ Y_{e2} \end{bmatrix} = \begin{bmatrix} K_{e1} & 0 \\ 0 & K_{e2} \end{bmatrix} \begin{bmatrix} Y_{01} \\ Y_{02} \end{bmatrix}. \quad (8a)$$

A similar partitioning applies to (9) but is not shown explicitly.

Zero Recycling

The mathematical structure corresponding to zero material recycling (essentially the structure now in existence in the industrialized nations) is obtained from (7a) and (8a) by simply setting $K_{22} = 0$, $K_{12} = 0$ and $K_{e2} = U$ to indicate that the externalities of production are transferred through the "recycling processes" without altering their form.

From the first expressions in (7a) and (8a) the environmental resource rates Y_{e1} required to supply product demand rates Y_{d1} is simply

$$Y_{e1} = K_{e1}(U - K_{11})^{-1} Y_{d1} \quad (14)$$

and the material rate inputs to the environment are

$$Y_{e2} = K_{21}(U - K_{11})^{-1} Y_{d1} + Y_{d2}. \quad (15)$$

The cost $X_{d1}{}^l$ of the product rates Y_{d1} to the human habitat as obtained from the partitioned form of (9) are

$$X_{d1}{}^l = (U - K_{11}{}^T)^{-1} K_{e1} X_{e1}{}^l + (U - K_{11}{}^T)^{-1} [F_1{}^l(Y_{01}) + G_1{}^l(Y_{01})] + (U - K_{11}{}^T)^{-1} K_{21}{}^T X_{e2}{}^l. \quad (16)$$

We now distinguish between two important classes of technologies; *material processing* technology and *production* technology (including transport). Changes in material processing technologies include, for example, developments of new genetic strains in agriculture and new agricultural and industrial products. They are reflected in the model as changes in the relative magnitude of the coefficient in the production matrices K and K_e. Production technologies including, specifically, *scale* of mechanization and automation are reflected in the model as changes in the processing and transport cost functions F^l and G^l. The central role of new material processing technologies in making new products available to the human habitat from environmental resources is clearly indicated by (14). But from (15) it is also clear that these increases in the yield of the environment is accompanied by the corresponding increase in the required waste processing capacity of the environment, the same multiplying factor $(U - K_{11})^{-1}$ appears in both expressions. Referring to (16) the cost of these products $X_{d1}{}^l$ to the human habitat depend upon both classes of technology. But the choice of production technologies depend upon the availability of the three forms of energy, (land, labor, and physical energy) and the extent to which these three forms are technically interchangeable. This implies that under certain conditions of population density, available resources and environmental conditions, the scale of mechanization in agriculture, for example, should be limited in the interest of increasing the total yield of the region or in the interest of social justice and for other social and cultural reasons [13].

The total energy cost $X_{d1}{}^1$ of products to the human habitat includes a weighted sum of the unit production and transport costs as given in the last term of (16). If these energy costs decrease monotonically with scale of operation, as they do for many agricultural and physical processes, then it is in the interest of overall energy efficiency to allocate the total productive capacity of a given product to a small number of productive units. The central impact of modern technology is that it has greatly extended these scale economies to human forms of energy through automation and large-scale machines. Introduction of computers extend them even further through mechanization of the management function of the firm. However, since the waste processing capacity of the environment depends not only on the total material discharge rates, but upon the spatial distribution and concentration of these wastes in relationship to the ecological characteristics of the environment, it eventually becomes necessary to limit size and spatial distribution of productive units for environmental reasons, with a corresponding *sacrifice* in energy efficiency. There appears to be ample evidence to indicate that major social and political stresses also develop with excessive concentrations of people and control of productive wealth. These stresses are not well understood and it is not clear which of the social or physical environmental factors are more limiting. Certainly there are synergetic effects between them.

The monetary cost of materials delivered to the human habitat $X_{d1}{}^4$ by a zero recycling system depend upon the monetary costs of production F^l and transportation G^l and the monetary cost $X_{e1}{}^l$ assigned the material exchange rates with the environment as given by (21) with $l = 4$. Economic competition between productive units in a free economy will over time generate *successional* changes in the technological and physical scale of the productive units in the direction of minimizing $X_{d1}{}^4$. At the present time in most industrialized nations the monetary costs $X_{e1}{}^4$, are based on short-term economic opportunities rather than long-term ecological considerations. Under these pricing mechanisms the monotonically decreasing functions in F^4 and G^4, made possible by modern technology, apparently are the dominating factor in successional changes. Approximately half of the productive capacity in the United States, for example, is now in the control of the 100 largest industrial corporations; a greater percentage than the 200 largest companies controlled 20 years ago [14]. The percentage of the population currently involved in agriculture is approximately 4.8 percent and is expected to drop to about 2 percent if present successional changes continue. These economically motivated successional changes in the technological and physical structure of the life-support system are of central concern. They can lead to what is perhaps appropriately called "successional instabilities," i.e., structures that fall outside the class of ecologically and sociologically viable systems.

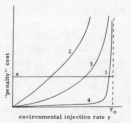

Fig. 7. Typical monetary cost associated with material exchange rates with environmental components.

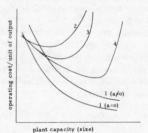

Fig. 8. Typical monetary cost curves as seen by production units for environmental costs indicated by Fig. 7.

It is easy to show from (15) that it is possible in principle to assign variable costs $X_{e1}{}^4 = H(Y_{01})$ (scale diseconomies) to the resources taken from the environment that will limit the scale of the productive units, but it cannot regulate their geographic location in relationship to other productive units and the natural ecological and physical features of the landscape. Spatial distribution of the production units can be regulated economically by assigning similar cost functions $X_{e2}{}^4$ to the waste discharges into the environment. For example, if the mass–energy exchange rates with the natural environment are well defined and undisputedly associated with the specific processes, as in the case with the industrial firms, then the costs in the form of taxes or the sale of discharge rights might be imposed [1], [6]. Such cost may vary with the exchange rate as illustrated in Fig. 7. In principle these costs are "calculated" to constrain the successional development of the life-support system to ecologically feasible structures. The upper bound of the exchange rate y_0 is specific to the recipient environmental component and might, for example, represent the input rate at which irreversible ecological damage to the environmental component is expected to occur.

The shape of the cost curve can be used to reflect varying degrees of uncertainty about the ecological limit, curve 2 reflects a much lower level of confidence than 4. All cost curves except number 1 indicate that society is not, under any circumstance, willing to risk permanent damage, but imputes little or no cost to very low exchange rates. In principle, the shape of the exchange cost curve can be adjusted to make the equilibrium state of the regional economy coincide with the mass–energy equilibrium state of the ecological structure. When discharge costs are imposed on waste inputs to the environment the cost curves, as typically seen by the designer of a plant site in the region, are illustrated in Fig. 8. Note that an inelastic environmental cost curve is ineffective against the technological economies of scale of man-made physical production processes.

Complete Recycling

The mathematical structure corresponding to complete material recycling is obtained from (7a) and (8a) by setting $Y_{e1} = Y_{e2}$ and $Y_{d1} = Y_{d2}$. It is easy to show that these conditions are satisfied only if the material processing technologies are constrained according to the matrix

$$[U - (K_{11} + K_{21})]K_{e2} = [U - (K_{22} + K_{12})]K_{e1}. \quad (17)$$

Given the required material processing technologies, the required energy sources and the materials necessary to build the production facilities, such a system can, in principle, provide a class of structured materials to the human habitat at unlimited rates. The energy requirements as obtained from a partitioned form of (9) are

$$X_{d1}{}^l = (U - K_{11})^{-1}[K_{e1}K_{12}(U - K_{22})^{-1}K_{e1}] \begin{bmatrix} X_{e1}{}^l \\ X_{e2}{}^l \end{bmatrix}$$

$$+ (U - K_{11})^{-1}K_{12}(U - K_{22})^{-1} \begin{bmatrix} F_1{}^l(Yd) \\ F_2{}^l(Yd) \end{bmatrix}$$

$$+ (U - K_{11})^{-1}K_{12}(U - K_{22})^{-1} \begin{bmatrix} G_1{}^l(Yd) \\ G_2{}^l(Yd) \end{bmatrix}. \quad (18)$$

In addition to production and transport costs we now have recycling process and transport costs $F_2{}^l$ and $G_2{}^l$. Note in particular the product of the inverse of two matrices operating on these two terms. Although no generalization can be made as to the magnitudes of these factors, they can under certain conditions lead to a very high energy requirement. Irrespective of what the energy requirements are, they can now be minimized through large-scale processing centers without having to be concerned about environmental hot spots, since the environment is physically isolated from the life support process. Since the environment is not a scarce resource to production, the monetary value assigned to X_{e1} and X_{e2} can be taken as a constant, perhaps even zero. Under these conditions successive changes in technological and physical size of productive units will again take advantage of scale economies.

The sociological and political implications of these economically motivated successional changes, however, are potentially of no less concern than in a zero recycling structure.

VIII. Conclusions

It is perhaps worth noting once more that the objective of modeling in the context of ecosystem design is not to simulate what is going on now, but to make judgments as to what might be and how to achieve it. The material and energy processing rates of environmental components represent a definitive set of ecological constraints on in-

dustrial, agricultural, and urban developments. These constraints provide the mechanisms for retaining a preassigned regional environmental quality over a given planning horizon. They provide a definitive set of meaningful specifications to which the technical and spatial features of man-made processes must be designed as a life support system. An important factor in judging the relative merits of alternative ecostructures is the steady-state mass–energy performance.

Zero recycling structure and the complete recycling structure represent end points of a spectrum of alternative structures. Any operational system clearly must fall somewhere within this spectrum. Given unlimited sources of energy and the ability to dissipate this energy from the surface of the earth without undesirable environmental side effects, it is conceivable that the productivity of the surface of the earth to man could be increased significantly over the levels that now exist in the developed countries. The attainable levels depend ultimately upon many factors, not the least of which are the availability of nonfossil sources of energy and the extent to which the technologies can be coordinated by (17). But given the enormous range of alternatives in technology and physical structure of our life-support system, the complexity of the relationship between these factors and the fact that fusion sources of energy are not as yet scientifically feasible, there appears to be little scientific basis for predicting the ultimate yield of the earth to man or the actual bounds of economic growth.

The principles and concepts discussed in this paper provide an overall perspective for coordinating policy decisions relating to these factors at all levels of functional, spatial, and technological organization. The topological structure provides an effective language for displaying system structure and for developing reduced mathematical models as an explicit function of system structures. The theoretical framework provides effective computational procedures for evaluating the important classes of tradeoffs in a quantitative sense at levels of organization where such quantitative information is important to the decision-making processes and for moving systematically from one level of organization to another. A multilevel analysis capability is one of the fundamental tools of modern systems science that is frequently overlooked but which is absolutely essential to deal with the design and management problems of large scale systems.

In the final analysis there are only two ways in which the exchange rates between the life-support process and the natural environment can be regulated; by physically limiting flow rates or by assigning prices to the exchange rates. To assign a zero value to the environmental exchange rates as in the zero recycling system, is to assign a zero value to the natural environment, yet the life-support system is absolutely dependent upon the natural environment for closing the material loops. To restrict the exchange rates to zero as in a complete recycling system is to assign an infinite monetary value to the environment and eliminate it from productive use. In this respect zero and infinite prices represent the end points on a spectrum of economic regulation alternatives. Practical answers are to be found somewhere within the spectrum. It is not practical to assign ecologically consistent economic value to all exchange rates and it is not practical to put physical restrictions on all forms of discharge. However, to the extent that it is feasible to assign ecologically consistent economic values to the exchange rates, the economy can be used as a "computing" mechanism for allocating the material processing capacity of the environment as a scarce resource to competing processes. If it is not feasible to internalize these environmental costs, the planner or other social instruments of decision-making must assume the responsibility for allocating limited exchange rates. Such decisions must be made at all levels of functional and spatial organization. But it is at the lower levels of organization—the production plant site design, the agricultural production unit, the urban community, etc.—that the critical decision on technological and physical structure and the couplings to the environment are initiated. It is at this level that quantitative articulation of the tradeoffs is both feasible and of greatest value.

It is not to be implied that decisions relating to planning and economic policy can be based entirely on static models of the type discussed in this paper nor is it to be implied that the theoretical concepts used to develop these static models are restricted to such forms. One of the most serious and demanding problems in economic policy decision is which group or groups in the community are to bear the short- and long-term "burdens of adjustment" toward ecologically desirable states. No less demanding is the problem of resolving legitimate socioeconomic conflicts imposed on lower levels of functional and spatial organiza-tions by higher level regulations. Neither the questions relating to the economic and ecological transients imposed by changes in economic pricing mechanisms nor the dynamic nature of the successional changes in technological and physical structure are considered here.

REFERENCES

[1] H. E. Koenig, W. E. Cooper, and J. M. Falvey, "Engineering for economic, social, and ecological compatibility," *IEEE Trans. Syst. Man Cybern.*, vol. SMC-2, pp. 319–331, July 1972.
[2] W. T. Edmondson, "Changes in Lake Washington following an increase in the nutrient income," *Verh. Int. Ver. Limnology*, vol. 14, pp. 167–175, 1971.
[3] ——, "Nutrients and phytoplankton in Lake Washington," in *Proc. Symp. Nutrients and Eutrophication: The Limiting-Nutrient Controversy*, W. K. Kellogg Bio. Station, Michigan State Univ., pp. 172–193, Feb. 1971.
[4] T. C. Koopmans, "Analysis of production as an efficient combination of activities," in *Activity Analysis of Production and Allocation*. New York: Wiley, 1951, (Cowles Commission for Research Economics, Monograph 13).
[5] R. U. Ayres and A. V. Kneese, "Production, consumption, and externalities," *Amer. Econ. Rev.*, vol. 59, June 1969.
[6] A. V. Kneese, R. U. Ayres, and R. C. D'Arge, *Economics and the Environment*. Baltimore, Md.: The Johns Hopkins Press.
[7] W. Leontief, "Economic repercussions and economic structure: input–output technique," *Rev. Econ. Stat.*, vol. 52, p. 262, Aug. 1970.
[8] H. E. Koenig, Y. Tokad, and H. K. Kesavan, *Analysis of Discrete Physical Systems*. New York: McGraw-Hill, 1967.
[9] P. H. Roe, *Networks and Systems*. Reading Mass.: Addison-Wesley, 1966.
[10] S. Seshu and M. B. Reed, *Linear Graphs and Electric Networks*. Reading, Mass.: Addison-Wesley, 1961.
[11] H. Caswell, H. E. Koenig, Q. Ross, and J. Resh, "An introduction to systems science for ecologists," To appear in *Systems Analysis and Simulation in Ecology*, vol. II, B. C. Patten, Ed. New York: Academic Press.
[12] D. Meadows, D. L. Meadows, *et al.*, *The Limits of Growth*. New York: Universe Books, 1972.
[13] K. Nair, *The Lonely Furrow*. Ann Arbor, Mich.: Univ. Michigan Press, 1969.
[14] Donald M. Morrison, "The future of free enterprise," *Time* (Time Essay), Feb. 14, 1972.

Reprinted from *Science*, **177**, 139–142 (July 14, 1972)

Forest Fuel Accumulation—
A Growing Problem

Marvin Dodge

Fire fighters in the western forests are doing an effective job. They have not yet reached the goal of total exclusion of fire, but over 95 percent of the wild fires are extinguished while small. The 3 to 5 percent that get out of control cause 95 percent of the damage.

This small percentage does tremendous damage. The 1967 fire season in the Pacific Northwest and Northern Rockies was a blow to the national economy as well as to the local timber and recreation industries. The year 1970 saw a repeat of huge fires in eastern Washington during the summer. Southern California was declared a disaster area by President Nixon. The final toll for California was 14 lives lost, 800 homes and buildings destroyed, and 242,000 hectares (600,000 acres) of timber and watershed cover burned.

More disastrous fires may be expected in future years. A major factor contributing to this prediction is the accumulation of dead fuels taking place in the wildland areas of the western United States. The fire control agencies may be making the situation worse. William E. Towell, chairman of a fire study group for the American Forestry Association, said (*1*), ". . . a fire control agency's worst enemy will be its own efficiency. The longer forests go without burning, the greater the fuel accumulation and the greater the hazard."

Forest Fuels

The fuels that burn in a forest fire are generally separated into two classes, living fuels and dead fuels. The living fuels, consisting of leaves, twigs, and stems of growing plants, are difficult to ignite and do not burn readily by themselves. When the moisture content of living plants is reduced by drought and when they are further heated and dried by a fire in dead fuels beneath them, they can burn, sometimes very intensely.

The real problem is dead material, consisting of fallen leaves and needles, dead twigs and branches, dead stems either standing or fallen, and dead grass and weeds. Fires start easily in dry, dead fuels and spread readily. The fuels that contribute most to fire intensity and rate of spread are those about 1 centimeter in diameter and smaller, plus the outer few millimeters of larger branches and logs. Larger fuels do not contribute to intensity or rate of spread since they burn after the main fire front has passed. These larger fuels do cause the fire to persist and are difficult to extinguish. They may also provide a source of burning embers for further fire spread.

A general formula for fire intensity (*1*) was developed by Byram (*2*):

$$I = HWR$$

where H is the heat or energy value of the fuel, W is the weight of fuel per unit area, and R is the rate of spread. Thus, increasing the fuel loading increases the intensity of the fire. However, not all the relations for free-burning fires have been worked out. According to Hodgson (*3*), Australian fire researchers found that doubling the amount of fine fuels (such as leaves, twigs, and long strips of bark) doubled the rate of spread. This produces a fourfold increase in intensity. This relationship has not been verified by experiments in the United States, nor have the effects of large volumes been adequately explored.

Conditions for Fuel Accumulation

All plants produce yearly increments of dead material. Annuals leave the whole plant body as fuel at the end of the growing season. Perennials shed leaves or needles and dead twigs every year, some species throughout the year, and others seasonally. At greater intervals the perennials contribute dead branches that have been shaded out by subsequent growth and, eventually, the stems or trunks of the dead plants that have completed their life cycles.

The normal annual increment of dead fuel may be increased tremendously by events that kill a sizable proportion of the plants growing in the area. Insect and disease epidemics, blowdown during storms, timber harvesting, and wildfire all contribute great quantities of dead material.

Very few studies have been made of the annual accumulation of litter. Accumulation rates of 1.1 to 3.2 tons per hectare were found in the chaparral of southern California by Kittredge (*4*). Studies by Biswell and his colleagues (*5*) of several species in the central Sierra Nevada showed that 2.2 to 6.9 tons of litter per hectare are contributed each year to the forest floor.

Climate plays an important part in the accumulation of dead fuels. Both Kittredge (*4*) and Olson (*6*) have pointed out that in warm, moist climates, which provide an optimum environment for decomposition, there is little or no litter. Where conditions are less favorable for decomposition, as in a cool, moist climate, there is a greater accumulation of dead materials. A climatic pattern of warm, dry summers and wet, cool or cold winters leads to maximum accumulations of dead fuels or stored energy that may be released by wildfires. Thus we should expect the greatest potential for disastrous wildfires in the western United States.

There have been few measurements or estimates of the quantities of dead fuel under growing forests or brush fields. Weaver, discussing the ecology of the ponderosa pine (*Pinus ponderosa*), pointed out the great increase in fire hazard which results from fuel accumulation under almost complete fire protection (*7*). He found fuel quantities large enough to make control nearly impossible if a fire should start during critical weather conditions (*8*). In studies of the giant sequoia (*Sequoiadendron giganteum*) groves as much as 56 to 94 tons of dead fuel per hectare were found (*9*). Southern California has a reputation for devastating forest and brush fires, yet only two studies have been made of the amount of material fueling these fires (*10*).

Although the figures are sparse for California, there is an even greater lack

The author is a state forest ranger with the forestry division of the California Department of Conservation, Sacramento 95814.

of quantitative data for fuel loadings in the Northwest and the Rockies. The only published data are those obtained in a brief study of a small area in northern Idaho by Anderson (11). In 1967, during an extensive air and ground survey of forest conditions on a transect from western Montana across northern Idaho into eastern Washington, I found that nearly half the area has loadings of dead fuel ranging from 90 to 135 tons per hectare. And many sites have up to 225 tons of dead fuel per hectare beneath the overmature standing timber.

Critical fire weather, long periods of drought culminating in a few days of very low humidities and strong winds, only occurs on an average of once in 10 or 15 years in the Pacific Northwest and Northern Rockies. But when a fire does start in these conditions in an area with large quantities of dead fuels, it will be so intense that little or nothing can survive.

Fire History

Forest fire was a regular phenomenon in the western United States before the white man arrived. Show and Kotok (12), in one of the earlier studies of fire chronology for California, found that fires have occurred at intervals of about 8 years since 1685, which was as far back as they could reliably date the trees studied. They stated that forests persisted in spite of the numerous fires, which indicates that most were light surface fires. The pine forests of California, Oregon, and eastern Washington (13) and of the Southwest (14) all have long histories of fire. In the Southwest, areas of pine forest with periodic fires were still open and parklike, but well stocked with young pines, in 1951 (14).

Great arguments have raged over whether Indians set fires or whether fires were all started by lightning. Readers may wish to refer to articles by Burcham (15) and Stewart (16) for examples of opposing views. However interesting this history may be, it has little bearing on today's problems. The management practices (or lack thereof) of a hunter-gatherer culture were determined by conditions entirely different from those prevalent now. Today's management decisions must be made with regard to present economic and social conditions and must be based on sound principles and definite goals, not nostalgic recollections of a rather vague historic past.

Conditions changed drastically with the arrival of white men in the West. Extensive forest areas were cut for building materials for new towns and farms, mine timbers, and fuel wood. The western forests also supplied wood for the continued economic growth of the East and Midwest. Logging left great quantities of slash—treetops, limbs, and waste—on the ground. The opening of the crown canopy permitted greater insolation and drying of the increased amounts of dead fuel on the ground, and the removal of obstructing trees also produced higher wind velocities at ground level. The result of these more severe conditions was a series of devastating fires over the years. According to Davis, who compiled a list of the large forest fires in the past (17), "It is a significant fact that every one of the fires listed started in slash or other debris resulting from logging and land clearing and gained initial momentum from such fuels."

The destruction wrought by these early logging-slash fires undoubtedly created the sentiment and reasoning that led to the policy of total fire protection. While this concept of complete exclusion of fire was probably the only practical solution at the time, it has created other problems, some of which are just beginning to be recognized now. As discussed above, dead fuels have accumulated and present a serious hazard in many areas. Fire exclusion has also altered the ecological relationships of the forests, changing drastically the composition and growing conditions of many timber stands (18).

Fire has been essential for perpetuating valuable species (19). For example, Haig stated that Douglas fir (*Pseudotsuga menziesii*), western white pine (*Pinus monticola*), and lodgepole pine (*P. contorta*) have been maintained in more or less pure form by fire (20). He further observed that foresters have been reluctant to accept the idea that fire may aid regeneration. Observations of giant sequoias have indicated that reproduction occurs primarily on burns or mechanically disturbed areas (21). Since reproduction occurs on mechanically disturbed sites as well as on burned areas, we must consider fire as only one means of securing a mineral seedbed. However, many other ecological considerations have been studied very little. Hartesveldt *et al.* (21) referred to the occurrence of pathogenic fungi in soil and the beneficial effect of fire in sterilizing soil. Baker (22) suggested that leaving large quantities of

slash and debris may produce an unfavorable carbon/nitrogen ratio.

Greatly increased growth rates in fire-thinned patches of pine compared to unthinned stands were reported by Weaver (8). He suggested that the elimination of fire-thinning of sapling and pole-sized trees has led to greatly increased competition, subsequently weakening the trees and making them more vulnerable to insect attack (7, 14). Stocking rates, giving the optimum number of trees per hectare for best vigor and growth, support his views (Table 1). In addition, corrective measures, such as deliberate thinning, may increase the fire hazard. It was found that slash produced in thinning ponderosa pine in the Northwest greatly increases fire intensities and also resistance to fire-control efforts since the jack-strawed stems are difficult to clear for fire lines (23).

Present Approaches

Many fire control men recognize that a problem of fuel accumulation exists. Brown (24) stated that "The amount of fuel that is allowed to accumulate and its continuity are recognized by fire control men everywhere as fundamental in determining the cost of effective fire control and in fixing the losses that are bound to occur over a period of years." In the wildland research plan for California (25) it was pointed out that the stockpiles of dead fuels continue to accumulate, and Edward P. Cliff, then chief of the Forest Service, commented that (26) "Disposal of logging slash and reduction of fuel build-up through prescribed burning or other means constitute special challenges."

Even though the problem has been recognized, little has been done to measure its magnitude or extent. Proposed solutions have taken a number of directions. Efforts to find efficient chemical or biological digesters of dead fuels have not been successful. Mechanical chipping and shredding is quite expensive, with estimated costs running from $148 to $222 per hectare when prison labor is used. Efforts have been made to crush and chop slash and brush with heavy mechanical equipment (27). No cost figures have been published, but one administrative study by the California Division of Forestry showed costs of $73 to $99 per hectare.

The old concept of firebreaks permanently cleared on ridgetops has been extended to fuel breaks. These are strips

30 to 91 meters or more in width where the heavy fuels have been removed to break up large expanses of brush or timber (28). They are often seeded to grass to minimize erosion, with only a narrow strip in the middle cleared to mineral soil. The cost of clearing and maintaining them runs from $3500 to $5000 per kilometer. And fuel breaks do not eliminate the problem of fuel accumulation, they merely divide it up.

Another possible solution, prescribed burning, has been a subject of tremendous controversy. Weaver has often proposed the use of fire both as an ecological tool and for hazard reduction (7, 8, 14), and Biswell (29) has pointed out that the removal of dead fuel by prescribed burning reduces the damage from wildfire. These proposals have often been greeted by an emotional reaction and the view that all fires are raging monsters that destroy everything. The critics fail to recognize the differences between high-intensity wildfires that do destroy everything and low-intensity fires that may cause little or no damage.

Several studies of logging-slash flammability have been made, and there is a considerable amount of prescribed burning on clear-cut blocks where no vegetation remains (19, 30). However, these are all high-intensity fires, and less attention has been given to fuel accumulations outside of logging areas or to the use of low-intensity fires.

In two studies, low-intensity fires have been used in timber areas. Gordon's study in dense, stagnated stands of ponderosa pine saplings and poles is difficult to evaluate as few data are given on the sizes or number of trees (31). The photographs included in the report seem to indicate that the stands were extremely thick and that burning thinned the suppressed trees effectively. The costs on small plots (about 0.4 hectare) ran from $51.91 to $69.72. Schimke and Green (32) discussed the use of prescribed fire for maintaining fuel breaks in the central Sierra Nevada. They gave recommendations for weather conditions suitable for maintaining low-intensity fires and reported average costs of $10.45 per hectare.

Attempts by the National Park Service to restore natural ecosystems are encouraging. Houston (33) has reported on efforts with prescribed burning and natural fires in Sequoia and Kings Canyon national parks. Studies are being conducted in Yosemite National Park to obtain definite data on the effects of fire by using different weather and fuel

Table 1. Density of trees for full stocking (37). Diameter, average diameter at breast height; N, number of trees; ha, hectare.

Diameter (cm)	Density (N/ha)	
	Ponderosa pine	Douglas fir
10.2	4700	3110
15.2	2470	1865
20.3	1480	1300
25.4	990	956
50.8	247	296
76.2	106	173

conditions according to Schimke and Green's prescriptions (32). No results of these studies have been published to date, but my examination of several treated areas showed effective thinning and hazard reduction.

High-intensity fires have been used widely in brush areas for range conversion and hazard reduction. Since the passage of enabling legislation, nearly 800,000 hectares have been burned by private property owners in California (34). This activity is decreasing, probably because most of the areas in which it is economically feasible have been treated and because of the increasing financial liability associated with high-intensity fires. An economic study of controlled brush burning showed steadily decreasing costs with increasing size of the burn up to 178 hectares, where costs totaled $1.48 per hectare. Above that size, costs again rose slightly (35). These costs for high-intensity burns contrast sharply with those given for small plots and illustrate clearly the differences in scale. On a large scale, low-intensity fires would be considerably easier and cheaper to manage.

It must be recognized that there are many areas where high volumes of fuel would cause severe damage with any attempt at burning. However, little valid work has been done on the effects of low-intensity fires in the West. There are possible damages and disadvantages in the use of fire, but we may be forced to accept them as the premiums due for insurance for the whole forest. Evidence that a program of prescribed burning is compatible with modern economic and social objectives is given in Australia. Heavy losses in a series of disastrous fires prompted a study of management objectives. Their attempts at total fire exclusion had made changes in the nature of the forest and, while reducing the area of forest burned annually, had produced more severe and damaging bush fires (36). They concluded that complete fire protection is almost im-

possible to achieve and is undesirable ecologically. Hence, they have started a program of prescribed burning to reduce the fire hazard and restore the ecological conditions needed by the forest and wildlife.

Conclusions

Fire has been part of the western forests, probably for thousands of years. Apparently these frequent fires consumed much of the dead fuels and prevented large fuel buildups. Since fire protection agencies started their policy of total fire protection, dead fuels have gradually accumulated in many areas and represent a serious hazard. Studies and surveys should be started to obtain data on fuel volumes and distributions.

The large quantity of stored energy in accumulated fuels is released rapidly by fires, producing very high fire intensities. It is reasonable to predict that, where fuels are permitted to accumulate, fires will become more severe and more damaging and will be more difficult to control. More men and more fire trucks will not solve the problem of bigger and more damaging fires, although they may delay the ultimate results. The only big improvement possible in forest fire protection lies in the area of hazard reduction.

Intensified efforts to find economical and practical ways to reduce the fuel hazards are needed. The relatively neglected use of prescribed burning should be considered, particularly the use of low-intensity fires. Advantage should be taken of large wildfires that have reduced the fuel hazards, and vigorous efforts should be made to prevent future fuel accumulations.

References

1. W. E. Towell, *Amer. Forests* **75** (6), 12, 40 (1969).
2. G. M. Byram, in K. P. Davis, *Forest Fire: Control and Use* (McGraw-Hill, New York, 1959), pp. 61–89.
3. A. Hodgson, *J. Forest.* **66**, 601 (1968).
4. J. Kittredge, *Forest Influences* (McGraw-Hill, New York, 1948).
5. H. H. Biswell, R. P. Gibbens, H. Buchanan, *Calif. Agr.* **20** (9), 5 (1966); J. K. Agee and H. H. Biswell, *ibid.* **24** (6), 6 (1970).
6. J. S. Olson, *Ecology* **44**, 322 (1963).
7. H. Weaver, *J. Forest.* **41**, 7 (1939).
8. ———, *ibid.* **45**, 437 (1947).
9. H. H. Biswell, R. P. Gibbens, H. Buchanan, *Nat. Parks Mag.* **42** (No. 251), 16 (1968); J. K. Agee, thesis, University of California, Berkeley (1969).
10. Progress Report 5, Operation Firestop, California Division of Forestry and Cooperative Agencies (1955); L. R. Green, *U.S. Forest Serv. Res. Note PSW-216* (1970).
11. H. E. Anderson, *U.S. Forest Serv. Res. Pap. INT-56* (1968).
12. S. B. Show and E. I. Kotok, *U.S. Dep. Agr. Circ. 358* (1925).
13. W. W. Wagener, *J. Forest.* **59**, 739 (1961).

14. H. Weaver, *ibid.* **49**, 93 (1951).
15. L. T. Burcham, in *Proceedings of the Society of American Foresters*, San Francisco (1959), pp. 180–185.
16. O. C. Stewart, *Geogr. Rev.* **41**, 317 (1951); in *Man's Role in Changing the Face of the Earth*, W. L. Thomas, Ed. (Univ. of Chicago Press, Chicago, 1956), pp. 115–133.
17. K. P. Davis, *Forest Fire: Control and Use* (McGraw-Hill, New York, 1959).
18. M. Oberle, *Science* **165**, 568 (1969).
19. D. S. Olson and G. R. Fahnstock, *Univ. Idaho Forest Wildl. Range Exp. Sta. Bull. 1*, (1955).
20. I. T. Haig, *J. Forest.* **36**, 1045 (1938).
21. R. Hartesveldt, H. T. Harvey, H. S. Shellhammer, R. E. Stecker, *Science* **166**, 522 (1969).
22. F. S. Baker, in *Regional Silviculture of the United States*, J. S. Barrett, Ed. (Ronald, New York, 1962), pp. 460–502.
23. G. R. Fahnstock, *U.S. Forest Serv. Res. Pap. PNW-57* (1968); J. D. Dell and D. E. Franks, *Fire Control Notes* **32** (1), 4 (1971).
24. A. A. Brown, *J. Forest.* **45**, 342 (1947).
25. California Division of Forestry, "Wildland Research Plan for California" (California Division of Forestry, Sacramento, revised, 1969).
26. E. P. Cliff, *Amer. Forests* **75** (6), 20 (1969).
27. K. O. Wilson, *J. Forest.* **68**, 274 (1970).
28. U.S. Forest Service, "Progress Report on Fuel-Break Research" (U.S. Forest Service, Pacific Southwest Station, Berkeley, 1969).
29. H. H. Biswell, *Calif. Agr.* **13** (6), 5 (1959).
30. G. R. Fahnstock, *U.S. Forest Serv. Intermt. Forest Range Exp. Sta. Res. Pap. 58* (1960); E. R. De Silvia, *Proceedings of the Tall Timber Fire Ecology Conference*, Tallahassee, Florida (1965), pp. 221–330; R. W. Steel and W. R. Beaufait, *Univ. Mont. Mont. Forest Conserv. Exp. Sta. Bull. 36* (1969).
31. D. T. Gordon, *U.S. Forest Serv. Res. Pap. PSW-45* (1967).
32. H. E. Schimke and L. R. Green, "Prescribed Fire for Maintaining Fuel-Breaks in the Central Sierra Nevada" (U.S. Forest Service, Pacific Southwest Station, Berkeley, 1970).
33. D. B. Houston, *Science* **172**, 648 (1971).
34. California Division of Forestry, "Brushland Range Improvement Annual Report" (California Division of Forestry, Sacramento, 1969).
35. A. W. Sampson and L. T. Burcham, "Costs and Returns of Controlled Brush Burning for Range Improvement in Northern California" (California Division of Forestry, Sacramento, 1954).
36. R. G. Vines, *Aust. Sci. Teach. J.* (November 1968), reprint.
37. G. A. Craig and W. P. Maguire, *The California Pine Region Handbook* (California Division of Forestry, Sacramento, 1949).

8

Reprinted from *Proc. XI Intern. Grassland Congr.*, (University of Queensland Press), A131–A143 (1970)

A systems approach to grasslands

GEORGE M. VAN DYNE

Colorado State University, Fort Collins, Colorado, U.S.A.

Summary

This paper focuses on the role of mathematical modelling and analysis in grasslands research, management, and training. Examples are given of intraseasonal and interseasonal dynamics models of herbage biomass and of a total system energetics model. A generalized, computer-compatible notation is provided for modelling ecosystems. The complexity of grassland ecosystems imposes an interdisciplinary team approach for research and management. This complexity also imposes a modelling and systems analysis approach which requires a new approach in training. We now must train many grassland scientists and managers to work in interdisciplinary teams.

Introduction

The "systems approach" is not new, but unfortunately, it is utilized all too infrequently. It is a systematic consideration of problems, stressing analysis, and seeking optimal action by systematically examining objectives, costs, effectiveness, and risks of alternative management strategies — and designing additional strategies if original ones are insufficient. A systems approach to grassland problems includes (i) compiling, condensing, and synthesizing much information concerning the system components, (ii) detailed examination of the system structure, (iii) translating knowledge of components, function, and structure into models, and (iv) using models to derive new insights about management and utilization.

89

Grassland systems analysis

A system is an organization or structure that functions in a particular way. Analysis of grassland structure is treated thoroughly in plant management, ecological measurement, and range management books. Grassland functions of interest here include matter dynamics, energy flow, and nutrient cycling. Combinations of many specific functional processes account for these dynamics, e.g. photosynthesis, decomposition, herbivory, carnivory, parasitism, and symbiosis. To understand or manipulate wisely the overall systems functions of nutrient cycling and energy flow, researchers and managers must be concerned with these subprocesses.

Dynamic vs static models

Various models are necessary in studying grasslands. One starts with word models, converts to diagrammatic models, and then develops either static or dynamic models. My comments here are about dynamic models.

Let us use a set of differential equations describing the change in system components over a period of time:

$$\frac{dv_i}{dt} = \dot{v}_i = \sum_{j=1}^{n} f_{ij}\, v_j \qquad \text{for } i = 1, 2, \ldots n.$$

This is a linear, constant-coefficient model for a system of compartments where f_{ij} are "proportionality coefficients", and v_j are compartments referring to matter, energy, or nutrients in that compartment at time t. $\dot{v}_j$ refers to the change in amount of matter, energy, or nutrients in a compartment during a time interval. In matrix-vector notation (Van Dyne, 1969a) we have

$$\dot{V} = F\,V,$$ where $\dot{V}$ and V are n x 1 and F is n x n.

The constant-coefficient model is only a starting point in analysing grasslands systems. Usually, non-constant models involving time lags, probabilistic components, etc., are necessary for a realistic description. Examples follow.

Grassland succession

In some North American areas old, worn-out fields have been abandoned and native vegetation has reinvaded, resulting in soil, vegetation, and animal community changes over as many as 50 to 100 years as the system progresses towards the pre-cultivation state. Secondary succession includes a series of more-or-less recognizable transitory states, appearing with surprising regularity considering the differences in climate, grazing, and duration and type of cultivation. Bledsoe and Van Dyne (1969) describe a secondary succession on Great Plains grassland; the model includes six differential equations (Fig. 1).

There are appropriate stages of buildup of the vegetation part of the community from abandoned field to climax vegetation (Fig. 1a); one community is replaced by another, and each community contributes to the litter supply. Each community's importance is shown in relative units of abundance of organic material. In this matter-flow model a step function controls litter buildup.

The first community, characterized by annual weeds or grasses, is replaced in a few years by a short-lived perennial grass, dominating temporarily and in turn replaced by a second short-lived perennial grass. The next stage is dominated by a somewhat longer-lived grass, but is eventually replaced by two of the species that are climax dominants.

Intraseasonal herbage dynamics

In grasslands, organic material is translocated from tops to roots, and there is seasonal shattering of live vegetation, change to standing dead material, and decomposition of various plant parts. Concurrently, there is varying photosynthetic input. These relationships, shown in Figure 2a in a 6-compartment model, include data adapted from Kelly et al. (1969).

Photosynthetic input enters into the live vegetation foliage in amounts that vary seasonally. The algebraic equation for seasonal input is shown in Figure 2b; coefficient values were obtained by least-squares fitting to data on solar energy input in the experimental area. Difference equations for compartments 2 through 6 show change in grams per square metre of organic material each day. The seasonally varying coefficients in equations are largely empirical approximations, but they

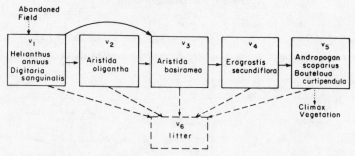

a. COMPARTMENT MODEL

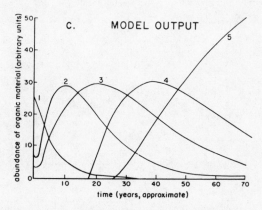

b. MATHEMATICAL MODEL

$\dot{v}_1 = -0.17\,v_1$

$\dot{v}_2 = 0.43\,v_1 - 0.87\,v_2$

$\dot{v}_3 = 0.17\,v_1 + 0.87\,v_2 - 0.06\,v_3$

$\dot{v}_4 = 0.13\,v_1 - 0.87\,v_4$

$\dot{v}_5 = 0.05\,v_4$

$\dot{v}_6 = \sum\limits_{i=1}^{5} f_{6i}\,v_i$ where $f_{6i}\begin{cases} = 0 \text{ if } v_6 \le c_i \\ = f_{6i} \text{ if } v_6 > c_i \end{cases}$ for $i = 1, 2, \ldots 5$

and $\begin{bmatrix} c_1 = 0 \\ c_2 = 2.7 \\ c_3 = 3.0 \\ c_4 = 6.5 \\ c_5 = 8.2 \end{bmatrix} > f_{61} = 0.01,\ f_{66} = -0.26,$ and $f_{6k} = 0.09$ for $k = 2,3 \ldots 5$

c. MODEL OUTPUT

abundance of organic material (arbitrary units)

time (years, approximate)

Fig. 1. An example of a compartment model (a), mathematical model (b), and model output (c) for old-field succession in a great plains grassland (adapted from Bledsoe and Van Dyne, 1970).

a. COMPARTMENT MODEL

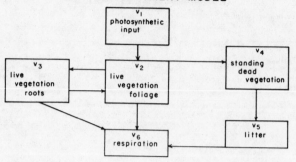

b. MATHEMATICAL MODEL

$$v_1(t) = \frac{3.0 + \cos\left\{3.1 + 2.0\left[\frac{\pi(t-1.0)}{365}\right]\right\}8.6}{1.4}$$

$$\frac{\Delta v_2}{\Delta t} = 1.0\,v_1 + 0.004\,v_3 - \left\{0.00027\,e^{0.0121\,t} + \left[0.002 + 0.002\sin(2t-0.7)1.4\right] + 0.0014\right\}v_2$$

$$\frac{\Delta v_3}{\Delta t} = \begin{cases} t \le 280 = 0.004\,v_2 - \left\{\left[0.0005 + 0.01\sin(t+2)\right]1.1 + 0.004\right\}v_3 \\ t > 280 = 0.004\,v_2 - \left\{\left[0.0005 + 0.01\sin(t+2)\right]1.1\left(\frac{365-t}{110.0}\right) + 0.004\right\}v_3 \end{cases}$$

$$\frac{\Delta v_4}{\Delta t} = 0.002\,v_2 - 0.001\,v_4$$

$$\frac{\Delta v_5}{\Delta t} = 0.00185\left[1.0 + \sin(2t-1.56)\right]v_4 - 0.002\,v_5$$

$$\frac{\Delta v_6}{\Delta t} = \frac{0.00185\left[1.0 + \sin(2t-1.56)\right]v_4 - 180.0 + v_5}{v_5} + 0.0014\,v_2 + 0.0007\,v_3$$

c. MODEL OUTPUT

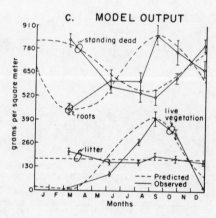

Fig. 2. An example of a compartment model (a), mathematical model (b), and model output (c) for intraseasonal dynamics of an *Andropogon virginicus*-dominated grassland. Observed means and standard errors are shown at measurement points (adapted from Kelly *et al.*, 1969).

could be converted, through additional mathematical and ecological study, to mechanistic functions of temperature, soil moisture, day-night ratios, and other field-measurable phenomena. Output from the model fits relatively well the observed values measured in the field (Fig. 2c).

Respiration losses and photosynthetic rate are predictable from the model. If the model is sufficiently detailed and describes accurately the real-life system, and if the field-measurable compartment values agree well with the model-predicted values for the same compartments, then considerable confidence could be placed in estimates derived from the model.

Multi-trophic level models

Examples in Figures 1 and 2 concerned only the vegetation, but grassland ecosystems contain more than one trophic level. The following model, utilizing difference equations synthesized from various experimental studies and assumptions but not representing any particular field experiment, is a 9-compartment ecosystem simulation (Fig. 3). Triangles in Figure 3 represent processes affecting energy flow among compartments. A tenth compartment is present, i.e. respiration losses.

The models of Figures 1 and 2 originally

COMPARTMENT MODEL

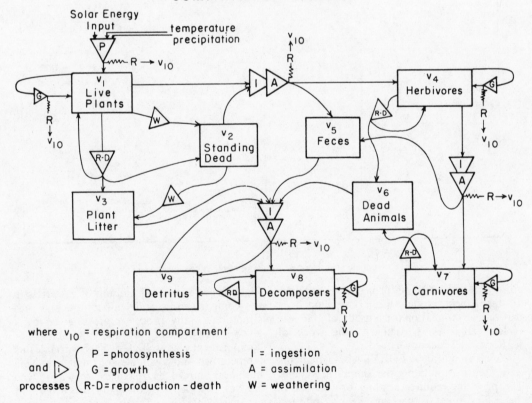

Fig. 3. A model diagram for a 9-compartment multi-trophic level grassland ecosystem. Energy flow is shown by arrows; compartments, in squares, are connected through processes, triangles. At many steps there is a respiratory energy loss, R, to the environment, v_{10}.

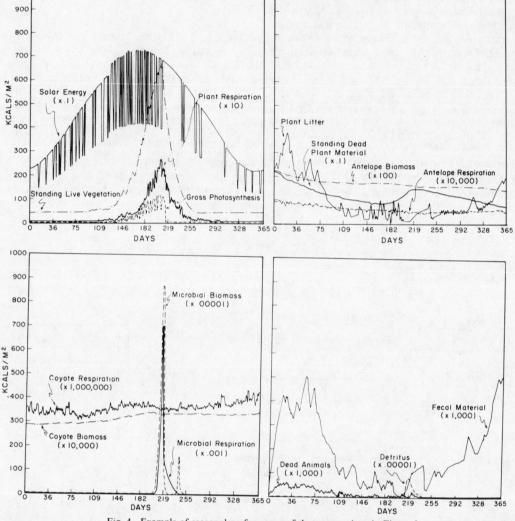

Fig. 4. Example of seasonal performance of the system given in Figure 3.

arose as Master's thesis projects, whereas the 9-compartment model (Fig. 3) developed as part of an examination for a group of graduate ecology students; particularly, the work of K. Redetzke is acknowledged.

More than 1,000 FORTRAN statements were used in simulating the simplified system. The program requires reading or calculating the driving forces or functions, i.e. precipitation and temperature. The remaining variables are calculated from time-dependent functions or from a combination of time-dependent

functions, the levels or amount of materials in the compartments, and temperature and precipitation. Many processes are affected by a given compartment, or the interaction of compartments, as well as by such variables as temperature and available soil moisture. This nonlinear model is detailed by Van Dyne (1969c). The activity of many processes and compartments is affected by the environmental driving forces. Many lag effects, upper and lower asymptotes, switch functions, and step functions are included to approximate seasonally varying phenomena. An example of the seasonal changes derived from the computer experiments in energy concentration in each compartment is given in Figure 4.

The model approaches the kind of complexity needed for simulating real-life ecosystems yet it only considers one generalized plant, herbivore, carnivore, and decomposer. There is a very simplified age-sex structure and generalized organisms are used. Even so, this represents a step in increasing complexity towards realism and interesting output may be obtained (Fig. 4).

Generalized concepts and notation for grassland models

A flexible scheme, compatible with computer implementation of models, is presented in Table 1, utilizing matrix algebra and set notation to denote ecosystem components or compartments. No distinction is made here between matter or energy flow. Major segments of the model in Figure 5 are ecosystem components (**V**), driving forces (**E**), parameters (**P**), processes (Q), and controls (S).

Basic macromodel—A compartment generally is an easily recognized component of the ecosystem in which we measure matter and energy in concentration units (Table 1). These are dependent variables; driving forces are independent variables. Parameters are measurable properties of the system whose time-fluctuation description is not the main object of the model; parameters interact with other dependent variables. Subscripted V's are used for intrinsic variables, subscripted E's for extrinsic variables, and subscripted P's for parameters. Processes control the transfer of material or energy between compartments. Major processes have been classified initially into 9 categories (Table 2).

Table 1. Conventions for a preliminary grassland macromodel (see Figure 5)

A vector of ecosystem components or compartments (variables)			*A vector of driving forces or input functions*		
	v_1	live plants, above ground		e_1	radiant energy
	v_2	live plants, below ground		e_2	wind
	v_3	standing dead vegetation		e_3	atmospheric temperature
	v_4	soil flora and fauna		e_4	atmospheric humidity
	v_5	humus	$E =$	$e_5 =$	atmospheric CO_2
	v_6	soil nitrogen compounds		e_6	external surface water
	v_7	litter, animal and plant		e_7	atmospheric O_2
$V = v_8 =$	v_8	litter flora and fauna		e_8	atmospheric N_2
	v_9	consumers			
	v_{10}	surface water			
	v_{11}	soil moisture		*A vector of parameters or properties*		
	v_{12}	available soil minerals				
	v_{13}	unavailable soil minerals				
	v_{14}	soil hydrogen ion	$P =$	$P_1 =$	soil temperature
	v_{15}	litter moisture		P_2	litter temperature

A set of control functions			*A set of processes*		
	s_1	plant community effects		q_1	photosynthesis
	s_2	animal community effects		q_2	ingestion
$S = s_3 =$	s_3	microbial community effects		q_3	assimilation
	s_4	human manipulation effects		q_4	growth
			$Q =$	$q_5 =$	reproduction-death
(s_i represent subsets)				q_6	decomposition
				q_7	emigration-immigration
				q_8	physical processes
				q_9	mobilization-immobilization

(q_i represent subsets)

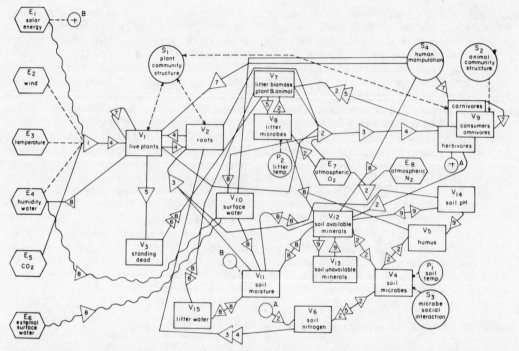

Fig. 5. A generalized macromodel of a grassland ecosystem showing driving forces (hexagons), compartments (boxes), processes (triangles), controls (large circles), and parameters (small circles). See the text for detailed description. Note that paired and letter-labelled circles represent connection points for flow lines. Thus + means going out of the plane of the paper and · means going into the plane.

Ecosystem function is not specified only by **V**, **E**, and **P**, but also depends upon structure. The position in space and density of some components has an effect on choice of co-efficients in the process functions which inter-connect compartments or on the form of these functions. I use a set of control variables, *S*, to show these relationships (Table 1). Three controls concern behavioural aspects of organisms such as animal, plant, and microbial interactions. Human intervention in ecosystem structure and function is a control variable. Parameters also can affect other compartments only in this way, since parameter variables represent properties of the system other than quantities of materials or energy.

Describing ecosystem function—**V**, **E**, **P**, *S*, and, to some degree, *Q* are changing through time, so each can be represented as a function of time as $V_{(t)}$, $E_{(t)}$, $P_{(t)}$, $S_{(t)}$, and $Q_{(t)}$. These properties or variables of the ecosystem are highly interconnected, but the degree of coupling varies widely, giving some first-order effects, second-order effects, etc.

Given initial values for **V** and given $E_{(t)}$ and $S_{(t)}$, we can calculate

$$P_{(t)} = g\ (E_{(t)}, V_{(t)}\ |\ S_{(t)}).$$

The set of control functions, *S*, affects the way in which **P** is determined by **E** and **V**. Some matrix of functions, **H** (*Q*), is determined by the interactions of **P**, **E**, and **V** as modified by *S*.

$$H(Q) = f\ (P_{(t)}, E_{(t)}, V_{(t)}\ |\ S_{(t)}).$$

Q, representing the processes of the ecosystem, is a set of relations which give these interactions

$$\dot{V} = H(Q)V.$$

This modelling scheme and notation differs greatly from those used before. Compartment models with constant-coefficients are the most common approach in modelling ecological phenomena. The equivalent here would be $\dot{V} = T\,V$, where $\dot{V}$ is the change in magnitude of the compartments V in a given time period, given simply by multiplication using the matrix of constants or transfer coefficients T. Perhaps this would approximate microcosms but it is inadequate for real-life field ecological phenomena where, as shown in the examples herein, non-constant-coefficient approaches are necessary.

Incorporating historical effects—A set of processes Q has been described (Table 2) but for each q_i the coefficients, or even the form of the function, may vary. The Q may be formed into a matrix $\overset{*}{Q}$ analogous to the F matrix in the constant coefficient model $\dot{V} = F\,V$. The particular $\overset{*}{Q}$ used depends upon system structure and history.

Implications of systems approaches

Grassland scientists and managers can greatly help man to understand nature so that he can live in harmony with it. Understanding productivity means understanding ecosystems. Detailed ecosystem studies are needed before existing theories can be applied fully to optimize biological productivity for man. Because of the complexity of ecosystems no one man encompasses all the required special knowledge to undertake the study of a complete grassland. This necessitates interdisciplinary teams and development of study procedures. Models provide a framework whereby individual studies can contribute to the understanding of the whole system.

As grassland science progresses from mythology through observation, measurement,

Table 2. General descriptions of families of processes used in the macromodel shown in Figure 5

Code	Description
1	*photosynthesis*—A family of processes applying to live green plants. Coefficients would vary for different species, nutrient conditions, etc. This process integrates solar input, atmospheric CO_2, etc., and is affected by such variables as wind speed, temperature, and humidity. The output of this process is net photosynthate.
2	*ingestion*—This family of processes applies to animals, especially those that search out their food.
3	*assimilation*—These processes apply to both plants and animals. For plants it represents primarily the transference of nutrients from the soil solution into the plant. For animals it represents primarily the transference from materials from the intestinal tracts into the animal. (Note: a process *egestion* is given by the difference between *ingestion* and *assimilation*.)
4	*growth*—This is a generalized and weakly defined process that applies to both plants and animals. It is concerned with the positive and negative weight changes of the organism, but not with subdivision of the organism to form new organisms. Thus, pregnancy in mammals and seed formation would be considered a kind of growth until parturition or seed fall, respectively. The growth process has as input the output products of assimilation.
5	*reproduction-death*—This group of processes applies to plants and animals and is concerned with the addition of new individuals to the population. Reproduction in concept includes both positive and negative changes in the population size. Thus, death is included as negative reproduction. The reproduction process is not shown throughout the model but it is implicit in the boxes representing plants and animals.
6	*decomposition*—This process is brought about by both micro-plants and micro-animals, i.e. all microbes, and some larger animals. Generally, decomposition has as input dead organic matter and the products are minerals usually returned to the soil solution.
7	*emigration-immigration*—These processes apply to both plants and animals and concern the addition of new individuals to the population from without the defined spatial range of the ecosystem under concern. This process may be under natural controls, or it may be due to man's manipulation of ecosystems by adding or deleting organisms from the system.
8	*physical*—This is a broad and weakly defined group of processes due primarily to abiotic agents. Examples would include compartmental transfers by weathering, respiration losses of energy, etc.
9	*mobilization-immobilization*—These processes are concerned largely with transfer of matter among living and nonliving compartments of the ecosystem. In general, many soil nutrient transformations are included here.

and understanding, we need to gain more knowledge for prediction and control. We must be able to predict the consequences of technological manipulation of grasslands. If we can mathematically model grassland ecosystems then we can better test our ability to predict consequences.

Ecosystem research and systems approaches

Much present grassland management and theory is based on knowledge of the relationships of a dominant plant species or vegetation to an edaphic situation, of a consumer to vegetation, or of one consumer to another (Coupland et al., 1969). But we have very little understanding of the processes in, nor optimal management for, the entire ecosystem.

Coupland et al. (1969) focus upon the entire ecosystem rather than upon components in their paper on grassland ecosystem research. When individual components are discussed it is in relation to the whole system.

There must be a great deal of sharing and collecting of basic data within a grassland ecosystem study, as shown by Figures 1-5. Scientists must participate as a team and must include services to others as an essential activity. The models also show that there must be a careful blend of field and laboratory studies in measuring system components and processes. The interaction of a large number of specialists requires a scientific committee to evaluate the rationale and management of the project. Natural grassland ecosystems are too complex for every organism and feature to be investigated. A scientific committee can help assign priorities. Priorities should be dynamic, shifting as knowledge increases. Systems analysts must be an integral part of the research team. Van Dyne (1969a) has emphasized the importance of early development of models in a grassland ecosystem study of the complexity found, for example, in the Canadian and United States grassland investigations in the International Biological Program. In the United States program much initial effort was in workshops in which information on grasslands was reviewed, condensed, discussed, and integrated into model framework (Bledsoe and Jameson, 1969). This synthesis procedure involved a large group of scientists of varying disciplines.

Members of the research group also got to know each other, which helped communication throughout the life of the program.

The main points about modelling in grassland ecosystem research are: (i) successive revisions of a model are essential in order to approach a realistic simulation, and no model is perfect; (ii) a good model does not only provide a graphic representation of the system but is also an instrument for manipulating the system and predicting changes under natural and artificial stresses; and (iii) the development of realistic mathematical models of grassland ecosystems is not a trivial nor short-term task, but a product of team effort that requires constant feedback and communication among field, laboratory, and "armchair" investigators (Van Dyne, 1969a).

Training grassland scientists and managers

Modifications of existing training patterns may be necessary for grassland scientists and managers before they can effectively participate in the types of interdisciplinary research and management outlined above (Van Dyne, 1969b). Future grassland scientists must be equipped with tools and techniques of studying ecosystems and an operational philosophy for contributing to, and operating in, interdisciplinary teams. Scientists have emphasized individual research and we will continue to need self-sufficiency in many research phases. However, to focus upon the challenging problems of ecosystem complexity the "loner" approach will not be adequate. Tomorrow's grassland scientist and resource manager must be psychologically conditioned to cope with the many interactions of team research and team management. Consider the different individuals and skills that would be needed to implement an ecosystem model of the type shown in Figure 5, remembering that the number of contacts between members of the total research program increases in proportion to $2^n - n$, and n is the number of people participating. The greatly increased number of contacts and interactions could easily overtax our communication, transportation, and psychological systems!

Our challenge is to train scientists to retain their individuality and yet to contribute to the team effort toward well-defined goals. Perhaps definition of goals should be emphasized in training grassland scientists and managers. Grassland scientists must become more precise in their terminology, for it is axiomatic that ambiguous use of terminology and ambiguous statement of problem, especially in an inter-disciplinary investigation, leads to ambiguity of thought. To examine grasslands requires conventional field and laboratory procedures and an increasing use of an array of sophisticated new tools and techniques. In addition to training in the basic sciences of chemistry, physics, mathematics, communications, and in the humanities, I feel the grassland scientist and manager should have training in the physiology and ecology of each of the trophic level disciplines involved. His undergraduate program should be integrated with courses in which systems structure, simulation, management, and analysis are stressed. This will require increased attention to analytical methods and computer applications, and should be undertaken in the context of real-life systems. It cannot be obtained solely in mathematics or engineering departments.

Most undergraduate students interested in grasslands will not have available to them analytically-orientated courses on systems techniques and concepts; these are available only in a few universities (Van Dyne, 1969*b*). There is increasing need for students to become acquainted with these tools in the classroom context. Simulation models can be constructed for classroom use, especially if remote computer terminals are available for the necessary man-machine interaction. The models need not be highly complicated; but they will be a computerized extension of logical thinking and scientific method. By such means students can examine the outcome of different alternatives in the management of resource systems.

Students must train themselves increasingly for one of three levels of scientific and management contribution. These are (Van Dyne, 1969*b*): (i) laboratory or field scientists largely working alone on a specific process or phase of the problem; (ii) leaders coordinating the efforts of a group of such scientists or managers; and (iii) scientists coordinating the efforts of the group leaders. In research, the third-level scientists will be "theoretical biolo-

gists" whose role in advancing knowledge is to synthesize into testable theories the results obtained in experimental studies. To do this, they must have broad knowledge of the entire grassland ecosystem, a penchant for searching and condensing literature, and skill in combining and evaluating information from field, laboratory, and library investigations. They must be systems ecologists (Van Dyne, 1966)

Systems implications for grassland management

Systems analysis has been successful in the study and management of complex systems, e.g. manufacturing, marketing, corporate structure. Grasslands offer a similar complexity, and knowledgeable management must involve simultaneous consideration of many variables. Consider especially the multiple-use concept and management of renewable natural resources. Assembling and studying the information needed for single-use management is difficult, but multiple-use management requires that a staggering array of problems be formulated and solved. Resource managers must assess quantitatively alternative management practices to be able to compare the marginal returns for investing funds in each activity. Changing and uncertain resource use values increase the difficulties of prediction. Analytical ecologists advising resource managers are resorting increasingly to computer tools for simulating large, complex resource systems; examples of applications are noted by Paulik (1970) and Van Dyne *et al.* (1969).

The manager must be able to make accurate forecasts of his system's responses to management input variables, at least on a short-term basis. Increasingly, routine measurements are being made on resource systems, using automatic monitoring and sensing equipment. There is a rapidly growing volume of resource system data on inputs and outputs which can be computerized and utilized in simulation models. The resource biologist must work in association with the modeller, the data collector, and the administrative manager. Coupled with conventional and time-tested resource management tools these new techniques and approaches can provide the decision-maker with a powerful array of methods for solving complex problems (Van Dyne, 1969*b*).

A142

GRASSLAND AS A SOIL-PLANT-ANIMAL SYSTEM

Conclusions and extensions

Systems analysis includes the orderly and logical organization of data on grasslands ecosystems into models. Word, picture, and mathematical models will all be useful, although only mathematical models are stressed in this paper. In grassland systems analysis we must determine the components or properties of the ecosystem, their functional interrelationships, and how these interrelationships connect to external or driving forces. Simplified system views will be taken initially and many interactions or interrelationships may be disregarded at first.

Some of the preceding models included only one of two trophic levels, or if all trophic levels were included, the model was greatly simplified. Final simulation models are complex in appearance, but they are built up from simple mathematical statements and statistical distributions which represent the functions, interrelationships, and values attributed to the realworld ecosystem. Numerical simulation is important; it is the only known technique which is capable of representing the complexities of real ecosystems.

The major values of models are (i) to provide a framework for the organization and development of a research program and the assessment of importance of different facets of field and laboratory investigation; and (ii) to permit inferences about the real system to be drawn, from operation and manipulation of the model, more easily than from observation of the real system itself. Examples of such inferences were made with respect to Figure 2.

A successful model can greatly improve the planning of long-term field research. This can be accomplished primarily through "sensitivity analysis", which is a method of determining the relative sensitivity of the model output to the changes in the coefficients, input functions, and forms of the equations of which the model is composed. If the model output remains essentially the same over a range of variation in one part, then the model is insensitive to that part, and less time and energy may be devoted to its accurate determination in field and laboratory. Alternatively, if the part in question is known to have profound effects on the output of the real system, yet sensitivity analysis does not reflect

this, then the structure of the model is suspect. Other systems analysis procedures that could be investigated include frequency and time responses of the model.

Some probabilistic components were included into the model shown in Figure 3. Stochastic models are much more complex than deterministic models and require more computing time. They demand more man-machine interaction and adequate computing facilities, including remote console terminals, for detailed model development and examination. Much of the modelling work in systems analysis of grasslands must be done on a day-to-day basis with analysts and experimentalists in close communication and cooperation.

Model output must be applied in research or management. Models are a useful framework on which to hang our knowledge about grasslands and the models are useful in making predictions, but the predictions must be validated or tested in the field and laboratory (Van Dyne, 1969a).

The number of equations and amount of computer code needed for a model, such as in Figure 5, will be large. A special problem exists in communicating information on simulation models. Our conventional journals and publications must allow inclusion of detailed lists of equations, FORTRAN or other computer code statements, and model output if we are to succeed in rapidly advancing the systems approach to grassland analysis.

Acknowledgment

Much of the modelling work reported herein was done in a team research project on grassland ecosystems sponsored by National Science Foundation Grant GB-7824. Further details for this paper are given by Van Dyne, 1969c, including FORTRAN listings for the model of Figures 3 and 4.

References

BLEDSOE, L. J., and JAMESON, D. A. (1969). Model structure for a grassland ecosystem, *The Grassland Ecosystem: a Preliminary Synthesis* (eds R. L. Dix and R. G. Beidleman), Range Sci. Ser. 2. Colorado State Univ.

BLEDSOE, L. J., and VAN DYNE, G. M. (1969). Oak Ridge Nat. Lab. TM-2414.

BLEDSOE, L. J., and VAN DYNE, G. M. (1970). A compartment model simulation of secondary succession, *Systems Analysis and Simulation in Ecology* (ed. B. C. Patten), Academic P., New York.

COUPLAND, R. T., ZACHARUK, R. Y., and PAUL, E. A. (1969). Procedures of study of grassland ecosystems, *The Ecosystem Concept in Natural Resource Management* (ed. G. M. Van Dyne), Academic P., New York.

KELLY, J. M., OPSTRUP, P. A., OLSON, J. S., AUERBACH, S. I., and VAN DYNE, G. M. (1969). Rep. Oak Ridge Nat. Lab. 4310.

PAULIK, G. J. (1970). Digital simulation modelling in resource management and the training of applied ecologists, *Systems Analysis and Simulation in Ecology*

(ed. B. C. Patten), Academic P., New York.

VAN DYNE, G. M. (1966). Rep. Oak Ridge Nat. Lab. 3957.

VAN DYNE, G. M. (1969a). Some mathematical models of grassland ecosystems, *The Grassland Ecosystem: a Preliminary Synthesis* (eds R. L. Dix and R. G. Beidleman), Range Sci. Ser. 2. Colorado State Univ.

VAN DYNE, G. M. (1969b). Implementing the ecosystem concept in training in the natural resource sciences, *The Ecosystem Concept in Natural Resource Management* (ed. G. M. Van Dyne), Academic P., New York.

VAN DYNE, G. M. (1969c). Range Sci. Ser. 3. Colorado State Univ.

VAN DYNE, G. M., FRAYER, W. E., and BLEDSOE, L. J. (1969). *In* Society for Industrial and Applied Mathematics, *Studies in Applied Mathematics: Symposium on Optimization.*

9

Copyright © 1972 by the American Geophysical Union

Reprinted from *Water Resources Res.*, **8**, 326–338 (Apr. 1972)

Watershed Management: A Systems Approach

L. M. EISEL*

*Division of Engineering and Applied Physics, Harvard University
Cambridge, Massachusetts 02138*

Abstract. A systems approach to land use management decision making in wildland areas is developed and demonstrated by using a hypothetical example. A major component of this systems approach is a chance-constrained programing model for investigating the effects of risk and uncertainty on land use management decisions. The solution of the chance-constrained model indicates the most efficient set of land use management activities for reaching specified goals subject to certain requirements and regulations. In addition to indicating the type and extent of land use management activities for a specific area, the systems approach provides additional results and conclusions concerning more general aspects of land use management decision making. The development and solution of the chance-constrained model indicate that the risk and uncertainty associated with the system physical and economic parameters can significantly affect the design of land use management activities. The model results further suggest that future outdoor recreational demand and benefits are important factors in the design of land use management activities. The solution of the chance-constrained model indicates that forest management practices to increase stream-flows may have only a minimal effect on the design and operation of downstream reservoirs.

The U.S. government is responsible through various federal agencies for the management of approximately 660 million acres of forest, grassland, and desert. Much of this land is in the western United States and is administered by the U.S. Forest Service and the U.S. Bureau of Land Management. These wildlands are the headwater areas of many western rivers and are important sources of outdoor recreation, timber, and grazing land. Rapidly increasing future demands for timber, grazing, and water are forecast, and even more rapid growth in the future demand for outdoor recreation is expected. Maintaining the necessary environmental quality required for the expanding recreational demand while satisfying the increased demand for other forest and range products will require sophisticated management of these wildland ecosystems.

A systems approach to land use management employing operations research methods may be necessary to meet future demands for forest and range products and still maintain a high level of wildland environmental quality for future outdoor recreation. Watershed manage-

* Now at Environmental Defense Fund, East Setauket, New York 11733.

ment recognizes the interdependence of soils, vegetation, the surface runoff phase of the hydrologic cycle, and management activities and provides a basis for developing a systems approach to the management of these wildland areas.

A systems approach to wildland management is developed in this paper. Major emphasis is on the development of a mathematical model to assist the wildland manager. This model is not intended to obviate the traditional functions of the land use management decision maker. Model development is intended to demonstrate how a systems approach to wildland management can supplement the decision process and help to meet the challenges of future land use management problems in wildland areas.

THE SYSTEMS APPROACH

To facilitate the attainment of the study objectives, a water resource system consisting of an upstream watershed and a proposed downstream irrigation reservoir is specified that illustrates the major characteristics of land use management activities (Figure 1). A proposed irrigation reservoir is included in the system to allow the investigation of the interdependence between watershed management activities and structural developments. This system is similar to many small river basins in the western United States that produce forest products, grazing, irrigation water, and outdoor recreational opportunities. For management purposes, the hypothetical watershed can be divided into three distinct units: (1) a 20-mi² forest area suitable for timber production, (2) a 40-mi² forest area suitable for pulp production, and (3) a 40-mi² grassland area suitable for grazing. The forest and grassland areas of the watershed have been depleted by overgrazing, burning, and inadequate management. In the simplified world of the hypothetical system, administrators have two alternatives for each square mile of land: (1) to permit existing practices to continue or (2) to apply a set of 'improved' land use management practices. The set of improved practices is designed to maintain or reestablish a quasi-steady state in the wildland ecosystem by preventing the degradation of the water quality, the soil, the vegetation, and the wildlife resources.

Sustained yield forest management practices have been recommended that prescribe a 90-year rotation for sawtimber and a 30-year rotation for pulpwood. These practices would not only improve the quality and quantity (in the long run) of forest products but would also preserve the soil resource and improve the environmental quality of the watershed for outdoor recreation. Employment of special timber-harvesting practices could increase the annual runoff from the watershed. Improved management of the watershed could result in reduced sedimentation in the proposed irrigation reservoir, which would then require a lower sediment storage capacity. Increasing the annual river flow by forest management would provide more irrigation water if adequate reservoir storage were available. The agency responsible for managing the upstream watershed and the agency responsible for designing and operating the reservoir have agreed to investigate these possibilities.

The management of the hypothetical wildland watershed demands the making of numerous decisions that would require administrators to consider not only the effects of specific practices on the soil–vegetation complex and the surface runoff component of the hydrologic cycle but also the quality and quantity of

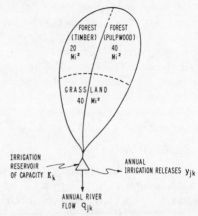

Fig. 1. A hypothetical water resource system consisting of an upstream watershed and a proposed downstream irrigation reservoir.

future demands for lumber, plywood, pulp, grazing products, water, outdoor recreation, and so forth. This decision process is complicated by (1) a lack of functional relationships between wildland system inputs and outputs, (2) an imperfect knowledge of an uncertain and stochastic environment, (3) the uncertainty of future demands and technology, and (4) the difficulty of quantifying benefits from outdoor recreation and beauty. A major objective of a systems approach to land use management is the incorporation of these factors into the land use management decision process.

Mathematical model. A major component of the systems approach to land use management is a mathematical programing model. The specification of a programing model permits an investigation of the system economic and physical responses to different modes and levels of management and to various assumptions concerning the quality and quantity of future demands.

The management of wildland areas is complicated by the uncertainty and risk associated with management activities. The physical effects and economic results of management activities cannot be predicted with certainty because of the uncertainty associated with imperfectly understood natural phenomena and the risk associated with stochastic factors (e.g., precipitation, streamflow, forest growth, insolation, and so forth). Explicit consideration of stochastic factors necessitates a stochastic mathematical programing model. The specification of a stochastic model could be accomplished by either linear programing under uncertainty [*Dantzig,* 1963] or chance-constrained programing [*Naslund,* 1965; *Charnes and Cooper,* 1963]. For the purposes of this study, a chance-constrained model is developed. The selection of a chance-constrained model is discussed further in *Eisel* [1972].

DEVELOPMENT OF THE SYSTEMS APPROACH

Management Objectives

A principal guideline for wildland management is the multiple use policy that prescribes management for '. . . outdoor recreation, range, timber, watershed, wildlife, and fish purposes' [*U.S. Congress,* 1960]. Other components of the multiple use philosophy require '. . . permanent good for the whole people . . .' and '. . . the greatest good for the greatest number in the long run' [*U.S. Congress,* 1960]. The National Environmental Policy Act of 1969 (Public Law 91-190) provides further guidelines for the federal wildland manager. Land use managers desire to develop management practices for the hypothetical system in Figure 1 that will maximize the present value of the time stream of benefits from timber, pulpwood, irrigation water, grazing, and outdoor recreation and still conform to multiple use policy. The specification of a chance-constrained programing model requires the translation of these management goals into a mathematical objective function and constraint set.

Outdoor Recreation

The development of this objective function and constraint set is complicated by a consideration of the benefits from outdoor recreation. Explicit consideration of outdoor recreation in a programing model could be achieved by two methods: (1) the introduction of environmental quality constraints for maintaining high quality wildland environments for outdoor recreation or (2) the quantitative evaluation of outdoor recreational benefits. In the first method, the dual variable value associated with each environmental quality constraint could be analyzed to determine the imputed costs of maintaining or reestablishing a specific level of environmental quality. These imputed costs could help decision makers to select a 'desirable' level of wildland environmental quality. This

process, however, requires an implicit quantification by agency decision makers of environmental quality and outdoor recreational benefits.

The second method was selected for this investigation on the basis that the explicit quantification of outdoor recreational benefits is preferable to their implicit quantification by agency decision makers. The user-day procedure is employed for estimating benefits from outdoor recreation because of its simplicity and the lack of information concerning the quantitative effects of specific land use management activities on outdoor recreation.

Estimates of recreation user-days were made for areas under various 'improved' and 'present' land use management practices. Guidelines are available for transforming these estimates into economic criteria [*Ad Hoc Water Resources Council,* 1964]. The practical and theoretical deficiencies of these guidelines and of the user-day metric are recognized [*Cichetti et al.,* 1969; *Knetsch and Davis,* 1966; *Krutilla,* 1967]. These deficiencies indicate that multiplying user-days by monetary benefit estimates will, in general, underestimate the social value from outdoor recreation in high quality natural environments because (1) consumer surplus is excluded, (2) option demand is excluded, and (3) the utility of natural environments to future generations is inadequately estimated. This underestimation must be considered in an analysis of model results.

Soil–Vegetation Complex Stability

The management of wildland areas in accordance with multiple use policy requires the maintenance of ecosystem stability. The soils and vegetation of the hypothetical watershed are an interdependent complex of the watershed ecosystem. For management purposes, this interdependent complex can be considered to be in a quasi-steady state. The evolution and adaptation of species are still occurring, soil formation is continuing, and so forth. However, the changes produced by these natural processes are insignificant within the time horizon of present land use management.

Watershed management in wildland areas involves land use management activities designed to maintain or reestablish a quasi-steady state in the soil–vegetation complex. A soil–vegetation system in a quasi-steady state normally insures the continual production of timber, grazing products, high quality water, and aesthetically pleasing environments for outdoor recreation. Natural catastrophes (e.g., forest fires and insect infestations) or man-induced perturbations from poor management practices may disturb the quasi-steady state of the wildland soil–vegetation complex. Such disturbances may result in reduced future production of forest and grazing products, eutrophication problems resulting from soil erosion, alteration of streamflow characteristics, increased erosion and reservoir sedimentation, and degraded environments for outdoor recreation.

To develop the mathematical model, it is necessary to select a quantifiable indicator of soil–vegetation complex stability. Several parameters could be used for this purpose: the rate of soil erosion, water quality parameters [*Borman et al.*, 1968], the composition and growth rates of vegetation, and so forth. For the purposes of this initial systems investigation, the rate of soil erosion is used as an index of soil–vegetation complex stability.

Prerequisite Assumptions

The rotation period for sawtimber is approximately 90 years. Let this 90-year period be the time horizon for the chance-constrained programing model. The division of this 90-year period into shorter time increments is necessary because timber growth rates, soil loss rates, and so forth for a specific land use management practice will not remain constant over the entire 90-year period. This continuous change must be transformed into incremental change for inclusion in a programing model. Therefore the 90-year period is divided into three 30-year periods corresponding to the length of pulpwood rotation. In the succeeding notation, the subscript $j = 1, 2, \cdots, 30$ will denote years, and $k = 1, 2$, and 3 will denote decision periods.

Consideration of forest management activities to increase the annual flow from the Figure 1 watershed complicates the chance-constrained programing model because estimates of the mean and variance of annual streamflow must incorporate the effects of land use management activities. Numerous small watershed investigations have studied the relationship between streamflow and vegetation [e.g., *Sopper and*

Lull, 1967] and have not produced general stochastic relationships that permit the prediction of flow changes from vegetative manipulation. In the absence of such empirical relationships, a simple linear stochastic relationship is postulated:

$$q_{jk}' = q_{jk} + \sum_i \theta_{ik} x_{ik} \tag{1}$$

where q_{jk}' is the annual streamflow (in 1000 ac ft) in year j of decision period k after vegetation management practices (a random variable), q_{jk} is the annual streamflow (in 1000 ac ft/yr) without vegetation management practices (a random variable), θ_{ik} is the streamflow change (in 1000 ac ft/mi^2) resulting from performing 1 mi^2 of land use management activity i in period k (a random variable), and x_{ik} is the land use management activity i (in mi^2) in decision period k (a decision variable). The specification of this stochastic relationship between land use management and streamflow is necessary for the development of the chance-constrained model.

CHANCE-CONSTRAINED MODEL

The Decision Problem

The land use management agency of the hypothetical system must either continue with existing management practices or implement 'improved' practices. For the simplified world of the Figure 1 system, let

x_{1k}, square miles of existing timber management practices;

x_{2k}, square miles of improved management practices;

x_{3k}, square miles of improved timber management practices to increase runoff;

x_{4k}, square miles of existing pulpwood management practices;

x_{5k}, square miles of improved pulpwood management practices;

x_{6k}, square miles of improved pulpwood management practices to increase runoff;

x_{7k}, square miles of existing grassland management practices;

x_{8k}, square miles of improved grassland management practices.

The x_{ik} are decision variables in decision period k for $k = 1, 2$, and 3.

In their decision the land use management agency must consider the effects of specific practices on soil loss and the consequent

330 L. M. EISEL

stability of the soil–vegetation complex. The effect of a specific land use management practice cannot generally be predicted with certainty. To incorporate the risk associated with the effects of land use management practices into the model, the soil loss effects of specific practices are assumed to be random variables, where d_{1k} is the soil loss (in ac ft/mi^2) from 1 mi^2 of existing timber management practices in decision period k (a random variable), $E(d_{1k})$ is the expected soil loss in decision period k from 1 mi^2 of existing timber management practices, $V(d_{1k})$ is the variance of d_{1k}, and so forth for $E(d_{ik})$ and $V(d_{ik})$ where $i = 2, 3, \cdots, 8$ and $k = 1, 2,$ and 3. These moments could be estimated by Bayesian statistical procedures. For examples of Bayesian estimates of parameters in forest management and forest fire system studies, see *Thompson* [1968], *Lindquist* [1969], and *Davis* [1969].

Land use management administrators desire to coordinate the management of the Figure 1 watershed with the design and operation of the proposed downstream irrigation reservoir. The design and operation of this reservoir require the specification of the initial reservoir capacity K_0 in 1000 ac ft (a decision variable) (the reservoir capacity in decision period k, K_k, will be less than K_0, owing to sedimentation), the annual irrigation target for decision period k, T_k, in 1000 ac ft (a decision variable), and a dimensionless reservoir operating policy parameter a_k [*Eisel*, 1972] (a decision variable).

The decision problem can be summarized. The land use management agency desires to coordinate the management of the hypothetical watershed with the design and operation of the downstream irrigation reservoir. The objective of this coordination is to optimize the economic benefits from the production of timber, pulpwood, grazing products, outdoor recreation, and irrigation water and still maintain the soil–vegetation complex in a quasi-steady state.

The Benefit Function

The objective function of the chance-constrained programing model can be specified. Maximize:

$$\sum_k B_k T_k - c K_0 + \sum_i \sum_k b_{ik} x_{ik} \qquad (2)$$

where B_k is the present value of 1000 ac ft of irrigation water delivered annually throughout period k (in \$1000/1000 ac ft), c is the capital costs and present value of operating and maintenance costs of reservoir storage (in \$1000/1000 ac ft), and b_{ik} is the present value of benefits from all products resulting from 1 mi^2 of land use management activity i in period k (in \$1000/mi^2).

Constraint Set

Constraint on soil–vegetation complex stability. The specification of a constraint requiring the maintenance of the soil–vegetation complex in a quasi-steady state is necessary under multiple use policy. Because the effects of land use management practices on the soil–vegetation complex cannot be predicted with certainty, this constraint is specified as a chance constraint, where soil loss is employed as an index of soil–vegetation complex stability:

$$P\{D_k X_k \leq t_k\} \geq \alpha_k{}^{(t)} \qquad k = 1, 2, 3 \quad (3)$$

where

t_k, the soil loss limit for decision period k (in ac ft), an exogenous variable specified by the land use management agency;

$D_k = [d_{1k}, d_{2k}, \cdots, d_{8k}]$;

$$X_k = \begin{bmatrix} x_{1k} \\ x_{2k} \\ \cdot \\ \cdot \\ \cdot \\ x_{8k} \end{bmatrix};$$

$\alpha_k{}^{(t)}$, the decimal percent reliability level for not exceeding the soil loss limit t_k in period k, $0 \leq \alpha_k{}^{(t)} \leq 1.0$;

P, the probability operator.

Inequality 3 requires the design of land use management practices to insure that soil loss from the hypothetical system will not exceed t_k ac ft in decision period k with probability $\alpha_k{}^{(t)}$. Inequality 3 can be reduced to a deterministic form if D_k is assumed to have a multivariate normal distribution. This assumption may not be restrictive in practice if only extreme fractiles are considered (e.g., $\alpha_k{}^{(t)} \geq 0.90$). The fractile level is the only characteristic of the distribution incorporated into the deterministic equivalent of (3). Little numerical difference exists between the extreme fractiles of Gaussian and some common skewed distribution (e.g., gamma). Consequently, the D_k distribution could be skewed, and little disparity would exist between its extreme fractiles and the corresponding fractiles of a multivariate normal distribution.

Reducing (3) to its deterministic equivalent yields

$$P\left\{ \frac{D_k X_k - E(D_k) X_k}{[X_k^T V(D_k) X_k]^{1/2}} \right.$$

$$\left. \leq \frac{t_k - E(D_k) X_k}{[X_k^T V(D_k) X_k]^{1/2}} \right\} \geq \alpha_k^{(t)}$$

where $V(D_k)$ is the variance-covariance matrix of D_k, and E is the expectation operator. This inequality requires

$$t_k - E(D_k) X_k \geq F^{-1}[\alpha_k^{(t)}][X_k^T V(D_k) X_k]^{1/2}$$

$$(4)$$

where $F^{-1}[\alpha_k^{(t)}]$ denotes the $\alpha_k^{(t)}$ fractile of a Gaussian distribution with mean 0.0 and variance 1.0. The set of feasible solution points satisfying (4) will be convex; this relationship can be demonstrated by investigating the Hessian matrix and associated quadratic form [*Charnes and Cooper*, 1963; *Van de Panne and Popp*, 1963]. If $V(D_k)$ is diagonal, (4) can be rewritten as two separable constraints:

$$t_k - E(D_k) X_k \geq {}_3 w_k \qquad (5)$$

and

$$\{F^{-1}[\alpha_k^{(t)}]\}^2 X_k^T V(D_k) X_k \leq {}_3 w_k^2 \qquad (6)$$

$$k = 1, 2, 3$$

The assumption of a diagonal $V(D_k)$ connotes statistical independence between the effects of land use management activities. This assumption presupposes that the effects of management activities on soil loss from a grassland area are independent of management activities in an adjoining forest. Examples exist to support and contradict this assumption. However, quantitative estimation of the interdependence of land use management activities is difficult because of the numerous factors affecting the results of management activities and their spatial variation in a wildland watershed. For the purposes of this limited initial investigation, independence is assumed. However, for more extensive investigations in existing watersheds, additional consideration should be given to this assumption and its analytical implications.

Additional land use management constraints. A functional relation between the reservoir capacity in decision periods k and $k + 1$ must be specified to account for reservoir sedimentation:

$$K_1 = K_0 - 1/2 \sum_i E(d_{i1})(x_{i1}) \qquad (7)$$

$$K_k = K_{k-1} - 1/2 \sum_i E(d_{ik})(x_{ik})$$

$$- 1/2 \sum_i E(d_{i,k-1})(x_{i,k-1}) \quad k = 2, 3 \qquad (8)$$

where linear interpolation between the beginning and the end of the decision period is used for determining the reservoir capacity in decision period k.

Upper bounds are required on the land use management activities in the hypothetical watershed of Figure 1:

$$x_{1k} + x_{2k} + x_{3k} \leq 20$$

$$x_{4k} + x_{5k} + x_{6k} \leq 40$$

$$x_{7k} + x_{8k} \leq 40 \qquad k = 1, 2, 3 \qquad (9)$$

$$\sum_i x_{ik} \leq 100 \qquad k = 1, 2, 3 \qquad (10)$$

Reservoir constraints. A set of reservoir operating constraints involving stochastic variables is necessary to insure the continuity of flow, limited spilling, and the meeting of the irrigation target release with specified probability. To specify these constraints as chance constraints, the moments of the $(q_{j+1,k}' + s_{jk})$ distribution must be estimated (where s_{jk} is the reservoir storage at the end of year j). The mathematical development of these reservoir operating constraints is presented in *Eisel* [1972] together with the convolution problem associated with moment estimation for the $(q_{j+1,k} + s_{jk})$ distribution. The estimation of the moments of the $(q_{j+1,k}' + s_{jk})$ distribution is a minor extension of this convolution problem and is described in *Eisel* [1970].

Model summary. The chance-constrained model of the hypothetical Figure 1 basin may be summarized as follows: maximize the objective function (equation 2) subject to the deterministic equivalent of the soil–vegetation complex constraint (equations 5 and 6), the additional land use management constraints (equations 7–10), and the three reservoir chance constraints.

SOLUTION AND RESULTS

The model of the Figure 1 system is a non-linear, separable convex programing problem. A linear programing algorithm with restricted entry can achieve a global optimal solution with the 'δ method' of separable programing [*Hadley*, 1964].

Obtaining the necessary economic and physical data proved difficult for even the hypothetical system in Figure 1. These difficulties in obtaining the necessary data were primarily responsible for limiting the objectives of this investigation to demonstrating a systems approach to land use management in a hypothetical and simplified water resource system. This limitation must be considered in interpreting the succeeding results and conclusions.

Three data sets (Tables 1 and 2) were developed to investigate how land use management decisions are affected by (1) future upward reevaluation of outdoor recreation benefits, (2) the increased future demand for water, and (3) time preference criteria (the discount rate). Data set 1 represents the static case, in which product demand and unit price remain constant, and data sets 2 and 3 represent the dynamic cases, in which preferences for water and outdoor recreation increase in relation to the demand for timber, pulpwood, and grazing products. Data set 2 assumes that agricultural use of water will be superseded by municipal and industrial use. This demand change requires

that the evaluation of 1 ac ft of water in the first decision period be increased by 50% over the data set 1 evaluation and that in the second and third periods by 100%. In data set 2, benefits from outdoor recreation in the first, second, and third decision periods are assumed to increase by 50, 100, and 200%, respectively, over the data set 1 evaluation. This increase is due both to the rising use of the outdoor recreational resource and to the upward reevaluation of outdoor recreational benefits by future societies.

The changes in relative preferences in data set 3 are the same as those in data set 2, but different discount rates are employed. In data sets 1 and 2, benefits from timber, pulpwood, grazing, irrigation water, and outdoor recreation are discounted at 6%, whereas in data set 3, benefits from outdoor recreation are discounted at 3% and those from all other sources at 7.3%. The U.S. Forest Service currently uses the 6% rate, and *Haveman* [1969] suggests 7.3% as a suitable rate for water resource developments. The 3% rate indicates high evaluation of future outdoor recreational opportunities and is included for comparison with data set 2, in which all the products of the hypothetical watershed, including outdoor recreation, are discounted at 6%.

The model solution results from the three data sets are presented in Table 3. The results and discussion presented here primarily concern land use management, the results and discussion of the chance-constrained reservoir model being presented elsewhere [*Eisel*, 1972].

TABLE 1. Necessary Economic Parameters for Data Sets 1–3

Parameter	Data Set 1			Data Set 2			Data Set 3		
	$k=1$	$k=2$	$k=3$	$k=1$	$k=2$	$k=3$	$k=1$	$k=2$	$k=3$
b_{1k}	70.7	2.2	1.1	76.9	4.3	1.9	81.2	14.6	9.4
b_{2k}	54.7	2.9	1.0	63.0	5.7	2.0	70.9	19.4	12.2
b_{3k}	50.2	2.9	0.9	58.5	5.8	1.9	66.6	19.4	12.2
b_{4k}	39.1	6.4	1.0	45.3	8.5	1.7	48.7	17.0	9.4
b_{5k}	30.0	5.9	0.8	38.2	8.6	1.8	46.2	21.1	12.2
b_{6k}	24.6	4.6	0.8	32.9	7.5	1.8	41.8	20.4	12.1
b_{7k}	16.0	2.0	0.3	17.3	2.5	0.5	17.4	4.2	2.0
b_{8k}	12.1	2.1	0.4	14.9	3.0	0.7	17.5	7.2	4.1
B_k	276.0	48.0	8.0	414.0	96.0	16.0	361.0	58.0	7.0
c	50.0			50.0			50.0		

Data set 1 represents the static case, and data sets 2 and 3 the dynamic cases, in which preferences for water and outdoor recreation increase in relation to preferences for timber, pulpwood, and grazing products.

The optimal solution of the model with any of the data sets specifies a reservoir of approximately 94,000 ac ft capacity and an annual irrigation target in the first 30-year period of about 59,000 ac ft. Approximately 5000 ac ft of reservoir sedimentation occurs in the 90-year time horizon in all cases. The marginal cost of preventing 1 ac ft of soil loss from the hypothetical watershed (i.e., the dual value of constraint 4) is $824 for the data set 1 solution (Table 3). This value indicates that increasing the soil loss limit t_k from 2500 to 2501 ac ft would reduce the net economic benefits from the watershed by $824. However, constructing an extra acre-foot of reservoir capacity to store this additional acre-foot of sediment would cost only $50. Data set 2 represents a situation in which the benefits from water and outdoor recreation are increased relative to those of other watershed products. In the data set 2 solution, all improved activities except x_{3k} (the improved forest management activity to increase streamflow) are employed to their respective limits, and the marginal cost of

TABLE 2. Necessary Physical Parameters for Data Sets 1–3

Parameter	$k = 1$	$k = 2$	$k = 3$
$E(d_{1k})$	27	3	3
$E(d_{2k})$	14	3	3
$E(d_{3k})$	14	3	3
$E(d_{4k})$	24	27	27
$E(d_{5k})$	14	14	14
$E(d_{6k})$	14	14	14
$E(d_{7k})$	50	54	60
$E(d_{8k})$	30	30	30
$V(d_{1k})$	9	1	1
$V(d_{2k})$	9	1	1
$V(d_{3k})$	16	1	1
$V(d_{4k})$	9	9	9
$V(d_{5k})$	16	16	16
$V(d_{6k})$	9	9	9
$V(d_{7k})$	9	25	25
$V(d_{8k})$	25	25	25
$E(\theta_{3k})$	0.032	0.0	0.0
$E(\theta_{6k})$	0.032	0.032	0.032
$V(\theta_{3k})$	0.0001	0.0001	0.0001
$V(\theta_{6k})$	0.0001	0.0001	0.0001
t_k	2500.	2500.	2500.
$V(q)$	400.0	400.0	400.0
$E(q)$	64.0	64.0	64.0
$F^{-1}[\alpha_k^{(t)}]$	1.28	1.28	1.28

The physical parameters are the same for all three data sets.

TABLE 3. Model Solution Results

	Data Set 1 Results			Data Set 2 Results			Data Set 3 Results			Data Set 4 Results		
	$k=1$	$k=2$	$k=3$	$k=1$	$k=2$	$k=3$	$k=1$	$k=2$	$k=3$	$k=1$	$k=2$	$k=3$
K_0	93.75			93.75			93.67			93.74		
K_k	92.64	90.42	88.39	92.64	90.63	88.81	92.64	90.72	88.89	92.65	90.65	88.83
T_k	59.34	57.37	57.67	59.34	58.41	57.81	59.72	58.44	57.84	57.80	57.14	56.53
α_k	0.90	0.868	0.838	0.90	0.871	0.845	0.90	0.872	0.846	0.90	0.871	0.845
x_{1k}	11.91	11.91	11.91	11.91	11.91	11.91	0.0	0.0	0.0	9.89	9.89	9.89
x_{2k}	0.0	0.0	0.0	0.0	0.0	0.0	0.0	0.0	0.0	10.11	10.11	10.11
x_{3k}	8.09	8.09	8.09	8.09	8.09	8.09	20.0	20.0	20.0			
x_{4k}	0.0	30.47	0.0	0.0	0.0	0.0	0.0	0.0	0.0	0.0	0.0	0.0
x_{5k}	0.0	0.0	0.0	0.0	0.0	0.0	0.0	0.0	0.0	40.0	40.0	40.0
x_{6k}	40.0	9.53	40.0	40.0	40.0	40.0	40.0	40.0	40.0			
x_{7k}	0.0	0.0	0.0	0.0	0.0	0.0	0.0	0.0	0.0	0.0	0.0	0.0
x_{8k}	40.0	0.0	0.0	40.0	40.0	40.0	40.0	40.0	40.0	40.0	40.0	40.0
Constraint 5 dual	824.1	1.1	0.0	235.0	0.0	0.0	0.0	0.0	0.0	923.2		

Data sets 1–3 are described in detail in Tables 1 and 2 and the text. Data set 4 is the same as data set 2 except that the land use management activities to increase streamflow x_{3k} and x_{6k} have been omitted. The net benefits of data sets 1–4 were $18,059,000, $30,369,000, $26,747,000, and $29,857,000, respectively.

preventing 1 ac ft of soil loss in period 1 has dropped to $235. Data set 3 portrays a case in which future opportunities for outdoor recreation in a high quality wildland environment are evaluated upward from those of data set 2, and the future economic benefits from the production of water, timber, pulpwood, and grazing are evaluated downward. In the data set 3 solution, all improved land use management activities are employed to their respective limits. As a result, the marginal cost of preventing 1 ac ft of soil loss has fallen to zero in all decision periods.

A traditional justification for watershed management activities has been the prevention of sedimentation in downstream reservoirs and irrigation canals. These results indicate that land use management activities for this purpose in the hypothetical watershed may be economically inefficient unless they preserve or restore the quality of natural environments for present and future outdoor recreational use.

The forest management practices designed to increase streamflow (x_{3k} and x_{6k}) were excluded from data set 2, and the model was solved with the resulting data set 4 (Table 3). A comparison of these results with those from the original data set 2 indicates that forest management practices for increasing streamflow had little effect on the design and operation of the proposed irrigation reservoirs. Only slight differences exist between optimal values for the initial reservoir capacity K_0, the irrigation target T_k, and the reservoir operating parameter a_k in the data set 2 and 4 solutions.

Risk and Uncertainty

Figures 2 and 3 present the results of parametric programing analyses for changes in the reliability level for meeting the soil loss target $\alpha_1^{(t)}$ and the irrigation target $\alpha_k^{(T)}$. Increasing the level of certainty for meeting the irrigation target (Figure 2) from 80 to 95% reduced the net economic benefits by more than $5 million (18%), whereas increasing the level of certainty for meeting the soil loss target (Figure 3) by

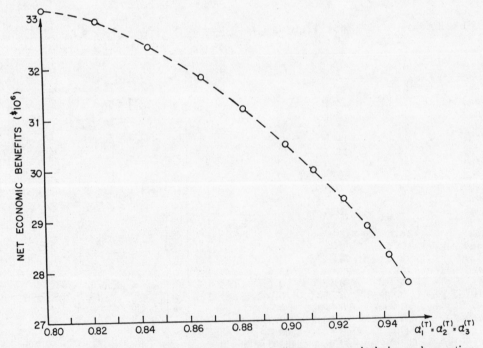

Fig. 2. Parametric investigation of reliability level effects on reservoir design and operating parameters. In $100\alpha_k^{(T)}\%$ of the years, farmers can expect delivery of the target level of irrigation water. If $\alpha_1^{(T)} = 0.90$, then farmers can expect delivery of the target level in 9 out of 10 years during the first decision period (see *Eisel* [1972] for detailed development of the irrigation target constraint).

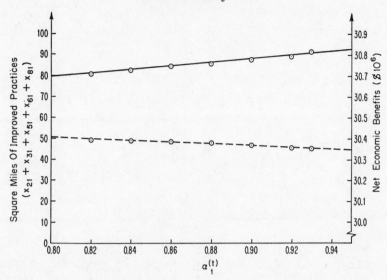

Fig. 3. Parametric programing analysis of the soil loss target reliability level $\alpha_1^{(t)}$. The solid line represents the square miles of recommended practices, and the dashed line the net economic benefits.

the same amount resulted in net economic benefit reductions of less than $100,000 (0.4%). Increasing the level of certainty from 80 to 95% indicates that attainment of the respective targets can be expected in 19 out of 20 years rather than in 4 out of 5 years. Further comparison of our Figure 3 with Figure 4 of *Eisel* [1972] suggests that the design of land use management activities in the hypothetical example is less affected by risk than reservoir design and operation are. These results indicate that the cost of increased certainty in maintaining a quasi-steady state in the hypothetical wildland ecosystem (and a high quality environment for future outdoor recreation) may be relatively low.

The prediction of future demands for outdoor recreation and the evaluation of outdoor recreational experiences for future generations involve uncertainty. The effect of this uncertainty on system design was investigated by parametric programing, and the results of this analysis are presented in Figure 4. The data set used in the preparation of Figure 4 resembles data set 1, except that the outdoor recreational benefits in decision periods 2 and 3 have been multiplied by factors of 2.0 and 3.0, respectively. The point at 1.0 and 84.0 in Figure 4a indicates that 84.0 mi² of improved land use management

activities are specified by the model in the optimal design. The point at 1.5 and 100 indicates that if the decision period 1 outdoor recreational benefits are increased by 50% (i.e., multiplied by a factor of 1.5), all improved land use management activities are specified in the optimal design. Even for higher discount rates (Figure 4b), outdoor recreational benefits must be increased by only approximately 300% before all improved land use management activities are specified in the optimal design. Discrepancies of this order of magnitude (50–300%) are not unlikely because of difficulties in assessing future benefits from outdoor recreation and in predicting future demand for outdoor recreation.

For simplicity, all benefits from outdoor recreation in the hypothetical watershed were evaluated at $3/user-day. This estimate is consistent with the $1.50–6.00/user-day range for 'specialized outdoor recreation days' suggested by the *Ad Hoc Water Resources Council* [1964] and with the benefit estimates by the *U.S. Forest Service* [undated]. Evaluation by future generations of outdoor recreation in high quality natural environments may substantially exceed these present estimates. A 300% increase in user-day benefit evaluation is a distinct possibility in future societies with increased

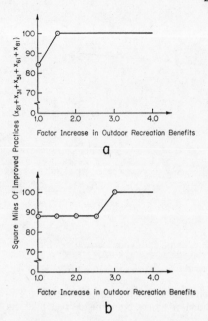

a

b

Fig. 4. Parametric programing analysis of outdoor recreation benefits. (*a*) Outdoor recreation benefits are discounted at 3%, and the economic benefits from production of irrigation water, timber, and grazing at 6%. (*b*) All benefits are discounted at 6%.

leisure time and disposable income. Figure 4 indicates that underestimating future benefits from outdoor recreation can significantly affect the provision of a high quality natural environment in the hypothetical watershed for future outdoor recreation.

Additional parametric programing results demonstrating the sensitivity of the system design to variation in the soil loss coefficients are presented in Figures 5 and 6. Figure 5 indicates that if estimates of all soil loss coefficients for the first decision period $E(d_{11})$, $E(d_{21})$, $\cdots$, $E(d_{81})$ are reduced by approximately 27% from their values in Tables 1 and 2, improved practices can be reduced from 88 to 40 mi², and net benefits increased by approximately \$117,000. Figure 6 indicates similar results for parametric variations of the soil loss coefficients for only the grassland, the principal area for soil loss in the hypothetical watershed.

Comparison of Figures 4, 5, and 6 indicates that the design of land use management activities for the hypothetical watershed is significantly affected by the uncertainty of predicting

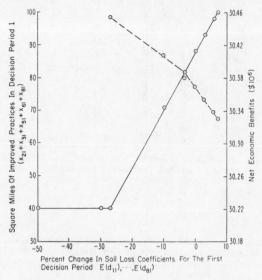

Fig. 5. Parametric variation of the soil loss coefficients for all areas in the first decision period. The solid line represents the square miles of improved land use management activities, and the dashed line the net economic benefits.

both the physical effects of land use management activities and the future outdoor recreational benefits. Numerous investigations have produced a vast literature detailing the physical effects of land use management practices on

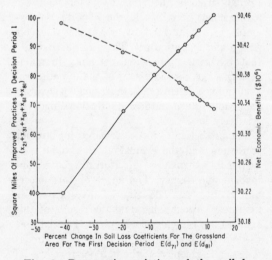

Fig. 6. Parametric variation of the soil loss coefficients for the grassland area for the first decision period. The solid line represents the square miles of improved land use activities, and the dashed line the net economic benefits.

wildland watersheds. In contrast to this situation, few studies and little information are available that permit the quantification of the effects of land use management practices on outdoor recreation in a specific area.

A comparison of the results from the solution of the chance-constrained model with data sets 2 and 3 (Table 3) indicates the importance of the discount rate in land use management decisions. A comparison of Figures 4a and 4b further denotes the significance of this parameter. No general agreement, however, exists on the 'correct' rates for discounting future benefits from the continued production of recreation, water, lumber, grazing, and so forth in wildland areas.

DISCUSSIONS AND CONCLUSIONS

A systems investigation of land use management in an existing wildland area will indicate the most efficient set of land use management activities for reaching specified goals subject to certain requirements and regulations. Such results would be far more detailed than those for the hypothetical watershed in Figure 1 with its eight land use management activities. Nevertheless, the results from the solution of the chance-constrained model of the hypothetical system indicate the design recommendations resulting from a systems approach to land use management.

An important aspect of the study was the integration of the upstream watershed into the downstream basin. Traditionally, these areas have not been incorporated into one system but have been managed separately by different agencies with individual policies. The investigation results indicate that land use management practices increasing streamflow and reducing reservoir sedimentation in the hypothetical watershed have minimal effects on the design and operation of the downstream irrigation reservoir (data set 4 results in Table 3). The traditional administrative division of the Figure 1 basin may, therefore, be adequate.

The justification for watershed management practices has traditionally been based on the preservation of the soil resource for sustained yield production of forest and grassland products, the prevention of reservoir sedimentation, the production of high quality water from wildland watersheds, and so forth. The dual values associated with the deterministic equivalent of the soil–vegetation complex constraint (inequality 5) and other results presented in Table 3 indicate that performance of land use management practices for these purposes may be economically inefficient in the hypothetical system unless these activities preserve or restore the quality of natural environments for present and future outdoor recreational use. These results further suggest that management emphasis in the hypothetical system should be transferred from traditional objectives (e.g., preventing soil erosion) to those of preserving the quality of natural environments for outdoor recreation. This revised objective will still necessitate soil erosion control, but the design of specific land use management activities should be in accordance with the revised objective of maintaining and restoring environmental quality in wildland areas rather than with soil erosion control per se.

The parametric programing results indicate that the design of land use management activities for the hypothetical Figure 1 watershed is significantly affected by the uncertainty in predicting the physical effects of land use management activities and the future outdoor recreational benefits. Numerous investigations have studied the physical effects of land use management on wildland ecosystems, but relatively few investigations have attempted a quantitative evaluation of the recreational benefits from specific land use management activities. These results suggest the necessity for developing procedures for estimating the recreational benefits from the projected physical effects of land use management activities.

Acknowledgments. The support and helpful criticism of Professor Myron B Fiering, Harvard University, have been appreciated. Mr. James Reid, U.S. Forest Service Southwest Forest and Range Experiment Station, Berkeley, California, assisted in obtaining data and provided important suggestions and criticism. Acknowledgment is made to Dr. Steven Marcus and Mr. John Gaudette, Harvard University, and to Mr. William Fleming, University of British Columbia, Vancouver, for their helpful discussions. Publication was assisted by Mr. Grant Schaumburg, Harvard University and V. H. Eisel. This investigation was supported by a U.S. Public Health Service Traineeship and grant 44-5325-2 from Resources for the Future. Publication was made possible by the Environmental Systems Program,

113

Harvard University, through National Science
Foundation grant GI-24.

REFERENCES

Ad Hoc Water Resources Council, Evaluation
standards for primary outdoor recreation benefits, supplement 1 to the President's Water
Resources Council policies, standards and procedures in the formulation, evaluation and review of plans for use and development of water
and related land resources, *Senate Doc. 97*,
9 pp., 87th Congress, 2nd Session, June 1964.

Borman, F. H., G. E. Likens, D .W. Fisher, and
R. S. Pierce, Nutrient loss accelerated by clear-
cutting of a forest ecosystem, *Science, 159*,
882–884, 1968.

Charnes, A., and W. W. Cooper, Deterministic
equivalents for optimizing and satisficing under
chance constraints, *Oper. Res., 11*(1), 18–39,
1963.

Cicchetti, C. J., J. J. Senecca, and P. Davidson,
The demand and supply of outdoor recreation,
report of the Bureau of Economic Research,
253 pp., Rutgers Univ., New Brunswick, N. J.,
1969.

Dantzig, G. B., *Linear Programming and Extensions*, 632 pp., Princeton University Press,
Princeton, N. J., 1963.

Davis, J. B., Application of bayesian decision
theory to large fire control systems, paper presented at Joint National Meeting, Amer.
Astronaut. Soc. and Oper. Res. Soc., Denver,
Colo., June 1969.

Eisel, L. M., Land-use management in a hydrologic system, Ph.D. thesis, Harvard Univ., Cambridge, Mass., 1970.

Eisel, L. M., Chance-constrained reservoir model,
Water Resour. Res., 8(2), 339–347, 1972.

Hadley, G., *Nonlinear Programming*, 484 pp.,
Addison-Wesley, Reading, Mass., 1964.

Haveman, R. H., The opportunity costs of dis-
placed private spending and social discount
rate, *Water Resour. Res., 5*(5), 947–957, 1969.

Knetsch, J., and R. K. Davis, Comparative methods of recreation evaluation, in *Water Research*,
edited by A. Kneese and S. Smith, pp. 125–142,
Johns Hopkins University Press, Baltimore,
Md., 1966.

Krutilla, J. V., Conservation reconsidered, *Amer.
Econ. Rev., 57*, 778–786, 1967.

Lindquist, J. L., Probability estimates of handline
construction rates, Forest Fire Laboratory
progress report, 36 pp., U.S. Forest Serv., Riverside, Calif., 1969.

Naslund, B., Mathematical programming under
risk, *Swed. J. Econ., 67*(3), 240–255, 1965.

Sopper, E., and H. W. Lull (Eds.), *Proceedings of
the International Symposium on Forest Hydrology*, Pergamon, London, 1967.

Thompson, E. F., The theory of decision under
uncertainty and possible applications in forest
management, *Forest Sci., 14*(2), 156–163, 1968.

U.S. Congress, hearings before the Subcommittee
on Forests of the Committee on Agriculture,
U.S. House of Representatives, on HR-10572,
National forests, multiple use and sustained
yield, 85th Congress, 2nd Session, March 1960.

U.S. Forest Service, A special PPB system study
of three selected areas in the Rocky Mountain
national forests, undated.

Van De Panne, C., and W. Popp, Minimum cost
cattle feed under probabilistic protein constraints, *Manage. Sci., 9*(3), 405–430, 1963.

(Manuscript received July 8, 1971.)

10

Reprinted from *Operations Res.*, **19**, 1698–1707 (Nov.–Dec. 1971)

SCHEDULING BIOPRODUCTION HARVEST

Tung Liang, Wen-yuan Huang, and Jaw-Kai Wang

University of Hawaii, Honolulu, Hawaii

(Received November 6, 1969)

Bioproduction is unique in its heterogeneous growth, unsteady product value, and production-rate-dependent harvesting cost structure. This paper develops a stochastic model for the harvesting process of a bioproduction, uses it to optimize a tropical-fruit-production harvesting operation, and conducts a sensitivity analysis with respect to the parameters in the model.

MOST bioproduction manifests the following characteristics: (1) heterogeneous growth, (2) unsteady fruit quality or value (the word 'fruit' is used here to represent bioproduction products), and (3) a fruit-density-dependent harvesting cost structure.

Harvesting cost can usually be broken down into: (1) travelling (which is independent of fruit density), (2) inspecting and searching for fruits (which demand a certain amount of time and labor even if no marketable fruit is available for collection), and (3) the activity of collecting fruits (which is definitely dependent on the density of marketable fruit on a tree— 'tree' being used here to refer to any individual bioproduction element that produces fruits, such as a fruit tree, a hen, etc.).

If one considered only the harvesting cost per fruit, it would be definitely advantageous to wait for a higher marketable fruit denisty before harvest such that the fixed-cost items (1) and (2) mentioned above can be shared by more fruits per harvest.

However, as a result of heterogeneous growth, fruits do not reach marketable stage at the same time. They also do not lose or gain values at the same rate. In fact, they never reach a steady value as a finished industrial product does. Their value is discrete and unsteady. The discrete nature of value is artificially created by marketing practice. A fruit may lose its value entirely if it is left unharvested too long. If harvested too early, it also loses a portion of its potential value. Therefore, harvesting should be managed so that both harvesting cost and fruit loss are carefully considered in the scheduling process and so that the production profit is optimized.

MODELING FRUIT RIPENING

THE OBJECTIVE of modeling fruit ripening ('ripening' is defined to be the growth process by which unmarketable fruits grow into various degrees of marketability and, finally, into overripe fruits without value) is to provide a means of estimating the number of fruits at different stages of ripeness for any fixed number of fruit trees and at any harvesting interval. The term 'harvesting interval' is defined to be the number of days between two consecutive harvesting operations. The time scale was chosen to be discrete and with the day as its unit, which is commonly used in scheduling harvesting operations.

The number of fruits at the ith stage of ripeness F_i that can be harvested from a field ('field' means a collection of bioproduction elements) on a given day, is a function of X_k, M_j, L, and N. Thus,

$$F_i = f(X_k, M_j, L, N), \qquad (K = 1, 2, \cdots, L, \text{ and } j = 1, 2, 3, 4) \quad (1)$$

where

F_i = total number of fruits at the ith stage of ripeness that are harvested
　　on a given day from a given field,

L = number of days between harvests,

R = ripeness,

X_k = number of fruit arrivals on the kth day after previous harvest
　　(fruit arrivals mean fruits have just reached the marketable stage,
　　this stage being denoted as the T stage, or turning stage; for example, X_1 is the number of fruit arrivals on the first day after the
　　previous harvest, and so forth),

M_1 = number of days for a fruit to ripen from the T stage to the $(\frac{1}{4})R$
　　stage,

M_2 = number of days from the $(\frac{1}{4})R$ stage to the $(\frac{1}{2})R$ stage,

M_3 = number of days from the $(\frac{1}{2})R$ stage to the $(\frac{3}{4})R$ stage,

M_4 = number of days from the $(\frac{3}{4})R$ stage to the $(\frac{4}{4})R$ stage,

N = number of trees in a field.

The degree of fruit ripeness varies discretely from the T stage, just reaching ripeness (marketable), to $\frac{4}{4}$ ripeness in five stages. The value of fruit at each stage may be different.

For any given field and given L and N, the X_k's and M_j's are random variables. Therefore, F_i is also a function of random variables. The expected number of fruit arrivals on a given tree on the kth day after the previous harvest can be expressed as

$$F_k = E(X_k). \qquad (k = 1, 2, \cdots, L) \quad (2)$$

The total number of fruit arrivals on a given tree between harvests can be expressed as

$$FT = \sum_{k=1}^{k=L} E(X_k). \qquad (3)$$

Let M^k denote the ripeness of a fruit that arrived on the kth day after the previous harvest and will be harvested at the next harvest; then M^k

can be determined in accordance with the conditions in Table I and expressed in the following set notations, for $k = 1, 2, \cdots, L$:

$$(M^k = 1) = (M_1 > L - k + 1) \tag{4a}$$

$$(M^k = 2) = (M_1 \leq L - k + 1) \cap (M_1 + M_2 > L - k + 1), \tag{4b}$$

$$(M^k = 3) = (M_1 + M_2 \leq L - k + 1 \cap (M_1 + M_2 + M_3 > L - k + 1), \tag{4c}$$

$$(M^k = 4) = (M_1 + M_2 + M_3 \leq L - k + 1)$$
$$\cap (M_1 + M_2 + M_3 + M_4 > L - k + 1), \tag{4d}$$

$$(M^k = 5) = (M_1 + M_2 + M_3 + M_4 \leq L - k + 1). \tag{4e}$$

TABLE I

RIPENESS CLASSIFICATION OF FRUITS ARRIVING BETWEEN HARVESTS

Ripeness stage	Conditions	Ripeness classification on harvesting day
T stage $M^k = 1$	$M_1 >$ (number of days between the kth day and the next harvesting day), or $M_1 > L - k + 1$	The T stage—includes all fruits between turning stage to $(\frac{1}{4})R$
$(\frac{1}{4})R$ stage $M^k = 2$	$(M_1 \leq L - k + 1)$ and $(M_1 + M_2 > L - k + 1)$	The $(\frac{1}{4})R$ stage—includes all fruits between $(\frac{1}{4})R$ to $(\frac{1}{2})R$
$(\frac{1}{2})R$ stage $M^k = 3$	$(M_1 + M_2 \leq L - k + 1)$ and $(M_1 + M_2 + M_3 > L - k + 1)$	The $(\frac{1}{2})R$ stage—includes all fruits between $(\frac{1}{2})R$ to $(\frac{3}{4})R$
$(\frac{3}{4})R$ stage $M^k = 4$	$(M_1 + M_2 + M_3 \leq L - k + 1)$ $(M_1 + M_2 + M_3 + M_4 > L - k + 1)$	The $(\frac{3}{4})R$ stage—includes all fruits between $(\frac{3}{4})R$ to $(\frac{4}{4})R$
$(\frac{4}{4})R$ stage $M^k = 5$	$(M_1 + M_2 + M_3 + M_4 \leq L - k + 1)$	The $(\frac{4}{4})R$ stage—includes all fruits from full-ripe to over-ripe

The probability of a fruit harvested at different stages of ripeness can be expressed as follows:

$$P(M^k = 1) = P(M_1 > L - k + 1), \tag{5a}$$

$$P(M^k = 2) = P((M_1 \leq L - k + 1) \cap (M_1 + M_2 > L - k + 1)), \tag{5b}$$

$$P(M^k = 3) = P((M_1 + M_2 \leq L - k + 1)$$
$$\cap (M_1 + M_2 + M_3 > L - k + 1)), \tag{5c}$$

$$P(M^k = 4) = P((M_1 + M_2 + M_3 \leq L - k + 1)$$
$$\cap (M_1 + M_2 + M_3 + M_4 > L - k + 1)), \tag{5d}$$

$$P(M^k = 5) = P(M_1 + M_2 + M_3 + M_4 \leq L - k + 1). \tag{5e}$$

117

The expected number of the fruits at various ripeness stages, given the fruit arrivals on the kth day, can be obtained as follows:

$$F_{k,i} = E[X_k P(M^k = i)], \quad (i = 1, 2, 3, 4, 5) \quad (6)$$

The total number of fruits per tree that can be harvested at the ith stage of ripeness on the next harvest, for $i = 1, 2, 3, 4, 5$, is

$$\begin{aligned} TF_i &= E[X_1 P(M^1 = i)] + E[X_2 P(M^2 = i)] + \cdots + E[X_L P(M^1 = i)] \\ &= \sum_{k=1}^{k=L} E[X_k P(M^k = i)]. \end{aligned} \quad (7)$$

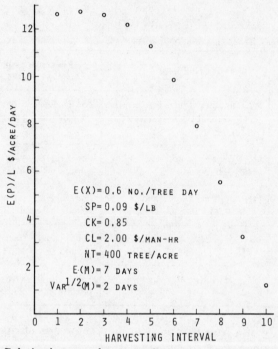

Fig. 1. Relation between the expected net return $E(P)/L$ and the harvesting interval L.

Therefore, the total number of fruits that can be expected at the ith stage of ripeness at the next harvest for a field of N trees is

$$F_i = N \times TF_i \quad (8)$$

OPTIMUM-HARVESTING-FREQUENCY MODEL

IN BIOPRODUCTION, maximum profit and minimum production cost are probably the most frequently used objectives in production policy making.

In this study, the objective is to determine the harvesting frequency that will maximize the net return per day. The expected net return per harvest can be expressed as:

$$E(P) = CK \times SP \times [E(NF1) - E(NF2)] - CO - CF(L). \quad (9)$$

Then the expected net return per day is

$$E(P)/L = \{CK \times SP \times [E(NF1) - E(NF2)] - CO - CF(L)\}/L, \quad (10)$$

where

$\quad E(P)$ = expected net return for one harvest, in \$/harvest,

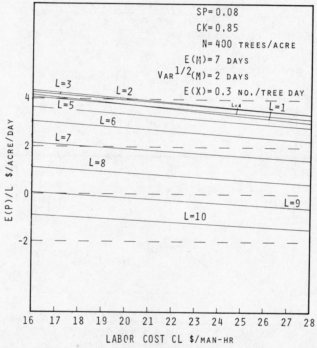

Fig. 2. Relation between the expected net return $E(P)/L$ and the labor cost CL.

$E(NF1)$ = expected number of marketable fruits (within the desirable ripeness), in no./harvest,

$E(NF2)$ = expected number of bad fruits (exceeding the desirable ripeness), in no./harvest,

$\quad CO$ = operating cost, in \$/harvest,

$CF(L)$ = shared fixed cost per harvest, in \$/harvest (it is a function of the harvesting interval L),

CK = conversion factor, in lb/fruit,
SP = sale price of fruit, in \$/lb.

The fruits exceeding the desirable ripeness are considered as a loss. Also, the fruits of different ripeness are sold at the same price if they are within the definition of marketable fruits. Equations (9) and (10) can be modified to accommodate the case where different prices are paid to different-ripeness fruits within the marketable range.

The number of harvested fruits, $E(NF1)$, that are in the desirable ripeness stages can be expressed as

$$E(NF1) = \sum_{i=1}^{i=J} F_i, \text{ for } J < 5. \tag{11}$$

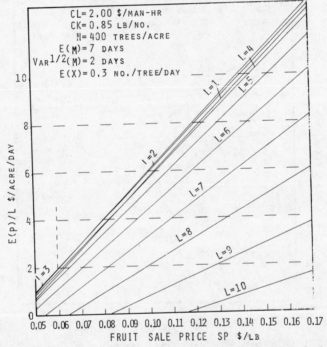

Fig. 3. Relation between the expected net return $E(P)/L$ and the fruit sale price SP.

Similarly, the expected number of discarded (bad) fruits is

$$E(NF2) = \sum_{i=J+1}^{i=5} F_i, \text{ for } i > 1, \tag{12}$$

where J = the upper stage of desirable ripeness. It depends on marketing requirements. The term $E(NF2)$ is identical to F_5 for a local-marketing requirement. Export or far-away markets may require J to assume values other than 5. By equation (8), $E(NF1)$ may be expressed as

$$E(NF1)=N\sum_{k=1}^{k=L} E[X_k P(\bigcup_{i=1}^{i=J} (M^k=i))].$$

The term $P[\bigcup_{i=1}^{i=J} (M^k=i)]$ may be simplified to

$$P[\bigcup_{i=1}^{i=J} (M^k=i)]=P(M_1+M_2+\cdots+M_J>L-k+1) \qquad (13)$$

and

$$E(NF1)=N\sum_{k=1}^{k=L} E[X_kP(M_1+M_2+\cdots+M_J>L-k=1)]. \qquad (14)$$

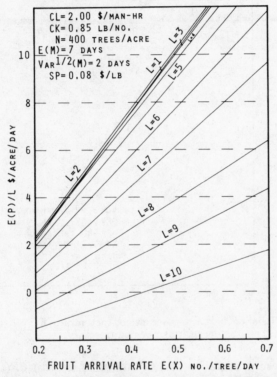

Fig. 4. Relation between the expected net return $E(P)/L$ and the fruit arrival rate $E(X)$.

Since the event of a harvested fruit beyond the jth ripeness stage is the complement of $\bigcup_{i=1}^{i=J} (M^k=i)$, equation (12) becomes

$$E(NF2)=N \sum_{k=1}^{k=L} E\{X_k[1-P(M_1+M_2\cdots+M_J>L-k+1)]\}. \qquad (15)$$

In general, the upper stage of desirable ripeness J is determined by the post-harvest handling system. A distant market requires the grower to pick his fruits at an early stage of ripeness, but a close market can allow the grower to pick the fruit at a later stage. A good post-harvest handling system can increase the shelf-life of the fruit and reduce the transportation time between the production field and the market. The system then can allow a grower to pick fruits at a riper stage and hence increase the marketable range.

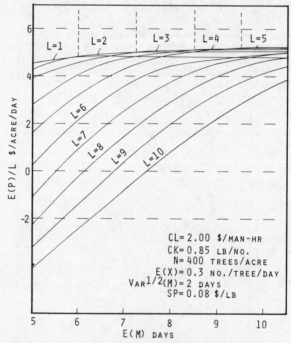

Fig. 5. Relation between the expected net return $E(P)/L$ and the mean fruit ripening rate $E(M)$.

The harvesting operation cost CO and the fixed cost $CF(L)$ in the objective function $E(P)/L$ can be evaluated for any given bioproduction. The term CO, in general, depends on the fruit density on the tree, which is a function of harvesting interval L. The fixed cost has to be shared by the annual total harvesting, which again is a function of the harvesting interval L. Therefore, the objective function is a function of one controllable variable L in addition to parameters such as CK, SP, etc., which are not affected by L. Since L can only assume very limited values, direct enumeration of $E(P)/L$ for all possible L will yield the optimal harvesting interval.

AN APPLICATION

THE OPTIMIZATION model was applied to papaya, a tropical tree fruit produced in Hawaii. It is now harvested by manual labor with some mechanical aid. Beside local consumption, fruits are airshipped to mainland markets. Fruits appear year-round on the trees, and they are harvested for a period of approximately two years. Harvesting cost constitutes a major portion of the production cost. This investigation was part of an effort to reduce the present harvesting cost.

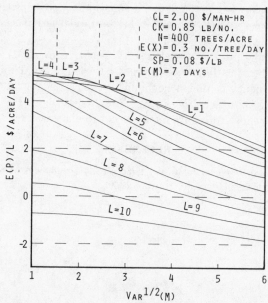

Fig. 6. Relation between the expected net return $E(P)/L$ and the standard deviation of the fruit ripening rate $\text{var}^{1/2}(M)$.

Computations for determining the optimum harvesting frequency were done on an IBM 360-65 computer. The relation between $E(P)/L$ and L is shown in Fig. 1. The net return $E(P)/L$ is at a maximum when L is 2 days for values of $E(X)$ (the mean arrival rate), $E(M)$ ($M = M_1 + M_2 + M_3 + M_4$), and other factors listed in the figure.

SENSITIVITY OF THE OPTIMUM MODEL

FOR A particular papaya-production area (defined by the various parametric values), the net return $E(P)/L$ is affected by the following parameters: (a) the number of harvested trees per unit area N, (b) the fruit weight conversion factor CK, (c) the fruit sale price SP, and (d) the labor cost CL.

A slight change in the value of any parameter may cause a significant change in the $E(P)/L$ and consequently may change the optimal harvesting policy. Of these parameters, labor cost and market price do change in reality; therefore, a sensitivity analysis with respect to the variations of these two factors is desirable. The objective function $E(P)/L$ is also a function of random variables X_k and M_j, which have a Poisson and normal distributions, respectively. Sensitivity analysis with respect to the variations of these variables yielded the results shown in Figs. 2 to 6.

The model is not sensitive (Fig. 2) to the change in labor cost. The result in Fig. 3 shows that the model is somewhat more sensitive to the sale price. The result in Fig. 4 shows that the model is not sensitive to the fruit arrival rate.

It was found that the model is very sensitive to $E(M)$, as shown in Fig. 5. However, the difference in profit between optimal and nonoptimal harvesting frequency is almost negligible.

The model is more sensitive to $\text{var}^{1/2}(M)$, as shown in Fig. 6. But loss is not great if the harvesting frequency is not too far away from the optimal.

REFERENCE

WEN-YUAN HUANG, *Optimizing Papaya Harvesting Frequency*, M.S. Thesis, University of Hawaii, Honolulu, Hawaii, 1969.

III

Air Quality

Editors' Comments on Papers 11, 12, and 13

11 **Savas:** Computers in Urban Air Pollution Control Systems
    Socio-Econ. Plan. Sci., **1**, 157–183 (Dec. 1967)

12 **Shieh, Halpern, Clemens, Wang, and Abraham:** Air Quality Diffusion Model; Application to New York City
    IBM J. Res. Develop., **16**, 162–170 (Mar. 1972)

13 **Seinfeld and Kyan:** Determination of Optimal Air Pollution Control Strategies
    Socio-Econ. Plan. Sci., **5**, 173–190 (June 1971)

In this section attention is turned to the subject of air quality. Three papers, arranged in order of increasing complexity, were selected from the growing body of air quality/air pollution literature. Paper 11 provides a general background concerning air quality and air pollution abatement measures. It is followed by a paper describing a Gaussian plume diffusion model and its application to New York City. The final paper of Part III, Paper 13, focuses on air pollution control strategies that minimize the cost of pollution control for a specified air quality. None of the papers deal with the chemistry of the air (e.g., photochemical reactions) because it was felt that such material might be too specialized for this book. However, the paper by Wayne et al., "Modeling Photochemical Smog on a Computer for Decision-Making," which is cited in the Air Quality subsection in the Bibliography, should serve as an initial guide to modeling that involves air chemistry.

Emanuel S. ("Steve") Savas, who in 1972 became Professor of Public Systems Management in Columbia University's Graduate School of Business after five years as Deputy City Administrator in New York City and eight years with IBM, is the author of Paper 11, "Computers in Urban Air Pollution Control Systems." The paper is clearly written and makes good use of control system block diagrams to depict systems for dealing with the maintenance of air quality. The approach Savas uses is one of systematic analysis and planning. His experience in urban planning leads him to describe the planning environment as one of "constant change" and technical problems "embedded within a larger legal and socio-political context."

To structure the generation of alternatives for air pollution abatement, Savas lists ten factors that influence air quality. In Table 3 he relates the ten factors to specific activities such as spaceheating, incineration, and transportation. The type of factor and activity breakdown used by Savas lends itself to the specification of air pollution control strategies. In concluding the discussion of abatement alternatives, Savas notes that political pressures must be taken into account in specifying, selecting, and tailoring the control policies to be implemented.

The evaluation of alternatives is the next topic discussed. To assist the planner in evaluating air pollution control actions, Savas suggests that use be made of computer simulation models. He describes in some detail "a simulation system that can be used for both long- and short-term planning and has the potential to be used for real-time prediction of air quality as well." Virtually all the important modeling considerations relevant to planning for air pollution abatement are summarized.

The second half of the paper (Part B) "describes the role of the computer and system analysis methods in some of the more immediate operational problems which face a municipal air pollution control department." Progressively better feedback control systems for dealing with air quality are outlined. A desirable system would provide sufficient information to allow the use of localized control measures rather than broad all-city or all-region emergency control measures. It would also relate air quality standards to the health effects of air pollution.

In Table 4 the unusual and unexplained abbreviation "Cohs" is used; it may relate to particulate matter. On page 148 the phrase "meso-meteorological monitoring network" is used. The adjective "meso-meteorological" might best be interpreted as "regional," since mesometeorology includes a larger airspace than micrometeorology and is concerned with the detection and analysis of the state of the atmosphere as it exists between meteorological stations rather than around a single point.

Paper 12, by IBM researchers, presents a model of air pollutant concentrations as functions of time and space, and compares the predictions of the model for sulphur dioxide in New York City to empirical measurements. Such an air pollution model may be used to determine the conditions under which a high concentration episode is likely to occur, to identify the major source contributors at any point, and to estimate the emission reductions necessary to avoid an episode.

The input variables are the characteristics of the pollutant sources (location, emission rate, and stack height) and the meteorological conditions (temperature, wind velocity, and direction). Output variables are particle concentrations. A Gaussian plume model is used to relate outputs to inputs. In this model it is assumed that the concentration of a pollutant is given by crosswind and vertical Gaussian (normal) distributions, where the standard deviation of each distribution is a function of the distance, in the downwind direction, from the pollutant source [see Equation (1)].

One problem common to virtually all environmental models is the appropriate degree of compromise between accuracy and model complexity. Usually, the simpler the model, the more gross are the predictions. The authors of this article confront this situation with regard to defining the location, emission rate, and stack height of the sources. A gridded segmentation of the area is used (refer to Figure 3), with a finer grid in the urban region and a coarser grid in the rural region. As it is too complex, in an urban area, to define each separate source, within each segment a uniform source—representing an average of the characteristics of the sources within that segment—is assumed. In this manner the model becomes tractable, although a compromise has been made with reality.

Since approximations are a part of every model, it is highly desirable to check the accuracy of the model by a comparison with experimental data. The authors calculate contours of constant SO_2 concentration (Figures 6 and 7) over the New York City area and compare their predictions with measurements made at monitoring stations located at different points in the city. The model did provide reasonable estimates for the actual cases that were considered.

John Seinfeld and Chwan Kyan look at the problem of minimizing the cost of achieving a specified air quality in Paper 13, "Determination of Optimal Air Pollution Control Strategies." In focusing on control measures, they assume the existence of a simulation model that dynamically handles atmospheric diffusion of pollutants (see Paper 12) and chemical reactions among substances in the air. The authors use a control systems approach and discuss the basic strategies of open-loop control and closed-loop control. They note that both these strategies are important in air pollution control efforts. Interestingly, their first reference is Paper 11 by Savas.

The focus is directed toward long-term air pollution control strategies, that is, control of air pollution over a period of several years. The authors note that perhaps the most difficult aspect of the problem is the selection of a criterion or criteria for evaluation of the relative merit of alternative control policies. The total cost of air pollution is a criterion that is proposed and discarded as unknowable for many years. Thus, instead of using a pollution cost function, the authors decided to minimize the cost of control measures to achieve a given air quality. The goal is "to provide the analyst with information on the relative costs of various degrees of clean air." Seinfeld and Kyan mention the difficulties of including the costs of social and economic consequences of each control policy and note that the political implications of control policies must be heeded. They also appropriately emphasize that "air quality" is most usefully considered as a vector with several components. Each component has a cost of control that may or may not be linked with the cost of control of other components.

A systematic methodology is presented for meeting the objective of a minimum cost of control for a given air quality. An interesting facet of the method is the decomposition of a high-dimension nonlinear problem into a high-dimension linear problem and a low-dimension nonlinear problem. Linear programming solution techniques can be applied to the linear subproblem. A simple airshed simulation model is developed as a prerequisite for the hypothetical example that illustrates the use of the analysis method.

An assumption that may prove to be a weakness in the methodology when applied to the long term is that each year is independent of other years, thus making the T-year control problem simply one of the combination of T one-year problems. Certainly industrial firms will not consider next year's control method(s) independent of this year's control method(s); in reality, the control devices used this year heavily influence the control devices used in the following year(s).

There are three minor typographical errors in Equation (6); the equation should read

$$z_m = f_m(t, \eta, b_1, \ldots, b_N) \qquad m = 1, 2, \ldots, M$$

For consistency and clarity, it also appears that the air quality constraints g_i^* should be g_{im}^*, although the need for a limit for each air component m for each type of air quality indicator i could be covered by an appropriately broad use of the index i on g_i^*. On page 176 the reference to Figure 4 should be to Figure 5, and in Figure 5 an arrow originating in cell 1 and terminating in cell 4 should be added. Finally, the value in the first row and first column of the **A** matrix in Equation (29) should be -0.4256×10^{-2} instead of -0.4522×10^{-2}.

11

Reprinted from *Socio-Econ. Plan. Sci.*, **1**, 157–183 (Dec. 1967)

COMPUTERS IN URBAN AIR POLLUTION CONTROL SYSTEMS

PART A. PLANNING FOR POLLUTION ABATEMENT

PART B. MONITORING, PREDICTION, AND CONTROL OF AIR QUALITY

EMANUEL S. SAVAS*

Deputy City Administrator, Office of the Mayor, City of New York, N.Y.

(*Received* 21 *September* 1967)

Abstract—Part A presents a brief overview of the entire waste-handling process, for solids, liquids, and gases, in order to place the air pollution control problem in perspective. Next, a systematic approach to long-term and short-term planning for air pollution abatement is described. Goals of the planning process, alternative plans, criteria for selection of alternatives and a procedure for conducting the evaluations are discussed. A comprehensive spectrum of abatement plans is presented.

The conceptual outline of a suitable simulation system to facilitate planning by local air pollution control agencies is portrayed and the elements of the system are identified. Several available diffusion models that might be employed in such a simulation system are described.

In Part B, a useful framework is developed for viewing the evolutionary stages of an integrated urban air pollution control system. Current off-line systems are characterized by manual reporting of air quality and manual inspection procedures for emission sources. On-line air monitoring systems, featuring remote, unattended sensor stations, enhance the responsiveness of the local air pollution control agency. Several factors are involved in selecting the sites for such stations.

A true pollution early-warning system requires improved meteorological prediction as well as improved air quality prediction. Given the former, the simulation system of Part A can provide the latter. There appears to be substantial spatial (as well as temporal) variability of the pollution pattern in an urban area. This suggests the need for small area prediction and offers the possibility of exercising specialized control action in local areas rather than gross, city-wide actions which may be almost catastrophic in economic terms.

A medical information subsystem, and on-line stack monitoring, may also be incorporated into the integrated system.

PART A. PLANNING FOR POLLUTION ABATEMENT

THE METABOLIC system of a city requires the daily input and output of prodigious amounts of materials. Food, water, fuel, durable and non-durable consumer goods, capital goods and materials enter a city via road, rail, ship, plane, pipeline, and aqueduct. The waste products of a city's metabolism are divided among solid, liquid, and aerial carriers and eliminated from the city. Thus, air pollution is one facet of the problem; water pollution is another, and the possible exhaustion of suitable sites for land fill is a third. Table 1 indicates the magnitude of some of these flow rates[1].

---

* This work was performed while the author was Manager of Urban Systems at the New York Scientific Center of the International Business Machines Corporation.

158   EMANUEL S. SAVAS

Trade-offs are possible. Waste particles from combustion processes can be released directly into the air, washed by scrubbers into a liquid waste stream, or trapped in dust collectors and carted off as solid waste. Food refuse can be carried away, incinerated, or ground up right in the kitchen and discharged into the liquid waste system. In general, interchanges between the different waste streams can be effected at the point of origin or at various other points in the series of collection, transportation, treatment, and disposal activities.

A municipal agency concerned with environmental protection would have to analyze this entire system and apply cost-benefit yardsticks in order to prepare suitable long-range plans for handling the total problem. Restricting our attention to the air pollution portion of the problem, it is clear that systematic analysis and planning is required in order to

TABLE 1. SOME OF THE METABOLIC INPUTS AND OUTPUTS FOR AN AMERICAN CITY OF ONE MILLION INHABITANTS [1]

Inputs	(Tons/day)	Outputs	(Tons/day)
Water	625,000	Sewage	500,000
Food	2000	Refuse	2000
Fuel	9500	Air pollutants	950

manage the air resources of a metropolitan region. It is no longer sufficient merely to point to the increased use of fuels with lower sulfur content, the advent of nuclear energy, the possibility of battery-powered vehicles, the installation of improved exhaust controls, and the upgrading of incinerators. Detailed, quantitative assessments are required of these and many other possible long-term abatement strategies by the responsible agency.

This is no simple matter when one considers that a metropolitan region, like any other vital organism, is undergoing constant change. Population growth and movements take place, land use and transportation patterns shift, and technological evolution occurs. The result of these long-term changes is that new pollution sources appear, old ones disappear, and others change their emission characteristics.

For the shorter term, it is evident that there exists a wide profusion of possibilities; abatement plans could be formulated with respect to fuels, combustion equipment, exhaust controls, type of emission source, location of emission sources, and scheduling of emissions.

Clearly, the planning of suitable policies for abatement of air pollution, in both the near term and the long term, is a complex technical problem. However, the technical problem is embedded within a larger legal and socio-political context, as is to be expected of any matter of such serious public import. This fact does not diminish the role of technical analysis, it accentuates it, for a technically impregnable position must be the basis for seeking legislative relief.

1. SYSTEM ANALYSIS APPROACH

A system analysis approach can be applied to air pollution abatement planning. This involves the following classical series of steps:

1. Defining the desired goals, in quantitative terms;
2. Proposing a comprehensive spectrum of alternative abatement policies;

130

3. Establishing the criteria by which the alternative policies (or combinations of policies) are to be evaluated;
4. Evaluating the alternative strategies in rigorous, quantitative terms from a cost/benefit viewpoint.

1a. Goals

The ultimate goals of an air pollution control program relate to the medical, aesthetic, and economic effects of air pollution. Practically speaking, however, there is insufficient knowledge at present to permit a firm, quantitative link between these ultimate factors and the level of air pollution; control standards cannot be expressed directly in such terms. For instance, one cannot say that the air must be of such purity that no more than M cases of lung emphysema per year can be attributed to it. Nor can one say that the air must be clean enough so that less than N dollars per year are required to clean buildings, streets, and clothing which are soiled by air-borne pollutants.

In time, it is reasonable to expect that goals will be stated as maximum permissible annual dosages of particular pollutants, e.g. in ppm-hr. Meanwhile, however, the goals must be expressed in terms of a secondary factor, air quality, which lends itself to relatively unambiguous and quantitative specification and measurement. In fact, air quality standards have already been established with respect to sulfur dioxide[2] and, in some cases, carbon monoxide[3]. For example, the standards for New York City[4] are shown in Table 2. Although such goals may be revised as knowledge increases, and additional goals may be established for other pollutants, such as hydrocarbons, ozone, nitrogen oxides, selenium, lead, phosphorus, and others the existing standards are the goals toward which any abatement plan must be directed, and against which alternative plans are to be tested as to their efficacy.

TABLE 2. AIR QUALITY STANDARDS FOR NEW YORK CITY[4]

Sulfur dioxide	
24-hr average:	only 1 day out of 100 can exceed 0.10 ppm
60-min average:	only 1 hr out of 100 can exceed 0.25 ppm
Carbon monoxide (tentative standards)	
8-hr sample:	85 samples out of 100 must never exceed 15 ppm;
hourly sample;	only 1 out of 100 may exceed 60 ppm

1b. Alternatives

When a systems analyst approaches the problem from an objective and fresh point of view, and ignores for the moment the ingrained attitudes about the subject, he can postulate a broad and inclusive spectrum of possible control actions, or policy alternatives, to be evaluated. Thus, control policies designed to improve air quality need not be the same for commercial buildings as for residential buildings, in summer as in winter, in the east side of the city as in the west side.

The full range of possible alternatives can be indicated by examining the numerous factors which contribute to the overall pollution picture; each factor could be affected by an appropriate control policy. The ten factors enumerated here are:

1. The activity which produces the pollution,
2. Type of emission source,
3. Ownership (public or private) of the emission source,
4. Size (pollution potential) of the emission source,
5. Kind of fuel burned,
6. Kind of combustion equipment employed,
7. Kind of treatment applied to the combustion exhaust products,
8. Specific pollutants emitted,
9. Location of the emission source,
10. Time when the pollution-producing activity occurs.

Table 3 illustrates an original attempt to structure the problem from the viewpoint of potential control action. Under each of the factors is listed a possible classification scheme for that factor.

The smallest possible unit to which a control strategy (or plan, or policy—these terms will be used interchangeably) can be applied is designated an 'emission group'. Such a group is defined by a set of ten elements, one element selected from each of the ten columns or factors of Table 3. For example, one emission group consists of the public schools smaller than five stories high which burn No. 4 oil (using a certain type of combustion and exhaust control equipment) for space heating purposes on winter mornings and thereby produce sulfur dioxide.

Obviously, not all possible emission groups are worthy of serious consideration. Nevertheless, the number which lend themselves to policy decision-making or are subject to foreseeable technological change remains very large, and it suggests that a wide choice of alternatives can be considered and evaluated. Examples of some alternatives (and mixtures of 'pure' alternatives) are the following:

—prohibiting or encouraging the use of one or more types of coal, oil and gas as fuel;
—enforcing certain standards for the efficiency of combustion in different types of combustion installations;
—enforcing certain standards for the treatment of combustion exhaust products;
—prohibiting or encouraging the use of certain fuels in certain parts of the city;
—prohibiting or encouraging the use of certain fuels for certain purposes (e.g. residential space heating, commercial space heating, etc.);
—prohibiting or encouraging the use of certain fuels in certain types of buildings (e.g. one- and two-family residential, greater than 10-storey commercial, greater than 50,000 ft² industrial, etc.);
—prohibiting certain activities (e.g. incineration) in certain types of structures in certain parts of the city;
—requiring that new buildings in certain parts of the city be heated from a central utility;
—prohibiting or encouraging the use of certain fuels for certain activities at certain times in certain zones of the city (e.g. no burning of No. 6 fuel oil during the heating season in residential buildings greater than six stories high);

TABLE 3. ANALYSIS OF POLLUTION FACTORS FOR ABATEMENT PLANNING*

(1) Activity	(2) Type of source	(3) Ownership of source (Administrative Control)	(4) Size (Emission Potential)	(5) Fuel	(6) Combustion Equipment	(7) Exhaust control equipment	(8) Pollutants	(9) Location of source	(10) Time of emission
Space heating	residential commercial -office -store -warehouse theatre/ auditorium institutional -school -hospital -religious industrial other	private public -federal -state -municipal quasi-public	number of floors 1–4 5–8 9–12 13+ floor space (in square feet) small medium large	*Type* gas oil -No. 1 -No. 2 -No. 3 -No. 4 -No. 5 -No. 6 coal -soft -hard *characteristics* chemical composition heating value	gas burner -type A -type B -etc. oil burner -type E -type F -etc. coal burner -hand fired -spreader stoker -underfeed stoker -pulverized coal	type A type B type C etc	sulfur dioxide particulates -by size distribution carbon monoxide hydrocarbons	*Block* (all sources located by block via geographic coordinates) for control purposes area sources may be aggregated by: square mile neighborhood zone borough sub-division borough etc. point sources may be treated as individual control points	seasonal -heating -non-heating weekly -workday -weekend/ holiday daily -morning (5–9am) -mid-day (9–3pm) -afternoon (3–7pm) -evening (7–11pm) -night (11–5am)
Incineration	(as above)		(as above)	food wastes paper plastics cloth other	type J type K type L etc.	type F type G type H etc.	particulates -by size distribution hydrocarbons sulfur dioxide	*Height* height above ground where emissions enter the atmosphere; for each source	
Transportation	bus truck automobile -taxi -other aircraft ship railroad		N.A.	gasoline -type A -type B -etc. diesel oil jet fuel etc.	N.A.	gasoline engine -type K -type L -type M -etc. diesel engine -type Q -type R -etc. etc.	carbon monoxide nitrogen oxides hydrocarbons metals	may be treated as area and line sources	
Power generation	public utility other		small large	(as under 'Space Heating' above)	type P type Q etc.	type U type V type W etc.	(as under 'Space Heating' above)	treated as point sources	
Industrial Processing	N.A.		small medium large	(as under 'Space Heating' above, plus process-specific fuels)	type S type T type U etc.	type Y type Z etc.	(as under 'Space Heating' above, plus process-specific pollutants)	treated as point sources	
Evaporation	architectural coatings industrial solvents		small large	N.A.	N.A.	N.A.	hydrocarbons	N.A.	
	gas storage					seals -type 1 -type 2 etc.		may be treated as point sources	N.A.

*Numbers in parentheses at column headings refer to similarly numbered factors in the text.
N.A.=not applicable.
Note: Wind-borne particles produced by the erosion of structures, soil, etc., are not considered in this analysis.

—establishing zoning regulations for the emission of pollutants;

—requiring the set-back of new buildings on certain streets in order to prevent the formation of 'canyons' with high carbon monoxide levels;

—encouraging the replacement of existing power plants with nuclear plants, either inside or outside of the city;

—encouraging electrical heating for certain types of new buildings;

—encouraging the use of electrically powered vehicles for certain services (e.g. bus, taxi);

—discouraging the use of certain packaging materials if they contribute appreciably to pollution upon incineration;

—securing the region-wide application of certain controls, instead of applying more stringent controls only in the city.

One notes that special interest groups throughout the country have begun to voice their concern as the first pollution control measures have been promulgated[5, 6]. Coal producers, petroleum refiners, automobile manufacturers, utilities, truckers, landlords, tenants, and housing administrators and sanitation officials in government, among others, all have a legitimate and proper interest in the solutions that evolve. Even foreign affairs are involved, for Venezuela's vital export is its crude oil, which happens to have a high sulfur content and whose market value is thereby threatened.

It is not difficult to foresee that many of these heavy pressures (and many more) will be transmitted to the local municipal scene, and that any planning effort aimed at air pollution abatement in a city will be subjected to them. (For example, owners in one part of the city will object strongly to having certain control measures applied to their buildings, where it may make little difference, merely because the measures are indispensable in another part of the city.) This is in no way a derogatory, cynical, or pessimistic appraisal, for this pluralistic situation will be wholly in keeping with the principles and practices of the democratic process. It does mean, however, that there will be a tendency toward accommodation and optimality, that is, that there will be exceptions, variations, and loopholes suggested to modify the initial alternatives that are offered, and that relatively complex and careful tailoring of control policies will have to be made to satisfy conflicting interests. This accentuates the need to consider the full spectrum of control actions which are indicated in Table 3, and to provide the planner with a suitable tool for evaluating the many relatively complicated control measures which will no doubt evolve from the political process.

1c. Criteria

From the foregoing discussion it would appear that there are very many possible control policies which theoretically could be applied by a municipal air pollution control agency to achieve the desired level of air quality. How does one make the selection in that case? The answer: by examining (1) the political acceptability, (2) the social desirability, and (3) the economic feasibility of each proposed alternative.

Rejection on the grounds of political unacceptability would occur, for example, for the hypothetical policy which would forbid space heating when the indoor temperature exceeds 45°F. It is not necessary to launch into an analysis of the economic consequences of the resulting illnesses to discard that particular alternative.

Social consequences may result from certain air pollution control policies, for example, restrictive zoning for high and low pollution activities. Residential areas near high pollution zones may be occupied by the more disadvantaged socio-economic groups and shunned by higher income groups. Attention must be given to this possible effect in studying the various plans.

Alternatives may be proposed or screened qualitatively with an eye towards their political and social acceptability, but the principal quantitative criterion which will determine the choice will be an economic one. Land use, industry costs, consumer costs, and public costs will be affected by air pollution control programs. Not to be neglected are the costs of inspection and enforcement activities, which will differ, in general, for different plans.

Capital and operating costs of space heating, incineration, power generation, industrial processing, and transportation would be affected as auxiliary equipment, new process units, and more expensive fuels are called for by abatement plans. Alternative policies must therefore be investigated as to their economic impact on these activities. If competitive manufacturing activities can be conducted at lower cost in neighboring areas, relocation (at a certain cost) can be expected to occur; therefore, it is necessary to consider this factor and to assure continued regional agreement on air quality standards. If a factory or industry cannot afford the price of required abatement action, among the possible trade-offs to be examined are the costs of subsidizing the activity, e.g. through tax credits, versus the costs of unemployment.

In many situations direct industry relocation will not occur, for increased costs could be passed on to the consumer in the form of larger utility bills or higher rents, for instance. (These may influence population relocation, however.) Again, the total economic effect on the public at large must be considered, as well as the immediate cost to the polluter of operating under more stringent control criteria. The concepts of planning-programming-budgeting systems (total capital and operating costs, long time horizons, etc.) and cost-benefit analysis[7] must be brought into play to analyze all these consequences to each affected interest group.

1d. Evaluation

At this point in the discussion it seems evident that the agencies responsible for air pollution abatement planning require a convenient tool with which to examine the effects of alternative control actions and potential technological developments on air quality. Simulation appears to be such a tool. Given an appropriate simulation model (described in the next section), the analyst/planner can do the following:

1. propose (or select) an abatement plan which is socially and politically acceptable;
2. 'dial in' this plan by 'setting the knobs' (i.e. the factors in Table 3, which define the plan) on the simulation model;
3. run the simulation model on the computer to determine the effect of the plan on air quality;
4. conduct an economic analysis, in accordance with the preceding section, if the improvement in air quality is substantial enough to warrant further consideration of the proposed plan;
5. repeat steps (1)–(4) for a variety of alternatives, and under a variety of assumptions about the future development of the region at different points in time;
6. apply cost-benefit analysis to determine the best composite plan (it is assumed here that a number of different abatement measures will be required to reach the desired air quality objectives) after a sufficient number of conditions have been examined.

When one views the large number of possible control policies which could be evaluated, along with the various technological improvements which can be anticipated, one is faced with the potentially awesome task of seeking the optimal abatement plan out of this myriad of possibilities. In this situation one is tempted to consider the application of an optimization technique, such as linear (or nonlinear) programming or some other mathematical programming tool which is designed to handle precisely this sort of situation. That is, one might link the simulation model to an optimizing algorithm in order to select the best abatement plan, in accordance with the established economic criteria.

Some work along these lines has been conducted[8] and could be applied at least to limited classes of alternatives; however, the widely varying form that the economic factors can be expected to assume for widely different plans tends to mitigate against this as a general approach. A more pragmatic approach in such cases is to let the analyst exercise his judgment to select a relatively limited number of alternative plans, to evaluate them as indicated above, to synthesize new plans on the basis of the results obtained in simulation runs, and ultimately to apply enlightened judgment based on cost benefit considerations.

2. A SIMULATION SYSTEM

From the foregoing discussion it appears that one of the most valuable tools for air pollution abatement planning is a simulation system that can be used for both long- and short-term planning and has the potential to be used for real-time prediction of air quality as well. The block diagram of Fig. 1 illustrates the conceptual approach to such a simulation system. The components of the system are identified in the figure and are discussed below in some detail.

2a. *Emission inventory*

A machine-processable emission inventory for a metropolitan region would include the location of each significant stationary pollution source in the region, by block and geographic coordinates, building class, height, and floor area, and type of combustion equipment[9]. In general, the information listed in Table 3, above, should appear in the emission inventory. However, it may not be necessary (or practical) to retain the same high level of detail for all types of sources. It is recognized that while such an inventory may be readily available for the major city in the region, comparable information from the suburbs and satellite cities may be difficult to obtain.

For the inventory of mobile sources, as distinguished from the stationary sources discussed above, information can sometimes be obtained from regional planning agencies concerning vehicular traffic volumes and speeds in the different area segments of the region at different times, by type of vehicle (bus, truck, taxi, private automobile, etc.)

With any data bank, such as this emission inventory, there is the question of updating; that is, how and when to enter into the inventory the changes in emission sources which actually take place, or whether in fact this is necessary at all. Given the rather slow rate of change expected in such matters, and the manner in which the model would be applied, it may be adequate to utilize a generally static inventory for several years.

In time, and given the current trend, a unified, comprehensive municipal data bank for real property may be in existence, and procedures could be developed to update the emission inventory routinely from such a data bank. Prior to that time, changes in major point sources in the metropolitan region, such as shutdowns or fuel changes affecting certain power plants, new municipal incinerators, etc., should be inserted into the inventory

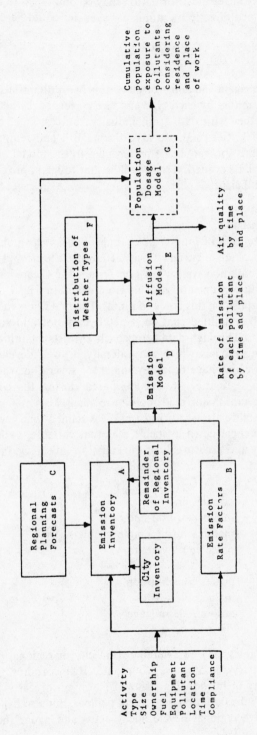

FIG. 1. A simulation system for urban air pollution abatement planning. Letters in blocks refer to similarly designated sections in the text.

whenever they occur, and however the information may be obtained. In addition, licensing and inspection data developed directly by municipal agencies could be used to update the inventory.

2b. Emission rate factors

It is necessary to supplement the emission inventory with quantitative data about the amount of each pollutant emitted by each type and size of source, by activity. Input from survey questionnaires together with data from industry sources, can be used to develop these elements of the model. Seasonal, weekly, diurnal[10], and temperature-dependent variations in the rates of emission should be represented in the model. It is by no means certain that detailed data will be needed; it may well be that rough approximations based on informed guesses will be sufficient.

2c. Regional planning forecasts

In order to study the effects of long-term changes in the metropolitan environment, such as population and housing shifts, new transportation patterns and changes in living habits and standards, the simulation system must be designed to accept long-term forecasts, and thereby to alter the emission inventory. This is indicated in Fig. 1.

Two methods of accomplishing this come to mind. The first offers more detail than the second, although it must be admitted that such detail may not add significantly to the overall planning capability. The high-detail approach consists of utilizing the (off-line) output from other planning models which may already be well developed, namely, (1) residential location models which have been developed for several metropolitan areas[11]; (2) automobile trip origin/destination models prepared during the course of in-depth transportation studies of several major metropolitan regions[12].

The second method is more straightforward but less refined than the first. It requires that the emission model be constructed in such a way that variables such as the following can be manipulated directly and used to adjust the emission inventory, in a simple proportional manner, for instance:

$$\left.\begin{array}{l} \text{population} \\ \text{number of households} \\ \text{manufacturing employment} \\ \text{office employment} \end{array}\right\} \text{by county}$$

$$\left.\begin{array}{l} \text{power consumption} \\ \text{incinerable waste} \\ \text{automobile mileage} \end{array}\right\} \textit{per capita}$$

Forecasts of such variables may be available from city or regional planning agencies [13], utility companies, industry and business groups, etc. By incorporating into the simulation system this flexible way of introducing regional planning forecasts, this same system could be utilized not only by the pollution abatement planner, but also by the urban planner who can examine alternative regional development plans and gauge their effects on air quality throughout the region[14].

2d. Emission model

By combining in an appropriate and convenient form the emission inventory, the emission rate factors, and the regional planning forecasts described above, one can build an emission model which predicts the rate of emission of each pollutant by time and place. It would be capable of predicting this for each 'emission group' (defined earlier), and of aggregating the data by any factor or combination of factors given in Table 3.

The model would be constructed so that the different control variables, or 'knobs' (the factors listed in Table 3 and indicated as inputs in Fig. 1) can readily be manipulated to facilitate evaluation of alternative strategies. As indicated in the figure, it is desirable to include the degree of compliance with a policy as a manipulated input, in order to take into consideration the effect of violations. Technological innovations, such as the introduction of electrically powered vehicles, or the replacement of a conventional power plant by a nuclear one, would be accommodated in the model simply by setting the appropriate 'knobs' to zero. (Radiation hazards, if any, could be considered outside the framework of the model.)

To a certain extent, this emission model could be utilized for planning purposes without further elaboration. For example, one could investigate a proposed control strategy and see what effect it would have on the rate of emission of a pollutant, e.g. tons of sulfur dioxide per average winter day. In this restricted form, however, the simulation system suffers from the following four shortcomings from the operational point of view:

1. It does not predict air quality (e.g. sulfur dioxide concentration in the air in parts per million by weight), although the goal for air pollution abatement planning is a given air quality standard. (See Table 1.)
2. It does not permit consideration of the transport of polluted air from a source area to another area.
3. It cannot be used to evaluate fully the effect of introducing, altering, or eliminating major point sources.
4. It cannot be used reliably for urban planning purposes, in terms of evaluating the effects of zoning for low pollution, etc.

Each of these points will be discussed in turn.

With respect to item (1), the failure to predict air quality, one might simply assume that the effects will be linear; that is, a long-term reduction of 50 per cent in the current rate of emission of a particular pollutant in an area will result in a 50 per cent drop in the average concentration of that pollutant in that area. This simplifying assumption (which corresponds to a simple diffusion model), and therefore this simple simulation system, may be adequate for coarse screening and may even be satisfactory for planning of gross alternatives, with respect to some area sources, but it fails as an overall approach due to shortcomings (2) and (3).

With respect to item (2), it is well known that polluted air can be transported across considerable distances and that an area with little local emission nevertheless may be afflicted with poor air quality[15]. If such transport (and diffusion) effects are neglected: and one considers only 'coarse' plans (i.e. those dealing with gross average emissions over large areas), one of two difficulties is likely to be encountered once such plans are applied, (i) some areas of the city will not have attained the target air quality objectives, due to the obscuring of certain weather conditions in the averages; or (ii) that 'overkill' was applied, along with an unduly heavy economic burden. To illustrate how 'overkill' could result from the use of a coarse averaging plan based solely on an emission model, consider the following hypothetical example: a city adopts broad measures which are intended to reduce the average pollution level, but do so, in effect, by penalizing the homeowner in a relatively 'clean' part of the city in a misguided attempt to counterbalance (arithmetically) the heavy emissions from more industrialized areas. The effect of the resultant super-clean air in the former part of the city is more likely to be felt in the courts and in the ballot boxes than in the city-wide average!

With respect to item (3), proposed changes in point sources cannot adequately be evaluated by an emission model because a buoyant plume which is released from a high stack may or may not contribute significantly to the concentration of pollutants near ground level in that locale[16].

With respect to item (4), an emission model is inadequate for evaluating the downwind effects of open spaces and low pollution zones, i.e. the transport and diffusion of clean air, just as it fails to consider the transport and diffusion of polluted air from its sources.

2e. Diffusion model

The four shortcomings of the emission model can be overcome, in principle, by linking to it a suitable diffusion model. The result would be a simulation system theoretically capable of predicting air quality (not merely rates of emission) at any point in the city under specified meteorological conditions.

Statistical models. Diffusion models may be statistical or theoretical. A statistical model is based on analysis of past data and consists of the observed statistical relationships among the variables. Needless to say, theoretical considerations enter into the selection of variables to be included, but do not lead to a precise specification of the functional form of the relationships. The greatest limitation of statistical methods is that they cannot be extrapolated with any confidence; one never knows how far beyond the conditions under which they were obtained the relationships will continue to hold. This is why direct statistical prediction of pollution levels is probably of little value for long-range planning.

Theoretical models. A theoretical diffusion model may be constructed by deriving the approximate differential equations governing the convection and diffusion of pollutants, and solving them subject to the appropriate boundary and initial conditions. All so-called 'diffusion models' use this approach, whether implicitly or explicitly. The models differ only in the nature of their approximations and simplifying assumption, and in the empirical constants that are introduced in order to cover up their theoretical deficiencies. The accuracy of predictions obtained by a theoretical model depends on:

(i) The accuracy of the theory used. (ii) The accuracy of the numerical implementation. (iii) The accuracy of the input data.

The last point is a vital one. All diffusion models developed to date require as inputs the predicted values of meteorological variables. At the present time, these predictions are unreliable for short-range prediction of air quality. However, the situation is different when it comes to long-range prediction. We may not know what the weather will actually be tomorrow, but we know what kinds of weather conditions we will have 10 yr from now. Hence, a sophisticated diffusion model would be directly useful for simulation studies, where the weather is to be postulated rather than predicted. Furthermore, only a truly theoretical model (which does not depend on many empirical constants fitted to present data) can be expected to yield accurate predictions on the effects of changes in the environment, i.e., the emission inventory.

Micro-scale models. At this point it is important to distinguish between diffusion models which are suitable for calculating the transport and diffusion of pollutants on a macro scale, and diffusion models which are suitable for micro-scale studies. On the macro scale, materials such as sulfur dioxide and particulate matter which enter the atmosphere at roof-top and chimney-top levels are subjected to considerable mixing in the turbulent air at that level. The result may be (subject to verification) the absence of pronounced vertical or lateral concentration gradients for sulfur dioxide or suspended particulate matter.

The situation is believed to be quite different for vehicular emissions, such as carbon monoxide, hydrocarbons, nitrogen oxides and certain metals. In canyon-like streets with high traffic volumes the diffusion effects are quite different from those acting upon higher level emissions, and are likely to be highly dependent on the geometry and micrometeorology of the particular 'canyon'. Furthermore, the main focus of interest with respect to vehicular emissions is also on a micro scale: the effect on policemen, workers, shop clerks, shoppers, pedestrians, and drivers who are exposed for long periods at ground level.

A micro-scale diffusion model for CO would permit better planning with respect to this facet of the problem. Without it, it is difficult to answer questions about the inter-relation between carbon monoxide concentrations and traffic volume, traffic speed, traffic pattern, street width, building set-back, building height, building frontage, block length, width of intersection, etc. The task of developing a model which incorporates all these complexities is probably impossible, but even a simpler model might be used to answer such questions as to whether a school or apartment house should be built over a highway; how far buildings should be set back on streets of a given width; what limitations should be imposed on traffic on a particular 'canyon' or expressway.

Macro-scale models. Several different models for sulfur dioxide have been reported. These are applicable for suspended particulates as well, and also describe the movement of vehicular emissions over broad areas once the gases rise above ground level . . . once they get into the mainstream, so to speak. The various major reported models are compared below, and the major assumptions inherent in each are classified.

(1) *Davidson's model*[17]. The city is divided into a grid of areas. An emission inventory is compiled for each area, and is augmented by estimates of diurnal and seasonal variations. Thus, an emission forecast for any hour in any area can be made. The aerial trajectory for the center point of each area is computed, using data on wind direction and velocity as functions of time, location, and elevation. The motion of the pollution emanating from each area is assumed to consist of translation along the trajectory plus diffusion according to an ever-spreading Gaussian distribution. Ground level and inversion height form lower and upper constraints on the spread by diffusion. Point sources are treated separately in an analogous manner. To obtain the pollution level at any point in space and time, one sums up the contributions from the emissions of all sources at all previous times.

(2) *Turner's method*[18]. Essentially similar to Davidson's, except that: (a) The emission from each area is treated as though emanating from a line source drawn cross-wind through the center of the area. (b) The wind is assumed constant throughout the city throughout any given time period. (c) Apparently, the path of the pollutants is not

141

traced beyond a single time period (2 hr). (d) The results are reduced to 24 hr averages. (e) Diffusion in the wind direction is neglected.

(3) *Clark's method*[19]. The area surrounding the sensing station is divided into sixteen circular segments, and each segment into four areas. The contribution of a unit source in the center of each area to the pollution at the station is computed for a representative set of wind velocities and conditions, using the point source diffusion equation, with only vertical diffusion taken into account. The pollution at the station is then estimated by summing the contributions from all the segments using a single wind observation for each time period.

(4) *Miller and Holzworth's method*[20]. Similiar to Turner's model, but sources are assumed to be uniformly distributed throughout the city, and are treated as a set of infinitely long parallel line sources across the wind.

2f. Weather types

Thorough evaluation of a contemplated control strategy requires studying its effect under various weather conditions, so that one can make probability statements about the expected air quality; the air quality standards themselves are expressed in such stochastic terms (see Table 2).

Meteorological conditions enter the simulation as input to the diffusion model, and are expressed in terms of the following variables: wind direction and wind velocity at different points in the city and at various altitudes, temperature, and vertical temperature gradient (or alternatively, the mixing depth). The concept of the 'stability' of the urban atmosphere has been introduced; this parameter has been found to be a useful descriptor for some purposes[18].

Clearly, it is necessary to compile a great many observations of these variables, and to group these data into appropriate classes of weather types. An analysis of the frequency of occurrence and duration of each type of weather would then facilitate the simulation studies, and permit meaningful conclusions to be drawn. Computer-based classification techniques and cluster analysis[21] may prove useful in the weather classification effort.

2g. Population dosage model

The final component of this simulation system is a population dosage model. This model would permit evaluation of alternative control policies from the epidemiological viewpoint, by estimating the exposure of the population to different pollutants. When air quality standards are related more clearly to physiological factors, the goals of an air pollution control program could be expressed in terms of population dosage standards, and this model would then assume great importance.

The dosage model would consist of three elements: (1) the zonal distribution of the population by place of residence, (2) the zonal distribution of employment, (3) interzonal journeys to work, i.e. the number of people living in each zone who work regularly in each of the other zones. By ignoring other activities, and introducing other simplifying assumptions, the model can compute the expected exposure of different population groups. Current information for these data items is available for those metropolitan regions which have been the subjects of recent transporation studies. Equivalent information for 1970 will be available via the census. Regional planning forecasts might be utilized to project future conditions.

Perhaps it is too naive to believe that some day, as people become more aware and knowledgeable about air pollution, they will consider their total pollution exposure (at home, at work, and on the road) when selecting their residential location, just as many today consider their tax exposure when locating in metropolitan area. It remains to be seen whether this factor will become significant enough to warrant the attention of the location theorists among the urban planners, that is, whether air quality, like housing cost and accessibility, will have a measurable influence on regional location and land-use patterns[14].

PART B. MONITORING, PREDICTION, AND CONTROL OF AIR QUALITY

The preceding section discussed a procedure for planning an air pollution abatement program, and described a computer-based simulation system which can be a valuable tool for the planner. This section describes the role of the computer and system analysis methods in some of the more immediate operational problems which face a municipal air pollution control department.

A conceptual frame of reference for viewing the control functions of an air pollution control department is the conventional feedback control diagram. A series of such diagrams is presented and discussed, to indicate a possible sequence of developmental stages through which an integrated system may evolve.

1. CURRENT SYSTEMS

The present mode of operation in many cities is indicated in Fig. 2. First, let us examine the inner loop which deals with the code standards for combustion equipment. Scheduled and unscheduled inspections and observations of emission sources by field personnel provide feedback and indicate code violations. Enforcement procedures are then applied to secure compliance with the code. At the present time, the entire feedback process is strictly a manual one; this is indicated in Fig. 2 by the dashed line. Conventional computer data-processing is often utilized for the associated record-keeping functions.

With respect to the outer control loop, the air quality control limits are not to be confused with the air quality objectives discussed in Part A. In those cities where they have been established, the control limits serve as the reference value to trigger control action when they have been violated, that is, when the measured air quality has deteriorated to a certain point. A representative set of control limits is displayed in Table 4[22]. In

TABLE 4. CONTROL LIMITS FOR AIR POLLUTION ALERTS[22]

	ppm SO_2	ppm CO	Cohs	Duration (hr)
	0.5	10	5.0	4
	0.7	10	7.5	2
	1.5	10	9.0	1

First alert
(1) If any two components exceed the indicated 4-hr levels.
(2) If SO_2 or Cohs exceed the indicated 2-hr or- 1hr levels.

Second Alert
If all three components exceed the limits indicated below and the thermal inversion is expected to continue for eight hours:

	ppm SO_2	pmm CO	Cohs	Duration (hr)
	1.0	20.0	7.5	2

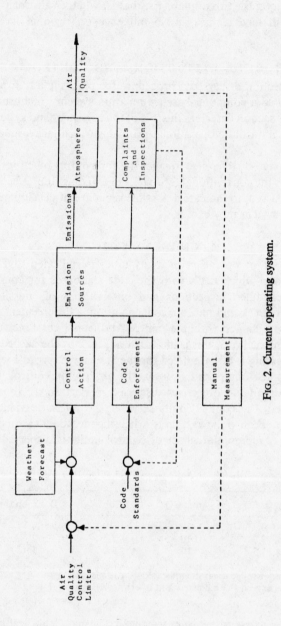

FIG. 2. Current operating system.

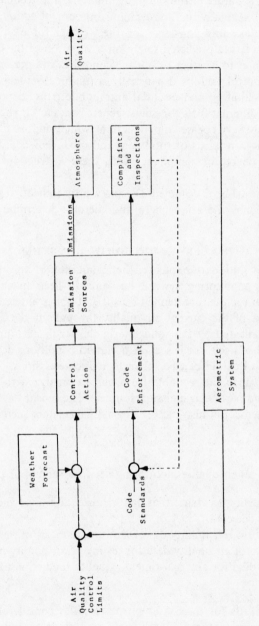

Fig. 3. Mode of operation after aerometric system is functioning.

many existing systems, air samples are taken and analyzed manually, by and large, and the results of the analyses are transmitted by mail, messenger, or phone to the decision-makers, who compare the observed air quality to the control limits and decide on what action, if any, to take. (The non-automatic nature of the sampling, analysis, and feedback is indicated by the dashed line in Fig. 2.) Actually, before deciding on control action, i.e. declaring an alert, the decision-makers take into consideration the weather forecast, such as it is. In other words, control action is not predicted solely on a violation of the control limits, but rather on a prognostication as to whether or not the condition will persist.

Certain emergency control actions are usually available to the responsible public officials. These include prohibiting non-essential automobile trips, banning incineration, requesting a switchover to different fuels for power generation, etc. It should be noted that at the present time these are very coarse control actions which impose uniform and heavy burdens throughout the city, in terms of total cost and inconvenience; furthermore, it is difficult to determine the degree of compliance with these instructions, or their overall effectiveness.

The remaining blocks in Fig. 2 complete the picture by indicating that the emission sources release pollutants into the atmosphere and thereby determine the observed air quality.

2. AIR QUALITY MONITORING SYSTEM

The next step in the evolutionary sequence, the installation of an on-line 'aerometric' system, i.e. an air quality monitoring system, has already been taken in several cities [23, 24]. Information from remote, unattended sensor stations is entered into a computer to permit rapid assessment of the current air quality throughout the area. The on-line nature of this system is indicated by the solid line in the feedback loop of Fig. 3. This knowledge is useful in and of itself, for if serious pollution conditions develop they can be detected rapidly and control action can be taken to reduce emissions. Furthermore, the data which is acquired in machine-readable form can be analyzed to find the relation between pollution levels and particular emission sources. The sensor data may include not only air quality measurements, but also meteorological parameters such as wind direction and velocity[24].

2a. Site selection

The sensor stations in an air quality monitoring network serve four primary purposes:

(a) To monitor air quality, to assure that dangerous conditions (whether predicted or not) do not go undetected.
(b) To supply a record of air quality data for future analysis of medical data.
(c) To supply a record of air quality data for testing of air quality prediction models.
(d) To provide input data for any real-time air quality prediction technique that may be in use.

The location requirements for these four purposes differ considerably. For (a) it is essential to locate the sensors in those spots which are known to suffer from relatively high pollution levels. Such spots could be pinpointed by means of mobile sensors and/or predictions based on a detailed diffusion model. For (b) the sensors are best located in high-density residential areas, to indicate the air quality to which large population masses are exposed. For (c) it would seem that locations having pollution levels close to the average of

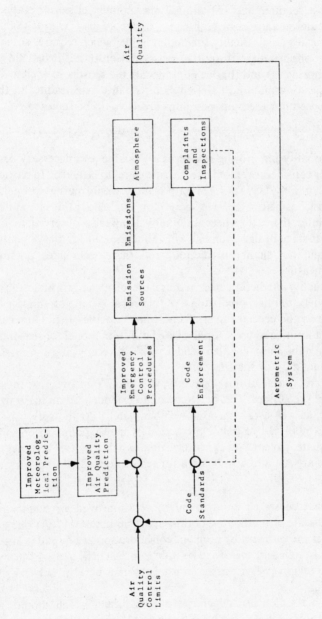

FIG. 4. An effective early-warning system for air pollution control.

the area in question are desirable. Purpose (d) can be discussed only if one knew how the data were going to be used in the predictions, for the problem is related to that of experimental design: what sensor location gives the maximum information for the purpose of the particular predictor in use.

To illustrate how requirements (b) and (d) may conflict, a sensor station located six floors above the ground in an area of high-rise apartments may be near the mean altitude of the population, but for an air quality prediction model which requires some wind data it might be a convenient opportunity to install the sensor station at the thirtieth floor in that area.

Similarly, by ignoring (a) and (b) one could locate the sensors in such a way as to give a favorable but misleading picture of the city's air quality—misleading in the sense that neither the most exposed nor the most populous areas would be represented.

3. AN EFFECTIVE EARLY WARNING SYSTEM

An on-line air quality monitoring system is a valuable and necessary first step, for it offers the prospect for considerably more effective measures to be taken. In particular, the prediction of future air quality (say, 6–48 hr in advance) would permit proper anticipatory action to reduce emissions before an 'alert' condition is reached; this is manifestly superior to merely reacting after the fact, when air quality is already observed to be very poor. Fig. 4 shows this next stage of development, which incorporates improvements in meteorological prediction and air quality prediction, and therefore makes possible improved emergency control measures.

Improved air-quality prediction may require use of an appropriate diffusion model. However, such models require predictions of meteorological variables, and these are unreliable at present in the crucial range of low wind velocities[19]. Under these circumstances there is little incentive to use a sophisticated diffusion model for real-time prediction. It is doubtful whether the most detailed model would give any more accurate short-term predictions than would a crude model.

If such prediction is deemed unsatisfactory, the initial improvement must come from better meteorological prediction. This may require an extended urban meso-meteorological monitoring network, as well as a wide-area weather watch. At the very least, better local weather prediction would lead to better prediction of major pollution episodes. At best, however, with adequate meteorological forecasting capability, the very same simulation system described in Part A could also be used for real-time prediction of air quality.

3a. Scale of prediction

One question that comes to mind in calling for improved air quality prediction is: What is the required scale of prediction, both spatial and temporal? At one extreme, must one be able to predict the moment by moment concentration at any point in the city, or, at the other extreme, it is sufficient merely to be able to predict the gross 24-hr average for an entire metropolitan region? The answer must lie somewhere in between, but it is not clear precisely where.

The requisite time scale must depend on medical data, which should determine the minimum time of exposure that is likely to be harmful. Fig. 5 illustrates two hypothetical situations where the average annual pollution concentrations are equal. If both situations are hazardous to health, then it becomes important to predict the short-term peak heights of Fig. 5b, clearly a more difficult task than to predict the extended pollution episodes of Fig. 5a.

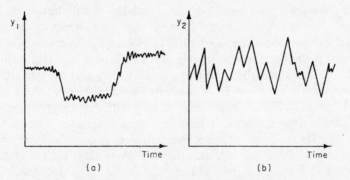

FIG. 5. Two different pollution patterns which exhibit the same average annual pollution concentrations.

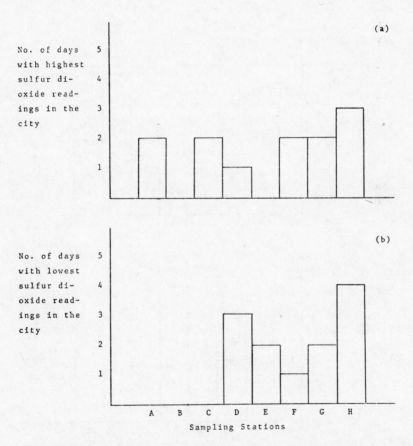

FIG. 6. Distribution of (a) highest and (b) lowest sulfur dioxide hourly readings among stations for 12 days.

149

The requisite space scale must depend on the variability of pollution levels from place to place in the city. Now it is well known that sulfur dioxide concentrations can vary many-fold throughout a city at the same moment[15]. However, it is commonly assumed that there are certain 'pollution pockets' in the city which have a rather consistently high pollution level; i.e. that although the absolute levels throughout the city will rise and fall with the weather, certain areas will be consistently worse relative to others.

In order to investigate this hypothesis, hourly data for an entire year from eight sulfur-dioxide monitoring stations in a large city were examined. Twelve high-pollution days were selected. The readings at each of the stations were compared and are summarized in Fig. 6. The height of the bar indicates the number of days (of the twelve) when the indicated station had the highest (Fig. 6a) and the lowest (Fig. 6b) readings in the city Surprisingly enough, the same location, station H, had the *lowest* reading in the city on four of the twelve days, and the *highest* on three of the twelve. The rather even distribution of the high (or low) readings among the eight stations certainly suggests that there is considerable variability in the pollution pattern throughout the city, i.e., that the spatial pattern itself varies with time. Similar results appear in Fig. 7 for a selected sample of 31 days in 1966. Note that station G had the highest readings on five of the days, and the lowest on five other days.

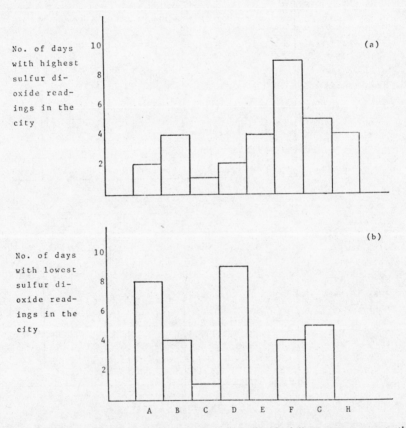

FIG. 7. Distribution of (a) highest and (b) lowest sulfur dioxide daily average concentrations among stations for 31 selected days.

At the very least this brief investigation fails to support the hypothesis that if one were anticipating a rather bad day, one could readily predict where in the city the high pollution levels would be found. It seems that small scale air quality prediction is necessary. This tentative finding, while dismaying on the surface, raises some interesting possibilities. On moderately bad days, it would appear that some parts of the city may be subjected to hazardous pollution levels (if no control action were taken) while other areas may be comparatively safe. At present, emergency control measures are applied rather generally and indiscriminately throughout the city. If one were able, however roughly, to predict localized air quality, this would afford the opportunity to apply control measures in a similarly localized manner, and thereby obviate the need for imposing economically drastic city-wide emergency measures. Furthermore, this capability would endow the control agency with the ability to act in marginal cases, when the situation is not serious enough to demand broad emergency measures, but still serious enough so that limited action could properly be taken to alleviate a potential local hazard.

3b. Emergency control actions

Given adequate meteorological and air-quality prediction methods which suffice to indicate the expected levels of pollutant concentrations in different areas of the city for a day or two in advance, the local control agency can issue directives for localized emergency control measures during the earliest stages of anticipated major and minor pollution episodes. When that capability is forthcoming, the agency can plan such localized control measures through a study which involves the following:

1. dividing the city into suitable areas within which the control measures would be applied;
2. listing in rank order of pollution potential the principal emission sources in each of the areas for each type of pollutant; this is done from the emission inventory.
3. estimating the response time for each major emission group (see Table 3); the response time is the total time required to communicate a directive from the air pollution control center to the controlling party of the emission source, and for the latter to take the requested action and actually reduce the emission rate to the desired value;
4. formulating realistic emergency control measures, estimating the degree of compliance with each measure, and using the simulation system and economic evaluation procedure of Part A to determine the most suitable and effective emergency measures; these could involve scheduling of power plants, selection of a minimum number of key emission sources (in each area) to be temporarily shut down, prohibition of traffic on certain arteries, etc.

Figure 8 illustrates, for hypothetical data, the fact that in some areas a very small fraction of the total emission sources are responsible for a major fraction of the emissions from that area. By preparing such a characteristic chart for each area of the city, and listing the specific sources, the local control agency is in a position to improve its emergency planning for each stage of an air pollution alert. It must be remembered, of course, that the chart for an area will differ according to seasonal and other time-dependent factors, as spelled out in Table 3.

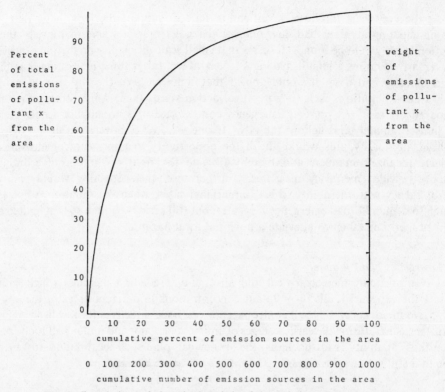

Percent
of total
emissions
of pollu-
tant x
from the
area

weight
of
emissions
of pollu-
tant x
from the
area

cumulative percent of emission sources in the area

cumulative number of emission sources in the area

FIG. 8. Cumulative emission characteristic of the emission sources in an area.

4. INTRODUCING HEALTH CONSIDERATIONS

Present air quality control limits are somewhat arbitrary, and are not rigorously founded on the medical effects of air pollution. As pointed out above, it is not yet known what threshold effects there are, if any, to duration of exposure, dosage levels, etc. The next step in the progression, therefore, is to embed the air quality control system within a larger and more meaningful system which explicitly considers the relationship between air quality and community health. The block diagram of Fig. 9 shows the system of Fig. 4 nested within a medical feedback loop. It illustrates the fact that the air, and its pollutants, acts upon both people and property, and that it has an impact on the physiological, economic, and aesthetic condition of the community. In such an ultimate system, health standards are established based on analysis of medical data, and these in turn are used to set the air quallity control limits. The latter may be dynamic, that is, it may be necessary to change them from time to time as feedback of medical information so dictates. As the dashed line of Fig. 9 indicates, this feedback need not be automatic, although conceivably it could be if it is a convenient byproduct of a city-wide, computer-based health information system that is developed and implemented for a broader purpose.

5. STACK MONITORING

A subsequent step that one can realistically envision for a municipal department of air pollution control in this progression of increasing effectiveness and increased mastery of the problem, is the use of automatic sampling, analyzing, and recording equipment to monitor

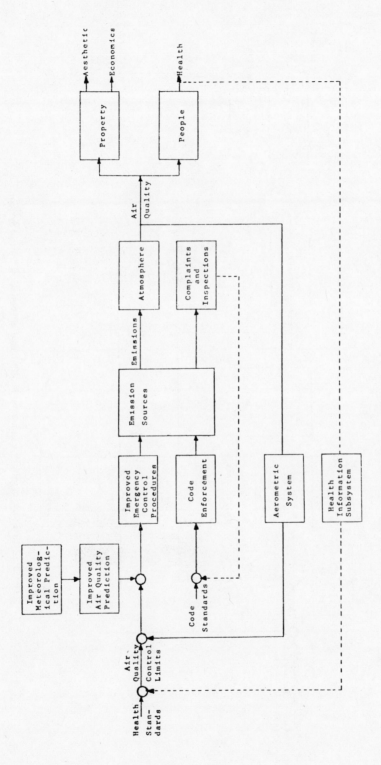

FIG. 9. Taking health aspects into consideration in an air pollution control system.

153

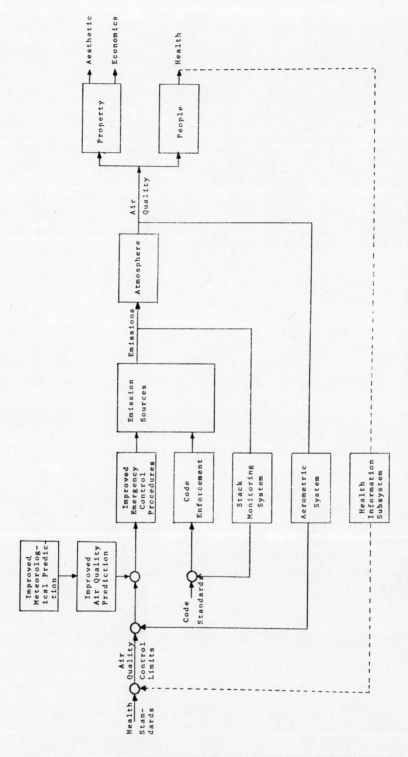

FIG. 10. Stack monitoring in a total health-based air pollution control system.

154

the smokestacks of major pollution producers. (See Fig. 10) This would tend to minimize inefficient 'smoke-chasing' activities by field personnel and would increase effective control over these emission sources. Such monitoring might be accomplished by installing on-site devices on the stacks themselves, and either telemetering the data to a central point for recording and analysis, or recording it locally in sealed recording units which can be examined periodically by authorized inspectors. Another approach is to use remote, telescopic, infra-red analyzers which can be focused, from a stationary platform or from a patrolling helicopter, on suspect smokestacks[25]. In either approach, computers would be used to analyze raw data and report on the findings. Conventional data processing would be applied to the licensing, inspection, and complaint-processing activities.

REFERENCES

1. A. WOLMAN, The Metabolism of Cities, *Scient. Am.* **213**, 179 (1965).
2. Air Quality Criteria for Sulfur Oxides, U.S. Department of HEW, Public Health Service, (1967).
3. Ambient Air Quality Objectives—Classifications System, Air Pollution Control Board, State of New York, 1965 (adopted December 11, 1964).
4. A. N. HELLER, (Remarks by) Commissioner of New York City Department of Air Pollution Control, at the New York–New Jersey Interstate Air Pollution Conference, sponsored by U.S. Department of HEW, New York City (January, 1967).
5. *Chem. Engng News* **45**, 20 (3 April, 1967).
6. *Chem. Engng News* **45**, 19 (15 May, 1967).
7. D. NOVICK, *Program Budgeting: Program Analysis and the Federal Government*, Harvard University Press (1965).
8. A. TELLER, Feasibility and Cost of Fuel Substitution, presented at 60th annual meeting, *Air Pollut. Control. Ass.* (June, 1967).
9. A. R. DAMMKOEHLER, *Inventory of Emission for the City of Chicago, J. Air Pollut. Control Ass.* **16**, 151 (1966).
10. D. B. TURNER, The Diurnal and Day to Day Variations of Emissions from Space Heating Sources, presented at 60th annual meeting, *Air Pollut. Control Ass.* (June, 1967).
11. I. S. LOWRY, *A Model of Metropolis*, Memorandum RM-4035-RC, The RAND Corp., Santa Monica, Calif. (1964).
12. Traffic Origin-and-Destination Studies, *Bull. Highw. Res. Bd* 253 (1960).
13. *The Next Twenty Years*, The Port of New York Authority, August (1966).
14. C. P. RYDELL, Air Pollution and Optimal Urban Form, presented at 60th annual meeting, *Air Pollut. Control Ass.* June (1967).
15. New York–New Jersey Air Pollution Abatement Activity, Sulfur Compounds and Carbon Monoxide, U.S. Department of HEW, Public Health Service, National Center for Air Pollution Control, Cincinnati, O. (January, 1967).
16. C. SIMON, and W. PROUDFIT, Some Observations of Plume Rise and Plume Concentration Distribution Over New York City, presented at 60th annual meeting, *Air Pollut. Control Ass.* (June, 1967).
17. B. DAVIDSON, A Summary of the New York Urban Air Pollution Dynamics Research Program, Paper 66–29, *Proc. 59th Meet. Air Pollut. Control Ass.* (1966)
18. D. B. TURNER, A Diffusion Model for an Urban Area, *J. appl. Meteor.* **3**, 83 (1964).
19. J. F. CLARKE, A Simple Diffusion Model for Calculating Point Concentration from Multiple Sources, *J. Air Pollut. Control Ass.* **14**, 347 (1964).
20. M. E. MILLER, and G. C. HOLZWORTH, An Atmospheric Diffusion Model for Metropolitan Areas, *J. Air Pollut. Control Ass.* **16**, 46 (1966).
21. P. SNEATH and R. SOKAL, *Principles of Numerical Taxonomy*. Freeman, San Francisco (1963).
22. Regional Air Pollution Warning System, New York–New Jersey Cooperative Committee on Interstate Air Pollution (1964).
23. D. A. LYNN and T. B. McMULLEN, Air Pollution in Six Major U.S. Cities as Measured by the Continuous Air Monitoring Program, *J. Air Pollut. Control Ass.* **16**, 186 (1966).
24. S. G. BOORAS, City of Chicago Air Quality Telemetering System, presented at Great Lakes Regional Meeting, *Am. Chem. Soc.* June (1966).
25. A. R. BARRINGER, Molecular Correlation Spectrometer for Remote Sensing of Gaseous Pollutants, presented at 60th annual meeting, *Air Pollut. Control Ass.* (June, 1967).

12

Reprinted from *IBM J. Res. Develop.*, **16**, 162–170 (Mar. 1972)

L. J. Shieh
P. K. Halpern
B. A. Clemens
H. H. Wang
F. F. Abraham

Air Quality Diffusion Model; Application to New York City

Abstract: An experimental multisource air pollution diffusion model based on the Gaussian plume formulation is described. The model incorporates point and area sources, time and space dependence of source strengths, and time and space dependence of meteorological variables. Numerical simulation of the SO_2 concentration distribution for New York City on January 11, 1971, agrees favorably with experimental measurements.

Introduction

The Palo Alto Scientific Center is developing an experimental "Gaussian plume" diffusion model for the prediction of air pollution concentrations. This model consists of a set of mathematical equations that can be solved on a digital computer for given rates of emission of pollutants into the atmosphere under a given set of meteorological conditions, principally wind vectors and atmospheric stabilities. The solution is a set of numerical values giving the concentration of a pollutant such as sulfur dioxide at spatial points in the region under consideration. Mixing and dilution in a diffusion model are ascribed to the random movement of turbulent eddies in the atmosphere, which carry the pollutants along with them. The rate at which this transport and diffusion take place is determined principally by wind speed, temperature gradient, and local topographic conditions.

The equations representing the model are partly phenomenological because some of the parameters must be evaluated from experimental observations. The use of such a model requires weather and pollution data from a real-time monitoring network and a complete catalogue of pollution emissions within the area (the "source emission inventory"). After the parameters in the model are evaluated using real data, the model must be run for a reasonable period of time for "validation," i.e., comparison of the predictions with the subsequent observations.

Turner[1] was the first to develop a multisource diffusion model for an urban area. The various diffusion models that had been developed up to 1968 have been critically reviewed by Moses[2] and Neiburger[3]. More recent models are the "Air Quality Display Model" developed by the TRW Systems Group[4], the Metropolitan New York Model" by Shieh[5], and the "Practical Multipurpose Urban Diffusion Model for Carbon Monoxide" by Johnson et al.[6].

Gaussian plume model

• *General discussion*

The physical process that transports the molecules of pollutant from one point to another is principally eddy diffusion, or turbulence, on a much smaller scale than the large-scale fluid motions of the mixing atmosphere. Turbulence generally refers to a collective random (or nearly random) motion involving a group of many molecules.

The description of diffusion of pollutants by turbulence, which is physically related to the temperature stratification of the atmosphere and to the wind field, leads to the concept of stability classes. The stable layers are those located in regions in which large-scale turbulence is suppressed, but small-scale turbulence, or eddy diffusion, still occurs. A mathematical theory of diffusion exists in which the parameter that measures the rate at which diffusion takes place is directly related to the scale and intensity of the turbulence that occurs. The theory is far from complete—many fundamental questions remain unanswered, but it is complete enough to be applied in a phenomenological sense inasmuch as the parameters that cannot be determined *a priori* can be estimated from experimental results. Such a diffusion theory can then be used, in conjunction with a description of the motion of the wind, to calculate the spread of pollutants through the atmosphere. At this point the theory is empirical, since it contains experimentally estimated parameters, and crude, because it can include the complicated geometry of the real world (buildings, mountains, etc.) in only a generalized sense.

With these qualifications in mind, however, it turns out that the diffusion models under static weather conditions, if carefully applied, can give good order-of-magnitude (and sometimes better) estimates of spatial distributions of time-dependent pollutant concentrations. The estimate can be made on either a short-term or a climatological basis. The diffusion model applied with day-to-day local weather forecasts is more useful in determining the normal background exposure of the population to pollution and the actual onset of an episode. One could even estimate the emission reductions necessary to suppress the episode or identify the major contributors to the concentration of pollutant at any given point.

One of the most accepted computational approaches is the Gaussian plume model for a single fixed source (see, e.g., Ref. 7). A Gaussian plume model assumes that if a pollutant is emitted from a point source, the resulting concentration in the atmosphere, when averaged over sufficient time, will approximate a normal statistical distribution in space. This is a good approximation for sufficiently long time averages. Thus the diffusion of airborne material can be described in terms of a set of diffusion coefficients that correspond to the standard deviation term in the Gaussian function. These coefficients are assumed to be functions of the atmospheric conditions and the distance downwind from the source. In addition, the entire diffusion cloud is assumed to be transported downwind while the cloud diffuses and spreads. Knowing the diffusion coefficients, one can calculate the steady state concentration associated with emission from an elevated, continuous-emission, point source for a given point in space by using the Gaussian plume formula (illustrated in Fig. 1). In general, the Gaussian plume formulation can be applied to the continuous-emission line and area sources, thereby making possible the construction of an urban diffusion model capable of handling sources of various types.

The Gaussian plume model requires three basic inputs when applied to multiple sources:

1. the spatial and temporal variation of the source emissions;

2. the prevailing average wind speed and direction; and

3. the time and space scales over which the pollutants are to be predicted.

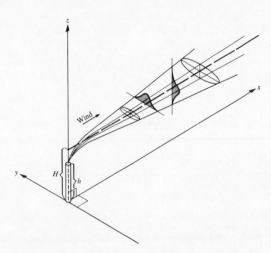

Figure 1 Representation of a continuously emitting point source in the Gaussian plume diffusion model; the function plotted is $\chi \bar{u}/Q$ [see Eq. (1)].

A simulation of concentration fields resulting from multiple sources becomes tractable if one assumes a steady state solution to the diffusion equation for a point source, which has the form of the Gaussian distribution function. By a superposition of single-source solutions, one can represent multiple sources that are spatially distributed and continuous in time.

• *Basic equations*

The surface concentration field $\chi(x,y,0)$, resulting from a fixed point source at effective height H emitting a gaseous pollutant at the rate Q, can be computed with the formula

$$\chi = \frac{Q}{\pi \bar{u} \sigma_y \sigma_z} \exp -\frac{1}{2}\left(\frac{y^2}{\sigma_y^2} + \frac{H^2}{\sigma_z^2}\right), \qquad (1)$$

where the x and y axes are the downwind and crosswind directions, respectively, $\bar{u}$ is the average wind speed, and $\sigma_y(x)$ and $\sigma_z(x)$ are the horizontal and vertical diffusion coefficients. (For a discussion see Ref. 8.) The ground surface is assumed to be a perfect reflector of the pollutants. The effective stack height H has to include the distance the plume rises due to buoyancy; i.e., $H = h + \Delta h$, where h is the physical stack height of the point source and Δh is the plume rise (Fig. 1). In general, Δh can be related to various stack parameters, such as physical configuration of the stack, exit velocity of the flue gas, heat emission rate Q_h, temperature difference between the flue gas at the top of the stack and the ambient air, mean wind speed $\bar{u}$, and local topography. We

163

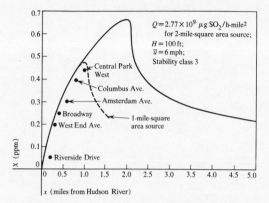

Figure 2 shows, chart with axes: y-axis χ (ppm) from 0 to 0.7; x-axis x (miles from Hudson River) from 0 to 5.0. Labels on chart:

$Q = 2.77 \times 10^9 \ \mu g \ SO_2/\text{h-mile}^2$
for 2-mile-square area source;
$H = 100 \text{ ft}$;
$\bar{u} = 6 \text{ mph}$;
Stability class 3

Central Park West
Columbus Ave.
Amsterdam Ave.
Broadway
West End Ave.
1-mile-square area source
Riverside Drive

Figure 2 Comparison of observed and computed sulphur dioxide concentrations along 79th Street in New York City for March 9, 1966, 0615 to 0825 EST.

use one of the more widely accepted empirical formulas[9],

$$\Delta h = k \, Q_h^r / \bar{u}^s, \qquad (2)$$

where k, r, and s are experimentally determined constants. (Currently we are using $k = 0.047$, $r = 0.5$, and $s = 0.78$.)

It is not practical to construct a diffusion model for a large region which includes many small sources by treating all of the pollution sources as point sources. To make the computation one combines the small sources within a chosen area into an "area source" with a uniform emission rate. In our model we adopt the integrated plume formula. The concentration at a given point x,y contributed by an area source is obtained by integrating Eq. (1) over the source region:

$$\chi = \int_{y'=-a}^{a} \int_{x'=0}^{2a} \frac{Q'}{\pi \bar{u} \sigma_y \sigma_z} \exp \left\{ -\frac{1}{2} \left[\frac{(y-y')^2}{\sigma_y^2} + \frac{H^2}{\sigma_z^2} \right] \right\} dx' dy', \qquad (3)$$

where σ_y and σ_z are functions of $x - x'$ and the atmospheric stability class, and Q' is the source strength per unit area. In the integration of Eq. (3) the area sources are oriented according to the mean horizontal wind direction.

Figure 2 shows an early use of Eq. (3) to calculate concentration from a continuous area source in New York City. The agreement with observational data is remarkably good. In particular, the ability to simulate the spatial gradient of concentration over a wide range of values portends great promise for using this method, which provides desired accuracy while improving computational efficiency. A detailed discussion of this method compared to other approaches currently used in diffusion modeling is given by Shieh and Halpern[10].

The parameters σ_y and σ_z in Eqs. (1) and (3) are the most controversial in any diffusion equation. It is generally accepted that these parameters depend on atmospheric thermal stability, dynamic wind field, aerodynamic effects of the surrounding boundaries, and travel distance of the plume. The σ curves as functions of plume travel distance are determined by observation for various atmospheric stability conditions and wind speed intervals.

Our model incorporates the σ_y and σ_z values of Pasquill[11] and Gifford[8,12] (especially Ref. 8, pp. 102–103, Figs. 3.10 and 3.11). However, we assume that an inversion layer at height H acts as a barrier to the dispersion of the pollutant so that Gifford's value of σ_z is replaced by σ_z', where

$$\sigma_z' = \min \ (\sigma_z, \tfrac{1}{2}H). \qquad (4)$$

• *Computational procedure*
Given the meteorological and source emission data, the model provides the pollutant concentration at a receptor as the sum of the contributions from the individual sources in the region. The usefulness of this "source-oriented" procedure is twofold. First, one may isolate individually the contribution of any source to the overall concentration field. Second, the procedure is computationally more efficient when the concentration field is to be computed over a relatively large area.

To reduce the computation time for real-time prediction of the concentration fields, an independent program calculates the $\chi \bar{u}/Q'$ function of Eq. (3), which is independent of x' *and* y' within each source area, for various source sizes, effective stack heights, and atmospheric stability classes. The specification of an inversion layer in the lower atmosphere is also considered as a variable. The results of these computations are stored in a peripheral memory device of the computer.

The main program has four parts: 1) analysis of source inventory data; 2) analysis of meteorological data; 3) calculation of surface pollutant concentration field; and 4) graphic display of computed results. The input requirements for the source inventory and meteorological data are discussed in succeeding sections. The surface pollutant concentration field is determined by the following sequence of computations: 1) area-source contributions; 2) point-source contributions; 3) sum of 1) and 2); 4) 24-hour average of 3).

SO$_2$ source emission data
The method of producing an inventory of sulphur dioxide emissions for a specific region is well documented. Studies such as those of NAPCA[13], Venezia and Ozolins[14], and Ozolins and Smith[15] describe in detail the procedure for obtaining an annual emission invento-

164

ry. However, they do not discuss procedures for estimating bihourly emissions from annual emissions for use as input to the air pollution diffusion model.

The diffusion model requires emission data at two-hour intervals in order to predict the average ground level concentration for that period. There are various approaches to obtaining this bihourly data from annual emission data. The method that we outline below has been successfully used in a study by Shieh[5].

In general, the source emissions are usually divided into contributions from industrial sources and from sources that generate space heating and hot water. For modeling, it is inefficient to consider each source as an individual point on the source grid. A logical approach to this problem is to represent all small sources within a reasonable geographic area as a single area source. The individual sources comprising the area source are largely a result of space heating.

The extent of the area sources and their size distribution over the model region are functions of the annual emission inventory and the classification of source types. Table 1 illustrates a typical classification of sources (the definition is not comprehensive).

The area-source and point-source emission data include the location of the sources with respect to a reference coordinate system. This system is defined in the model as a rectangular coordinate system with the origin located at the southwest corner of the geographic region of interest. The x axis represents east-west directions, the y axis, north-south directions, and the z axis, the vertical direction. The x and y coordinates and the physical source height H are required for point sources. The area sources are represented by the geometric center and horizontal extent of the source grid. Each area source grid of a given mesh size has only one source-height parameter H.

• *Area sources*

Size
The source emission inventory is usually made on a square grid system, using annual data for each grid square. There is no fixed dimension for the mesh size. The techniques used in such a survey are reviewed by Ingram, Kaiser, and Simon[6].

The diffusion model currently accepts various sizes of area source. In areas where SO_2 emission is small, or in areas far from the receptor, the refinement given by a small-mesh grid is not required to obtain a reasonable concentration distribution. In regions of strong SO_2 output, the use of smaller source-grid mesh sizes enables the model to predict concentration fields more accurately. The specification of area sources, i.e., their distribution and dimensions over the model region, is based on the

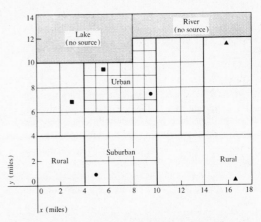

Figure 3   Schematic representation of point sources (● power plant; ■ incinerator; ▲ refinery) and area source grid.

Table 1   Examples of pollutant sources.

Point
  Power generator
  Incinerator
  Manufacturing plant (SO_2 is an indirect product)
  Industrial combustion (SO_2 is a direct product)
  Chemical refinery
  Oil refinery
  Mineral smelter
  Hospital

Area
  Commercial: stores, office buildings, hotels, laundries, etc.
  Domestic: space heating, hot water heating, cooking, etc.

receptor location relative to the sources and the distribution of source strengths according to the available source inventory.

The definition of an area source assumes that within the area the emission inventory is approximately uniform. However, it is important to account for any geographic (spatial) gradient of source strength in forming the area source grid map. This map should also reflect large source-free areas such as lakes or rivers. Figure 3 shows a schematic diagram of a typical area source grid map. The one-mile-square mesh is in an urban area where the SO_2 emission is large and the spatial gradient is strong. The two-mile-square mesh is in a suburban area, and the four-mile-square mesh is in a rural area.

Height
The model is capable of handling different source emission heights for each area source mesh size. Since the computational procedure can be made more efficient if   **165**

fewer heights are specified, the current model treats only one emission height per area source. The emission height is the average building height within the source area.

Emission rate
One method of subdividing the annual basic area source emission data into two-hour emission periods is to relate daily emission patterns to the average daily temperature. Area source emissions result mainly from fuel combustion for space heating and hot water heating. The amounts consumed are highly dependent on the average daily temperature. Using such a relationship, Ingram et al.[16] suggested the following formula for determining daily output:

$$Q_d = \gamma_1 Q_a / 365 + \gamma_2 DD Q_a / TDD, \qquad (5)$$

where Q_d is the daily and Q_a is the annual total output of SO_2. The degree-day DD is defined as the difference between 65°F and the daily mean temperature if the latter is 65°F or less; otherwise DD is zero. The long-term climatological average of the annual total degree-days is TDD, and γ_1 and γ_2 are the proportions of sources due to hot water heating and to space heating, respectively $(\gamma_1 + \gamma_2 = 1)$.

An alternative to Eq. (5) is based on heating, nonheating, and transient seasons. The calendar year is divided into three specified seasons and the fractions of the annual source emission for these seasons are denoted by α_1, α_2, and α_3, respectively $(\alpha_1 + \alpha_2 + \alpha_3 = 1)$. The emission data obtained using this method are not as accurate as those by the former method.

Experience has shown that when the daily mean temperature is greater than 60°F, there is no significant emission variation in such data. However, different diurnal emission "patterns" are observed when the daily mean temperature lies in the range 30 to 40°F or 50 to 60°F. By assuming that within each 10°F interval of the daily average temperature there is a specific pattern of emissions, a simple statistical analysis can be used to obtain 12 two-hour-period coefficients for each temperature range, i.e., the β_i in Eq. (6):

$$Q_i(2h) = \beta_i Q_d, \quad \sum_{i=1}^{12} \beta_i = 1. \qquad (6)$$

The accuracy of pollutant concentrations predicted by the model is strongly dependent on the accuracy of the two-hour source emission data.

Small point source in an area source
To reduce the total number of sources handled by the model, we adopt the following criteria for considering a point source as part of an area source: 1) The annual emission from point sources must be less than one tenth of the annual area source emission; 2) the physical stack height of the point source must be within 20 m of the emission height of the area source; and 3) the point source must have no appreciable plume rise.

• *Point sources*
The sources not classified as area sources are point sources. Associated with point-source input data are the stack parameters mentioned previously as necessary to calculate the rise of the plume above the physical stack height due to buoyancy. Various formulas are available to compute plume rise, e.g., see Ref. 17. One must, however, decide which formula is appropriate for the particular locality. To simplify the computation of plume rise, it is possible to estimate Q_h as a function of $Q(SO_2)$ for each point source or to establish it on an industrial basis.

As in the case of area sources, the model requires bihourly emission data for each of the point sources. If such data are not readily available, the daily emission pattern may be derived from total annual output by classifying the point sources according to industry with an operational mode assigned to each industry. The daily output is assumed to follow a pattern for each industry. For power generation plants, the daily output pattern can be associated with seasonal variation.

The diurnal emission pattern may be formulated analogously to provide the two-hour point source emission data. These are a function of daily output on an industry wide basis. For example, one diurnal emission pattern would be associated with an industry that has a 24-hour operational shift and another pattern would characterize an eight-hour operational shift.

Meteorological data
The diffusion model requires observations of wind speed and direction, averaged over two-hour periods, with a spatial density of observation points dependent on local geography and the observed concentration field. The meteorological data network should be sufficiently dense for a meteorological grid to be defined. (This grid is not necessarily the same as the source grid.) Since the meteorological stations do not necessarily coincide with the grid line intersections, data must be interpolated between the observation points and the meteorological grid, which requires knowledge of the x and y coordinates of the observation locations with respect to the source grid system.

The diffusion model computation uses the atmospheric stability classes defined by Pasquill[11] and used by Turner[1]. These stability classes are based on the vertical thermal stratification. The model requires a stability classification for each two-hour interval.

It is quite common for the region to be covered, in whole or in part, by an above ground temperature inver-

166

sion layer. The height of this inversion, if present, is included in the input data.

Example

The simulation experiments discussed in this section demonstrate the capability of the diffusion model. Figure 4 is a map of the New York City area showing the locations of the air quality and meteorological observation stations in a four-mile-square grid system. Figure 5 shows the source emission inventory grid with about 500 area sources for the New York metropolitan area. The simulation area is 10.5 miles × 18 miles and the area source mesh sizes are one-half-, one-, and two-mile squares. The effective areas of simulated sources are related to the geographic gradients of emissions. The two-hour-average emissions for the area sources were obtained with the assistance of the New York City Department of Air Resources by using the annual emissions inventory and Eqs. (5) and (6). Point source data were not used in the example.

The simulation period is the 24 hours of January 11, 1971. The choice of this date was based on two considerations: 1) There was a high sulphur dioxide pollution incident, and 2) a low wind speed condition posed a severe test for the computational method of the model. On that day only five of the ten telemetering stations were operating (1,3,5,14, and 18 in Fig. 4), a not unusual situation in an air quality monitoring network. Four of the operating stations are in the simulation area and station 14 is not far outside it.

A simulation of the SO_2 concentration field for the hours 0200 to 0400 eastern standard time (EST) is shown in Fig. 6(a). The wind for the modeled area is from the southwest at 7 mph. The stability class used is 4. The peak concentration occurs on the west side of mid-Manhattan. There is a very steep concentration gradient across the Hudson River. The general isopleth pattern in Brooklyn, Queens, and the Bronx shows elongation of the closed isopleths in a direction parallel to the mean wind direction.

The observations of ground level sulphur dioxide concentration at station 5 in Manhattan and at stations 1 and 3 in the Bronx verify the simulation for these two areas. In Brooklyn (station 18) the simulation indicates a relative maximum. The 0.08 ppm isopleth falls just to the right of the observed 0.09 ppm. However, in Queens at station 14 the simulation underestimates the concentration by a factor of seven. This is due to two factors: the point source emissions and the new housing in the Queens area—for at least four years, the source inventory has not been updated with these contributions.

The 0800 to 1000 EST simulation is shown in Fig. 6(b). The mean wind direction has changed to northwest, the speed to 4.5 mph, and the stability class to 3.

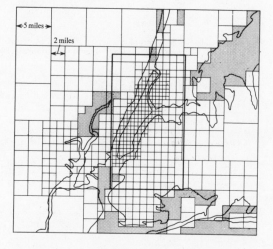

● Manual — Air quality data
▲ Telemetering — Air quality and meteorological data
■ National Weather Service — Meteorological data

Figure 4 Monitoring stations and meteorological grid. The area simulated in the January 11, 1971, example is outlined in the center.

Figure 5 Area source grid for New York City; the largest mesh is 5 miles square, the smallest, ½ mile square; shading indicates no-source areas.

There are three distinct areas of maximum concentration, two in Manhattan and one in the southeastern Bronx. All three areas are elongated in the direction of the mean wind. The change in wind direction has resulted in the movement of the previous Manhattan maximum from the eastern shore of the Hudson River southeastward to the East River. A tongue of high concentration extends from the East River into Queens. A second **167**

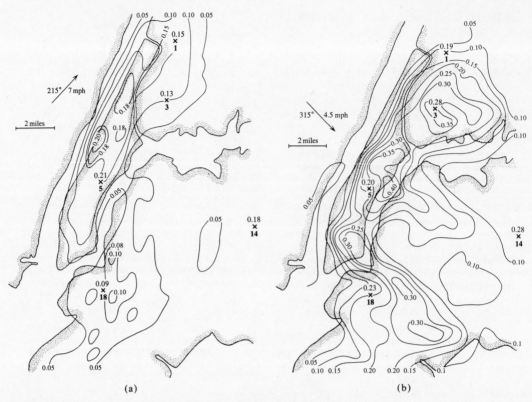

Figure 6 Two-hour simulations of the sulphur dioxide concentration field in New York City on January 11, 1971; observation points and readings are indicated by ×: (a) 0200 to 0400 EST, stability class 4; (b) 0800 to 1000 EST, stability class 3.

Table 2 Six- and 24-hour pollutant concentration (ppm) averages for four stations, January 11, 1971.

Time (EST)	Bronx High School (1) Observed	Computed	Morrisania Health Center (3) Observed	Computed	Central Park Arsenal (5) Observed	Computed	Brooklyn Public Library (18) Observed	Computed
0000–0600	0.20	0.19	0.17	0.22	0.22	0.19	0.10	0.09
0600–1200	0.17	0.18	0.14	0.38	0.22	0.24	0.19	0.15
1200–1800	0.14	0.11	0.16	0.18	0.15	0.14	0.09	0.08
1800–2400	0.41	0.30	0.29	0.29	0.25	0.26	0.14	0.14
0000–2400	0.22	0.19	0.22	0.25	0.21	0.21	0.12	0.11

maximum appears in the southeastern part of Manhattan. Here again a region of high concentration extends southeastward to Brooklyn with two relative maxima there. The regions of high concentration are directly related to upwind emissions from areas in Manhattan that have the largest annual emission of SO_2 in all of New York City. The northwest wind has also shifted the early region of maximum SO_2 gradient along the Hudson

River into the western shore line of Manhattan. The observations at stations 1, 3, and 5 verify the simulation quite well. In Brooklyn (station 18) and in Queens (station 14) underestimations by the simulation occur for the reason previously stated.

Figure 7 shows the isopleths of the simulated 24-hour-average surface concentration of SO_2 on January 11, 1971. The peak concentrations occur in Manhattan be-

168

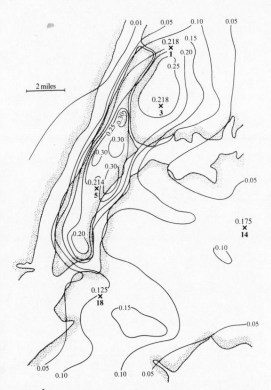

Figure 7 Twenty-four-hour-average simulation of the sulphur dioxide concentration field in New York City on January 11, 1971.

tween station 3 and station 5. A secondary maximum occurs between station 10 and station 37 near the southern end of Manhattan. The southwesterly prevailing wind during the 24-hour period results in a sharp gradient in concentration along the western shoreline of Manhattan. The relatively low-emission sources upwind and the significant density of emission sources downwind (in Manhattan) result in the concentration isopleths' being almost parallel to the shoreline.

The secondary maximum in the southeastern Bronx can be attributed to the high-emission sources in this area and the transport of SO_2 by the wind from Manhattan. There is a large area in Brooklyn and Queens with less than 0.05 ppm concentration. This is also a consequence of the prevailing wind direction and the low emission rate in that area.

The averaged observations of ground level concentration compare favorably with values from the simulation. Station 3 in the southeastern Bronx recorded 0.22 ppm, which is in good agreement with the simulation isopleth of 0.25 ppm. Station 5 in Manhattan agreed with the

0.25 ppm simulation isopleth. The 0.10 ppm isopleth in Brooklyn falls just to the left of station 18, which showed a reading of 0.13 ppm. The simulation in the Queens area near station 14 again underestimated the concentration of SO_2, thus being consistent with our previous comparisons.

The bihourly variations on January 11, 1971, of the sulphur dioxide concentration are presented in Fig. 8 for four stations. In Table 2, six- and 24-hour averages of the concentrations for the same period are listed for these four stations. In general, there is good agreement between the observed and computed concentration variations for this period.

These numerical experiments indicate that this model has the capability of predicting the general pattern of the ground-level SO_2 concentration in an urban region on a time scale of about two hours. Furthermore, limited comparisons of the simulated values and the observed data in the New York City area give a positive indication of the useful potential of this model. We are presently extending our study to include 1) the spatial dependence of the meteorological input, 2) the point source inventory, and 3) the complete systematic evaluation of the various physical approximations used in the model.

The computer program[18] for this diffusion simulation is written in FORTRAN IV(H) and requires 200K bytes of core storage. For an IBM System/360 Model 91 computer, the CPU execution time for a two-hour simulation was 15 seconds.

Acknowledgments

The authors are grateful to the New York City Department of Air Resources for their cooperation in supplying the observational air quality, meteorological, and source emission data that were used in the simulation experiments. Special thanks are acknowledged to former assistant commissioner C. Simon and to senior meteorologist H. Nudleman for their constructive comments. The acknowledgments also extend to the IBM Research Division for making their facilities available to the authors.

References and note

1. D. B. Turner, "A Diffusion Model for Urban Areas," *J. Appl. Meteorol.* 3 (1), 83 (1964).
2. H. Moses, "Mathematical Urban Air Pollution Model," presented at the 62nd annual meeting of the Air Pollution Control Association, June 1969, *Paper 69-31*, Air Pollution Control Association, 4400 Fifth Avenue, Pittsburgh, Pennsylvania 15213.
3. M. Neiburger, "Diffusion Model of Urban Air Pollution," presented at the Symposium on Urban Climates and Building Climatology, *CLU Document 35*, World Meteorological Organization, Brussels, 1968.
4. TRW Systems Group, "Air Quality Display Model," report prepared for the U.S. Dept. of Health, Education and Welfare, Public Health Service, National Air Pollution Control Administration, Washington, D. C., 1969.

169

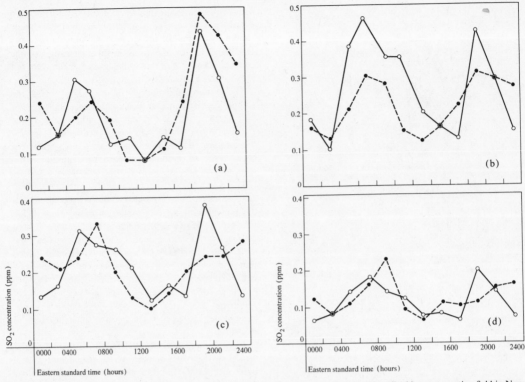

Figure 8 Comparison of the observed (●) and computed (○) bihourly variations of the sulphur dioxide concentration field in New York City on January 11, 1971; (a) station 1, Bronx High School; (b) station 3, Morrisania Health Center; (c) station 5, Central Park Arsenal; (d) station 18, Brooklyn Public Library.

5. L. J. Shieh, "A Multiple Source Model of Turbulent Diffusion and Dispersion in Urban Atmospheres," Research Grant AP-00328-04 Final Report in "Mathematical Models of Urban Air Pollution Dynamics," Vol. 2, *Report TR-69-11*, New York University, 1969.
6. W. B. Johnson, F. L. Ludwig, A. E. Moon, and R. L. Mancuso "A Practical Multipurpose Urban Diffusion Model for Carbon Monoxide," *Project Report 7874*, Stanford Research Institute, Menlo Park, Calif., 1970.
7. O. G. Sutton, "A Theory of Eddy Diffusion in the Atmosphere," *Proc. Roy. Soc. (London), Series A* **135**, 143 (1932).
8. F. A. Gifford, Jr., "An Outline of Theories of Diffusion in the Lower Layers of the Atmosphere" in *Meteorology and Atomic Energy*, edited by D. H. Slade, U. S. Atomic Energy Commission, 1968, p. 66.
9. *The Calculation of Atmospheric Dispersion from a Stack*, Concawe Publications, The Hague, The Netherlands, 1966.
10. L. J. Shieh and P. K. Halpern, "Numerical Comparison of Various Model Representations for a Continuous Area Source," *Report G320-3293*, IBM Scientific Center, Palo Alto, California, 1971.
11. F. Pasquill, *Atmospheric Diffusion*, D. Van Nostrand Company, Inc., Princeton, New Jersey, 1962.
12. F. A. Gifford, Jr., "Use of Routine Meteorological Observations for Estimating Atmospheric Dispersion," *Nuclear Safety* 2(4), 47 (1961).
13. "Guidelines for the Development of Air Quality Standards and Implementation Plans," U. S. Dept. of Health, Education and Welfare, Public Health Service National Air Pollution Control Administration, Washington, D. C., 1969.
14. R. Venezia and G. Ozolins, "Air Pollutant Emission Inventory; Interstate Air Pollution Study, Phase II," Vol. II of project report, U.S. Dept. of Health, Education and Welfare, Public Health Service National Air Pollution Control Administration, Washington, D. C., 1966.
15. G. Ozolins and R. Smith, "A Rapid Survey Technique for Estimating Community Air Pollution Emission," *Public Health Service Publication 999-AP-29*, U.S. Dept. of Health, Education, and Welfare, Washington, D. C., 1966.
16. W. Ingram, E. Kaiser, and C. Simon, "Source Emission Inventory for SO_2 in the New York Metropolitan Area," Dept. of Civil Engineering report, New York University, 1965.
17. C. A. Briggs, "Plume Rise," *C.F.S.T.I. TID-25075*, Division of Technical Information, U.S. Atomic Energy Commission, Washington, D. C., 1969.
18. This is an experimental computer program developed for research purposes only and is not available outside IBM.

Received July 2, 1971; revised November 17, 1971

The authors are located at the IBM Data Processing Division Scientific Center, 2670 Hanover Street, Palo Alto, California 94304.

170

Reprinted from *Socio-Econ. Plan. Sci.*, **5**, 173–190 (June 1971)

DETERMINATION OF OPTIMAL AIR POLLUTION CONTROL STRATEGIES

JOHN H. SEINFELD and CHWAN P. KYAN

Department of Chemical Engineering, California Institute of Technology, Pasadena, California 91109

(*Received* 17 *December* 1970)

A general theoretical framework for the determination of long-term air pollution control strategies for an airshed is presented. The problem is stated as determining the set of control measures to be exercised on airshed source emissions such that the total cost of control is minimized subject to the constraint of maintaining a certain level of air quality. It is assumed that a dynamic airshed simulation model exists that is capable of describing atmospheric diffusion and chemical reaction, thus relating total mass emissions to atmospheric concentrations. The problem is based on the typical daily air pollution syndrome with air quality being defined as some function of the daily pollutant concentrations. It is shown that the overall optimization problem can be decomposed into two problems, the first, one of linear programming, the second, one of a dynamic optimization. The theory is illustrated by means of detailed hypothetical example of selecting air pollution controls for 1 yr.

INTRODUCTION

AN AIR pollution control strategy specifies requirements for the abatement and regulation of air pollution in a given airshed. The control strategy should be the result of a comprehensive analysis of the total air pollution problem in the airshed, including determination of desired levels of air quality, delineation of alternative abatement policies and establishment of criteria by which the alternative policies are to be evaluated.

Desired air quality will ultimately be related to levels of concentrations injurious to health and plant life, unpleasant from an aesthetic point of view and costly in an economic sense, although current knowledge seldom affords a quantitative relationship between airborne concentrations and these factors. Nevertheless, current air quality standards are related to persistent levels of airborne pollutant concentrations, for example, maximum hourly average parts-per-million (ppm) concentrations.

Many alternative air pollution control strategies may exist for a given region, although the principal element is legally enforceable emission standards. It is necessary to establish criteria by which alternative strategies are to be evaluated, perhaps the most difficult task of all. The criteria should include not only the direct and indirect costs of air pollution control to the polluter and the public but also the damages incurred by the general public as a result of polluted air.

We have on one hand the various emission control strategies and on the other hand air quality based on airborne pollutant concentrations. Therefore, there must exist an airshed simulation model, which incorporates source and meteorological information to predict contaminant concentrations as a function of time and location in the airshed. Presumably the criteria by which strategies are judged will include both source and airborne concentration factors. Thus, the formulation of emission control policies requires knowledge of source

173

emission rates, control technologies and their costs for various sources, wind patterns and topography of the airshed, rates of chemical reactions that may take place in the atmosphere and anticipated growth in pollution-related activities.

An airshed may be envisioned as a dynamic system with uncontrollable inputs of meteorology and controllable inputs of source strengths and locations. The output of the dynamic airshed system is air quality, concentrations of pollutions by time and location. The principal unit of time for the airshed is the day since in most airsheds the air pollution syndrome is repeated daily, or during conditions of prolonged stagnation, over a period of several days.

The meteorology of an airshed is generally constant from year to year, that is, each year there can be expected to be a certain fraction of days conductive to air pollution episodes. Thus, whether we are concerned about air quality tomorrow or in 10 yr, we must always deal with the daily cycle of source emissions and meteorology. This will also be the case when assessing air quality, since injurious pollutant levels whether expressed on an hourly, daily, or long-term dosage basis can be computed from the airshed behavior for a typical day. The simulation model will describe the pollutant concentrations as a function of time and location.

There are two basic strategies for controlling a dynamic system: open-loop control and closed-loop control. In open-loop control the control policy to achieve a desired objective is determined on the basis of the initial state of the system and any expected inputs during the evolution of the system, i.e. open-loop control is predetermined and not altered during the evolution of the system. In feedback closed-loop control the control policy is determined at each time during the evolution of the system by comparing the actual output of the system and the desired output and manipulating system inputs to make the actual output match the desired output. Both open- and closed-loop control play important roles in an overall air pollution control strategy.

Open-loop control is important in a day-to-day as well as a year-to-year capacity. If weather predictions indicated the possibility of forthcoming adverse meteorological conditions, e.g. low inversion and light winds, control measures could be announced that would have to be instituted by various sources during the affected period. Also, the determination of time-tables for emission controls over the next decade or more is an exercise of open-loop control. Figure 1 illustrates the elements of the determination of open-loop air pollution control strategies.

Closed-loop control is most important when combined with an air quality monitoring system, from which measurements of air quality made during the day can be used to put into

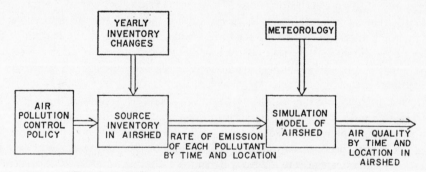

FIG. 1. Open-loop control of air pollution in an airshed.

operation rapid control actions when pollutant concentrations begin to exceed specific warning levels. These warning levels would normally be somewhat lower than those considered to be injurious to health because of the inherently sluggish response of the entire airshed or portions thereof to source emission changes. The system of smog alerts existing in Los Angeles is an example of such control. Savas [1] and Croke and Booras [2] have discussed the role of feedback control in an urban air monitoring system. Figure 2 illustrates closed-loop control of an airshed. In both types of control it is clear that an airshed simulation model is an integral component of the control system.

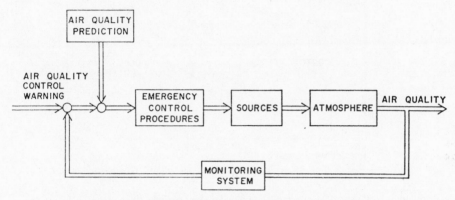

FIG. 2. Closed-loop control of air pollution in an airshed.

We will henceforth confine our attention to the determination of long term air pollution control strategies, that is, the systematic determination and evaluation of emission abatement programs over a many year period. Let us assume that we can forecast the control technologies that will be available over the time period of interest. This would include, for example, techniques for cleaning stack gases, emission controls for motor vehicles and methods of operation of power generating combustion equipment.

As we have noted, the most difficult part of our overall problem is the selection of a criterion or criteria upon which to evaluate alternative control strategies. Most optimization problems seek to minimize a single scalar criterion which in some manner reflects the total cost of the control. The total cost of air pollution can be considered a sum of two opposing costs: (1) the cost of control, i.e. the direct cost to the polluter of control equipment and abatement measures; and (2) the costs of polluted air to the general public, i.e. decreased property values, loss of visibility, increased medical expenses, deterioration of paint and rubber, etc. Both costs are shown as functions of air quality in Fig. 3. If both curves could be determined accurately as functions of air quality then a conceivable solution would be to select the air quality at the minimum of the total cost curve. However, to adopt this approach for air pollution control is, unfortunately, a gross oversimplification. First, the assignment of dollars of cost to pollution levels is at best, guesswork. In fact, we probably can not even catalogue all the adverse affects of air pollution, the result of which is that the pollution cost curve in Fig. 3 is basically unknown. For example, atmospheric pollution is believed to be unhealthy, but how to ascertain the effects of long, continued exposure to sublethal concentrations of pollutants on health has proved to be an enormous problem. Thus, we will not attempt to determine a pollution cost function, but will cast the optimization problem as determining a minimum cost combination of control measures to meet a specified level of air quality. The solution of this problem produces one point of the control cost curve of Fig. 3. The problem can then be solved repeatedly for various levels of air quality.

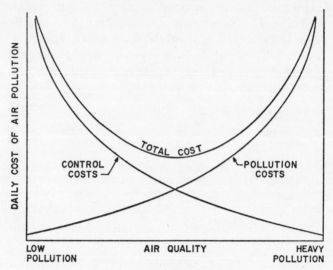

FIG. 3. Costs of air pollution.

We have not really told the whole story yet, however. It will be difficult for the cost of control function to include the social and economic consequences of various strategies. It will also be difficult to compare incommensurable strategies, such as using large land areas for solid waste disposal vs. incineration of solid waste. Finally, it is necessary to include the political implications of control policies. Since control strategies must be enacted into law, a politically unpalatable policy might just as well have been eliminated from the start.

Lest we become so discouraged that we drop the problem let us make the following compromises. We will determine that set of control policies over a stated time period that minimizes the total cost of control, as best it can be determined, subject to the constraint that a given level of air quality is maintained. It should be understood that our primary objective is to chart the shape of the control cost curve, to provide the analyst with information on the relative costs of various degrees of clean air.

We have, for illustrative purposes, been referring to the general term, "air quality". Of course, this must be carefully and precisely defined for a quantitative analysis of the problem. In the United States the five most common air pollutants in tons/yr emitted are CO, SO_2, hydrocarbons, nitrogen oxides and particulates. Mere consideration of weight emitted may not be a true measure of the destructive potential of these primary pollutants. Some may react chemically with other substances in the air to produce secondary pollutants that have far more severe effects than the original primary pollutants. The most important atmospheric chemical reactions are those that take place when emitted hydrocarbons and oxides of nitrogen are irradiated with sunlight. The resulting mixture, termed photochemical smog, consists of ozone, organic nitrates, aldehydes, ketones, etc. Thus, "air quality" is actually a vector of several quantities, and the cost curves in Fig. 3 will be surfaces in the cost–concentration space. The actual air quality measures can take several forms, such as maximum daily average or maximum hourly average concentrations.

There are two approaches to the specification of air quality. First, we could use as measures of air quality only the quantities of the various primary pollutant emissions. Clearly, this approach avoids the necessity for a simulation model, since we desire only to reduce primary emissions without regard to the resulting airborne concentrations. Such an approach may be valid for a pollutant which does not undergo transformation in the atmosphere, such as CO and SO_2, since airborne concentrations are probably linearly related to total emissions. The second approach, which we have been discussing above, bases air quality on airborne concentrations and requires an airshed simulation model. When atmospheric chemical reactions produce secondary pollutants, such as in photochemical smog, this approach must be used. Nevertheless, a simulation model represents a difficult undertaking, so there is much practical merit in the first approach in cases in which CO and SO_2 are the main air pollution components. Such an analysis was carried out by Kohn [3–5] for the St. Louis airshed for the year 1975. He chose values for the reductions in primary emissions desired in lb/yr and determined by linear programming the combination of control measures minimizing the total control cost corresponding to the chosen level of reduction. Farmer et al. [6] have also discussed formulation of emission control strategies for SO_2, and particulates for St. Louis. Burton and Sanjour [7] have considered the choice of control strategies for SO_2 and particulates for Kansas City and Washington, D.C. utilizing a diffusion model and yearly averages. Herzog [8], in an interesting study, has outlined how an urban air diffusion model can be incorporated into zoning decisions.

In this work we develop a systematic methodology for the determination of regional air pollution control strategies. We will assume that emission control policies are enacted on a yearly basis. The problem is to determine control strategies over a given period of time such that the total cost of air pollution control is minimized subject to the constraint of a specified level of air quality. We will develop a general theoretical framework for control strategy evaluation and illustrate the application by means of a hypothetical example.

AIR POLLUTION CONTROL STRATEGY EVALUATION

Let us first consider the problem of determining an optimal air pollution control strategy for a single year. In general terms the optimization problem is to determine the minimum cost combination of emission controls while maintaining a given level of air quality. The various aspects of the problem are depicted in Fig. 4. We make the following definitions:

d_{ij} = number of units of control method j per unit activity of source i, $i = 1,2,......,p$, $j = 1,2,......,q_i$;

s_i = daily units of activity of source i, $i = 1,2,......,p$;

$e^0{}_{ik}$ = daily mass emission of primary pollutant k from unit activity of source i with no control, $i = 1,2,......,p$, $k = 1,2,......N$;

e_{ik} = daily mass emission of primary pollutant k from unit activity of source i with control, $i = 1,2,......,p$, $k = 1,2,......N$;

r_{ijk} = reduction in the emission of primary pollutant k per unit of control method j per unit activity of source i, $i = 1,2,......,p$, $j = 1,2,......,q_i$, $k = 1,2,......,N$;

b_k = total daily mass emission of primary pollutant k, $k = 1,2,......,N$;

z_m = atmospheric concentration of pollutant m, a function of time and location in the airshed, $m = 1,2,......,M$;

$g_i(z_m) =$ functions of the atmospheric concentrations that serve as measures of air quality, for example, the average daily concentration of species m or the maximum instantaneous concentration of species m reached during the day, $i = 1,2,\ldots,L$. g_i^* is the maximum allowable values of the $g_i(z_m)$.

The total quantities of primary emissions are related to source emission and control levels by

$$b_k = \sum_{i=1}^{p} s_i e_{ik} = \sum_{i=1}^{p} s_i e_{ik}^0 - \sum_{i=1}^{p} \sum_{j=1}^{q_i} r_{ijk}\, d_{ij}\, s_i \qquad (1)$$

To formulate the cost criterion we let:

$c_{ij} =$ daily cost of one unit of control method j per unit activity of source i, $i = 1,2,\ldots$, $p, j = 1,2,\ldots,q_i$;
$C_i^* =$ maximum daily cost of control to be imposed on source i, $i = 1,2,\ldots,p$.
The total daily cost of control is thus

$$J = \sum_{i=1}^{p} \sum_{j=1}^{q_i} c_{ij}\, d_{ij}\, s_i \qquad (2)$$

One constraint which must exist on the choice of the d_{ij} relates to the net value of the source activity. For example, if we minimize J by choice of the d_{ij}, it is possible that a small firm with a net profit of $2000 per day would be required to undertake control measures costing $5000 per day, whereas a large firm with a net profit of $100,000 per day might also be required to install control measures at a cost of $5000 per day. Obviously, the small firm would be forced out of business. Thus, we introduce the constraint

$$\sum_{j=1}^{q_i} c_{ij}\, d_{ij}\, s_i \le C_i^* \qquad i = 1, 2, \ldots, p \qquad (3)$$

which provides a maximum daily control cost for each source.

It is also necessary to incorporate any constraints which may exist on the d_{ij} which are related to the manner in which the control methods are employed. For example, there may exist a maximum number of units of control method j that may be applied per unit activity of source i,

$$d_{ij} \le d_{ij}^* \qquad (4)$$

and there may exist a maximum number of total units of control, d_i^*, that may be exercised per unit activity of source i,

$$\sum_{j=1}^{q_i} d_{ij} \le d_i^* \qquad (5)$$

For example, to control SO_2 emissions from an industrial furnace (called source 1) burning 1 per cent sulfur coal, three possible control techniques are:

$d_{11} =$ pounds of carbonate powder injected per ton of coal burned;
$d_{12} =$ tons of 0.5% sulfur coal replacing 1% sulfur coal;
$d_{13} =$ tons of 0.25% sulfur coal replacing 1% sulfur coal.

Then, the constraints for the d_{ij} are $d_{11} \le d_{11}^*$, where d_{11}^* is the maximum number of pounds of carbonate that can be injected without affecting combustion or causing excessive particulate emission and $d_{12} + d_{13} \leq d_1{}^*$, where $d_1{}^*$ is the total number of tons of coal burned per day. Clearly we cannot replace more tons of coal than will be burned per day.

The airshed simulation model can be represented by the functional relationship,

$$z_m = f_m(t,\eta,b_l,\ldots,b_n) \qquad m = 1, 2, \ldots, m \qquad (6)$$

It is important to realize that equation (6) merely expresses the dependence of the solution of the simulation model on time t, location η and the b_k. The simulation model will, in general, be a set of nonlinear ordinary or partial differential equations. We will not, of course, be able to obtain an explicit functional relationship between the z_m and the b_k. This relationship can only be obtained by solution of the differential equations of the airshed model.

The constraints on air quality can be denoted by

$$g_i(z_m) \leq g_i{}^* \qquad i = 1, 2, \ldots L \qquad (7)$$

where, as noted, $g_i(z_m)$ is a set of air quality indicators. For example, $g_1(z_m)$ might represent the maximum hourly average concentration of species m over the entire airshed. Then, $g_1{}^*$ would represent the maximum allowable value of $g_1(z_m)$.

The single year optimal control problem can be stated as follows: choose the d_{ij} to minimize J subject to equations (1)–(7). This problem can be solved in principle using techniques from optimization theory, such as the method of steepest descent. However, the large dimensionality of the d_{ij} and the nonlinearity of the airshed model make the problem computationally formidable. To make the problem tractable we will decompose the problem into two subproblems.

From Fig. 4 the variables involved are related by the following flow diagram:

$$\begin{matrix} & & & & \text{airshed} \\ & & & & \text{model} \\ {}_{i=1}^{P}\{d_{ij}\}_{j=1}^{q_i} \longrightarrow {}_{i=1}^{P}\{s_i e_{ik}\}_{k=1}^{N} \longrightarrow \{b_k\}_{k=1}^{N} \longrightarrow \{z_m\}_{m=1}^{M} \end{matrix}$$

The large dimensionality associated with the possible control techniques d_{ij} can be separated from the nonlinearity associated with the dynamic airshed simulation model by the decomposition into two optimization problems:

$$\begin{matrix} {}_{i=1}^{P}\{d_{ij}\}_{j=1}^{q_i} \longrightarrow {}_{i=1}^{P}\{s_i e_{ik}\}_{k=1}^{N} \longrightarrow \{b_k\}_{k=1}^{N} & | & \{b_k\}_{k=1}^{N} \longrightarrow \{z_m\}_{m=1}^{M} \\ \text{problem 1} & & \text{problem 2} \end{matrix}$$

Problem 1

For any given set of total mass emissions b_k, $k = 1,\ldots\ldots,N$, determine the optimal set of control actions d_{ij} to be imposed on the p sources such that the cost of air pollution control is a minimum. Let x_k denote the reduction in the daily mass emissions of species k,

The $x_k = b_k{}^0 - b_k$, where $b_k{}^0 = \sum_{i=1}^{p} s_i e_{ik}^0$. The problem can be stated formally as: choose d_{ij} to minimize J subject to equations (2)–(3) and

$$\sum_{i=1}^{p} \sum_{j=1}^{q_i} r_{ijk} \, d_{ij} \, s_i \geq x_k \qquad k = 1, 2, \ldots, N \qquad (8)$$

which states that the reduction in emissions from all sources should at least be equal to x_k. By solving this problem for several different values of the b_k, and hence the x_k, we can determine J as a function of b_k, that is, we will know the minimum cost combination of controls corresponding to any level of total mass emissions. Problem 1 corresponds to the formulation of Kohn [3–5] and can be solved by linear programming. Having the minimum control cost as a function of the b_k we now formulate the second problem.

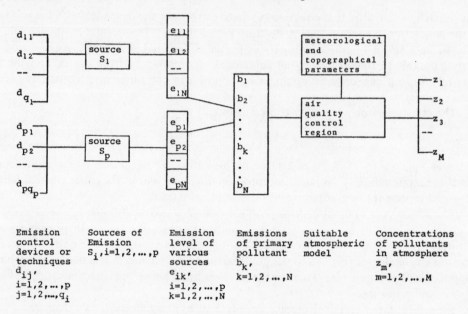

Emission control devices or techniques d_{ij}, i=1,2,...,p j=1,2,...,q_i	Sources of Emission S_i, i=1,2,...,p	Emission level of various sources e_{ik}, i=1,2,...,p k=1,2,...,N	Emissions of primary pollutant b_k, k=1,2,...,N	Suitable atmospheric model	Concentrations of pollutants in atmosphere z_m, m=1,2,...,M

FIG. 4. Schematic diagram of the relationship between control measures, source emissions and air quality.

Problem 2

Determine the b_k that minimize J subject to equations (6) and (7), where $J = \psi (b_k)$ is the result of Problem 1.

The solution of this problem will require dynamic optimization techniques because of the airshed model. The importance of the decomposition lies in the uncoupling of the high dimensional part of the problem and the nonlinear section of the problem. The large dimension of the d_{ij} will not affect the linear programming solution. The second problem, which is inherently more difficult, is of much lower dimension than Problem 1. We will discuss the solution of this problem later.

The entire single year problem is based on airshed behavior for a typical day, and all costs are assumed to be expressed on a daily basis. The final result will be the minimum control cost J as a function of various levels of air quality, g_i^*, $i = 1,2,......,L$.

For a multi-year period the problem can be stated as: determine the number of units of control method j per unit activity of source i in year τ, $d_{ij}(\tau)$, $\tau = 1,2,......,T$, to minimize the total cost of air pollution control, $J = J(1) + J(2) + ... + J(T)$, subject to the constraints that the air quality each year is less than or equal to $g_i^*(\tau)$, $i = 1,2,......,L$. The result of the single year problem was the minimum value $J(1)$ and the corresponding $d_{ij}(1)$ for $g_i^*(1)$.

From year to year there will be changes in source activities, uncontrolled source emissions and costs of various control methods in accordance with industrial growth and expansion and inflation. We can represent these changes by means of increments to the previous year values,

$$s_i(\tau) = s_i(\tau - 1) + \Delta s_i(\tau)$$
$$e^0{}_{ik}(\tau) = e^0{}_{ik}(\tau - 1) + \Delta e^0{}_{ik}(\tau)$$
$$c_{ij}(\tau) = c_{ij}(\tau - 1) + \Delta c_{ij}(\tau)$$
$$C^*{}_i(\tau) = C^{*'}{}_i(\tau - 1) + \Delta C^*{}_i(\tau) \tag{9}$$

Thus, we assume that for a given span of years we can forecast the above changes, including the removal or appearance of sources. Conceptually we can consider each year as being independent of the other years, in that the level of control activity for a particular year is not necessarily dependent on previous levels of control activity. Therefore, the multi-year air pollution control problem can be viewed conceptually as a combination of T single year problems.

The result of the multi-year problem would be the total cost J as a function of the various yearly air quality levels, $q_i^*(\tau)$. It is clear that if L and T are both large, the representation of J as a function of the LT variables g_i^* will be cumbersome. Thus, for a multi-year analysis it would be best to select a single scalar measure of air quality ($L = 1$). Then J will be a function only of the T levels, $g^*(\tau)$, $\tau = 1,2,......,T$.

A SIMPLE AIRSHED SIMULATION MODEL

Our primary purpose is not to consider atmospheric simulation, thus, we will only consider this subject in enough detail to make clear its relationship to the economic problem. The necessary components of an urban airshed model are:

1. The transport and diffusion model

This is really the overall model, the major descriptive aspect of which is the atmospheric transport and dispersive processes. This model will include:

2. The reaction kinetics model

This describes the rates of reactions occurring in the atmosphere as a function of concentration, intensity of radiation, temperature, etc.

3. The emissions model

This includes a complete source inventory of the airshed describing mass emissions of pollutants as a function of time and location.

A rigorous approach to urban diffusion modeling is direct integration of the three-dimensional, time-dependent partial differential equations of continuity for each species [9, 10]. However, a somewhat simpler approach can be adopted, based on the concept of well-mixed cells.

Assume the airshed has been divided into an array of L cells, each of which is considered as a well-mixed reactor. The volumes of the cells, which need not be equal, are $v_1,......,v_L$. The concentration of species i in cell j is z_{ij}. In each cell there is a time-varying source of each pollutant, the rate of emission of species i into cell j being S_{ij}. Also, there exists the possibility that pollutants can be formed by chemical reaction at a rate R_{ij}, or removed by deposition, the rate of deposition being D_{ij}. Finally, the volumetric rate of airflow from cell j to cell k is q_{jk}.

Thus, a dynamic material balance for species i in cell k, when the volume v_k can vary with time, is

$$v_k \frac{dz_{ik}}{dt} = -z_{ik} \frac{dv_k}{dt} + \sum_{j=0}^{L} q_{jk}\, z_{ij} - z_{ik} \sum_{j=0}^{L} q_{kj} + S_{ik} - D_{ik} + R_{ik} \tag{10}$$

$$z_{ik}(0) = z^0{}_{ik} \tag{11}$$

Normally, dv_k/dt is set equal to $A_k(dh_k/dt)$, where A_k is the area of the base of a cell having vertical sides and h_k is the height of the base or an inversion of a convenient mixing height. In effect, the cell is a box with permeable walls and a movable lid. The subscript zero on q_{kj} and z_{ij} relates to flows into and out of the airshed. If we divide the airshed into L cells and consider M components, LM ordinary differential equations of the form of (10) will be required to describe the system. Such a model has been introduced for airshed modeling by Ulbrich [11].

The advantages of this approach are as follows:

1. Aspects of complicated topographical variations can be easily handled.
2. Changing inversion levels can be easily handled.
3. The model is conceptually easy to understand and implement.

However, this approach has several drawbacks:

1. In the absence of an inversion, the concept of a mixing cell is somewhat artificial.
2. The assumption that pollutants are instantaneously mixed throughout the entire cell may be a poor one. If vertical mixing is slow, as under stable meteorological conditions, strong vertical concentration gradients can develop and the well-mixed assumption will not hold.

In spite of its potential drawbacks, the well-mixed cell model represents a reasonable compromise between the complexity of a rigorous partial differential equation diffusion model and the statistical Gaussian plume formulation [8], inapplicable when chemical reactions are occurring.

The second component of the airshed model is the reaction kinetics model, which, in the cell model, appears in R_{ik}. A discussion of atmospheric chemistry is beyond our scope and intentions. For a qualitative model of atmospheric reactions involving nitrogen oxides and hydrocarbons the reader may consult Friedlander and Seinfeld [12].

The third component of an air pollution model is the sources. Emission magnitudes must be specified as a function of time and location. Sources can be divided conveniently into mobile sources (motor vehicles, aircraft, etc.) and fixed sources (power plants, refineries, factories, etc).

The following information is required to estimate motor vehicle emissions in an airshed:

α_i = fraction of total motor vehicles of model year i, $i = 1,2,\ldots\ldots,X$;

Y_{ij} = fraction of motor vehicles of model year i manufactured by company $j, j = 1,2,\ldots\ldots,$
J;

$\tilde{e}_{ijk}$ = volumetric emissions of species k from motor vehicles manufactured by maker j in model year i (ppm or vol. %);

w_{ijl} = fraction of motor vehicles of model year i manufactured by maker j of weight class $l, l = 1,2,......,L'$;

μ_{ij} = fraction of motor vehicles of model year i manufactured by maker j with standard transmissions;

μ'_{ij} = fraction of motor vehicles of model year i manufactured by maker j with automatic transmissions;

K_{kl} = conversion factor to convert volumetric emissions to g/mile for species k, weight class l and standard transmissions (based on standard driving cycle);

K'_{kl} = conversion factor to convert volumetric emissions to g/mile for species k, weight class l and automatic transmissions (based on standard driving cycle);

$\beta_{k\tau}$ = fraction of total airshed driving activity in cell k during time period τ;

M_k = total daily vehicle miles travelled in cell k.

Then, the total grams of species i emitted into cell k during time period τ is

$$Q_{ik} = M_k \beta_{k\tau} \sum_{l=1}^{X} \alpha_l \sum_{j=1}^{J} Y_{lj} \tilde{e}_{lji} \left[\sum_{v=1}^{L'} w_{ljv} (\mu_{lj} K_{iv} + \mu'_{lj} K'_{iv}) \right] \qquad (12)$$

S_{ik} is obtained by converting Q_{ik} to a ppm value using the cell volume v_k. The total magnitude of contaminant emissions from motor vehicles in an urban area is not easily estimated. Pollutant magnitudes and concentrations from a motor vehicle depend on the size, age and make of the car, its condition, the particular driving mode (e.g. accelerate, cruise, decelerate, idle) and the presence or absence of a smog control device. Obviously it is imposssible to calculate precisely the magnitude of motor vehicle emissions in an airshed. Certain averages will have to be used, for example, average emissions per mile or gallon as a function of vehicle age, make and size, averaged over a typical driving cycle.

The estimation of fixed source emissions is somewhat easier than mobile emissions. Emission magnitudes from fixed sources will depend on the type of fuel burned, the type of combustion unit, stack height and evaporative losses.

Having an airshed model we can now discuss the solution of Problem 2. We write (10) in the general vector form

$$\frac{dz(t)}{dt} = f[z(t), t, b] \qquad (13)$$

$$z(0) = z^0 \qquad (14)$$

The air quality constraint (7) can be written in the vector form

$$g(z) \le g^* \qquad (15)$$

The result of Problem 1 is $J(b)$. Then Problem 2 can be stated as: find b to minimize $J(b)$ subject to (13)–(15). Since we have already determined the range of b of interest, all that is necessary is to integrate (13) numerically for the b values and compute $g(z)$. Then when we select g^* we can refer to these results to find the set of b yielding $g(z) = g^*$.

A HYPOTHETICAL EXAMPLE

To illustrate the concepts we have developed, the determination of the relationship between air quality and minimum control cost for a hypothetical airshed for 1 yr is now presented. Even though the example is not based on an actual airshed, we have included all the elements that would appear in the application to a real urban area. In fact, an analysis of the Los Angeles basin, following along the same lines, is currently in progress. For simplicity, we confine our attention to one primary pollutant ($N = M = 1$) which does not participate in any chemical reactions.

We assume that there are two types of major sources ($p = 2$) with the following data for each:

$s_1 = 8.0 \times 10^6$ source units/day;
$e^0{}_{11} = 2640$ g pollutant/source unit;
$C_1{}^* = \$0.50$/source unit;

$s_2 = 8.7 \times 10^6$ source units/day;
$e^0{}_{21} = 3000$ g pollutant/source unit;
$C_2{}^* = \$0.40$/source unit.

We assume that there are two control techniques available for each source ($q_1 = q_2 = 2$):

$r_{111} = 800$ $c_{11} = \$0.30$/unit of d_{11};
$r_{121} = 650$ $c_{12} = \$0.25$/unit of d_{12};
$r_{211} = 1350$ $c_{21} = \$0.24$/unit of d_{21};
$r_{221} = 1500$ $c_{22} = \$0.20$/unit of d_{22}.

For source 1, there is a maximum of one unit of d_{11} or d_{12} that may be applied per source unit. Thus, $d_{11}{}^* = d_{12}{}^* = 1$, and

$$d_{11} \leq 1, \, d_{12} \leq 1$$

Such a constraint is typical if the control methods 1 and 2 represent installation of devices, such as precipitators or stack gas cleaning equipment. Then, the choice is to install control method 1 or 2 or both or nothing, but we would not use more than 1 unit of control method 1 or 2 per source unit.

For source 2, we assume that the total control effort should not exceed $d_2{}^* = 1$,

$$d_{21} + d_{22} \leq 1$$

This type of constraint is typical if d_{21} and d_{22} represent, for example, thousands of barrels of high sulfur fuel oil replaced by low sulfur fuel oil and natural gas, respectively. The $d_2{}^* = 1$ would indicate that the total equivalent fuel oil consumption should never exceed 1000 barrels/day, the maximum consumption. Thus, if $d_{21} + d_{22} = 1$, it would mean we are replacing all the high sulfur fuel oil with low sulfur fuel oil and natural gas.

The airshed is represented by four well-mixed cells with intercell flows as shown in Fig. 5. The arrows represent the prevaling flow directions during a typical day, assumed to be constant. We let m_{1k} and m_{2k} be the fractions of the total emissions of pollutant in cell k from sources 1 and 2, respectively. Table 1 presents the size and source distribution information for the cells, and Table 2 presents the volumetric flow rates of air from cell j to cell k. The inversion heights, or upper mixing depths, have been taken as constant and equal to 300 m. The volumetric flow rates in Table 2 correspond to the arrows in Fig. 4 and no entry in Table 2 indicates no flow between those two cells. In order to preserve continuity of mass, the total volumetric flow of air into the airshed equals the total volumetric flow out of the airshed, i.e. $q_{01} + q_{02} + q_{04} = q_{20} + q_{30} + q_{40}$.

We let $\beta_{1\tau}$ and $\beta_{2\tau}$ represent the fraction of the 24-hr source emissions from sources 1 and 2, respectively, that occur in hour τ. We let $\tau = 1$ correspond to the period 6–7 a.m.,

$\tau = 2$ to 7–8 a.m., etc., and consider only the period 6 a.m. to 6 p.m. Table 3 presents the assumed hourly source activity distribution for the two sources for this time period. The source distribution is taken to be the same in each of the cells.

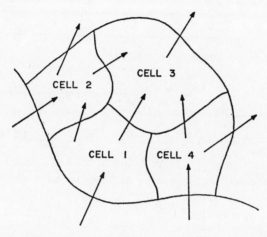

FIG. 5. Hypothetical airshed divided into four cells with intercell flows.

TABLE. 1. SIZE AND SOURCE STRENGTH DISTRIBUTION FOR CELLS

Cell No. k	Cell Area A_k (m² × 10^{-10})	m_{1k}	m_{2k}	Inversion Height h_k (m)
1	2.6	0.25	0.40	300
2	1.87	0.21	0.10	300
3	3.74	0.18	0.35	300
4	2.25	0.36	0.15	300

TABLE 2. VOLUMETRIC FLOWRATES OF AIR FROM CELL j TO CELL k q_{jk} × 10^{-10} M³/HR

j k	0	1	2	3	4
0		3.32	1.53		5.1
1			1.32	1.5	0.5
2	1.1			1.75	
3	5.55				
4	3.3			2.3	

TABLE 3. HOURLY SOURCE ACTIVITY DISTRIBUTION $\beta j\tau$

τ	1	2	3	4	5	6	7	8	9	10	11	12
Hour	6–7 a.m.	7–8	8–9	9–10	10–11	11–12	12–1 p.m.	1–2	2–3	3–4	4–5	5–6
$\beta_1\tau$	0.02	0.09	0.09	0.05	0.05	0.05	0.05	0.05	0.05	0.09	0.09	0.09
$\beta_2\tau$	0.04	0.04	0.05	0.05	0.05	0.05	0.05	0.05	0.05	0.07	0.07	0.07

The airshed is described by the simplified form of (10),

$$v_k\frac{dz_k}{dt} = \sum_{j=0}^{4} q_{jk}z_j - z_k \sum_{j=0}^{4} q_{kj} + S_k(t) \tag{16}$$

$$z_k(0) = z_k^0 \qquad\qquad k = 1,2,3,4 \tag{17}$$

where $z_k(t)$ is the concentration in ppm of pollutant in cell k at time t, t having units of hr. The total emissions of the pollutant into cell k in ppm/hr., S_k, is computed from

$$S_k = 800 \left[\frac{\beta_{1\tau}m_{1k}s_1e_{11}}{v_k} + \frac{\beta_{2\tau}m_{2k}s_2e_{21}}{v_k} \right] \tag{18}$$

$$k = 1,2,3,4$$
$$\tau = 1,2,......,12$$

where 800 is the assumed conversion factor from grams of pollutant to volume in cubic meters (computed based on the pollutant being carbon monoxide).

The air quality criterion $g(z)$ is taken to be the average hourly concentration in the airshed,

$$g(z) = \frac{1}{48}\sum_{k=1}^{4} \int_0^{12} z_k(t)dt \tag{19}$$

We first solve the linear programming problem that relates the minimum cost of control to the total mass emissions and determines the optimal control allocations corresponding to each minimum. This is what we have termed *Problem* 1. If x is the required reduction in total grams of pollutant emitted, the linear programming problem is the following:

$$\text{Min } J = 2.4 \times 10^6 d_{11} + 2 \times 10^6 d_{12} + 2.088 \times 10^6 d_{21} + 1.74 \times 10^6 d_{22} \tag{20}$$

subject to

$$1.23d_{11} + d_{12} + 2.26d_{21} + 2.28d_{22} \geq \frac{x}{5.2 \times 10^9} \tag{21}$$

$$d_{11} \leq 1 \tag{22}$$

$$d_{12} \leq 1 \tag{23}$$

$$d_{21} + d_{22} \leq 1 \tag{24}$$

$$1.2\, d_{11} + d_{12} \leq 2 \tag{25}$$

$$1.2\, d_{21} + d_{22} \leq 2 \tag{26}$$

$$d_{11}, d_{12}, d_{21}, d_{22} \geq 0 \tag{27}$$

Conditions (22)–(24) have already been explained. Constraints (25) and (26) are the particular forms of the general cost constraint (3).

The result of the linear programming problem is shown in Fig. 6 as the minimum control as a function of the total daily emissions of pollutant. If no control action is taken $J = 0$ and $b = 4.5 \times 10^{10}$g/day. Since only a discrete number of control possibilities exist, the curve of J vs. b is not smooth but consists of several straight line segments. With the given constraints, the minimum attainable emissions is about 2.16×10^{10}g/day at a cost of 6.13×10^6\$/day.

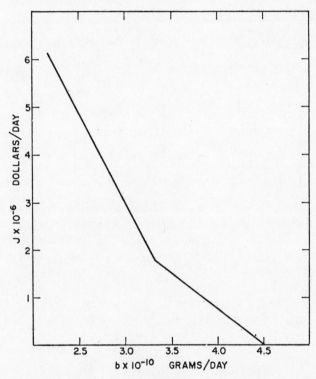

FIG. 6. Total minimum control cost (\$/day) as a function of the total mass emissions of pollutant (g/day).

TABLE 4. OPTIMAL CONTROLS d_{ij} FOR SELECTED VALUES OF b

$b \times 10^{-10}$	2.1655	2.55	3.3	3.85	4.512
d_{11}	1	1	0.05	0	0
d_{12}	1	0.25	0	0	0
d_{21}	0	0	0	0	0
d_{22}	1	1	1	0.56	0
e_{11}	1190	1677.5	2600	2640	2640
e_{21}	1500	1500	1500	2140	3000
$J \times 10^{-6}$	6.13	4.66	1.84	0.97	0

We now proceed to the second problem, the determination of the relationship between total mass emissions and air quality. The airshed model (16) can be written in the form

$$\frac{d\mathbf{z}}{dt} = \mathbf{Az} + \mathbf{S}(\tau) \qquad \begin{array}{l} 0 \le t \le 12 \\ \tau = 1, 2, \ldots, 12 \end{array} \qquad (28)$$

where $\tau = 1$ when $0 \le t \le 1$, $\tau = 2$ when $1 \le t \le 2$, etc., and from the data given previously,

$$\mathbf{A} = \begin{bmatrix} -0.4522 \times 10^{-2} & 0 & 0 & 0 \\ 0.235 \times 10^{-2} & -0.5075 \times 10^{-2} & 0 & 0 \\ 0.1335 \times 10^{-2} & 0.1558 \times 10^{-2} & -0.494 \times 10^{-2} & 0.205 \times 10^{-2} \\ 0.0741 \times 10^{-2} & 0 & 0 & -0.829 \times 10^{-2} \end{bmatrix} \quad (29)$$

$$\mathbf{S}(\tau) = \begin{bmatrix} 2.052 \times 10^{-4}\beta_{1\tau}e_{11} + 3.571 \times 0^{-4}\beta_{2\tau}e_{21} \\ 2.395 \times 10^{-4}\beta_{1\tau}e_{11} + 1.241 \times 10^{-4}\beta_{2\tau}e_{21} \\ 1.028 \times 10^{-4}\beta_{1\tau}e_{11} + 2.172 \times 10^{-4}\beta_{2\tau}e_{21} \\ 3.42 \times 10^{-4}\beta_{1\tau}e_{11} + 1.548 \times 10^{-4}\beta_{2\tau}e_{21} \end{bmatrix} \quad \tau = 1, 2, \ldots, 12 \qquad (30)$$

where we have assumed $z_0 = 0$, i.e. the air entering the airshed is free of pollutant. For a specified value of b, e_{11} and e_{21} are found from problem 1. Since $\mathbf{S}(t)$ is piecewise constant, the solution of (28) is straightforward, although tedious. We will not present the analytical solutions here because they are rather lengthy. The evaluation procedure, however, can be summarized:

1. Select a value for b, corresponding to which will be a J (b) from Problem 1.
2. Determine the emission levels e_1 and e_2 corresponding to b from the solution to Problem 1 from

$$e_{i1} = e_{i1}^0 - \sum_{j=1}^{2} r_{ij1}d_{ij} \qquad i = 1, 2$$

3. Compute $\mathbf{S}(\tau)$ from (30) and Table 3.
4. Evaluate $\mathbf{z}(t)$ and $\mathbf{g}(\mathbf{z})$ from the solution of (28) and (19).
5. Repeat 1–4 for various values of b.

Figure 7 shows $g^*(\mathbf{z})$ given by (19) as a function of b. As b increases $g^*(\mathbf{z})$, the average hourly concentration over the airshed, increases. With no control and $b = 4.5 \times 10^{10}$ g/day, the average hourly concentration is 10.4 ppm. With full control, $b = 2.16 \times 10^{10}$g/day, and $g = 5$ ppm. Figure 8 shows the total daily minimum control cost as a function of air quality $g^*(\mathbf{z})$, our ultimate desired result. Thus, for example, to achieve an average hourly concentration of 6 ppm for a day with the meteorological conditions of the example for our hypothetical airshed it would cost a minimum of 4.5×10^6 \$/day for control.

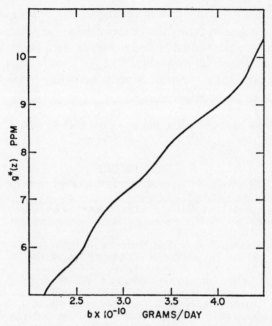

FIG. 7. Maximum hourly average pollutant concentration (ppm) as a function of the total mass emissions of pollutant (g/day).

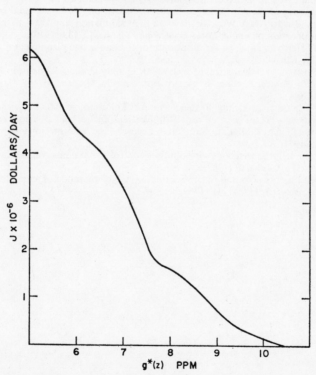

FIG. 8. Total minimum control cost ($/day) as a function of the maximum hourly average pollutant concentration (ppm).

SUMMARY

We have presented a general theoretical framework for the systematic determination of air pollution control strategies using a dynamic airshed model. The result of the study is the minimum cost of various levels of air quality. The decision of what levels of air quality to accept or legislate in anticipation of must be made separately on the basis of the damages incurred by various levels of air pollution.

Acknowledgement—This work was supported in part by a grant from the John A. McCarthy Foundation.

REFERENCES

1. EMANUEL S. SAVAS, Computers in urban air pollution control systems, *Socio-Econ. Plan. Sci.* **1**, 157–183 (1967).
2. EDWARD J. CROKE and SAMUEL G. BOORAS, The Design of an Air Pollution Incident Control Plan, APCA Paper No. 69–99, presented at the Annual Meeting, Air Pollution Control Association, New York (1969).
3. ROBERT KOHN, A Linear Programming Model for Air Pollution Control in the St. Louis Airshed, Ph.D. Thesis, Department of Economics, Washington University, St. Louis (1969).
4. ROBERT KOHN, A linear programming model for air pollution control: A pilot study of the St. Louis Airshed, *J. Air Pollut. Control Ass.* **20**, 78–82 (1970).
5. ROBERT KOHN, Abatement Strategy and Air Quality Standards, in *Development of Air Quality Standards*, edited by A. ATKISSON and R. GAINES, pp. 103–123. Charles E. Merrill (1970).
6. JACK R. FARMER, PHILLIP J. BIERBAUM and JOSEPH A. TIKVART, Proceeding from Air Quality Standards to Emission Standards, APCA Paper No. 70–85, presented at the Annual Meeting of the Air Pollution Control Association, St. Louis, June 14–19 (1970).
7. ELLISON S. BURTON and WILLIAM SANJOUR, A simulation approach to air pollution abatement program planning, *Socio-Econ. Plan. Sci.* **4**, 147–159 (1970).
8. HENRY W. HERZOG, JR., The air diffusion model as an urban planning tool, *Socio-Econ. Plan. Sci.* **3**, 329–349 (1969).
9. JOHN H. SEINFELD, Mathematical Models of Air Quality Control Regions, in *Development of Air Quality Standards*, edited by A. ATKINSON and R. GAINES, pp. 169–196. Charles E. Merrill (1970).
10. JOHN H. SEINFELD, Modeling Problems in Air Pollution, in *Mathematical Models of Public Systems—Proceedings of the 1970 National Invitational Seminar on Advanced Simulation*, edited by G. BEKEY. Simulation Councils, Inc. and the University of Southern California (1971).
11. E. A. ULBRICH, Adapredictive air pollution control for the Los Angeles Basin, *Socio-Econ. Plan. Sci.* **1** 423–440 (1968).
12. SHELDON K. FREIDLANDER and JOHN H. SEINFELD, A dynamic model of photochemical smog, *Env. Sci. Technol.* **3**, 1175–1181 (1969).

IV

Water Management

Editors' Comments on Papers 14 Through 17

14 **Leopold:** Hydrology for Urban Land Planning—A Guidebook on the Hydrologic Effects of Urban Land Use
U.S. Geol. Surv. Circ. 554, 1–18 (1968)

15 **Wolman:** The Nation's Rivers
Science, **174**, 905–918 (Nov. 26, 1971)

16 **Smith and Morris:** Systems Analysis for Optimal Water Quality Management
J. Water Pollution Control Federation, **41**, No. 9, 1635–1646 (Sept. 1969)

17 **Burke:** A Water Pricing and Production Model
IEEE Trans. Sys., Sci. Cybern., **SSC-6**, 272–275 (Oct. 1970)

Water management is an environmental area that received attention long before the blossoming of environmental awareness in 1970. However, the main concerns in water management have broadened in recent years from water impoundment (dam building) and sewage treatment to include aesthetics, the effect on biological organisms (both the bountiful and the scarce species), and ground subsidence due to removal of underground water. The four papers in Part IV were chosen to highlight different aspects of today's spectrum of water management interests. The first paper selected gives a geologist's view of the hydrologic effects of urbanization. In the second paper a geologist/geographer discusses water pollution and the management of river resources. The third and fourth papers were written by practitioners who represent the newer fields of systems analysis, operations research, and management science (all largely synonomous terms in practice), which grew out of engineering, mathematics, and statistics. These last two papers in Part IV describe, respectively, the use of systems analysis to optimize the management of water quality and the employment of a decision model in setting prices for water from two different sources.

The purpose of Paper 14, "Hydrology for Urban Land Planning—A Guidebook on the Hydrologic Effects of Urban Land Use," is to help the land-use planner incorporate information concerning the environmental impacts of urban development into the planning process. In particular, the problems considered are the hydrologic effects of urbanization on the Brandywine Creek basin in Pennsylvania.

The author, Luna Leopold, is a geologist with the U.S. Geological Survey. He is widely known for his studies of fluvial (river) geology. Although Leopold does not explicitly state the steps required for planners to incorporate environmental impact information, the steps may be summarized from his discussion, as follows:

1. Identify the important environmental characteristics of concern to the planner. The hydrologic factors are peak and total runoff, water quality, and the appearance of waterways.

2. Specify the means for measuring these factors. Runoff, for example, is measured by number and by characteristics of rise in streamflow.

3. Relate the measure to land use. Runoff is related to the percentage of area made impervious, which, in turn, is dependent upon the lot size. For an area in the Brandy-

wine Creek basin that is to be 50 percent sewered and 50 percent impervious, Leopold predicts a fourfold increase in the number of flows equaling or exceeding channel capacity in a given time period.

4. Consider corrective measures. Leopold suggests reservoir storage as a means for ameliorating the flooding problem.

The author indicates that the data required for determining environmental impacts are often sparse or inconclusive. It would have been helpful to have some additional discussion of how to proceed under these circumstances.

This paper is a very useful attempt to relate earth-science information to land-use planning. Too often the scientist and planner operate in different worlds, to the detriment of everyone.

An eminent geographer, M. Gordon Wolman, is the author of Paper 15, "The Nation's Rivers." This paper deals primarily with water pollution and the management of river resources. The selection of criteria for assessing the quality of water is discussed at the outset. Wolman emphasizes the need for data gathered over long periods (e.g., several decades) so that one is in a position to tell whether there are trends in the data concerning water quality parameters. However, Wolman notes the scarcity of good long-term hydrologic data and clearly describes various situations (e.g., changes in analysis techniques over the years) that can lead to incorrect inferences if the researcher is not careful in the selection and analysis of data.

The major variables generally used to characterize water and river quality are discussed. Interesting historical data are presented and interpreted for dissolved oxygen, dissolved solids, sediment, temperature, radioactivity, pesticides, and debris. Lack of understanding of "the precise causes of the changes in a parameter" is mentioned as a major difficulty in evaluating water management alternatives. Nonetheless, Wolman comments that some of the data cited illustrate "the possibilities of maintenance and, perhaps, enhancement of river quality." He concludes by stressing the need for systematic observation (data collection) programs that include more biological variables and other parameters that can be used to give a better idea of the probable effectiveness of expenditures intended to preserve or enhance river quality.

Ethan T. Smith, the first author of this article, is Chief of the Data Systems Branch of the U.S. Environmental Protection Agency in New York City. In his professional capacity he is concerned with both the economic and the managerial aspects of environmental protection. Paper 16, "Systems Analysis for Optimal Water Quality Management," considers various management alternatives for achieving specified water quality objectives. As one might predict, the system that is simplest to manage, where each pollutor is required to remove the same percentage of effluence, is less economically efficient than a system where overall cost is minimized without regard to equality of treatment. The problem then is to decide how to balance economic efficiency against complexity of management, and the authors suggest a compromise approach wherein the area is divided into zones and within each zone the pollutors receive uniform treatment.

The Delaware Estuary is the subject of the analysis. Several obstacles common to most other environmental systems studies are noted at the outset:

1. It is difficult to define goals for water quality.
2. It is difficult to model the environment.
3. It is difficult to combine a variety of goals into a single objective function.

Dissolved oxygen (DO) is used as the primary indicator for water quality, and rate equations are presented for the time rate of change of DO and BOD (biochemical oxygen demand). This model was tested in the steady state for the actual system, and a satisfactory verification was obtained. Linear programming and search techniques are used to minimize cost in the presence of constraints on DO.

Benefits (e.g., to recreation and fisheries) are mentioned, but additional discussion would have been helpful. It is noted that there are nonquantifiable benefits, but apparently these are ignored in the analysis. The problem of benefit definition is certainly one which requires further consideration, not only for water quality management but for most other environmental systems as well.

The notation for the equations on page 221 might be somewhat confusing. The left-hand side is written as a differential, whereas the right-hand side is a difference (see R. V. Thomann, *Systems Analysis and Water Quality Management,* listed in the Bibliography, for the development of such equations). Definitions are given for the parameters Q, ξ and E with a single subscript, whereas double subscripts should have been used to conform to usage in the equations. For example, the definition for Q should be

$$Q_{i-1,i} = \text{net flow from section } i-1 \text{ to section } i, \text{ l/day.}$$

In Paper 17, "A Water Pricing and Production Model," Emmett Burke considers a systems analysis approach to the determination of price and production levels for two possible sources of water in Santa Clara County, California. The water may be imported and treated at the county plant, or it may be obtained by pumping water from underground aquifers. An objective function is defined as the revenue to the county, minus operating costs, minus costs associated with land subsidence that results from the removal of underground water. Constraints are imposed which (1) define mass conservation for water percolation, (2) specify the capacity of the water treatment plant, (3) guarantee meeting demand, and (4) limit cost to the consumer to an amount no greater than current cost.

The decision variables (i.e., those variables that may be controlled) are the prices and quantities of each water source. Burke finds that minimizing ground water pumping (by pumping only enough to make up the difference between demand and the capacity of the water treatment plant) can save the county an average of two thirds of a million dollars per year. Random variation in two hydrologic variables, runoff and infiltration, vary the savings to the county in any particular year. The effect of this natural variation is indicated in Figure 1 by the vertical range between the dashed lines.

This paper has been included because it illustrates a rather common type of systems analysis problem. Specifically, the situation often arises where there are various methods for meeting demands, and the problem is to determine the best combination of use of these methods. The approach is to define and optimize an objective function in the presence of relevant constraints.

In three instances the typesetting of equations may be misleading. In Equations (1) and (7) and in the unnumbered equation preceding Equation (7), the variables over which the expression between braces ($\{\ \}$) is to be maximized are listed at the end of the equation after the closing brace instead of under the word "max." For example, in Equation (1) the variables Q_g, Q_t, P_g, P_t listed at the end of the line are the ones over which the expression

$$\{P_g Q_g + P_t Q_t - [C_p(Q_p - Q_v) + C_o Q_t + C_s (L + Q_g - Q_p)]\}$$

is to be maximized.

14

Reprinted from *U.S. Geol. Surv. Circ. 554*, 1–18 (1968)

HYDROLOGY FOR URBAN LAND PLANNING—A GUIDEBOOK ON THE HYDROLOGIC EFFECTS OF URBAN LAND USE

By Luna B. Leopold

This circular attempts to summarize existing knowledge of the effects of urbanization on hydrologic factors. It also attempts to express this knowledge in terms that the planner can use to test alternatives during the planning process. Because the available data used in this report are applied to a portion of the Brandywine Creek basin in Pennsylvania, this can be considered as a report on the basic hydrologic conditions of the Brandywine Creek basin prior to the beginning of major urbanization. Because the available data are not yet adequate, this report can be considered as a compilation of tentative suggestions in the form of an explanatory, not a definitive, handbook.

The application of current knowledge of the hydrologic effects of urbanization to the Brandywine should be viewed as a forecast of conditions which may be expected as urbanization proceeds. By making such forecasts in advance of actual urban development, the methods can be tested, data can be extended, and procedures improved as verification becomes possible.

PLANNING PROCEDURES AND HYDROLOGIC VARIABLES

A planning document presented to a community for adoption must always be more suggestive than coercive. This is true not only because the planner is unable to foresee the innumerable complications of actual development, but also because there are many detailed alternatives which would accomplish generally similar results. The planner is particularly concerned with both the constraints and the opportunities offered by the principal physiographic characteristics of the area, especially the location of hillslopes, soils, and streams. The existing pattern of land use and the accompanying distribution of woods and agriculture are parameters which over a period of years may actually change, albeit slowly. Roads, villages, industries, and other manmade features are more or less permanent and exert their greatest influence in their effect on further development, especially through land values.

Of particular concern to the planner are those alternatives that affect the hydrologic functioning of the basins. To be interpreted hydrologically, the details of the land-use pattern must be expressed in terms of hydrologic parameters which are affected by land use. These parameters in turn become hydrologic variables by which the effects of alternative planning patterns can be evaluated in hydrologic terms.

There are four interrelated but separable effects of land-use changes on the hydrology of an area: changes in peak flow characteristics, changes in total runoff, changes in quality of water, and changes in the hydrologic amenities. The hydrologic amenities are what might be called the appearance or the impression which the river, its channel and its valleys, leaves with the observer. Of all land-use changes affecting the hydrology of an area, urbanization is by far the most forceful.

Runoff, which spans the entire regimen of flow, can be measured by number and by characteristics of rise in streamflow. The many

1

rises in flow, along with concomitant sediment loads, control the stability of the stream channel. The two principal factors governing flow regimen are the percentage of area made impervious and the rate at which water is transmitted across the land to stream channels. The former is governed by the type of land use; the latter is governed by the density, size, and characteristics of tributary channels and thus by the provision of storm sewerage. Stream channels form in response to the regimen of flow of the stream. Changes in the regimen of flow, whether through land use or other changes, cause adjustments in the stream channels to accommodate the flows.

The volume of runoff is governed primarily by infiltration characteristics and is related to land slope and soil type as well as to the type of vegetative cover. It is thus directly related to the percentage of the area covered by roofs, streets, and other impervious surfaces at times of hydrograph rise during storms.

A summary of some data on the percentage of land rendered impervious by different degrees of urbanization is presented by Lull and Sopper (1966). Antoine (1964) presents the following data on the percentage of impervious surface area in residential properties:

Lot size of residential area (sq ft)	Impervious surface area (percent)
6,000	80
6,000–15,000	40
15,000	25

The percentage decreases markedly as size of lot increases. Felton and Lull (1963) estimate in the Philadelphia area that 32 percent of the surface area is impervious on lots averaging 0.2 acre in size, whereas only 8 percent of the surface area is impervious on lots averaging 1.8 acres.

As volume of runoff from a storm increases, the size of flood peak also increases. Runoff volume also affects low flows because in any series of storms the larger the percentage of direct runoff, the smaller the amount of water available for soil moisture replenishment and for ground-water storage. An increase in total runoff from a given series of storms as a result of imperviousness results in decreased ground-

water recharge and decreased low flows. Thus, increased imperviousness has the effect of increasing flood peaks during storm periods and decreasing low flows between storms.

The principal effect of land use on sediment comes from the exposure of the soil to storm runoff. This occurs mainly when bare ground is exposed during construction. It is well known that sediment production is sensitive to land slope. Sediment yield from urban areas tends to be larger than in unurbanized areas even if there are only small and widely scattered units of unprotected soil in the urban area. In aggregate, these scattered bare areas are sufficient to yield considerable sediment.

A major effect of urbanization is the introduction of effluent from sewage disposal plants, and often the introduction of raw sewage, into channels. Raw sewage obviously degrades water quality, but even treated effluent contains dissolved minerals not extracted by sewage treatment. These minerals act as nutrients and promote algae and plankton growth in a stream. This growth in turn alters the balance in the stream biota.

Land use in all forms affects water quality. Agricultural use results in an increase of nutrients in stream water both from the excretion products of farm animals and from commercial fertilizers. A change from agricultural use to residential use, as in urbanization, tends to reduce these types of nutrients, but this tendency is counteracted by the widely scattered pollutants of the city, such as oil and gasoline products, which are carried through the storm sewers to the streams. The net result is generally an adverse effect on water quality. This effect can be measured by the balance and variety of organic life in the stream, by the quantities of dissolved material, and by the bacterial level. Unfortunately data describing quality factors in streams from urban versus unurbanized areas are particularly lacking.

Finally, the amenity value of the hydrologic environment is especially affected by three factors. The first factor is the stability of the stream channel itself. A channel, which is gradually enlarged owing to increased floods caused by urbanization, tends to have unstable and unvegetated banks, scoured or muddy

2

channel beds, and unusual debris accumulations. These all tend to decrease the amenity value of a stream.

The second factor is the accumulation of artifacts of civilization in the channel and on the flood plain: beer cans, oil drums, bits of lumber, concrete, wire—the whole gamut of rubbish of an urban area. Though this may not importantly affect the hydrologic function of the channel, it becomes a detriment of what is here called the hydrologic amenity.

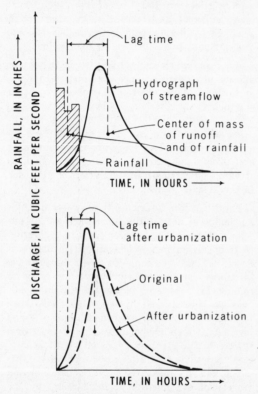

Figure 1.—Hypothetical unit hydrographs relating runoff to rainfall, with definitions of significant parameters.

The third factor is the change brought on by the disruption of balance in the stream biota. The addition of nutrients promotes the growth of plankton and algae. A clear stream, then, may change to one in which rocks are covered with slime; turbidity usually increases, and odors may develop. As a result of increased turbidity and reduced oxygen content desirable game fish give way to less desirable species. Although lack of quantitative objective data on the balance of stream biota is often a handicap to any meaningful and complete evaluation of the effects of urbanization, qualitative observations tend to confirm these conclusions.

AVAILABILITY OF DATA AND THE TECHNIQUE OF ANALYSIS

Basic hydrologic data on both peak flow and volume of runoff may be expressed in terms of the characteristics of the unit hydrograph, that is, the average time distribution graph of flow from a unit or standard storm. The unit hydrograph shows the percentage of the total storm runoff occurring in each successive unit of time. The standard storm may be, for example, a typical storm which produced 1 inch of runoff (fig. 1). Such data are derived from the study of individual storms and the associated runoff graphs measured at gaging stations.

One factor stating the relation between the storm and the runoff is lag time. This is defined as the time interval between the center of mass of the storm precipitation and the center of mass of the resultant hydrograph. Lag time is a function of two basin parameters—the mean basin slope and the basin length. These factors empirically correlate with lag time if expressed in the form of the basin ratio (basin length L divided by the square root of the mean basin gradient, s). This basin ratio is also related to drainage area. As drainage area increases, the basin length increases and the average value of slope generally decreases. Thus, natural basin characteristics can be translated into flood-flow characteristics.

Lag time may be materially altered by the effects of urbanization on the basin surface. Water runs off faster from streets and roofs than from natural vegetated areas. This tends to decrease the lag time. The construction of artificial channels, especially storm sewers, also decreases lag time. As the time required for a given amount of water to run off shortens, the peak rate of runoff (flood peak) increases.

In addition to the basin ratio and lag time, the regimen of a stream, however, can be described in many other ways, including flood

3

frequency, flow duration, mean annual flood, discharge at bankfull stage, and frequency of bankfull stage. This is evidenced in past studies of the effects of urbanization on the hydrology of an area. Many different techniques of relating rainfall to runoff have been used, along with various parameters to measure the degree of urbanization. In order to evaluate our present knowledge, it is necessary to express the results of these studies in some common denominator.

Most reports on hydrologic effects of urbanization present the conclusions in a form which is more useful to the hydrologist than to the urban planner. This circular will attempt to interpret the hydrologic conclusions of these studies in terms that are meaningful to the planner. Two forms of presentation will be used.

The first is a slight modification of a method previously used by several investigators, especially D. G. Anderson (1968) and L. D. James (1965). The percentage of an area sewered is plotted against the percentage of the area rendered impervious by urbanization; isopleth lines (lines of equal value of the ratio) on the graph show the ratio of peak discharge under urbanized conditions to the peak discharge under rural or unurbanized conditions. Such a graph will be different for different drainage area sizes and for different flow frequencies.

The second method utilizes a relationship between the degree of urbanization and the frequency at which the original channel capacity would be exceeded.

Table 1 is an interpretation and summary of the effects of urbanization on peak discharges based on previous studies. Results of the studies were interpreted and extrapolated to a common denominator of 1 sq mi (square mile), a practical unit of size for planning.

Carter (1961) developed a technique that followed the reasoning previously used by Snyder (1938) and that showed lag time as a function of basin characteristics. For 20 streams in the vicinity of Washington, D.C., Carter developed this relation for natural

basins, for partly sewered, and for completely sewered basins. As in most studies the difficulty comes in translating these descriptive terms to quantitative measures of urbanization. From data presented by Carter, values were read for a basin ratio of 0.12 representing a 1-sq-mi area having an estimated length of 1.2 miles and an average slope of 100 feet per mile. It was further assumed that in Carter's study, "partly sewered" is equivalent to 50 percent sewered and 20 percent impervious. These conditions provide some of the data shown in table 1.

Table 1.—*Increase in discharge as a result of urbanization in a 1-square-mile area*

[Discharge is mean annual flood; recurrence interval is 2.3 years. Data are expressed as ratio of discharge after urbanization to discharge under previous conditions. Data from James (1965) have no superscript]

Percentage of area served by storm sewerage	Percentage of area made impervious			
	0	20	50	80
0	1.0	[1]1.2	[1]1.8	[1]2.2
		[2]1.3	[2]1.7	[2]2.2
		1.3	1.6	2.0
20	1.1	[3]1.9	1.8	2.2
		1.4	—	—
50	1.3	[4]2.1	[1]3.2	[1]4.7
		[1]2.8	2.0	2.5
		[5]3.7	—	—
		[6]2.0	2.5	[3]4.2
		1.6	—	—
80	1.6	1.9	—	3.2
100	1.7	[1]3.6	[1]4.7	[4][5]5.6
		2.0	2.8	[1]6.0
		—	—	3.6

[1] Anderson (1968).
[2] Martens (1966).
[3] Wilson (1966).
[4] Carter (1961).
[5] Wiitala (1961).
[6] Espey, Morgan, and Masch (1966).

As an indication of the change in impervious area resulting from urbanization, Harris and Rantz (1964) showed that an area near Palo Alto, Calif., changed from 5.7 percent to 19.1 percent impervious in a 10-year period.

One of the most complete analyses of urbanization effects was made by D. G. Anderson (1968) in his study of the urbanization in Fairfax County, Va., near the metropolitan complex of the Nation's Capital. Anderson's analysis follows the procedure suggested earlier by Carter, but Anderson included a larger array of data from 64 gaging stations. Anderson closely confirmed the conclusions of Carter, but he carried the analysis further in a plot of

4

the ratio of peak discharge to the mean annual flood for different percentages of basin imperviousness and for flood flows exceeding the mean annual flood. For table 1, data from Anderson's study were read directly from his graph at the 2.33-year recurrence interval and expressed two separate conditions of sewerage. The first condition was expressed as "main channels natural, upstream drainage sewered"; this was assumed to be 50 percent sewered. The second condition was expressed as "completely sewered drainage basin"; this was assumed to be 100 percent sewered.

Wiitala (1961) presented data on urbanized versus rural conditions for a medium-sized watershed in Michigan. His data were translated into a ratio of peak discharges and it was assumed from his report that the urbanized condition represented 20 percent impervious area and 50 percent sewered area.

Martens (1966) reported on three small drainage basins in and near Charlotte, N.C. Using flood-frequency curves from long-term records at gaging stations in the State, he constructed a graph similar to that of Anderson; that is, ratio to mean annual flood for various degrees of basin imperviousness. As before, the difficulty lies in ascertaining the relation of Martens' urbanized condition to the degree sewered. In reading from Martens' graph for recurrence interval 2.33 years, it is assumed that the conditions he discussed include no sewerage and represent changes in impervious area only.

Wilson (1966) presented data on flood frequency for four drainage basins of 1.1 to 11.2 sq mi near Jackson, Miss. He presented his analysis in the form of discharge of mean annual flood plotted against drainage area size, and he interpolated lines to represent the percentage of the basin having storm sewers and improved channels. It is assumed that his description "20 percent of basin with storm sewers and improved channels" would be equivalent to 20 percent impervious and 20 percent sewered. Similarly, his value of 80 percent was assumed to be 80 percent sewered and 80 percent impervious.

Espey, Morgan, and Masch (1966) analyzed runoff data from urban and rural areas in Texas. To utilize this study, data were used corresponding to a basin length of 5,500 feet and a slope of 0.02. It was also assumed from his description of the area that "urban" could be expressed as 50 percent sewered and 20 percent impervious.

James (1965) analyzed runoff data from a 44-sq-mi basin south of Sacramento, Calif., within which 12 sq mi had been urbanized. From the basic data on flow, he obtained empirical coefficients used to route a series of synthetic flows by using a mathematical model expressed as a digital computer program. The results were plotted in a series of curves which separated the effects of flood frequency, drainage area, and degree of urbanization. Though the derived curves do not present field data, they also were incorporated into table 1.

Thus in table 1 are compiled, with certain necessary assumptions, the data for seven published and unpublished references which report measurements of the effect of urbanization on peak flow. Although interpretations were necessary to express the degree of urbanization in quantitative terms, there is considerable agreement among the data.

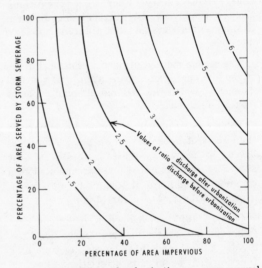

Figure 2.—Effect of urbanization on mean annual flood for a 1-square-mile drainage area. (Based on data from table 1.)

5

Data from table 1 have been transposed into the graph shown in figure 2. The ratios of peak discharge of urbanized to rural areas are presented for different percentages of sewerage and impervious area; lines of equal values of the ratio are drawn through the data. Briefly, these data show that for unsewered areas the differences between 0 and 100 percent impervious will increase peak discharge on the average 2.5 times. For areas that are 100 percent sewered, peak discharge for 0 percent impervious will be about 1.7 times the mean annual flood and the ratio increases to about eight for 100 percent impervious areas. Figure 2, then, reduces the basic data to the same units applicable to a 1-sq-mi drainage basin and to the mean annual flood.

A basin produces big flows from large and intense storms and smaller flows from less intense but more frequent storms. The great or catastrophic event is rare, and the storm of ordinary magnitude is frequent. These events can be arranged in order of magnitude and counted. For example, all discharge events exceeding 400 cfs (cubic feet per second) can be tabulated from the record at a stream-gaging station and arranged in order of magnitude; the values in the array can be plotted as a discharge-frequency curve. This has been done for the gaging station on West Branch Brandywine Creek at Coatesville, Pa., for 9 years of record (fig. 3). The theory and practice of constructing such flow-frequency curves is well known. The plotting position or frequency often used is defined as

$$R = \frac{n+1}{m}$$

where R is the recurrence interval in years, n is number of years of record, and m is the rank of the individual event in the array.

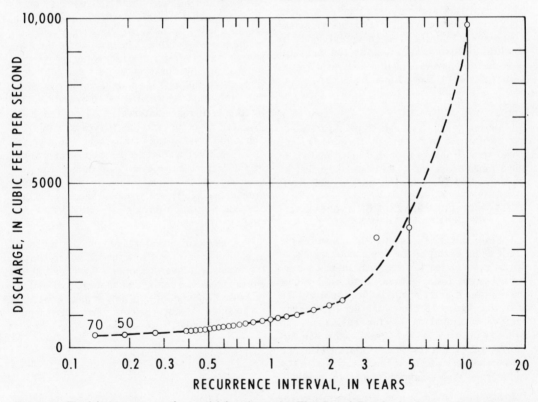

Figure 3.—Flood-frequency curve for partial-duration series, West Branch Brandywine Creek at Coatesville, Pa., based on data for 1942, 1944–51.

6

Note in figure 3 that the largest flow in the 9-year record was nearly 10,000 cfs. The number 50 printed on the graph means that there were 50 flows equal to or exceeding 500 cfs. Once a year, on the average, a discharge value of about 900 cfs will be equalled or exceeded.

A slightly different result would be obtained if, instead of using the peak flow for each storm, only the largest flow in each year were included in the array. The principle involved is similar. The arithmetic mean of the peak flows for the nine annual events is the "average annual flood." The statistics of this array are such that the recurrence interval of this average annual flood is the same regardless of the length of record, which specifically is 2.3 years. That is to say, a flood of that magnitude can be expected to be equaled or exceeded on an average of once in 2.3 years, or 10 times in 23 years.

Studies of river channels have shown that rivers construct and maintain channels which will carry without overflow a discharge somewhat smaller than the average annual flood. In fact the recurrence interval of the bankfull stage in most rivers is a flow having a recurrence interval of about 1.5 to 2 years.

Urbanization tends to increase the flood potential from a given basin. The channel then will receive flows which exceed its capacity not just once in 1.5 to 2 years on the average but more often. It is now proposed to estimate how much more often and to indicate the effect of this increased frequency on the channel itself.

EFFECT OF URBANIZATION ON INCREASING FREQUENCY OF OVERBANK FLOW

Taking the East Branch of Brandywine Creek as an example, the flow-frequency curve can be constructed for a typical subbasin having a 1-sq-mi drainage area. Figure 4A shows the relation of average annual flood to drainage area, and figure 4B shows the flood-frequency curve for annual peaks for basins in the Brandywine area. The diagrams shown in figure 4 are similar to those published in the nationwide series of flood reports, U.S. Geological Survey Water-Supply Papers 1671–1689.

From these curves a discharge-frequency relationship is developed for a drainage area of 1 sq mi. The average annual flood is read from the upper graph of figure 4 as 75 cfs, and the lower graph is used to construct the frequency curve in figure 5 pertaining to a 1-sq-mi basin marked "unurbanized."

The arithmetic for the construction of the curve is as follows:

Recurrence interval of annual flood [1] (years)	Ratio to mean annual flood [2]	Discharge [3] (cfs)	Recurrence interval duration series [4] (years)
1.1	0.55	41	0.4
1.5	.75	56	.92
2.0	.90	68	1.45
2.3	1.0	75	1.78
5	1.45	110	4.5
10	1.9	145	9.5

[1] Only the highest flood each year.
[2] From figure 4B.
[3] Obtained by multiplying ratios by 75 cfs from figure 4A for a drainage area of 1 sq mi.
[4] All peaks during the year. The values in this column are mathematically related to those in the first.

The graph marked "unurbanized" in figure 5 is constructed on semilogarithmic paper from the data listed in the third and fourth columns of the preceding table. The ordinate is the discharge, and the lower abscissa is the recurrence interval in the duration series. An auxiliary scale gives the average number of floods in a 10-year period (calculated as 10 years divided by the recurrence interval). Thus, the flow expected to occur once in 10 years would be about 145 cfs and the fifth largest would be 75 cfs. The latter would also be the average value of the largest flows each year during the 10-year record and thus would be the "average annual flood." It would plot, therefore, at an abscissa position approximately at 2.3-year recurrence interval.

The effect of urbanization on the average annual flood is shown in figure 2, which shows the increase in average annual flood for different degrees of urbanization as measured by the increase in percentages of impervious area and area served by storm sewers. For convenience these are tabulated as follows:

Percentage of area sewered	Percentage of area impervious	Ratio to average annual flood
0	0	1
20	20	1.5
40	40	2.3
50	50	2.7
80	60	4.2
100	60	4.4

7

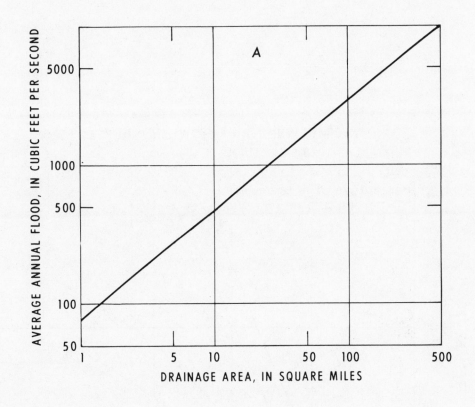

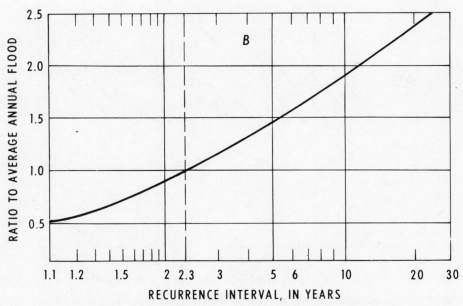

Figure 4.—Regional flood-frequency data for the Brandywine Creek basin, Pennsylvania. *A*, Relation of average annual flood to drainage area. *B*, Flood-frequency curve for annual peaks.

8

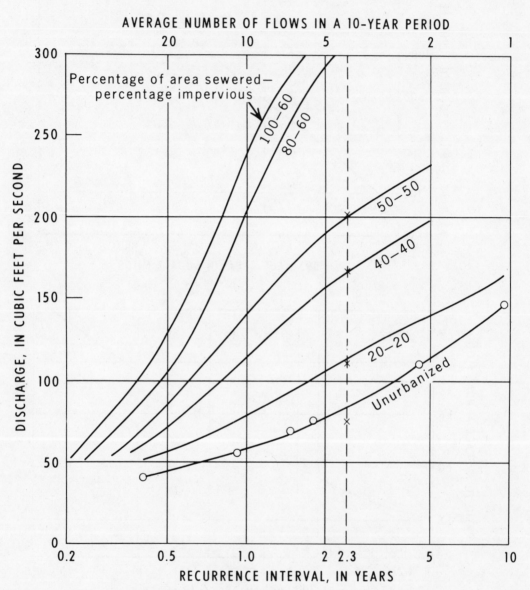

Figure 5.—Flood-frequency curves for a 1-square-mile basin in various states of urbanization. (Derived from figures 2 and 4.)

9

The average annual flood of 75 cfs was then multiplied by these ratios and plotted as shown in figure 5 at the 2.3-year interval. These values form the basis of a series of frequency curves for combinations of sewered area and impervious area. The shapes of the curves are guided by the principle that the most infrequent floods occur under conditions that are not appreciably affected by imperviousness of the basin.

The most frequent flows are therefore increased by smaller ratios than would be the average annual flood. Also, the most frequent flows are decreased in number because low flows from an urbanized area are not sustained by ground water as in a natural basin. The

curves representing urbanized conditions therefore converge at low flow values.

Obviously the frequency curves in figure 5 are extrapolations based on minimal data and require corroboration or revision as additional field data become available.

The flood-frequency curve under original (unurbanized) condition passes through a value of 67 cfs at a recurrence interval of 1.5 years. At bankfull condition natural channels generally can carry the flow having that recurrence interval. If one assumes that this flow approximates the capacity of the natural channels, the intersection of the estimated curves for different degrees of urbanization with the discharge value of 67 cfs can be used to estimate the in-

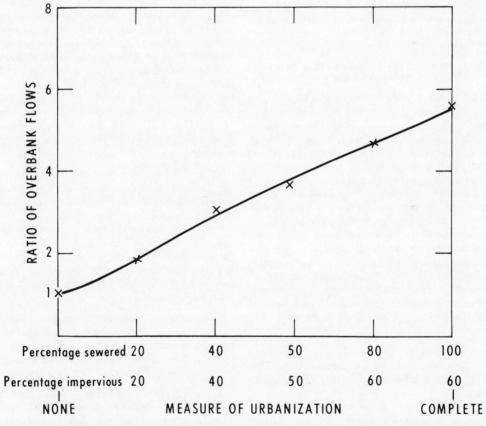

Figure 6.—Increase in number of flows per year equal to or exceeding original channel capacity (1-square-mile drainage area), as ratio to number of overbank flows before urbanization, for different degrees of urbanization. (Derived from figure 5.)

10

crease in number of flows equal to or exceeding natural channel capacity. An auxiliary scale is shown at the top of figure 5 to facilitate this.

For example, under natural conditions it is expected that a 10-year record would show about seven flows equal to or exceeding 67 cfs, or channel capacity. But if the average annual flood were increased 1.5 times (from 75 to 112 cfs) corresponding to 20 percent sewered and 20 percent impervious, the new frequency curve indicates that 14 flows of 67 cfs or greater would occur in a 10-year period, or a twofold increase in number of flows. Similarly, the ratio of number of flows exceeding bankfull capacity was read from the intersection of the other curves in figure 5 with the ordinate value of 67 cfs to obtain the ratios plotted in figure 6.

Figure 6 shows that with an area 50 percent sewered and 50 percent impervious, for example, the number of flows equal to or exceeding bankfull channel capacity would, over a period of years, be increased nearly fourfold.

LOCAL STORAGE TO COMPENSATE FOR PEAK FLOW INCREASE

Urbanization tends to increase both the flood volume and the flood peak. But the increase can be compensated so that the discharge through channels downstream is maintained to any degree desired within the range which existed prior to urbanization. It is obvious that reservoir storage is installed on a river in order to reduce the magnitude of peak discharge by spreading the flow over a longer time period. Channels themselves provide temporary storage and act as if they were small reservoirs. Overbank flooding on to the flat flood plain is a way that natural rivers provide for temporary storage and thus decrease flood peaks downstream. This effect of storage has been fully investigated and described (for example see Leopold and Maddock, 1954, especially p. 36–49).

The provision of flood storage upstream, then, will decrease flood peaks and compensate for the increase caused by urbanization. This storage could take many forms including the following:
1. Drop inlet boxes at street gutter inlets.
2. Street-side swales instead of paved gutters and curbs.

3. Check dams, ungated, built in headwater swales.
4. Storage volumes in basements of large buildings receiving water from roofs or gutters and emptying into natural streams or swales.
5. Off channel storage volumes such as artificial ponds, fountains, or tanks.
6. Small reservoirs in stream channels such as those built for farm ponds.

Various types of storage volumes could be used simultaneously in various mixes. The effectiveness depends on the volume of storage relative to the volume of inflow during a storm peak period. Design criteria to guide city engineers and developers are needed.

SEDIMENT PRODUCTION

The basic data available for analyzing the effect of urbanization on sediment yield, though sparse, have been summarized to some extent in the literature. Especially valuable is the report by Wolman (1964) who summarized not only the data obtained from sediment sampling stations in streams in Eastern United States but also studied the sediment yield from building construction activities. Sediment yields from urbanized or developing areas ranged from 1000 to more than 100,000 tons per square mile per year.

It should be recognized that sediment yield per square mile decreases with increasing drainage area, but nevertheless it is apparent that unurbanized drainage basins yield 200 to 500 tons per square mile per year, on the average. These figures are slightly higher for the farmed Piedmont lands, which may be expected to produce sediment yield of 500 tons per square mile per year, such as the Watts Branch basin near Rockville, Md.

The data on urbanized areas studied by Wolman are plotted in figure 7 together with data from suspended load sampling stations of the U.S. Geological Survey as summarized by Wark and Keller (1963).

In the graph (fig. 7) three bands or zones are labeled *A, C,* and *UC.* Wolman and Schick (1967) differentiated the following types of activity: *A,* agricultural or natural; *C,* under-

11

198

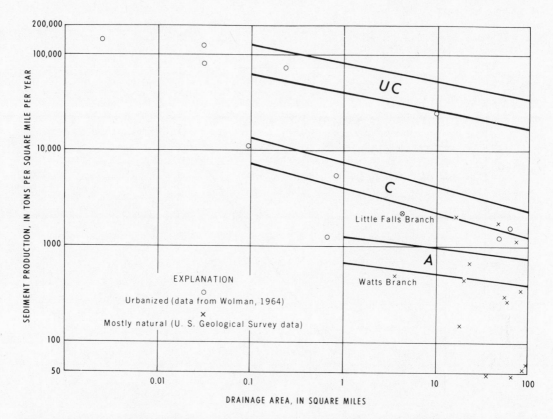

Figure 7.—Annual sediment production per square mile for urbanized and natural areas. Zones: *A*, agricultural; *C*, under construction; *UC*, under construction and undiluted.

going building construction, but highly diluted before reaching channels; and *UC*, undiluted sediment yields delivered to stream channels from construction sites.

They found that when building sites are denuded for construction, excavations are made, and dirt is piled without cover or protection near the site, the sediment movement in a rill or stream channel is very large in terms of tons per year immediately downhill from the construction site. If the channel contains little water except during storms (an ephemeral stream), there is no chance for dilution; during storm flow the sediment movement is great. If the construction debris gets into perennial channels, or for other reasons is distributed along a channel or dispersed over a wide area, the dilution lowers the yield per square mile

per year. Thus, Wolman and Schick drew the distinction between agricultural, construction, and construction-undiluted.

For very small areas, Wolman (1964) said, "Because construction denudes the natural cover and exposes the soil beneath, the tonnage of sediment derived by erosion from an acre of ground under construction in developments and highways may exceed 20,000 to 40,000 times the amount eroded from farms and woodlands in an equivalent period of time."

Figure 7 shows the data as a relation between annual sediment yield per square mile and drainage basin size. The usual suspended load station is on a basin of more than 10 sq mi in area. Seldom is urbanization complete for basins of this size.

12

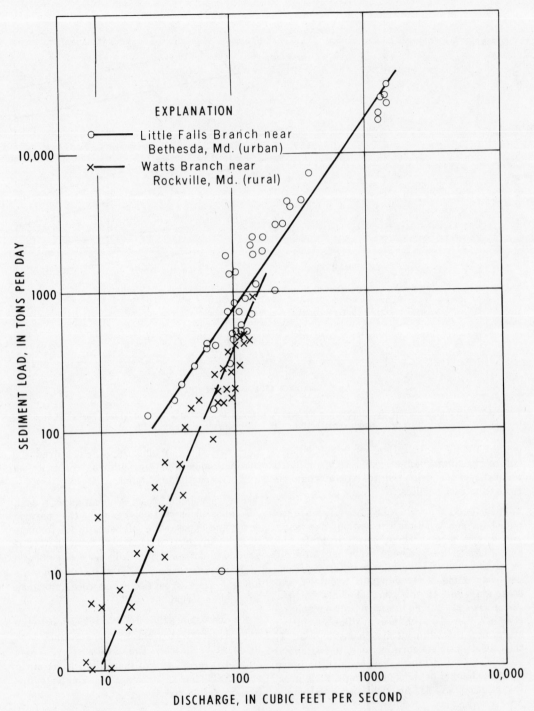

Figure 8.—Relation of sediment yield and discharge for an urban and a rural or unurbanized area.

13

The data measured or estimated by Wolman (1964) in small urbanizing, developed or industrial areas show clearly that the sediment yield is larger by 10 to 100 times that of rural areas. Guy and Ferguson (1962) observed an increase of 250 times in an area near Kensington, Md.

To illustrate the difference in sediment samples obtained during storm flow, actual data for two stations are shown in figure 8. The sediment rating curve, which is a plot of the discharge at any moment in time against the concurrent rate of sediment transport, gives an indication of the order of magnitude of the increase in sediment production from developed, as against rural, areas. The sediment rating curves in figure 8 are for stations near Washington, D.C. Watts Branch drains an area primarily used for farming though urban influences have recently extended into the basin. Little Falls Branch near Bethesda drains a nearly completely urbanized community, consisting of Bethesda and parts of Chevy Chase, Md.

Note that the sediment rating curves tend to converge at high discharges. One might suppose that at those discharges the urbanized areas are actually contributing no more sediment than the unurbanized ones. This is not the case, however, owing to the fact that as a result of urbanization, the number of high flows increases materially. Because most of the sediment during the year is carried during periods of high flow, the result is that urbanized areas yield on the average larger sediment loads than the unurbanized ones.

The difference in drainage basin size between Watts Branch (3.7 sq mi) and Little Falls Branch (4.1 sq mi) is not alone sufficient to explain the larger discharges in the latter basin. For about the same number of sample storms, note that Little Falls Branch data include discharges varying from 20 to 1500 cfs. In contrast, Watts Branch data (unurbanized) include flows ranging from 7 to 150 cfs. At least some of this difference is probably due to the effect of urbanization on increasing peak flow from a storm of given size, as discussed earlier. The two basins are only 10 miles apart and storms are comparable.

Keller (1962) compared the sediment rating curves for Northwest Branch of Anacostia River near Colesville, Md., a relatively unurbanized basin, and the Anacostia River basin near Hyattsville, Md., which is partly urbanized. He found the sediment production to be about four times greater in the urbanized area.

Most sediment carried by a stream is moved by high flows. In Brandywine Creek, for example, about 54 percent of the total sediment transported annually by the river (drainage area 312 sq mi) is carried by flows that occur, on the average, about 3 days each year.

In the tabulation below, a comparison is made between sediment yield from Watts Branch, a rural landscape, and Little Falls Branch, an urban one. These basins are of the size and type represented in East Branch Brandywine Creek.

	Drainage area (sq mi)	Tons per year	Tons per year per sq mi
Watts Branch in Rockville, Md. (rural)	3.7	1,910	516
Little Falls Branch near Bethesda, Md. (urban)	4.1	9,530	3,220

Sediment production is importantly related to land slope. Using multiple correlation techniques for a large variety of data from experimental watersheds, Musgrave (1947) developed a multiple correlation in which the rate of erosion is found to be proportional to the 1.35 power of land slope and to the 0.35 power of the slope length. The same conclusion had been derived theoretically by Horton (1945) and verified by comparison with the percentage of area eroded in the Boise River basin, Idaho. Sediment yield, therefore, is more highly sensitive to land slope than to length of slope but is positively correlated with both.

Some idea, however, can be obtained of the difficulty in keeping steep slopes stable after the original vegetation has been disturbed, particularly during construction. If, for example, land slopes of 5 and 10 percent are compared, the doubling of the slope would increase the erosion rate by 2.3 times.

Increased slope length does not have such a large effect on erosion rate. Doubling slope

14

length would increase the erosion rate by only 22 percent.

Because a slope of 10 percent drops 10 feet in a 100-foot horizontal, temporary storage in the form of depressions which might hold silt would be nearly absent. For land slopes above 10 percent, stream channels also would tend to be nearly devoid of areas or depressions which could hold up sediment during its passage downhill. From a practical standpoint, therefore, a figure of about 10 percent probably would be a physical and economic limit beyond which construction would be especially harmful insofar as sediment production is concerned. Any such limiting slope, however, would have to be determined by detailed economic studies.

Wark and Keller (1963) related the average annual sediment discharge in the Potomac River basin to percentage of forest cover and, separately, to the percentage of land in crops. Average annual sediment yield increased from 50 to 400 tons per square mile per year, or eightfold, as forest cover in the basin declined from 80 percent to 20 percent. Sediment yield increased from 70 to 300 tons per square mile per year, or fourfold, as land in crops increased from 10 to 50 percent.

EFFECT OF INCREASED PEAK FLOWS ON SEDIMENT YIELD

It has been pointed out in the comparison of sediment rating curves for urban versus rural areas that the rating curves do not appear to be as much different as the values of sediment yield on an annual yield basis. It has been mentioned that a slight increase of sediment concentration can make a large difference in total annual sediment yield owing to the fact that urban areas produce a larger number of high flows. If the number of flows above bankfull stage is increased by urbanization, the banks and bed of a channel in erodible material will not remain stable, but the channel will enlarge through erosion. Computation indicates the seriousness of this factor.

For example, assume that a channel is capable of carrying 55 cfs at bankfull stage. In the Brandywine area this represents a channel draining a basin slightly less than 1 sq mi in area. The channel necessary to carry 55 cfs

at bankfull stage would probably have a velocity of slightly less than 2.5 feet per second and would be about 2 feet deep and 11 feet wide. In figure 2, urbanization might cause a flow of this frequency to increase 2.7 times, or 150 cfs. If this channel had to adjust itself to carry a flood of 150 cfs at bankfull stage, it is estimated that the new velocity would be about 2.5 feet per second, and the necessary depth and width would have changed respectively to about 3 feet and 20 feet. In other words, this stream would deepen about 50 percent and increase in width a little less than twice its original size. If such erosion takes place through at least one-fourth mile of channel length in a drainage basin of 1 sq mi, the amount of sediment produced by this erosion would be 50,000 cubic feet. At 100 pounds per cubic foot, this amounts to 2,500 tons.

This amount can be compared with the mean annual sediment yield for Watts Branch, an unurbanized area near Rockville, Md. Annual sediment yield of Watts Branch is 516 tons per square mile. Thus, the channel erosion alone under the assumptions made would produce as much sediment as 5 years' usual production from an unurbanized area of the same size. Therefore, one can visualize that as urbanization proceeds, not only does construction activity have the potential of increasing sediment loads many thousands of times while construction is in progress, but also the result of the urbanization through its increase in peak flow would produce large amounts of sediment from channel enlargement as well. This emphasizes the need to provide temporary storage far upstream to counteract the tendency of urbanization to increase the number and size of high flows.

WATER QUALITY

There is little doubt that as urbanization increases, particularly from industrial use of land and water, the quality of water decreases. However, quantitative data to support this observation are sparse. There are two principal effects of urbanization on water quality. First, the influx of waste materials tends to increase the dissolved-solids content and decrease the dissolved-oxygen content. Second, as flood peaks increase as a result of the increased area of imperviousness and decreased lag time, less

15

water is available for ground-water recharge. The stream becomes flashier in that flood peaks are higher and flows during nonstorm periods are lower.

A recent study on the Passaic River at Little Falls, N.J., by Anderson and Faust (1965) provides quantitative data on the effect of urbanization and industrialization on water quality. Seventeen years of data for the flow and chemical quality of the 760-sq-mi drainage basin were analyzed. During these 17 years, diversions of water for domestic and industrial supplies increased more than 30 percent between 1950 and 1963. Returns of waste waters into the basin became as much as 10 percent of the water withdrawn. Analysis of the data showed that at relatively low discharge the dissolved-solids content increased about 10 ppm (parts per million) between 1948 and 1955 but increased 75 ppm between 1955 and 1963. That is, during the period of greatest population growth the dissolved-solids content increased nearly 40 percent in a period of 8 years.

A long-term change in the average content of dissolved oxygen was also noted. Between 1950 and 1964 the dissolved-oxygen content dropped from an average of 78 percent of saturation to 62 percent of saturation. Further, the analysis demonstrated that these average changes in water quality occurred in all seasons of the year.

An aspect of population growth not generally appreciated is the large segment of population using septic tanks for disposal of sewage. In a given area this segment often becomes large before community water and sewerage systems are built. For the planner it should be important to know how septic-tank installations can affect water quality in streams and in the ground. In the upper East Branch of Brandywine Creek, a basin of 37 sq mi, the population in 1967 was 4,200. As of that date, there were no community water or sewerage systems; all the population was served by individual wells and septic tanks. Population projections indicate that the basin will have 14,000 persons by the year 1990. During the initial part of this projected growth at least, the number of wells and septic tanks can be expected to increase materially.

The soil, containing as it does a flourishing fauna of micro-organisms, tends to destroy or adsorb pathogenic bacteria. Effluent draining from the seepage field of a septic tank tends therefore to be cleansed of its pathogens. McGauhey and Krone (1954) showed that the coliform count was reduced by three orders of magnitude in moving from an injection well a distance of 50 feet through sand and gravel. In 100 feet the count was reduced to a small number. As for rate of movement, Mallmann and Mack (1961) showed that bacteria introduced into a permeable soil by a septic-tank seepage field moved 10 feet in 2 days and 20 feet in 3 days and appeared in a well 30 feet away after 10 days.

Both the rate and effectiveness of the process of pathogen reduction depend on the type of soil as has been summarized by Olson (1964), who emphasized that position of the ground-water table is a critical factor in the transmission of pollutants.

Studies by Wayman, Page, and Robertson (1965) of the changes in primary sewage effluent through natural materials in conditions of saturated flow showed that "most soils removed over 90 percent of the bacteria from sewage within a few feet of travel * * * [but there was] severe clogging in the finer-grained soils." They found, however, that "dissolved solids moved through the columns [of soil] virtually unaffected * * *."

The same authors report on infiltration of polluted river water through sandy loam. "ABS [synthetic detergent] and coliform bacteria are significantly reduced by infiltration through the unsaturated zone; dissolved solids do not seem to be removed * * *. Once a pollutant gets into the ground water (saturated flow) little additional change in removal of ABS or dissolved solids, even for movement over extensive horizontal distances, is to be expected. This result is in agreement with the data * * * for flow of sewage effluent through various soil columns (saturated flow)."

The data are not definitive regarding the minimum distance a septic-tank seepage field should be separated from a stream channel, but the application of data cited above with

16

general principles does indicate some tentative rules of thumb which might be useful to the planner. A perennial stream represents the intersection of the saturated zone (water table) with the earth's surface. The observations indicate that, for soil cleansing to be effective, contaminated water must move through unsaturated soil at least 100 feet. Owing to the gentle gradient of the water table near the perennial stream and the fact that seepage water moves vertically as well as toward a nearby channel, it would seem prudent that no septic tank should be as close to a channel as about 300 feet, if protection of the stream water quality is to be achieved. The distance should probably be greater from a perennial than from an ephemeral channel. (An ephemeral stream is one which contains flowing water only in storm periods.) In general, it might be advisable to have no source of pollution such as a seepage field closer than 300 feet to a channel or watercourse.

Even this minimum setback does not prevent the dissolved materials (nitrates, phosphates, chlorides) from enriching the stream water and thus potentially encouraging the proliferation of algae and otherwise creating a biotic imbalance.

The only detailed study of the effect of urbanization on water temperature is that of E. J. Pluhowski (1968), some of whose results are summarized here. He chose five streams on Long Island for detailed analysis and found that streams most affected by man's activities exhibit temperatures in summer from 10° to 15°F above those in an unurbanized control. Connetquot River, the control stream, flows through one of the few remaining undeveloped areas of central Long Island. Temperatures in reaches most affected by ponding, realinement, or clear cutting of trees are significantly higher in summer, but winter temperatures are 5° to 10°F colder than those observed in reaches unaffected by man.

Solar radiation is the predominant factor in the energy balance determining a stream's thermal pattern. The more solar energy a stream absorbs, the greater its temperature variation diurnally as well as seasonally. By greatly increasing the surface area exposed to the sun's radiation, the construction of ponds and lakes has profoundly affected stream temperature regimen. On Long Island, Pluhowski found that ponds having mean depth of about 2 feet or less substantially increase downstream diurnal temperature fluctuations whereas ponds deeper than 2 feet exhibit a dampening effect on daily temperatures. For example, during the period October 31 to November 2, 1967, the mean daily range of temperatures at Swan River, in south-central Long Island, varied from 9°F in a reach immediately below a shallow pond (mean depth, 0.5 foot) to 3°F below Swan Lake (mean depth, 3 feet). In reaches unaffected by man's activities, the mean daily temperature fluctuation was about 4°F.

Under natural conditions, less than 5 percent of the streamflow on Long Island originates as direct surface runoff. With the conversion of large areas of western Long Island from farmland to suburban use during the last 20 years, the proportion of streamflow originating as surface runoff has increased sharply. As a direct consequence, streams most affected by street runoff may exhibit temperature patterns that are markedly different from those observed in streams flowing through natural settings. During the period August 25 to 27, 1967, a series of heavy rainstorms overspread Long Island. Throughout this period, temperatures at each of the five observation sites on Connetquot River showed little day-to-day change. In contrast, temperatures in the upper reaches of East Meadow Brook, which drains highly urbanized central Nassau County, increased steadily in response to the relatively warm street runoff. Pluhowski found that by August 27, water temperatures had risen 10° to 12°F above prestorm levels and were 15°F higher than concurrent temperatures in the control stream.

17

204

SELECTED REFERENCES

Anderson, D. G., 1968, Effects of urban development on floods in northern Virginia: U.S. Geol. Survey open-file rept., 39 p., 5 figs.

Anderson, Peter W., and Faust, Samuel D., 1965, Changes in quality of water in the Passaic River at Little Falls, New Jersey, as shown by long-term data, *in* Geological Survey research 1965: U.S. Geol. Survey Prof. Paper 525-D, p. D214-D218.

Antoine, L. H., 1964, Drainage and best use of urban land: Public Works [New York], v. 95, p. 88-90.

Carter, R. W., 1961, Magnitude and frequency of floods in suburban areas, *in* Short papers in the geologic and hydrologic sciences: U.S. Geol. Survey Prof. Paper 424-B, p. B9-B11.

Espey, W. H., Morgan, C. W., and Masch, F. D., 1966, Study of some effects of urbanization on storm runoff from a small watershed: Texas Water Devel. Board Rept. 23, 109 p.

Felton, P. N., and Lull, H. W., 1963, Suburban hydrology can improve watershed conditions: Public Works, v. 94, p. 93-94.

Guy, H. P., and Ferguson, G. E., 1962, Sediment in small reservoirs due to urbanization: Am. Soc. Civil Engineers Proc., HY 2, p. 27-37.

Harris, E. E., and Rantz, S. E., 1964, Effect of urban growth on streamflow regimen of Permanente Creek, Santa Clara County, California: U.S. Geol. Survey Water-Supply Paper 1591-B, 18 p.

Horton, R. E., 1945, Erosional development of streams and their drainage basins, hydrophysical approach to quantitative morphology: Geol. Soc. America Bull., v. 56, no. 3, p. 275-370.

James, L. D., 1965, Using a computer to estimate the effects of urban development on flood peaks: Water Resources Research, v. 1, no. 2, p. 223-234.

Keller, F. J., 1962, The effect of urban growth on sediment discharge, Northwest Branch Anacostia River basin, Maryland *in* Short papers in geology and hydrology: U.S. Geol. Survey Prof. Paper 450-C, p. C129-C131.

Leopold, L. B., and Maddock, T., Jr., 1954, The flood control controversy: New York, The Ronald Press Company, 275 p.

Leopold, L. B., Wolman, M. G., and Miller, J. P., 1964, Fluvial processes in geomorphology: San Francisco, Calif., W. H. Freeman and Co., 522 p.

Lull, H. W., and Sopper, W. E., 1966, Hydrologic effects from urbanization of forested watersheds in the northeast: Upper Darby, Pa., Northeastern Forest Expt. Sta., 24 p.

McGauhey, P. H., and Krone, R. B., 1954, Report on the investigation of travel of pollution: California State Water Pollution Control Board Pub. 11, 218 p.

Mallmann, W. L., and Mack, W. N., 1961, Biological contamination of ground water: Robert A. Taft Sanitary Eng. Center Tech. Rept. W61-5, p. 35-43.

Martens, L. A., 1966, Flood inundation and effects of urbanization in metropolitan Charlotte [North Carolina]: U.S. Geol. Survey open-file rept., 54 p.

Musgrave, G. W., 1947, Quantitative evaluation of factors in water erosion—First approximation: Jour. Soil and Water Conserv., v. 2, no. 3, p. 133-138.

Olson, G. W., 1964, Application of soil survey to problems of health, sanitation, and engineering: Cornell Univ. Agr. Expt. Sta. Mem. 387, 77 p.

Pluhowski, E. J., 1968, Urbanization and its effect on stream temperature: Baltimore, Md., Johns Hopkins Univ., Ph. D. dissert. (in preparation).

Snyder, F. F., 1938, Synthetic unit hydrographs: Am. Geophys. Union Trans., v. 19, p. 447-454.

Swenson, H. A., 1964, Sediment in streams: Jour. Soil and Water Conserv., v. 19, no. 6, p. 223-226.

Wark, J. W., and Keller, F. J., 1963, Preliminary study of sediment sources and transport in the Potomac River Basin: Interstate Comm. on Potomac River Basin, Washington, D.C., Tech. Bull. 1963-11, 28 p.

Wayman, C., Page, H. L., and Robertson, J. B., 1965, Behavior of surfactants and other detergent components in water and soil-water environments: Federal Housing Adm. Tech. Studies Pub. 532, 136 p.

Wiitala, S. W., 1961, Some aspects of the effect of urban and suburban development upon runoff: U.S. Geol. Survey open-file rept., 28 p.

Wilson, K. V., 1966, Flood frequency of streams in Jackson, Mississippi: U.S. Geol. Survey open-file rept., 6 p.

Wolman, M. G., 1964, Problems posed by sediment derived from construction activities in Maryland—Report to the Maryland Water Pollution Control Commission: Annapolis, Md., 125 p.

Wolman, M. G., and Schick, P. A., 1967, Effects of construction on fluvial sediment, urban and suburban areas of Maryland: Water Resources Research, v. 3, no. 2, p. 451-462.

18

U. S. GOVERNMENT PRINTING OFFICE : 1972 O - 468- 019

15

Reprinted from *Science,* **174**, 905–918 (Nov. 26, 1971)

The Nation's Rivers

Problems are encountered in appraising
trends in water and river quality.

M. Gordon Wolman

Definitions of water pollution are often rightly subjective. Thus pollution is sometimes defined as any impairment of water which lessens its usefulness for beneficial purposes, or anything the public does not like, or even that which is getting worse. Each implies both a value judgment and a subjective perception of the resource. Because such definitions do describe public attitudes toward environmental quality, they provide incentives to public action.

To evaluate alternative ways of combating observed or perceived conditions of water and river deterioration, however, requires objective measurements that describe the perceived conditions and can be related to technological and other measures appropriate to management of the water resource. Over the years in assessing water and river quality, a number of parameters have been observed which might provide objective criteria that could be used to study trends in the condition of the nation's rivers. Two principal questions can be raised about these observations: (i) whether there are measurable trends in the observed data and (ii) whether the variables being measured are those that provide the best measure of the perceived condition or qualities in which the public or society at large may be interested.

Relatively few studies of the quality of the nation's rivers have been directed toward determining changes in specific parameters over long periods of time (1). This is perhaps not surprising because a number of disabilities interfere with a truly adequate statistical analysis of such a series. First, hydrologic records in the United States are relatively short. There are few continuous records for periods as long as 50 or

The author is professor of geography at the Johns Hopkins University, Baltimore, Maryland 21218.

60 years. Second, techniques of observation and of analysis have changed over the years. Analytical techniques, in particular, have become more sophisticated, and routine measurements of exceedingly small quantities of contaminants are now possible which, only a few years ago, were considered impractical. Thus some comparisons reflecting changes in techniques of detection rather than real trends may be misleading. Third, changing the location or frequency of observations of water quality may distort the record. Observations of water quality are often made in the vicinity of metropolitan areas adjacent to the intakes of city water supplies. From time to time the intakes are moved to avoid sewer outfalls. While the intake may be moved upstream only a few hundred yards, the new record differs completely from the previous record, which was essentially monitoring the relation between the quality of the river and the inflow from the outfall. Fourth, adequate comparisons of specific variables related to water and to river quality require systematic correlation with hydrologic behavior. Such correlations are rarely available. Fifth, a knowledge of the "natural background" or temporal variability of a given parameter is often essential in detecting and measuring a trend. Statistically, a trend cannot be discerned unless it is possible to discriminate between the variability of the phenomena as it might occur unaffected by the influences that one wishes to measure, in this case so-called pollution, and the variability normally associated with diurnal, annual, and significant secular climatic variations that occur in the hydrologic record over any period of time. Lastly, explanation of changes in water quality parameters requires a knowledge not only of the natural background but of the economy and land use of the area.

This report could not be exhaustive. I have analyzed examples of changes in specific parameters at a few sites to illustrate the magnitude of changes in some variables over time as well as the presence or absence of trends which in some cases can be reasonably well documented. Diligent search, however, has revealed a limited number of similar comparisons, and examination of the available data suggests some tentative conclusions and poses a number of questions.

Selection of Variables

Variables that are used to describe water and river quality may be separated into two types, those associated with the water itself—transport characteristics—and those associated with the river or river site. Transport phenomena such as dissolved oxygen, dissolved solids, biochemical oxygen demand, suspended sediment, pH, and temperature are among the common measures of what is known as water quality. It is important to note that the phrase "water quality" does not adequately describe river quality or the quality of the river site itself. While the variety and associations of stream and shoreline biota are sometimes considered among the measurements of water quality, periodic or long-term observations are rare; site characteristics such as channel morphology, bottom conditions, trash, and shoreline vegetation are even more rarely recorded. The parameters chosen for illustrations in this article are those most generally observed, ones for which data are most readily available over the longest periods of time.

Dissolved Oxygen

Dissolved oxygen, expressed as percentage of saturation, for the Hudson River between Troy, New York, and the Battery below New York City, was measured in the summer of 1933, many times in the interim, and in August of 1965. A comparison between the summer data for 1933 and a curve drawn for 22 to 24 August 1965 is shown in Fig. 1. The comparison indicates that dissolved oxygen in the Hudson River

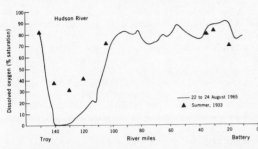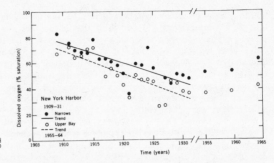

Fig. 1 (left). Comparative levels of dissolved oxygen in the Hudson River in 1933 (54) and in 1965 (3). Fig. 2 (right). Comparative levels of dissolved oxygen in the vicinity of New York Harbor. Note the broken scale 1933–55. All points are averages of summer values. Data and curves 1909–31 from Phelps (2). See text for discussion.

in 1965, despite the fact that flow in 1965 was quite low, is roughly the same over most of this length as it was in 1933, except in the vicinity of the metropolitan area below Albany and Troy. There the summer curve shows a lower sag in 1965 than was apparent in 1933. Observations of the river made at a number of intervals between 1933 and 1965 show wide variations in dissolved oxygen in this reach, with the winter curve showing considerably less sag than the summer curves both for 1933 and 1965. The data in Fig. 1, as well as curves for the interim period (available from the New York State

Health Department) suggest that, despite large increases in population in the New York metropolitan area as well as in the vicinity of Albany and Troy over much of the length of the river, roughly 100 miles (160 kilometers), the dissolved oxygen has remained essentially constant.

Of perhaps even greater interest are figures for dissolved oxygen at the narrows and the upper bay of New York Harbor (Fig. 2). Data for the period 1909 to 1931 (2) show the decline in the dissolved oxygen from 1909 to 1931. Phelps (2) noted, however, that much of the decline in the dissolved

oxygen actually took place between 1909 and 1920, as is shown by the distribution of points on Fig. 2. Three later points are shown for the years 1955, 1960, and 1965 from the Public Health Service report of 1965 (3). In themselves the three points cannot be used, of course, to define a trend, but it appears that there has been little or no change in dissolved oxygen between 1930 and 1965.

On a stretch of the nation's largest river, the Mississippi River at Minneapolis–St. Paul, dissolved oxygen has remained approximately constant since 1938 in the reach below the metropolitan area (Fig. 3). Between 1932 and 1938 dissolved oxygen increased as a result of construction of sanitary works, including major sewage treatment plants. The cost of these facilities including treatment plants ($4.8 million) and intercepting sewers ($11.6 million) totaled $16 million between 1934 and 1937. Oxygen levels attained in 1938 have been substantially maintained to the present time with the construction of additional treatment facilities between 1963 and 1966 at a cost of $27 million, as the population in the Twin Cities areas has grown from about 250 thousand to 1.5 million (Table 1). Not shown in the comparison in Fig. 3 are the effects of projected expenditures to 1971, which include an additional $33 million to reduce the biochemical oxygen demand by 90 percent [cost data from (4)]. The low value of the dissolved oxygen for the winter period is presumably related to the frozen condition of the river. A mean of the two curves suggests that the dissolved oxygen level has remained approximately the same over the last 30 years.

While below Philadelphia on the Delaware River the oxygen sag curve

Table 1. Approximate populations and expenditures on major works including sewage treatment, interceptors, and pumping stations in several regions.

River	Region	Population		
		Year	Numbers	Reference
Hudson	New York metropolitan	1930	7 million	(61)
		1964	11.3 million	
	Albany-Schenectady-Troy	1930	127,000	(61)
		1965	697,000	(3)
Mississippi	St. Paul–Minneapolis	1940	796,000	(4)
		1964	1.6 million	(4)
Potomac	Washington, D.C., metropolitan	1930	487,000	(61)
		1940	908,000	(62)
		1967	2.6 million	(56)
Delaware	Philadelphia metropolitan	1950	3.6 million	(62)
		1960	4.3 million	(63)
Ohio	Basin area	1950	17.6 million	(16)
	Municipal		9.3 million	
	Municipal		11.3 million	

River	Expenditures ($10⁶)	Period	Item	Reference
Hudson	12	1963–68*	Sewers only	(64)
	180	1968–70	Sewers only	
	1400	1965–76	Treatment plants	
Mississippi	16.4	1934–37	Treatment plants and sewers	
	27	1963–66	Treatment plant	(4)
	3.3	1968	Annual operation and maintenance	
	33	1969–71	Treatment plant	
Potomac	60	1950–60		(44)
Delaware	300	Since 1946		(65)
Ohio	1000	1952–61		(66)

* Per year.

falls to levels of 5 percent saturation or less, in a limited reach (5) for the period 1950 to 1963 some miles above Philadelphia and at central Philadelphia dissolved oxygen fluctuated around a value of 75 percent saturation, remaining essentially constant (6).

Where dissolved oxygen has been maintained in waters adjacent to major metropolitan centers, it has required large expenditures on sewers and on sewage treatment works to accommodate large increases in population (Table 1). The "Red Queen" process of running to stay in place, illustrated in St. Paul and in part for New York, may not be viewed as progress by some, but may still be viewed as an accomplishment often unrecorded in the current popular literature. The Potomac estuary in the vicinity of Washington, D.C., presents a different picture. Here the impact of large metropolitan growth on a relatively small river and sluggish estuary is evident. In contrast to the Mississippi, the Hudson, and the Delaware, the relative size of the metropolitan population of Washington, as contrasted with the flow of the river, has placed a burden on the capacity of the river which is insufficiently compensated by sewage treatment works built in the area. Despite expenditures upward of $70 million from 1938 through 1965, in recent years dissolved oxygen during the summer months has retreated to the position occupied by similar curves in 1932 before major treatment works were installed in 1938 (Fig. 4). The oxygen sag curve has not only declined but has moved downstream to include a larger area. The oxygen sag curves for 1954 to 1967 and for 1960 to 1967 for consecutive minimum flows of 28 days show declines below the figures for 1932 before major treatment works were installed. While major improvements in treatment processes and in the volume of wastes that can be treated are planned for the metropolitan area of Washington, the demands on the river, posed by sewage and other wastes, have exceeded the assimilative capacity of the Potomac, so that there results a burgeoning algal population with associated odors and esthetic problems.

In 1967, Torpey (7) suggested that the dissolved oxygen in such major estuaries as the New York Harbor and the estuary of the Thames in London was controlled primarily by organic activity associated with nutrients promoting the growth of algae. Torpey's curve, reproduced in Fig. 5, indicates that, in estuaries such as these, the dissolved oxygen declined rapidly once the input, expressed as biochemical oxygen demand (BOD) per acre of surface per day (1.0 pound per acre is equivalent to 1.12 kilograms per hectare) exceeded a specific threshold. He suggested that the rapid decline in dissolved oxygen occurred in a range of instability associated with large organic loading. This range or zone for New York appears to be defined roughly for the period from 1890 to 1916. Thereafter, a kind of plateau was maintained in the dissolved oxygen as a result of the production of oxygen by algae. Despite a large increase in population, from 1937 to 1965 loading decreased as a result of the installation of sewage treatment works. Gould (8) noted, however, that, while the pollutional load "decreased about 38 percent, the saturation deficit of dissolved oxygen decreased only 11 percent." If the data is applied to the Thames estuary at London, Torpey's estimate (Fig. 5) shows a dissolved oxygen in the Thames in 1910 of approximately 25 percent saturation, whereas in 1913 the figure was near zero or approximately 5 percent. The precipitous decline from 1910 to 1913 represented another zone of instability accompanied by wide fluctuations in dissolved oxygen.

In contrast to the figures for the dissolved oxygen in New York Harbor and the Thames, dissolved oxygen in the Potomac River at Washington, approximately 25 percent saturation at a loading rate of approximately 45 pounds of BOD per acre per day, falls well below the general curve defined on the basis of the New York Harbor and Thames (Fig. 5) data. This may result from the fact that the flushing time of the Potomac, including the effect of the relatively small river flow, is considerably longer than that either for New York Harbor or for the Thames in London. Torpey (7) analyzes these and the historical relationships for the Hudson and the Thames. The curve represents a complex of hydrodynamic, chemical, and organic interactions and hence cannot be simply generalized to other estuaries. A comprehensive paper on the Hudson by Howells et al. (9), for example, points out the complexity of the relation between biota and nutrients and the potential for rapid deterioration in oxygen levels and organisms despite the fact that "serious fouling and deoxygenation have so far been avoided for most of the river." Ketchum (10) also notes the very high phosphorus concentration in the Hudson estuary.

Whether or not the curve suggested by Torpey is indeed a general phenomenon remains to be verified. However, Torpey (7) and later Gould (8) have pointed out that, if the postulated "plateau" in dissolved oxygen does in-

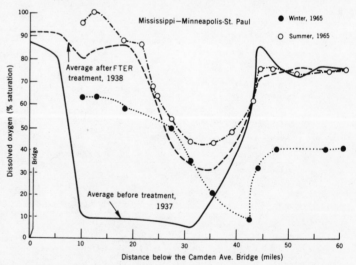

Fig. 3. Comparative levels of dissolved oxygen in the Mississippi River below Minneapolis–St. Paul. The winter data represent a survey during January, February, and March. The summer data were obtained during July, August, and September [data for 1938 from (55), for 1965 from (56)].

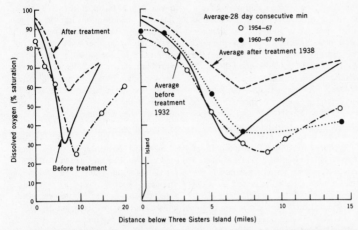

Fig. 4. Comparative levels of dissolved oxygen summer flows, Potomac River below Washington, D.C., 1932, 1938, 1954–67 inclusive, and 1960–67 [data for earlier period from (54), for 1932 and 1938 from (57)].

deed exist in a number of estuaries, large increases in sewage treatment works could have a relatively small impact on the dissolved oxygen of such estuaries controlled by biological activity where treatment levels still operated within a range of dissolved oxygen represented by the "plateau" from roughly 30 to 40 percent saturation. Large expenditures of funds for treatment works might be reflected in only modest improvements in the quality of the water as measured by dissolved oxygen. Thus the effort to reverse a trend may be exceedingly difficult in reaches subject to the complex biological phenomena likely to be found in a number of rivers in critical estuaries on the coasts of the United States.

Inability to understand the precise causes of the changes in a parameter such as the dissolved oxygen also poses problems from the standpoint of the evaluation of alternatives directed toward the improvement of river quality. Because of the magnitude of the expenditures required to correct the conditions of such estuaries and rivers, it becomes imperative that understanding be improved to the point where returns per dollar invested can be more easily evaluated if the public is to be able to choose those alternatives which it feels are worth the expenditures required.

Dissolved Solids

While information on the dissolved solids carried by rivers in the United States is more extensive than that for the dissolved oxygen or other parameters, few examples of comparative changes over a period of time have been analyzed which include controls of hydrologic variation as well as accurate measurements of dissolved materials. Curves relating streamflow and dissolved solids for the Passaic River at Little Falls, New Jersey, in the New York–New Jersey metropolitan area show a progressive increase in dissolved solids from 1948 through 1955 to 1963. At a discharge of 200 cubic feet per second (5.7 cubic meters per second), for example, the dissolved load has increased from approximately 130 parts per million (ppm) in 1948 to 150 ppm in 1955 and to approximately 220 ppm in 1963 (11). The significance of changes in flow in relation to dissolved load, however, is suggested perhaps by the increase in dissolved load between the years 1945 and 1960 on the Delaware River at Trenton (12). An increase in concentration is confounded by a decline in annual discharge in the same period. Records since 1964 suggest a reversal if not a downward trend in concentration of dissolved solids.

Analyses of long-term trends in dissolved materials for several rivers and a station in Lake Michigan constitute one of the most extensive analyses (1) of such records thus far published (Table 2). Beginning in the late 19th century, although the records are broken in places, the data show significant increases in chloride, sulfate, nitrate, and total dissolved solids at specific sites on the Illinois, Mississippi, and Ohio rivers and in Lake Michigan. Increases since about 1900 range from about 50 percent in total dissolved solids to 300 percent in chlorides, amounts that must be viewed as signif-

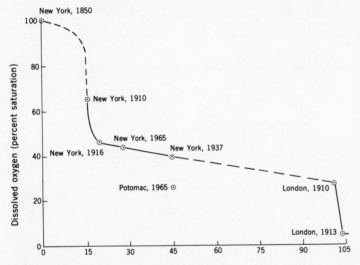

Loading rate (pounds of biochemical oxygen demand per acre of surface per day)

Fig. 5. Sequence of changes in dissolved oxygen in the Hudson and Thames River relating dissolved oxygen to BOD loading. There was a precipitous decline with intensive loading around 1900 in New York Harbor and near zero dissolved oxygen in the Thames River with high loading. The data was taken from (7). The Potomac point was added by M. G. Wolman.

icant despite the absence of controls for variations in discharge and the broad scatter in the values of each of the variables.

Acid waste from coal mines has long been recognized as a pollutant in the eastern United States where underground and open pit mining have been pursued. In some instances the amount of acid appears to be directly related to the rate of coal production. However, despite repeated efforts at mine sealing, particularly during the 1930's, relatively little reduction in acidity can be seen in rivers affected by discharge from mines. Thus, from 1947 to 1965 the sulfate concentration in the Schuylkill River at Berne, Pennsylvania, has remained constant despite installation of control measures particularly on effluents from active mines (13).

An increase in salinity is associated with return flows from major irrigation projects in the western United States. For example, it is estimated that the weighted average concentration of 501 ppm at Lee Ferry on the upper Colorado River is higher by 238 ppm as a result of man's activities (14).

An illustration of changes in the concentrations of dissolved organic matter is provided in a brief record on the Illinois River (15). Before about 1955 synthetic detergents were little used in the United States. After their introduction, detergents posed major problems in streams in many if not most regions here and abroad. The early "hard" detergents were not readily degradable by organisms within the streams. Before detergents were introduced, background levels in the Illinois River were about 0.5 milligram per liter measured as methylene blue active substance.

From 1959 to 1963 all of the detergents were alkyl benzene sulfonate (ABS). Between 1963 and 1965, ABS was being replaced by linear alkylate sulfonate (LAS), a more rapidly biogradable detergent. By 1965 to 1966 LAS was the sole surfactant in the Illinois River. From 1959 to 1965 mean monthly concentrations ranged from lows of about 0.4 mg/liter to a maximum of 9 mg/liter. During the winter the concentrations exceeded 0.6 mg/liter. By 1965 to 1966 concentrations of surfactants ranged from about 0.2 to 0.3 mg/liter. In general, high concentrations were associated with high input although monthly mean flow rates were somewhat higher in 1965 to 1966.

The detergent case illustrates the reversibility of certain pollution effects, in this case a reversibility achieved as a result of the compulsory introduction of a new material. The new products did not eliminate phosphorus from the river systems, but surfactant levels in the Illinois River decreased from 1959 through 1966 even though the amount of detergent used was increasing at a rapid rate. Similar results were obtained in Great Britain, in Germany, and elsewhere in Europe.

Perhaps the best set of analyses of trends in river quality—albeit over a relatively short period of time—is for the Ohio River, for the period since the introduction of the major cleanup efforts inaugurated by the creation of the Ohio River Water Sanitation Commission (Orsanco). During this period efforts to maintain the quality of the Ohio River have been successful in that the levels of a number of indicators have remained roughly constant from 1952 to 1964 (Fig. 6). Sulfate concentrations exceeded the standard at two points roughly 18 percent of the time (16). A detailed study by Orsanco (17) showed that in the 40-year period between 1914 to 1915 and 1952 to 1954 chloride concentrations increased 50 percent at Weirton and 60 percent at Cincinnati. Regression lines relating flow and chloride concentrations showed significant differences at all flows. However, these statistical analyses apparently show smaller increases in chlorides in this reach of the river than are suggested by the more limited data shown

for the Ohio at Cairo, Illinois in Table 2. Chloride concentrations are lower today than in 1952 at many points, with the exception of the reach below the confluence of the Muskingum which contributes large amounts of chloride into the main river. Limited chloride control has been achieved in part by scheduling the rate of chloride discharge in accordance with the capacity of the river and the desired quality objectives.

In the Ohio high pH at mile 16 and at mile 161 is attributed to the presence of considerable growth of algae (16). Throughout the river, temperatures are about the same as they were a decade earlier, and nitrate concentrations have remained constant at roughly 1.5 to 12.3 mg/liter for the past 12 years (16). In addition, phosphate (PO_4) levels have remained at approximately 0.06 to 1.15 mg/liter since measurements were first taken in 1954. Canalization of the Ohio and construction of dams up to 50 feet (15 meters) high have altered the oxygenation pattern along the river. Where sedimentation occurs in some pools, algae growth has occurred and reoxygenation is slowed at the greater depths. In contrast, higher detention times have reduced the amounts of coliform bacilli present.

In addition to improvements in control and treatment of industrial wastes, more than $1.2 billion have been spent on sewage treatment works for municipalities and now 99 percent of the urban population of the Ohio River Basin is served by sewage treatment facilities.

Table 2. Long-term observations, dissolved solids [Illinois; data from Ackerman et al. (1)].

Chloride		Sulfate		Nitrate		Total dissolved solids	
Year	Amount (mg/liter)	Year	Amount (mg/liter)	Year	Amount (mg/liter)	Year	Amount (mg/liter)
Lake Michigan off Waukegan							
1861	0.6 to 1.7	1900	5	1900	1.0	1861	136
1897	3.0	1926	14.4	1924	1.3		
1969	7.7	1969	25.7*	1930	0.5		
				1969	1.0*	1969	160
Illinois River at Peoria							
1889	33			1900	6.2	1895	318
1898	21.5	1907	46.6	1945	10.0	1900†	240
1908	15†	1945	100	1952–62	Decline		
1968	50‡	1968	120	1968	16	1968	417
Mississippi River at Alton							
1898	8			1900	4	1889	181
1925	13			1920	7	1897	220
1969	27§			1931	4.2		
				1969		1969	329
Ohio River at Cairo							
1910	11			1910	2.5 to 4.1		
1913	15			1915–20	3.5		
1919	10	1951	45	1950	3.5	1954	176
1966	23.6	1966	61.4	1966	5.4	1966	219

* Rise since 1948.   † Increased dilution.   ‡ Rise after 1950.   § Rise after 1960.

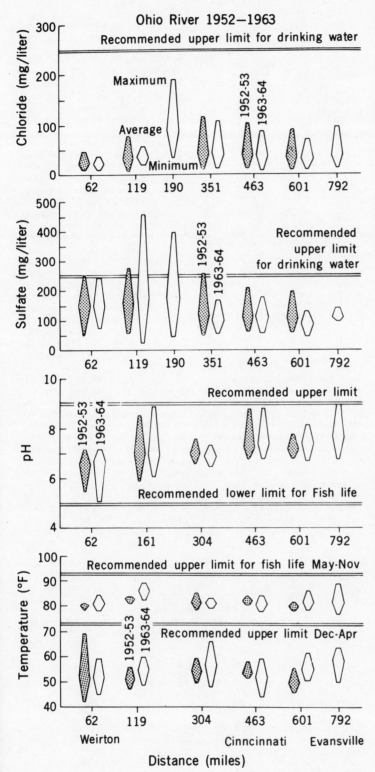

Of the total expenditure, $300 million has been spent in the Ohio Valley. Cleary (18, p. 115) estimates that, ". . . a price tag of $100 per capita would represent a fair average expenditure to cover the cost of treatment works, interceptor sewer installation, site purchase, and the associated engineering and legal fees." At places, for example Cincinnati, the river has improved sufficiently to permit the establishment of very large recreation facilities, including large marinas and many pleasure boats. In any event, the fact that the river has been rehabilitated for use in a number of places and that a number of measurable parameters have remained constant despite a fourfold increase in population from 1952 through 1963 (Table 1) illustrates the possibilities of maintenance and, perhaps, enhancement of river quality.

Sediment

The concept of sediment as a pollutant is of relatively recent origin. Even during the decade of the 1930's when problems associated with soil erosion and the need for soil conservation were first clearly faced—although the importance of reservoir sedimentation was recognized—emphasis was placed on the value of the land lost to production rather than on the offsite effects of erosion and sedimentation.

Problems posed by sediment as a pollutant are primarily a function of climate, associated vegetation, and land use. The significance of geologic erosion has not always been appreciated. Early predictions that Lake Mead behind Boulder Dam would fill within a period of 50 years, in part because of accelerated erosion due to overgrazing, gave way after detailed study (19) to estimates of 350 to 400 years. It was also recognized that the accelerated erosion resulting from poor land use, while of major consequence to husbandmen and to the local landscape, represented a minor contribution on top of an already high rate of erosion characteristic of semiarid regions. Where sediment yields are normally high in the Middle West and Southwest, the effects of man have probably

Fig. 6. Comparative levels of selected ions, pH and temperature in the Ohio River from 1953 to 1963 [data from (16)].

211

not exceeded the effects brought on by the vicissitudes of nature (*20*).

In the eastern United States, sediment yields have varied from about 100 tons per square mile per year (166 tons per square kilometer) or less in presettlement time (Table 3) to 600 to 800 tons per square mile per year where land is in crops. In contrast, in metropolitan regions undergoing development construction, activities increase sediment yields to rates of 1000 to 5000 tons per square mile or higher. Upon completion of construction, an urbanized cover of concrete, asphalt, and rooftops reduces sediment yields to levels probably less than those of forested areas before settlement. For the most part then sediment as a pollutant in the eastern United States is principally a function of land use. The association of sediment pollution with urban expansion in the humid east concentrates the problem in major population centers, centers often adjacent to estuaries.

In the western United States, changes in the sediment load have been associated not only with changes in land use, but perhaps even more significantly with regulation of stream flow by impoundments. The effect of large reservoirs that may trap as much as 99 percent of the inflowing sediment is illustrated by data from two sampling stations on the Colorado River (Fig. 7). The annual sediment load of the Colorado River at Grand Canyon upstream from Lake Mead fluctuates widely from year to year; in some instances variations are four- to sixfold. In contrast, the station at Topock, Arizona, 115 miles downstream from Boulder Dam, accurately reflects the effects of the major impoundment. Closure of Boulder Dam took place in 1933. Thereafter the annual suspended load at Topock declined progressively during the period of record shown here to virtually zero in 1939. On the Mississippi, a significant change in the average annual sediment load of the river over a long period of time did not occur until the advent of major regulation of the Missouri main stem (*21*). Reduction of sediment downstream from major reservoirs has often resulted in progressive degradation in the reach below the dam and in some instances in redeposition well downstream.

Whether river regulation enhances or degrades water quality depends in part on details of operation as well as

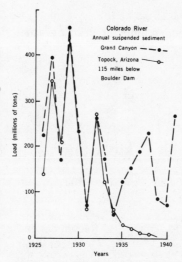

Fig. 7. Variability of annual suspended sediment load on the Colorado River and reduction in sediment load after closure of Hoover Dam in 1933 as reflected at Topock, Arizona, in contrast to upstream station at Grand Canyon [data from Howard (*58*)].

on the purposes of regulation. Clear water may reduce the cost of water treatment, provided that the water does not require the addition of solids for filtration. The clarity of the water may also increase algae production. Changes in gradient may result in sluggish flows preventing rapid dilution of waste from outfalls, and flow regulation may enhance growth of riparian vegetation. In that more than 150 major reservoirs have been constructed throughout the Middle West and western United States on virtually every major river, one can expect changes

in river sedimentation and in water quality in all of these rivers. In some instances, regulation of the flow will provide additional water for downstream reaches which will lead to the availability of water for dilution of wastes and hence, in the classical sense, a decrease in the probability of pollution. In others, the results may be exceedingly complex.

Sediment in the Schuylkill River in Pennsylvania has been significantly reduced by a major effort begun in 1950 (*13*). The reduction of sediment transported to tidewater reversed a trend in the Schuylkill initiated in the middle of the 19th century with the development of coal mining in the basin. The annual discharge of suspended sediment has been reduced about tenfold at Landingville in the upper basin, and downstream at Philadelphia a threefold reduction has been achieved (Table 4). While the program for coal mine reclamation and control has been unable to reduce the sulfate concentration in the river (see above), it took roughly 2 years at the upstream station for the accumulative curve of sediment yield to show a progressive reduction, and the effects of the program were felt downstream at Philadelphia roughly 3 to 5 years later. The total cost of the sediment control program—including the cost of three major desilting basins, the operating and maintenance costs required to remove accumulated sediment, and the costs of channel restoration involved in dredging 24 million cubic yards (18.5 million cubic meters) of fine sediment or culm associated with coal mine wastes—was originally estimated at $26 million (*22*). Actual expenditures have probably been higher.

Table 3. Sediment yield from drainage basins under different land use, eastern United States.

Drainage area (mi²)	Yearly sediment (tons/mi²)	Land use	Characterization
		Broad Ford Run, Maryland	
7.4	11	Forested: entire area	Presettlement
		Gunpowder Falls, Maryland	
303	808	Rural: agricultural 1914–43; farmland in county 325,000 to 240,000 acres	Intensive agriculture
	233	Rural: agricultural 1943–61; farmland in county 240,000 to 150,000 acres	Urban fringe rural
		Building site, Baltimore, Maryland	
0.0025	140,000	Construction: entire area exposed	Undergoing construction
		Little Falls Branch, Maryland	
4.1	2,320	Construction: small part of area exposed	
		Stony Run, Maryland	
2.47	54	Urban: entire area	Urban

Table 4. The effects of the restoration on sedimentation in the Schuylkill River (13).

Location	Drain- age area (mi²)	Trap efficiency of desilting basin (%)	Average annual suspended-sediment discharge*			
			Before restoration		After restoration	
			Tons	Tons/ mi²	Tons	Tons/ mi²
Landingville†	133		609,000	4,580	51,800	390
Desilting basin: Auburn	157	91				
Auburn	157		581,000	3,700	4,700	30
South Tamaqua† ‡	65.7		679,000§	10,000	62,400	950
Desilting basin: Tamaqua	67	86				
Desilting basin: Kernsville	340	49				
Schuylkill River at Berne	355		1,260,000	3,550	14,000	39
Schuylkill River at Philadelphia (Manayunk)	1,893		1,750,000	920	656,000	350

* Based on long-term flow duration and transport curve for indicated period. † Above desilting basin system. ‡ Called little Schuylkill River. § Calculated by subtracting suspended-sediment discharge at Auburn from Berne.

This record on the Schuylkill River indicates again that reclamation and reversal of trends are possible where the requisite control procedures exist and where society is willing to bear the cost.

Temperature

The use of water for cooling in industrial processes (particularly the large quantities demanded by modern power-generating stations), the increase in the numbers of impoundments—and the recognition that water temperature changes are associated with land use have all stimulated concern for the potential effects of higher temperatures in river water. Some examples of the magnitudes and effects of these changes have been documented.

In Long Island streams (23), urbanization produced changes in runoff and in temperature through changes in infiltration, in clearance of vegetation from the channel, in small impoundments in different reaches, and in land use adjacent to the channel. These changes have been accompanied by significant alterations in the temperature regime. On the forested Connetquot River, for example, the variability of both winter and summer temperatures is exceedingly small. On East Meadow Brook, which drains an urban watershed (Table 5), variations in temperature are large. Only a small length of channel of East Meadow Brook is bounded by shade trees, in contrast to the virtually continuous shade on the Connetquot. In addition, between 1937 and 1962 direct runoff increased by 270 percent on the urban watershed. Thus a seasonal temperature range on East Meadow Brook of 28°F (15.6°C) contrasts with 17°F (8°C) on the Connetquot, and mean temperatures in June

1967 were 6°F (3.3°C) higher on the urban stream. Removal of brush and trees along streams in the absence of urbanization was shown by Brown and Krygier (24) to increase the average maximum monthly temperature by 14°F (7.8°C) and to increase the annual maximum temperature from 57° to 85°F (14° to 30°C).

Analyses by Jaske and Goebel (25) of the effect of the large reservoir system on the Columbia River indicate that temperature variability has decreased with an accompanying small decrease in mean temperatures in upstream reaches of the river. However, below Hanford, Washington, no decrease in the mean temperature is apparent. Thus, the effect of the reservoirs has been to reduce temperature variability, much as flow variability is reduced by regulation. These observations of course do not describe the local effects of release of cold water, for example, from the bottom layers of stratified reservoirs, an effect seen downstream from a number of reservoirs in the southeastern United States.

The effects of urbanization and of reservoir regulation, while ubiquitous and perhaps growing as the urban areas grow, are probably small in contrast with the potential for temperature rises posed by heated discharges of cooling water from power plants. For this reason, the potential effects of such discharges have received much attention, and there have been several reviews of the subject (26).

A few illustrations are given here from the experience of the Tennessee Valley Authority (TVA) because they are suggestive of the nature and magnitude of the problem. The Widow's Creek steam plant on the Tennessee River, for example, has a capacity of 540 megawatts and the river has a flow of

47,000 cubic feet per second (Table 6). The cooling water discharges in a characteristic plume with higher temperatures at the surface decreasing in intensity with distance and with depth. The maximum temperature difference at the discharge point is 10.3°F (5.7°C) and a temperature elevation of about 1°F (0.6°C) exists at a distance of about 1 mile downstream from the plant. At three plants, as Table 6 shows, the maximum increase in temperature is from 10° to 12°F (5.6° to 6.7°C). At the Paradise plant on the Green River, essentially a pool, the maximum temperature is actually controlled by operation of the plant so as not to exceed the value shown.

Of particular interest is the response of the biota as observed at three sites reported in Table 6. On the Green River, at the Paradise plant, the temperature of the river water rises rapidly to about 98°F (37°C) immediately below the outfall. Initially the temperature rise is associated with a marked decline in zooplankton volume, but there is subsequent progressive recovery in the downstream direction. Recovery of all the biota is complete approximately 15 miles downstream from the plant (Table 6).

These few examples of temperature changes associated with cooling water discharges do not support a conclusion that major biota degradation will automatically accompany such discharges. Indeed, under some circumstances, the effects of thermal effluents on aquatic life appear minimal (27). However, where mixing appears to be limited and where the flow capacity of the river is low, a number of changes in aquatic life have been observed, including growth of epiphytic organisms on submerged surfaces near the plant, some reduction in the seasonal abundance of fish, especially in summer periods close to outfall areas, and a change (increase) in growth rates of some species. Beyond a relatively limited perimeter, temperature increases after mixing appear to be on the order of 1° to 5°F (0.5° to 3°C), the amount obviously influenced by the volume, location, temperature differences, and mixing characteristics encountered at each locality. While such results may not be unexpected, it is important to note that, in some instances, regulations have established upper bounds of temperature that may not be exceeded.

While temperatures on the Connecticut, the Ohio, and the Columbia rivers

213

appear to show little change over the past decade, rivers in urban areas and those subjected to large thermal effluents have apparently experienced some increase in temperature. That trends are exceedingly difficult to detect is suggested by preliminary study of mean temperatures for July and August on the Delaware River at Trenton from 1945 to 1965. Mean temperature corrected for discharge variations apparently increased 1.2°F (0.7°C). However, the great variability in the temperature indicates that the apparent trend is not statistically significant. Although discerning a trend in water temperature is difficult, evaluating potential biological effects is even more difficult. Unfortunately, empirical studies, which up to now suggest relatively modest effects on the biota, do not help to establish the level at which cumulative heating may exceed a biologically significant threshold. Such studies suggest simply that, in a number of places, if a threshold exists, it has not yet been exceeded.

Radioactivity

Radioactive materials have been released to the waters of the country from bomb testing in the atmosphere, from mining and milling, and from the use of radionuclides in industry, research, and commerce. For example, concentrations of strontium-90 in rivers, as measured at 128 stations in the United States, rose from a low in 1960 of less than 1 picacurie per liter to almost 4 pc/liter late in 1963 after a period of weapons testing (28). (Drinking water standards permit a maximum concentration of 10 pc/liter). From the cessation of testing in 1963, the level returned to about 1 pc/liter by late 1965.

For the mining industry a particularly interesting illustration is provided by a study of changes in gross alpha and beta radioactivity in a reach of the Colorado River in the vicinity of two mills (29). Samples of water, algae, and mud from the bed of the river were taken above and below two milling sites. As Fig. 8 shows, gross beta radioactivity rises rapidly immediately downstream from each site. The magnitude of the rise is greatest in the algae, and successively less in the mud and in the drinking water. Thus the relative concentration in the biota is highest and the beta activity is greater than the alpha. The increased radioactivity of

Table 5. Temperature variations and watershed characteristics of Long Island streams showing the effect of urbanization [data from Pluhowski (23)].

Item	Rivers	
	East Meadowbrook	Connetquot
Drainage (mi²)	31	24
Character	Urban	Forested
Vegetation along stream	Low	High
Temperature (°F)		
Mean of five stations		
June	66	60
January	42	42
Variation along stream		
June	10	0.5
January	2.0	1.0
Maximum range	17	28
Mean discharge (ft³/sec)	~16.6	~38.8
Runoff	270%*	3% †

* Direct runoff increase 1937–1962; † As stormwater, which was 3 percent of the total discharge.

the water exceeds that permitted by the drinking water standard of 1000 pc/liter at one location immediately downstream from a mill. Concentrations of beta activity in the algae decline more slowly and over a greater distance than they do in either the water or the bottom mud. While absolute amounts are low, the river travels 50 miles before the concentrations of radioactive materials are reduced to the background level.

No comprehensive figures which can be used to appraise the total content or extent of radioactive materials in the rivers of the country or the magnitude of changes in radioactivity in past decades are available. Data are available at several industrial locations, however.

Studies of the uptake of these nu-

clides by water, bottom mud, plankton, fish, and vascular plants indicate that removal of strontium-90 from the water is very limited in the Mohawk River downstream from the Knolls Atomic Power Laboratory. However, the amounts of cesium-137, zirconium-95, and strontium-90 in the bottom muds decrease by several orders of magnitude in a pool at a distance of about 2 miles, and at locations 12 to 15 miles downstream concentrations were lower by factors of 10 to 100 as compared to maximum values at the outfall (30). Cesium concentrations in algae also declined in the downstream direction.

On the Clinch and Tennessee river systems in the vicinity of Oak Ridge National Laboratory concentration of most radionuclides in bottom muds declined downstream by 4 or 5 orders of magnitude over a distance of 20 to 50 miles, except in reservoir deposits near dams. Quantities of radioactive materials in bottom muds of the Clinch River are directly related to the deposition of fine-grained sediments, particularly clay minerals, just as radioactivity is associated with fine sediments in the Columbia River below the source locations and above McNary Dam (31). On the Clinch River, according to Carrigan (32, pp. 15 and 17), "during 19½ years of waste disposal 64 percent of the stream bed has become covered with radioactive sediment, and, in the 22-mile-long reach, at least 20 percent of the cesium-137 and rare earths released to the river are retained in the bottom sediment." Present levels are not a hazard. Whether retention poses an eventual hazard depends both on future releases, now declining, and on

Table 6. Thermal effects: some TVA power plants (67).

Total capacity (Mw)	River-flow (ft³/sec)	Velocity (ft/sec)	Cooling water discharge (ft³/sec)	Temperature increase (°F)		Effect on aquatic life
				Maximum (°F)	After mixing (°F)	
Widows Creek, Guntersville Reservoir						
540	47,000	1.75	2,200	10.3	1	Diversity and abundance same above and below
Colbert, Pickwick Reservoir						
800	55,000	0.9	1,900	12		Discharge to Cane Creek; no significant effect on bottom organisms
Paradise, Green River above dam						
1,380	1,300	0.25	550	11 (controlled)		Relative abundance of fish reduced Summer periphyton growth rate down; winter, some rise Recovery 15 miles downstream Slime growth greatest near plant

Table 7. Pesticides in California rivers [from (36)].

Area	Application rate (pound/acre)	River	Location	Pesticide concentration (μg/liter)	
				Water	Mud
Sacramento Valley	0.75–1.0	Feather	Nicolaus Bridge	0.009	24
		American	Sacramento	.120	
		Sacramento	Walnut Grove	.113	94
Southern San Joaquin Valley	1.25 DDT	San Joaquin	Vernalis	.229	31
	2.40 other	San Joaquin	Antioch	.101	212
Imperial Valley	1.25 DDT	Alamo		.421	
	2.40 other				

maintenance of safe disposal practices. Concentration in fish has generally followed the food chain, with highest concentrations in plankton feeders and the lowest in game fish. Concentration factors for strontium-90 in fish ranged from 10 to 100 (30).

These examples confirm the fact that industrial activities are increasing the burden of radioactivity on specific rivers. Under existing controls, hazardous levels are not experienced at most sites. The extent to which such materials may accumulate or be dispersed in the river environment is a function of the half-life of the radionuclides, absorption by biota, and the rate of transport of sediments. Work remains to be done on the long-term relations among these variables, including the role of benthic organisms in concentrating radioactivity from both inorganic and organic sediments. Because radioactivity is carried upward in the food chain and because of the long half-life of many radionuclides, continuing vigilance appears warranted. Such vigilance should be coupled with recognition that thus far processes have not apparently resulted in widespread spatial dispersion of high concentrations of radionuclides in rivers.

Pesticides

Expansion of pesticide use is a post–World War II phenomenon, and warnings of potential damages beyond the point of application were issued in the 1940's. By 1962, a sampling program of 101 stations revealed DDT in 32 samples from nine rivers in the United States (33). Subsequent studies have added to the number of locations at which pesticides have been observed in river water and to the knowledge of modes of dispersal and accumulation (34). Schafer et al. (35) reported detectable quantities of dieldrin in 40 percent and of endrin in more than 30 percent of 500 samples of finished water derived from the Mississippi and

Missouri rivers. Smaller amounts of other pesticides were also found.

Because of the extensive agricultural enterprise in California, approximately 20 percent, as has been estimated, of all pesticides used are applied in that state. Thus the careful documentation of pesticide distribution in some California rivers by Bailey and Hannum (36) provides an excellent illustration of the relation between heavy pesticide application and its behavior in river systems. Where pesticide applications are heavy, such as in the southern San Joaquin and Imperial valleys, concentrations are correspondingly high in river waters (Table 7). For the Sacramento River system as a whole, high values of 0.229 micrograms per liter in the San Joaquin near Vernalis, 120 miles above Golden Gate, decline to 0.110 μg/liter at Antioch 66 miles downstream. The decline continues to San Francisco Bay because of "inflow dilution, including tidal flushing and uptake by aquatic organisms and sediments" (36). Concentration decreases at a rate of about 0.0016 μg/liter per mile. Pesticide concentrations in sediment exceeded those in water 20 to 100 times, with concentrations being proportionately higher as the sediment becomes finer. Even on the Feather River, at an upstream station where water samples averaged only 0.009 μg/liter, the average of ten samples of sediment contained a concentration of 24 μg/liter.

In contrast to the high concentrations observed in the agricultural areas of California, lighter applications in forestry suggest more rapid disappearance of the pesticide with surface and river runoff. On several experimental sites in Oregon, for example, endrin from endrin-coated Douglas fir seed was detected in streamflow 2 hours after seeding and during a high flow 6 days later. The total, however, amounted to only 0.12 percent of the endrin used, and during the storms later in the year concentrations averaged 0.02 μg/liter (37). The results were roughly similar in a study of concentrations of the herbicide

amitrole sprayed on a forested watershed. Immediately below the site a maximum concentration of 155 parts per billion, measured 30 minutes after application, decreased to 26 ppb after 2 hours, and none was detectable after 5 days. Similarly, no amitrole was detected at a station 1.8 miles downstream (38).

Although the association of fine-grained sediments and pesticides has been established, studies in the lower Mississippi River area failed to show significant concentrations at predictable depositional loci but rather concentrations were low and dispersed and no buildup was observed where pesticides were derived from farm use. Concentrations were greater near manufacturing and formulating plants (39).

Many of the very high concentrations of pesticides experienced in rivers or streams resulting in major fish kills, for example, have been associated with direct discharges, spills, or direct spraying from the air to the watercourse. The number of fish kills from pollutants, about 400 to 500 per year (40), has changed relatively little since reporting began. For other pollutants, as well as pesticides, however, each year is characterized by one or more excessive or exceedingly high mortalities associated with a particular accident. Statements such as (40)

The dam of a temporary slush pit broke and its contents (sodium chloride and/or bentonite, a drilling mud) entered the creek 100 yards from the well site. The pollution gave the creek a "milky" appearance and traveled the entire 39 miles as a "slug" . . . (1,200,000 fish). or, Potato field north of a creek was sprayed by plane . . . Area is stocked with trout . . . Water appeared normal, no other damage apparent (600,000 fish). . . .

in the reports of fish kills clearly indicate that the increase in manufacturing industries and ubiquitous use of insecticides is inevitably accompanied by accidents during manufacture and application of the product. In general, industrial and municipal activities account for about 50 to 75 percent of the reported fish kills and insecticides for only about 10 percent (40). The total number of reported incidents does not appear to be large, but presumably the number of accidents will continue to be some proportion of the total volume of use. While active campaigns promoting more careful handling will tend to reduce accidents, if increasing use of highly toxic products continues, one can expect increasing reports of the presence of pesticides as well as increasing numbers of significant fish kills from specific accidents.

Pesticides of different composition, of course, behave quite differently. Thus DDT with a lower solubility in water is likely to be found in sediments or biota and at lower concentrations in the river water itself. In addition, a number of the phosphorous compounds are degradable and short-lived in contrast to the persistent life of chlorinated hydrocarbons. In many ways, the chlorinated hydrocarbons pose much the same problem as radioactive materials do; they are persistent and concentrations increase from water, to sediment, and thence to biota with progressively higher concentrations higher in the food chain.

Experience shows that biota in river systems may be decimated by biocides. Effects of specific doses will depend not only on the strength and composition of the biocide, but on the seasonal and feeding cycles of the organisms. At the same time, populations may readily recover, either by drift or from remnant resident populations (41). In contrast,

subtle or delayed effects of pesticides accumulated in sublethal doses in various organisms have also been documented and may pose continuing hazards so long as persistent pesticides continue to be used. There is some evidence that public concern with the effects of such pesticides may be leading to a reduction in their use in the United States. However, even with marked reduction in use, many years may elapse before residuals accumulated in soils, vegetation, and other organisms are flushed to rivers and, in turn, disappear from the river systems themselves. With improved detection systems for pesticides, their reported presence in river water, bottom muds, or biota may continue to increase for some time to come. Thus, while one might suggest that thus far the magnitude of the effect of biocides on rivers has been quite limited, little can be said on the distribution and long-term alterations which may accompany increased use of biocides.

Trash and Debris

In the eyes of many, trash and debris are ubiquitous features of rivers and bays, particularly in metropolitan regions. Photographs of tires, market carts, tangled lumber, and tin cans commonly appear in newspapers and magazines. While sequential sets of ground or aerial photographs which might provide a record of changes in the quantity of debris at a given site are

unavailable, paired photographs at two points in time are graphic but not quantitative. There is reason to believe, however, that debris may be one of the most significant elements influencing the public's perception of pollution.

The inner harbor of the port of Baltimore (north branch of the Patapsco River) includes an area of about 1000 acres (400 hectares). A principal tributary, Jones Falls, drains 46 square miles, of which about 20 are in urban land uses. Several large storm drains also empty directly into the harbor.

From 1961 through 1969 approximately 600 tons of debris per year were removed from the inner harbor by the debris recovery program (42). While a marked increase appeared from about 300 to 600 tons between 1963 and 1964, data are insufficient to define a significant trend. (The estimated cost of removal is roughly $15 per ton.) Observations suggest that Jones Falls, the storm drains, illegal disposal of dunnage, and rotting derelicts are the principal source of debris. Monthly figures show that the maximum tonnage usually occurs in the summer months and is presumably related to rapid runoff from intense storms. This is corroborated by limited data from a trash boom sometimes in operation on Jones Falls. In one September storm, at least 7 tons of debris were trapped by the boom, and it is estimated that an equal amount escaped earlier when the boom failed. Each of several smaller storms contributed 4 tons of debris, and in one event 52 tires were retrieved. Observations of

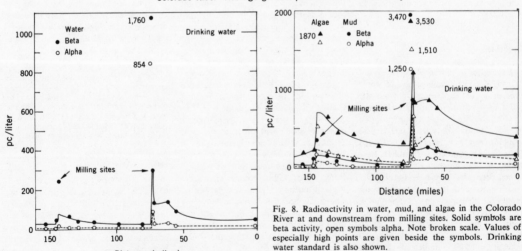

Fig. 8. Radioactivity in water, mud, and algae in the Colorado River at and downstream from milling sites. Solid symbols are beta activity, open symbols alpha. Note broken scale. Values of especially high points are given beside the symbols. Drinking water standard is also shown.

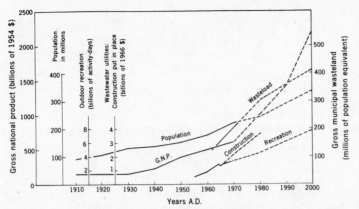

Fig. 9. Measures of demand on water resources including gross national product, population, municipal waste loads, and outdoor recreation, as compared with proposed construction of waste-water utilities to parallel waste loads. Estimated construction expenditures should approximate $4 to $5 billion per year. Actual expenditures approximately $1.5 billion per year (59, 60) [data from (59) and (60)].

urban rivers in Baltimore suggest that individual reaches may contain many cubic yards of debris per city block (43). This material appears to be periodically flushed to the estuaries by large storm flows.

Comparable quantities of debris (150 tons in 6 months) are being retrieved from the Potomac River near Washington, D.C., and doubtless equal and larger amounts from other areas (44). Data for the establishment of trends are not available. Of more importance, it may well be that volumes and weights are particularly inadequate as measures of the impact of debris on pollution.

Conclusions

All measures of growth whether population, gross national product, waste loads, or demand for recreation suggest that demands on the country's water resources are increasing at a rate that exceeds the rate of installation of waste treatment facilities (Fig. 9). Presumably it is these demands which constitute the customary measure of "pollution" and upon which prophesies are made regarding the condition of the resource and its future prospects. Pressures posed by population and growth must have an effect on the river systems which, even with regulation, have finite assimilative or carrying capacity for wastes of different kinds. At the outset, then, one might assume, even in the absence of proof, that the very large increase in society's activities should have recognizable conse-

quences. Unfortunately, it is easier to estimate the potential demand on a river system than it is to measure the effect. Certainly many rivers of the United States, like the rivers in Illinois, Colorado, and other places cited here, are not as they were 70 years ago. Nevertheless, while we "know" that conditions must be getting worse, we are hard-pressed to determine precisely the relation between the pressures posed by society and the responses of the river system.

To a large extent observations of the condition of river systems have been confined to measures of the quality of the flow itself and exclude descriptions of the bed, banks, and environs of the river. In some instances the parameters observed, for example, dissolved oxygen or coliform organisms, are surrogates for broader effects of greater interest. Although useful, they provide only partial measures of the river's condition or assimilative capacity (45). Thus in the Hudson River, while it can be shown that the change in dissolved oxygen is exceedingly small since 1922, we have some reason to believe that dissolved oxygen inadequately describes accumulating sludge on the bottom of the river as well as increasing growth of algae. Many years ago Henderson (46) demonstrated that a portion of the Shenandoah River was virtually devoid of life despite high oxygen levels and other quality standards of the water. Intended to surmount this disability, current attempts to quantify biological observations may not in fact do so unless they are associated with true bio-

logical surveys (47). Observation programs continue to emphasize measures of water quality to the virtual exclusion of measurements of river quality (48).

Observational programs appear to be particularly weak with regard to the detection of subtle initial changes from a natural to a polluted condition. This is not to suggest that no such observations are being made, for example on wild rivers, but rather that there is no systematic program for the detection of initial changes.

The initiation and progression of change are often considered impossible to prevent and inexorable in character. This leads further to the view that virtually all changes from a pristine to a used condition are irreversible. Such does not appear to be the case. Experience on the Ohio River described above, the reversal of changes in surfactant concentrations in the Illinois River, and the marked reduction of sediment in the Schuylkill River illustrate that it is possible to reverse or to improve the conditions of some rivers with respect to important pollutants. Fish may migrate rapidly to formerly polluted areas when either industrial shutdowns or large cleanup programs occur. The Thames River near London is reported to have fish for the first time in very many decades as a result of a major pollution control program. Indeed, a brief report suggests "that the condition of rivers in England and Wales . . . may well have improved since 1958" (49).

Because the evidence up to now does indicate that the pressure on the water resources posed by population and industrial growth significantly exceeds the rate of investment in control facilities, much higher cost must be incurred if even a rough parallelism between development and control is to be achieved. More important, the kind of technology that will be needed is not only more expensive, but may be distinctly different in kind from secondary treatment that is now projected for most municipal areas.

Even if treatment or pollution control were to maintain parity with pollution pressure, there is good reason to believe that many would continue to hold to the definition that "pollution is that which is getting worse." In the eyes of the beholders (50), the sights and smells of rivers today may be considered unsatisfactory, and while the "quality" of the rivers may be unchanged, they will be viewed as being polluted and becoming more so, as the interests of those living in more crowded areas turn toward rivers to satisfy their recreational and esthetic needs. Many observers have noted historic changes in values in the environment, and there is reason to believe that these play no small part in the current scene (51).

The paucity of information and the handful of investigators concerned with evaluating trends in the quality of the rivers of the United States suggest some specific conclusions. First, none of the observational programs were designed specifically to measure the quality of rivers or the river environment. The sampling programs emphasize the measurement of specific characteristics primarily related to water use by industry and municipalities. The new National Water Quality Network should improve on this single objective orientation (52). Few observational programs combine the necessary hydrology with measurements of water quality, river characteristics, and biology. While some long-term observations exist, the lack of coordinate observations makes long-term comparisons virtually impossible. For this reason, one must resort to the selected or case method described here. In addition, as Dworsky and Strandberg (53) emphasize, interpretation is "the vital part of the task of water quality assessment." Such interpretation requires the knowledge and skill of analysts familiar both with the data and with the changing characteristics of the land use and economy of the drainage basin. The new emphasis on quality of the environment demands continuing assessment and interpretation.

A second conclusion from the available data suggests that surrogate measures of river and water quality as well as a multiplicity of measurements of easily measured parameters may shed little light on the dynamics of the processes active in river systems and hence such measures may be of limited use in estimating the likelihood of reversing specific observed trends in the absence of a knowledge of their causes. Additional attention must be given to the measurement of parameters related to models of river behavior and to estimates of inputs based on budgets of materials derived from industrial outputs and land use.

Third, while hydrologists have long been concerned with variability of the flow of natural rivers, because of the difficulty of observation, much less attention has been given to the variability of biological activity as well as physical variability associated with natural variations and cycles in rivers. Many measurements of biological effects are done during low and summer flows where measurement is easy, organisms often flourish, and concentrations of various substances in the flow are high. The effect of winter flow on the growth of slimes on the bottom of rivers, for example, and the special significance to the flora and fauna of periodic floods are not well documented. Significantly, however, among the most common trends in river management is the progressive regulation of flow through the provision of storage. Conceivably regulation rather than pollutants alone may have the most far-reaching effects on the character of many river systems. To date, observations have not been designed to measure these effects.

Because the demands on the waters of the rivers of the country are increasing, the concept of threshold and irreversibility must be studied on (i) pristine waters to disclose the nature of the initial, presumably biological, changes which take place and (ii) specific rivers where large scale control or cleanup programs have been initiated. It may well be that observations designed to detect "polluters," that is, observations designed to support the enforcement of standards, may not in themselves provide satisfactory measures of thresholds, trends, or reversals of trends. If one is to judge the effectiveness of the expenditure of large sums of money, observational tools must be designed to evaluate the response of the rivers to these expenditures.

References and Notes

1. W. C. Ackerman, R. H. Harmeson, R. A. Sinclair, Trans. Amer. Geophys. Union 51, 516 (1970).
2. E. B. Phelps, Sewage Works 6, 998 (1934).
3. Division of Water Supply and Pollution Control, Pollution of the Hudson River and Its Tributaries (U.S. Department of Health, Education and Welfare, Public Health Service, Washington, D.C., 1965), 102 pp.
4. Metropolitan Sewer Board of the Twin Cities Area, personal communication, October, 1970.
5. Federal Water Pollution Control Administration (FWPCA), Delaware Estuary Comprehensive Study, Preliminary Report and Findings (Department of the Interior, Washington, D.C., 1966), p. 94.
6. W. B. Keighton, U.S. Geol. Surv. Water Supply Pap. 1809-0, 1–57 (1965).
7. W. N. Torpey, J. Water Pollution Control Fed. 39, 1797 (1967).
8. R. H. Gould, Proc. Amer. Soc. Civ. Eng. J. San. Eng. Div. 94, 1041 (1968).
9. G. P. Howells, T. J. Kneipe, M. Eisenbud, Environ. Sci. Technol. 4, 35 (1970).
10. B. H. Ketchum, in Eutrophication, G. A. Rohlich, Ed. (National Academy of Sciences, Washington, D.C., 1969), p. 197.
11. P. W. Anderson and S. D. Faust, U.S. Geol. Surv. Prof. Pap. 525-D, 214 (1965).
12. L. T. McCarthy, Jr., and W. B. Keighton, U.S. Geol. Surv. Water Supply Pap. 1779-X, 42 (1964).
13. J. E. Biesecker, J. B. Lescinsky, C. R. Wood, Commonw. Penn. Water Res. Bull. No. 3 (1968), 193 pp.
14. W. V. Iorns, C. H. Hembree, G. L. Oakland, U.S. Geol. Surv. Prof. Pap. 441, 1–370 (1965).
15. W. T. Sullivan and R. L. Evans, Environ. Sci. Technol. 2, 194 (1968).
16. Ohio River Valley Water Sanitation Commission, 17th Annual Report (1965), p. 28.
17. ———, Cloride Control Considerations for the Ohio River, unpublished report 1957.
18. E. J. Cleary, The Orsanco Story (Johns Hopkins Press, Baltimore, 1967).
19. H. E. Thomas, U.S. Geol. Survey Circ. 346, 27 (1954).
20. R. V. Ruhe and R. B. Daniels, J. Soil Water Conserv. 20, 52 (1965).
21. P. R. Jordan, U.S. Geol. Surv. Water Supply Pap. 1802, 1–89 (1965).
22. U.S. Army Corps of Engineers, Report on Schuylkill River, Pennsylvania, House Document No. 529, 79th Congress, 2nd Session.
23. E. J. Pluhowski, U.S. Geol. Surv. Prof. Pap. 627-D, 1–109 (1970).
24. G. W. Brown and C. A. Krygier, Water Resour. Res. 6, 1133 (1970).
25. R. T. Jaske and J. B. Goebel, J. Amer. Water Works Ass. 59, 935 (1967).
26. L. D. Jensen, R. M. Davies, A. S. Brooks, C. D. Meyers, The Effect of Elevated Temperature upon Aquatic Invertebrates: A Review of the Literature Relating to Fresh Water and Marine Invertebrates (Department of Geography and Environmental Engineering, Johns Hopkins Univ., Baltimore, 1969).
27. D. Merriman, Sci. Amer. 222, 42 (1970).
28. Water Pollution Surveillance System, Annual Compilation of Data 1962–63, Public Health Serv. Publ. No. 663 (1963), p. 5.
29. J. M. Morgan, Jr., A Stream Survey in the Uranium Mining and Milling Area of the Colorado Plateau, Colorado and Gunnison Rivers (Johns Hopkins Univ., Baltimore, 1959).
30. A. G. Friend, A. H. Story, C. R. Henderson, K. A. Busch, "Behavior of certain radionuclides released into fresh-water environments," Ann. Rep. 1959–60 Publ. Health Serv. Environ. Health Ser. (1965), 89 pp.
31. J. L. Nelson and W. L. Haushild, Water Resour. Res. 6, 130 (1970).
32. P. H. Carrigan, Jr., U.S. Geol. Surv. Prof. Pap. 433-I, 1–18 (1969).
33. A. W. Breidenbach and J. J. Lichtenberg, Science 141, 899 (1963).
34. H. P. Nicholson, ibid. 158, 871 (1967).
35. M. L. Schafer, J. T. Peeler, W. S. Gardner, J. E. Campbell, Environ. Sci. Technol. 3, 1261 (1969).
36. T. E. Bailey and J. R. Hannum, Proc. Amer. Soc. Civ. Eng. J. San Eng. Div. 93, 40 (1967).
37. R. B. Marston, R. M. Tyo, S. C. Middendorf, Pestic. Monit. J. 3, 167 (1969).
38. R. B. Marston, D. W. Schults, T. Shiroyama, L. V. Synder, ibid. 2, 123 (1968).
39. W. F. Barthel, J. C. Hawthorne, J. H. Ford, G. C. Bolton, L. L. McDowell, E. H. Grissinger, D. A. Parsons, ibid. 3, 8 (1969).
40. FWPCA, Annual Reports, 1960–1968, Pollution-Caused Fish Kills (U.S. Department of Health, Education, and Welfare–U.S. Department of the Interior, Washington, D.C., 1960–1968).
41. H. B. N. Hynes and T. R. Williams, Ann. Trop. Med. Parasitol. 56, 78 (1962).
42. Data from Maryland Port Authority, personal communication.
43. M. G. Wolman, Geogr. Ann. Ser. A 49, 385 (1967).
44. A. Wolman, J. C. Geyer, E. E. Pyatt, A Clean Potomac River in the Washington Metropolitan Area (Interstate Commission on the Potomac River Basin, Washington, D.C., 1957). 63 pp.
45. A. L. Downing and R. W. Edwards, Institute of Water Pollution Control Annual Conference (Water Pollution Research Laboratory, Stevenage, England, 1968), reprint No. 552.
46. C. Henderson, Progr. Fish Cult. 11, 217 (1969).
47. H. B. N. Hynes, in "Symposium on Environmental Measurement (Pub. Health Serv. Publ. No. 999-AP-15 (1964), pp. 289–98).
48. R. H. Langford and G. H. Davis, Proc. Amer. Soc. Civ. Eng. J. Hydr. Div. 96, 1391 (1970).
49. M. Owens, P. Maris, H. Rolley, New Sci., 2 April 1970, p. 25.
50. For a recent confirmation, see E. L. Davis, Water Resour. Res. 7, 453 (1971).
51. D. Lowenthal, Geogr. Rev. 58, 21 (1968).

52. W. T. Sayers, *Environ. Sci. Technol.* **5**, 114 (1971).
53. L. B. Dworsky and W. B. Strandberg, *Cornell Univ. Water Resour. Center Publ.* **16**, 71 (1967).
54. E. B. Phelps, *Stream Sanitation* (Wiley, New York, 1944).
55. U.S. National Resources Committee, Special Advisory Committee on Water Pollution, *Water Pollution in the United States, House Document No. 155*, 76th Congress, 1st Session, (1939), 165 pp.
56. Minneapolis–Saint Paul Sanitary District, 33rd Report, 1965, 88 pp.
57. J. G. Geyer, J. H. Carpenter, D. W. Pritchard, C. E. Renn, D. C. Scott, M. G. Wolman, "A research program for the Potomac River," report, Department of Sanitary Engineering, D.C., and Bureau of Environmental Hygiene, Maryland State Department of Health (1965), 145 pp.
58. C. S. Howard, *U.S. Geol. Surv. Water Supply Pap.* **998**, 1–65 (1947).
59. U.S. Water Resources Council, *The Nation's Water Resources*, part 1 (1968), 300 pp.
60. U.S. Department of Commerce, Business and Defense Services Administration, *Regional Construction Requirements for Water and Waste-Water Facilities* (1967).
61. U.S. Department of Commerce, U.S. Census Population 1940, p. 32.
62. ——, U.S. Statistical Abstract, 55 (1950).
63. ——, U.S. Statistical Abstract, 18 (1965).
64. New York City, Environmental Protection Administration, personal communication, July 1971.
65. Philadelphia Water Commission, *Clean Streams for Philadelphia* (1970).
66. Ohio River Valley Water Sanitation Commission, *Annual Report* (1968).
67. M. Churchill and T. A. Wojtalik, *Nuclear News* **12**, 80 (1969).
68. I thank colleagues at Johns Hopkins University for review and criticism of the manuscript; and Edward Cleary, C. H. J. Hull, William Ackerman, and Leonard Dworsky for comments on the manuscript; R. J. Dougherty of St. Paul, Minnesota; E. Cleary of Orsanco; and Captain Robert Wilcox at the Port of Baltimore, for unpublished data.

16

Reprinted from *J. Water Pollution Control Federation*, **41**, No. 9, 1635–1646 (Sept. 1969)

SYSTEMS ANALYSIS FOR OPTIMAL WATER QUALITY MANAGEMENT

Ethan T. Smith and Alvin R. Morris

In late 1961, the Delaware Estuary Comprehensive Study was undertaken by the Federal Water Pollution Control Administration* in cooperation with the Delaware River Basin Commission, the State of Delaware Water Pollution Commission, the New Jersey Department of Health, the Pennsylvania Department of Health, and the City of Philadelphia Water Department.

The study objectives were:

1. Develop methods of water quality management, including techniques for forecasting the variations in water quality caused by natural or man-made causes.

2. Determine the cause and effect relationships between pollution from any source and the present deteriorated quality of water in the estuary.

3. Prepare a program for the improvement and maintenance of water quality in the estuary, including the waste removal and other control devices necessary to manage the quality of water in the estuary for municipal, industrial, and agricultural water use, and for fisheries, recreation, and wildlife propagation.

In developing the estuary study, an attempt was made to define strictly controls for water pollution, i.e., 'the management of man's environment through a set of operational procedures

* Its predecessor Agency, the Division of Water Supply and Pollution Control, U. S. Public Health Service.

Ethan T. Smith and Alvin R. Morris are, respectively, Chief, Program Management Section; and Chief, Planning Branch, Delaware Estuary Study, Hudson-Delaware Basins Office, Federal Water Pollution Control Administration, Edison, N. J.

(regulating waste discharges, flow regulation, etc.) to achieve a desired water-use goal in an optimal fashion.

The definition obviously requires an expression of water-use goals—usually a difficult attainment. A second problem is that the definition presupposes the existence of a useful description of the environment in an engineering or scientific sense. Manipulation of a complex system is impossible without knowledge of the major cause and effect relationships involved. The results of any actions taken on the various major components of the physical environment should be able to be predicted with reasonable confidence. A third problem involves the resolution of the first two in an optimal manner. With defined goals plus cause and effect relationships in hand, criteria for selection of a particular program still must be developed, i.e., "optimum" must be given specific characteristics for the situation being analyzed.

Mathematical Model of the Estuary

The initial effort was to embody the cause and effect relationships of the physical environment in a mathematical model which could be programmed for a digital computer. One of the most significant cause and effect relationships in water pollution control is that between biochemical oxygen demand (BOD) and dissolved oxygen (DO). BOD is a measure of the quantity of dissolved oxygen removed from the water by the metabolic activity of microorganisms oxidizing materials in the water and can be expressed in lb/day (kg/day) of oxygen demanding load.

The DO present in the stream often is used as an index of water quality. This is because fairly high levels (i.e., >4.0 mg/l) of oxygen are required to maintain a desirable population of organisms, including the aerobic bacteria that consume waste material in the water. However, low levels of DO (i.e., <1.0 mg/l) mark the transition to anaerobic conditions, which are denoted by odors of hydrogen sulfide, organic sulfides, blackening of the water, and destruction of the desirable fish species and many other aquatic organisms. The large loads imposed on the Delaware Estuary deplete DO resources in some locations to critical levels (i.e., <0.5 mg/l) during the summer months.

Mathematical descriptions of the relationships in the physical estuary system were combined with the primary inputs to the estuary, forming a mathematical representation, or model, of the estuary.

The model utilized the following primary forcing functions (i.e., inputs):

(a) water temperature, and
(b) waste loads.

The system components that transform or operate on these inputs to produce DO output include:

(a) freshwater inflow to the estuary,
(b) magnitude of tidal turbulence,
(c) size of the estuary in terms of the volume of specific reaches,
(d) reaeration characteristics, and
(e) decay characteristics of organic matter.

The basic system for DO is composed of two subsystems—the BOD system and the DO system.

The mathematical representation of the estuary is divided into 30 sections with the lengths chosen representing geographical areas which essentially are homogeneous in water quality. The 30 sections model the area of the estuary from Trenton, New Jersey, to below Wilmington, Delaware. Each section of the model represents a distance of either 10,000 or 20,000 ft (3,048 or 6,096 m) of the real estuary. Within a section, all parameters are considered uniform; however, the parameters can be varied from section to section. For each of these sections a mass-balance equation was written for the BOD system, and another for the DO system. This resulted in two linear differential equations based on the physical, hydrological, and biochemical characteristics of the section. These equations are as follows:

System 1

$$V_i \frac{dL_i}{dt} = Q_{i-1,i}[\xi_{i-1,i}L_{i-1} \\
+ (1 - \xi_{i-1,i})L_i] \\
- Q_{i,i+1}[\xi_{i,i+1}L_i \\
+ (1 - \xi_{i,i+1})L_{i+1}] \\
+ E_{i-1,i}(L_{i-1} - L_i) \\
+ E_{i,i+1}(L_{i+1} - L_i) \\
- d_iL_iV_i + f_i..1$$

System 2

$$V_i \frac{dc_i}{dt} = Q_{i-1,i}[\xi_{i-1,i}c_{i-1} \\
+ (1 - \xi_{i-1,i})c_i] \\
- Q_{i,i+1}[\xi_{i,i+1}c_i \\
+ (1 - \xi_{i,i+1})c_{i+1}] \\
+ E_{i-1,i}(c_{i-1} - c_i) \\
+ E_{i,i+1}(c_{i+1} - c_i) \\
+ V_ir_i(c_{s_i} - c_i) \\
- d_iL_iV_i + P_i..2$$

where;

V_i = volume, l,
Q_i = net flow, l/day,
L_i = BOD level in the section, mg/l,
ξ_i = tidal mixing parameter, dimensionless,
E_i = dispersion coefficient, l/day,
d_i = decay rate, l/day,
f_i = BOD discharged to the section, mg/day,
c_i = dissolved oxygen level in the section, mg/l,
r_i = reaeration rate, l/day,
c_{s_i} = saturation level of DO, mg/l, and
P_i = other sources or sinks of DO acting on the section, mg/day.

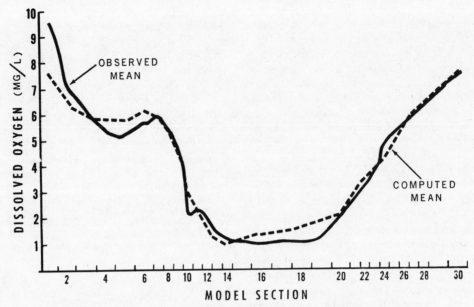

FIGURE 1.—Steady-state verification was performed for the June through
August 1964 period.

Modeling of all 30 sections yields two systems of 30 simultaneous equations each.

The details of this model are described by Thomann (1). In the steady-state application of this model, a set of transfer functions can be obtained from the equations. The set of transfer relationships details the transformation of a waste load input in any section to the stream quality output in any other section. For example, effluent BOD input in one section can be translated to output DO in another. This constitutes the simulation of the primary cause-effect relationship.

This steady-state mechanism was formulated as a digital computer program for the CDC 1604 and served as the basic program on which various techniques of load reduction and redistribution were superimposed later.

When the appropriate input data were supplied to the model for the summer months, it was possible to produce a satisfactory verification of the observed record. Steady-state verification was performed for the June through August 1964 period (Figure 1).

The verification obtained is inferred to support the theoretical approach of using a mass-balance model to simulate the estuary.

Water-Use Goals

The water-use and water-quality goals used in developing the estuary pollution control program were ascertained through a technical, quasi-political decision-making process involving the community of water users in the region; appropriate representatives of these were organized as the Water Use Advisory Committee (WUAC). They were queried about possible swimming areas, fishing locations, community desires on water withdrawal from the estuary, and industrial desires on water use. Based largely on their response, the many alternatives for improving water quality were reduced to five combinations. These were termed "Objective Sets," and gave alternative quality levels starting with "Objective Set V," that is, present water quality, and ranging to "Objective Set I," the maximum feasible enhancement of the

Delaware River using present technology. Each objective set is composed of 12 water quality parameters, of which DO is the most important.

Control Programs Based on Cost Allocation Models

Once a set of alternative goals is specified in terms of DO and the estuary's input-output characteristics can be reliably specified, the question of how a particular goal may be secured optimally can be asked.

Generally speaking, it is possible to determine the maximum allowable waste that can be discharged to the estuary under any given objective set. But how should this total quantity of load, and its associated costs, be allocated among the individual sources located along the length of the river? For example, in the case of objective set III, a maximum of 500,000 lb/day (227,000 kg/day) load may be discharged by all sources acting together. One possibility is to require that the largest sources remove the maximum possible amount of their load, and that only subsequently would smaller sources be required to act. This method, as employed in the past, leaves a great deal to be desired. No cognizance is taken of the relative costs of removal between sources, nor is any attempt made to treat sources in an equitable fashion. The problem that remains is how to select a rational set of treatment levels for the various sources along the estuary. The selection should balance the apparent equity of the solution to the individual source, the economic cost to the region, and the ease of administering the management program. The selected levels also must result in attainment of the DO goals which are used as an index of overall water quality.

The importance of cost is well illustrated by considering alternative waste removals in association with the cost-functions of the sources. For example, even in a simple situation involving only two sources, the best policy to follow is far from clear. Because of the influence of tidal hydraulics, the magnitude of the waste load removals which are required depends on the location of the sources on the estuary. However, the costs of removal depend on the economics of the individual sources which are extremely variable. Which source should be chosen? Should a solution be compounded of partial abatement at each source; if so, how much abatement and at which source?

Facilities to secure various water quality objectives require large investments for construction, operation, and maintenance. Thus a careful comparison of the various alternatives is necessary to ensure that the selected program best satisfies the needs and desires of the region. Comparison of alternative programs on the basis of cost allows two important comparisons:

1. The different costs for obtaining a given objective set are good criteria on which to base decisions as to type of control to be used in the abatement program.

2. Comparison of the costs to attain different objective sets are helpful in selecting a final set.

The cost of attaining various water quality objectives was determined under each of the following allocation programs:

1. Uniform Treatment—Each source to remove identical percentage of its raw load (i.e., the load before any reduction has taken place).

2. Cost Minimization—Using an optimization technique of linear programming, a program was formulated which would reach the desired quality goals at a minimum cost to the region. It selects effluent modifications based on greatest waste removal per dollar of treatment cost and simultaneously considers where in the estuary water quality must be improved.

3. Zoned Optimization—This solution combines elements of the uniform

and least cost approaches. The estuary was divided into a series of treatment or management zones, the treatment level in each zone chosen so as to achieve the water quality goal at minimum overall cost. Several constraints common to all three of the above methods are imposed on the various solutions:

(a) In no case will the DO in any area be permitted to decrease below its present level, even if present levels exceed the legal standard in the area.

(b) No effluent source may discharge any load above that discharged at present. When a particular treatment level is chosen, sources currently treating their wastes to a higher level may not lower the degree of treatment.

Each of these cost allocation programs now will be examined in turn. Some of the methodology will be described, and the implied efficiency and equity of each program will be pointed out.

The basic computer input data for each of the programs consisted of the following:

1. The mathematical representation of the physiochemical relationships in the estuary which are called transfer functions, or input-output coefficients for the estuary.

2. The differences between the present DO values in each section and the DO values defined for each objective set.

3. The waste load in lb/day (kg/day) discharged from each of the major sources.

4. A cost function for each of the major sources which gives the cost of removal as a function of the degree of removal.

The Uniform Treatment Cost Allocation Model

The uniform treatment program requires that an identical percentage of the raw oxygen-demanding load from each source be removed before discharge. Thus it is necessary to find a treatment level that is high enough to achieve the desired DO.

The general formulation for a minimization of costs program is as follows:

$$\text{MIN} \sum_{k=1}^{M} C_k(f_k) \ldots\ldots\ldots 3$$

subject to,

$$\sum_{j=1}^{N} A_{i,j} f_j \geq B_i, \quad i = 1,2,3,\cdots N \ldots 4$$

where;

$$f_j = \sum_{k\epsilon j} f_k \ldots\ldots\ldots\ldots 5$$

$$0 \leq f_k \leq U_k, \quad k = 1,2,3,\cdots M \ldots 6$$

$$f_k = \begin{cases} (S - P_k)T_k, & \text{for} \quad P_k < S..7 \\ \\ 0, & \text{for} \quad P_k \geq S..8 \end{cases}$$

In this case:

$C_k(f_k) =$ the cost function for each source k (dollars),

$f_k =$ the additional removal from each source which is summed over the proper sources to yield f_j, the input section removal (kg/day),

$A_{ij} =$ the steady-state transfer functions (mg/l/kg/day),

$B_i =$ the DO increment required in output section i(mg/l),

$U_k =$ the upper bound, or maximum waste that can be removed for source k(kg/day),

$P_k =$ present percent removal for each source k(dimensionless),

$T_k =$ raw waste load that k produces (kg/day), and

$S =$ the uniform percent removal (dimensionless).

The desired program is determined by the value of S that will satisfy the required DO goal.

Equation 3, by requiring the minimum cost, asks for the value of S that just satisfies the goals, since the cost of treatment increases with increasing percent removal.

Equation 4 requires that the load removed be enough to obtain at least the desired DO improvement in each section. Equation 6 limits the effluent modification to the maximum amount the source can remove. Equations 7 and 8 test to see if the present fraction of waste removal, P_k, at source k is equal to or greater than the level of uniform treatment being considered. If it is, source k is omitted from the solution. If it is not, the contribution from source k is described by Equation 7.

The approach for solution of this program was a search technique in which a low treatment level was chosen and its resulting DO profile calculated. If this did not satisfy the goal, the treatment level was increased by five-percent increments until the goal was reached in all places. Since costs increase with percent removal, the objective of minimum cost is met by choosing the treatment level that just satisfies the goal. It should be noted that the cost of effluent removal at each source need not be a linear function of the effluent removal. For simplicity in computing, however, the cost function was approximated by a series of linear segments. The algorithm is not dependent on the cost information given, but rather on the physical environment. A removal level of 85 percent to the effluent source will cost the same no matter what flow or temperature conditions exist in the estuary. However, the DO resulting in the estuary after 85-percent removal is very much a function of the inherent physical conditions at any given time.

Uniform treatment is the commonly used approach in existing management situations. Its primary advantage is administrative simplicity; however, it is economically inefficient in accomplishing the goal. Many sources are required to increase treatment in noncritical areas because the goal is not met in the more critical locations. The result is a large surplus of DO in noncritical areas, thus providing a measure of the inefficiency of the solutions. While each source is required to remove the same percent of raw load, no allowance is made for differences in unit cost of waste removal from source to source. In spite of apparent equity, the solution is actually quite inequitable. It at best can treat each source identically in terms of treatment level without regard for cost. It does not attempt to equate costs between sources or to treat sources similarly because of similar locations or types of operation. Thus, any gain in administrative simplicity is offset by the program's inefficiency and inequity plus its substantially higher costs to the general public.

The Cost-Minimization Cost-Allocation Model

The basic question answered by cost minimization is: What combination of waste effluent modifications along the estuary should be made to secure desired DO goals at least overall cost for the region? The key here is that no source is assumed to act in conjunction with any other source. Mathematically this can be formulated as a classical linear programming problem with upper bound constraints. This formulation and its application to the Delaware Estuary originally were accomplished by Thomann and Sobel (2).

$$\text{MIN} \sum_{k=1}^{M} C_k(f_k) \ldots \ldots \ldots 9$$

subject to,

$$\sum_{j=1}^{N} A_{i,j} f_j \geq B_i, \qquad i = 1,2,3,\cdots N \ . \ . \ 10$$

where;

$$f_j = \sum_{k \epsilon j} f_k \ldots \ldots \ldots \ldots 11$$

$$0 \leq f_k \leq U_k, \qquad k = 1,2,3,\cdots M \ldots 12$$

Equation 9 represents the minimization of the sum of all costs associated with waste removals at each of the k sources. Equation 10 represents the basic physical model as discussed previously and relates the environment through the input-output functions, $A_{i,j}$, to the waste load, f_j, and desired quality, B_i. Equation 12 specifies that in no case will a waste source treat less than its present level of treatment, and requires that the solution remove only up to some maximum level; in this case the maximum waste that can be removed at the source under present technology.

The optimum program consists of those values of f_k that satisfy the constraint Equations 10, 11, and 12 and minimize the objective function Equation 9. Note that this formulation is similar to that of uniform treatment except that some constraints have been dropped. Thus a solution to this problem will, at the most, cost the same as the uniform treatment scheme and for most cases will cost much less.

The computer solution of the linear programming problem most often is accomplished by the "Revised Simplex Method" which systematically searches for the "best" solution, thereby eliminating a complete enumeration of all possible solutions to find the optimum.

The technique is referred to as cost minimization because the least cost solution is subject to constraints which have associated costs. Thus the final cost of the program can be altered by changing the constraints. For example, placing of a lower bound constraint may be considered, with the plan that each discharge must have 35-percent minimum of primary treatment. The subsequent least-cost solution with this constraint will be greater than the cost without the constraint; the cost differential is the price for including the additional requirement in the problem.

Because of the nature of the specific problem of raising water quality by waste removal, some special modifications had to be made in computing the solutions to the above formulation of the cost minimization problem. These center around the problem that the cost-of-removal vs. the load-removed relationship is not linear. An investigation of this relationship shows it is generally a convex curve above primary treatment. A convenient method is available for solving convex nonlinear

models by linear programming. It involves approximating the cost curve by a series of linear segments, each of which has a cost per unit and a load that may be removed at that unit cost. The algorithm for solving the problem considers each segment as a separate source of effluent. There is no danger of any segment being taken out of order since each segment of the curve for a particular source has a higher unit cost than the one preceding; the algorithm, in looking for the least cost, will choose the lowest unit cost available to it.

Least-cost solutions show a number of interesting characteristics, aside from their guarantee of an absolute minimum investment. The linear programming model considers such items as location of the source with respect to the lowest part of the DO profile, the marginal cost to remove additional amounts of waste at each source, the maximum amount of waste that could be removed, and the relative proximity of other, cheaper waste loads. Thus the solutions show a number of trade-offs between individual sources. It is not unusual, for example, for the solution to indicate a high degree of removal at one source while its neighbor is not required to increase treatment beyond 35-percent removal.

The least cost solution is very efficient in allocating the treatment levels to be attained by each source to meet a specified goal. No unnecessary treatment is called for and only those removals are required which produce an increment of quality at the lowest cost. The DO surplus, i.e., the amount in excess of the goal, is smaller for this formulation than for any other.

The least-cost solution is equitable in the sense that a source causing no damage (i.e., not lowering water quality below desired goal) incurs no costs. The solution is likely to be extremely inequitable in the sense of not treating industrial competitors in a like manner. Two dischargers on opposite banks of the estuary can be expected to cause equal marginal damages. Yet, if waste treatment costs at one firm are low while at the other firm they are very high, it is likely that the source with the low-cost removal capabilities will be required to treat its waste to extremely high removals and the other discharger to provide no additional removal at all.

This solution would be difficult to implement administratively. Each source would have to be considered individually by a basinwide authority and waste loads allocated according to the least cost criterion. Because of the unequal treatment of some dischargers which may be alike in many relevant respects, great antipathy can be anticipated on the part of many sources. However, the problems are not insoluble.

The Zoned-Optimization Cost-Allocation Model

The optimization model for this program is formulated as a combination of the uniform treatment and the cost minimization programs. This results in uniform treatment levels for groups of sources within zones with cost trade-offs possible between zones that yield a set of levels having minimum cost. This can be expressed as follows:

$$\text{MIN} \sum_{k=1}^{M} C_k(f_k) \dots \dots 13$$

subject to,

$$\sum_{j=1}^{N} A_{i,j} f_j \geq B_i, \qquad i = 1,2,3,\cdots N \,.\,.\,14$$

where;

$$f_j = \sum_{k \epsilon j} f_k \dots \dots \dots \dots 15$$

$$0 \leq f_k \leq U_k, \qquad k = 1,2,3,\cdots M \dots 16$$

For each zone z:

$$z = 1,2,3,\cdots D$$

$$f_k = \begin{cases} (S_z - P_k)T_k, & \text{for} \quad P_k < S_z \,.\,. \, 17 \\ \\ 0 & , \quad \text{for} \quad P_k \geq S_z \,.\,. \, 18 \end{cases}$$

where;

$$k \epsilon j$$
$$j \epsilon z$$

The model requires that treatment level, S_z, be specified for each zone, z, so that the goal is reached at a minimum overall cost to the estuary community. Equations 17 and 18 relate the required treatment level in each zone to the effluent improvement necessary for source k. Equation 17 deals with the case of a source, k, having a present percent removal, P_k, below the required treatment level in the zone, S_z. If this is so, an effluent modification, f_k, must be made to raise the treatment level to S_z. Equation 18 states that if the present treatment level is above the proposed zone treatment level, no modification is needed. Equation 14 specifies that the effuent changes must raise the DO to at least the desired quality improvement. Equation 16 limits the effluent improvement at each source to the maximum available. Equation 13 shows the optimization criteria under which the final set of treatment levels, S_z, are chosen, yielding the minimization of the overall cost to the region for this particular management program. The value of f_k for each source in a zone is determined uniquely by the value of S_z. Non-negativity of f_k is implied through Equations 17 and 18 which will not specify a negative value of f_k.

The treatment zones need not be geographical in nature but may be used to group waste sources according to type of waste discharge, type of effluent source, or other category. If only one zone is chosen, the problem becomes the uniform treatment problem. If a zone is specified for each source, the problem becomes the cost minimization problem.

It would appear at first that a linear programming formulation would be possible for this problem, as well as for cost minimization. Unfortunately, this does not appear feasible because of nonlinearities in the constraint functions. Discontinuities were found to appear at points where new sources enter the problem. The method of solution adopted involves construction of a band of possible solutions between a set of lower and upper levels. Within this band a search technique is applied that yields all feasible solutions which satisfy the input DO goal. The technique involves significantly less effort than complete enumeration. It is possible to compute not only the zone treatment levels resulting in the minimum cost, but also other feasible levels that cost more but might be more desirable administratively. Any number of these more costly alternatives can be computed, depending on how much more costly a result will be accepted.

As might be expected, the results of this program are intermediate between those of the first two. The great surplus of DO that leads to inefficiency in the uniform treatment program is avoided. Surplus DO does exist within zones, but it is smaller than before, since in this case zones are chosen to encompass areas of similar water quality. Four zones are specified in the Delaware Estuary, which cover the area from Trenton, New Jersey to the beginning of Delaware Bay.

A measure of equity is attained in that sources located near each other, and adjacent to similar water quality, are treated similarly. Similar type sources also could be treated in equal fashion, although this was not presently superimposed on the geographical breakdown. The sort of marginal equity found in the linear programming approach is lost here, because the method does not consider the derivative of the cost function in selecting sources.

The zone approach is looked on with a certain amount of favor from the administrative viewpoint. It is nearly as easy to implement as the uniform treatment method, since it requires only locating sources within management zones. In addition, the equities of similar treatment for sources located near one another tend to reduce the objections of individual dischargers regarding their situation as compared to that of their neighbors.

Resulting Costs and Benefits

Applying each of the cost allocation models to each of the "Objective Sets" yields a complete tabulation of the different methods of attaining a given water quality goal.

The program costs increase as water quality increases. The DO objective for set I can be reached only by 92- to 98-percent removal of all carbonaceous waste sources plus in-stream aeration and removal of benthic deposits at an estimated cost of $695 mil (Figure 2). Since only experimental plants have attained such high removals so far, some doubt exists as to whether present technology, in fact, can achieve set I objectives on the large scale basis that would be necessary.

Many sources would have to make improvements to avoid degrading water quality below present levels, i.e., set V. This alone is estimated to cost $140 mil for the period 1964 through 1975. This cost would be incurred through facilities construction necessary to offset the effect of regional growth.

Figure 2 presents a breakdown of cost by objective set and allocation model. Sets I, II, III, and IV were selected by the Water Use Advisory Committee to the Delware Estuary Study. Between sets II and III, the Delaware River Basin Commission (DRBC) defined a set of goals that are intermediate in degree of stringency. For any given objective set the cost minimization model yields the smallest cost and the uniform treatment model yields the largest. This is in agreement with the efficiency concepts discussed earlier for each of these models. The exception to this rule is set I, which stringently demands the highest levels of treatment under all conditions.

Certain monetary benefits can derive from increased use of the estuary, once its water quality is improved. Numerous benefits are intrinsic in water quality enhancement programs. These are realized by a more economic utilization of natural resources, preservation of

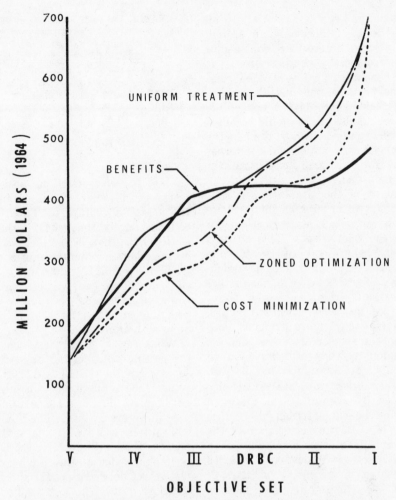

FIGURE 2.—Graphical presentation of the three economic allocation models
is plotted on the same axes with results of the benefit analysis.

fish and wildlife, and protection of regional health and welfare. One of the basic aims of the Delaware Estuary Study has been to better define and quantify the benefits of enhancing water quality in the estuary.

Quantification of the benefits is an essential part of any feasibility study. However, in the water pollution control field, the state of the art is new and much methodology currently is being developed. The Delaware Estuary Study proceeded with an analysis of the anticipated benefits for several water uses under each of the objective sets. It was not expected that all the benefits could be quantified. The re-

sults included positive benefits for recreation and commercial fisheries based on the increased use attributable to improved water quality.

A negative (i.e., a cost) benefit was estimated for some industries along the estuary. This is primarily because of increased corrosion rates within circulating water systems caused by higher DO levels in the intake water.

Comparison of program costs and benefits may be carried out in several ways. Perhaps the most explicit is the simple graphical presentation (Figure 2). Here the costs obtained from the three economic allocation models are plotted on the same axes along with the results of the benefits analysis. In all cases, the abscissa is only an ordinal scale. Some interesting conclusions can be reached by comparison of the graph, benefit/cost ratios (Table I), and marginal costs vs. benefits (Table II).

The benefit/cost (B/C) concept is that every dollar spent should be offset by a dollar gained through the benefits of implementing a particular alternative program. In such a case, a ratio of ≥ 1.0 is desired.

In many of the alternatives a B/C ratio equal to or greater than 1.0 results; however, all ratios decrease for the higher objective sets. This is caused by the occurrence of a benefits "plateau" (Figure 2). Once a particular level of use is reached, no further gains result from increased water quality, at least with respect to that use. Therefore, as ever more stringent quality goals are specified, costs increase tremendously while benefits tend to level off at a maximum value plateau.

Further insight into program choice is gained by examining the marginal costs and benefits of each alternative (Table II). In the case examined, the increments in cost or benefit for each objective set are somewhat analogous to the slopes of the curves in Figure 2.

In theory attempts should be made to proceed to higher objective sets as long as marginal benefits at least equal marginal costs. The functions shown in Figure 2 require a careful interpretation of this rule. The slope of the benefits function is generally greater than

TABLE I.—Total Benefit/Cost Ratio

Objective Set	Uniform Treatment Model	Zoned Optimization Model	Cost Minimization Model
I	.69	.69	.69
II	.83	.85	.97
DRBC	.96	.98	1.09
III	1.05	1.24	1.39
IV	.89	1.11	1.19
V	1.16	1.16	1.16

that of the cost functions until "Objective Set III" is reached. Comparing the marginal values in Table II, it could be argued that set III is the highest level that can be justified using the approach of incremental returns. That is, above set III, values move along the benefits plateau, whereas costs continue to increase. In addition, the benefit/cost ratios for set III are all greater than one.

What reason could be given, then, for seeking yet a higher objective set, specifically the DRBC set? It is of course true that all the benefits could not be quantified and hence do not appear in Figure 2 at all. The DRBC set will obtain more of these intangible benefits than will objective set III. It is emphasized that the ranking of the DRBC set between sets III and II was based on cost. The cost resulted from additional waste treatment measures designed to facilitate the occurrence of benefits, whether or not they could be quantified. In all probability the DRBC objective set represents that creature known as a reasonable alterna-

TABLE II.—Marginal Costs and Benefits (Million 1964 Dollars)

Objective Set	Costs			Benefits	Marginal Benefits / Marginal Costs		
	Uniform Treatment	Zoned Optimization	Cost Minimization		Uniform Treatment	Zoned Optimization	Cost Minimization
I	186	202	263	62	0.33	0.31	0.24
II	70	63	45	0	0	0	0
DRBC	56	104	96	17	0.30	0.16	0.18
III	64	70	53	121	1.89	1.73	2.28
IV	179	116	98	121	0.68	1.04	1.23
V	140	140	140	162	1.16	1.16	1.16

tive. It possesses some rationale from the standpoint of the economic analyst; note that the benefit/cost ratios are close to 1.0 for any of the economic allocation models. In the midst of the powerful conflicting forces that must be considered in such a public service decision, it probably represents a combination of logic and necessity.

The Decision-Making Process

The results of the Delaware Estuary Study as described here have been presented in the form of a preliminary report (3). The agency utilizing these results to the greatest degree has been the Delaware River Basin Commission. The DRBC was created in November 1961 on enactment of concurrent legislation by congress and by the respective legislatures of the states of Delaware, New Jersey, New York, and Pennsylvania, as an agency of the federal government and the signatory states. The Commission consists of the Governors of the states plus the Secretary of the Interior and has the authority to develop plans, policies, and projects related to the water resources of the basin.

As the primary decision-making agency within the basin, the DRBC is responsible for the selection of water quality goals which then serve as guidelines for the regulatory agencies of the constituent states in formulating their own plans for water quality improvement.

The DRBC held public hearings on water quality improvement programs for the estuary, which culminated in the annual Commissioner's Conference of March 1967, at which a decision was reached. At this meeting the four Governors and the Secretary of the Interior selected "Objective Set II" as the goal to be sought for the Delaware Estuary. Subsequent adjustments produced the DRBC objective set, intermediate between II and III. It was decided to use the zoned approach to estuary management, which would allow a certain degree of flexibility to be maintained, to meet future conditions. The water quality standards as selected by the DRBC are now in the process of being implemented along with state standards. The selection of a particular objective set and an allocation model has resulted in load allocations for individual firms and municipalities. These allocated loads are the maximum waste discharge loads that will be permitted under the given conditions. If the waste loads are maintained at their allocated magnitudes, water quality in the estuary will be maintained at an enhanced level.

This is an example of the way in which some of the more recently developed analytical techniques have been utilized in a real situation. The use of systems analysis methods and the advent of the digital computer have made possible the formulating of rational alternatives for consideration by policy making agencies. Seldom are all recommendations of the scientist and engineer accepted. But it is possible to employ these modern tools so that the good and bad points inherent in alternative decisions are readily distinguishable. It thus is possible to minimize illogical and inconsistent actions while attempting to maximize the returns for necessary expenditures. In a multi-faceted field such as water resources, optimization may not mean getting the best of all possible worlds; rather it will mean making the best possible use of the world as it is.

References

1. Thomann, R. V., "Mathematical Model for Dissolved Oxygen." *Jour. San. Eng. Div., Proc. Amer. Soc. Civil Engr.,* 89, SA5 (1963).
2. Thomann, R. V., and Sobel, M. J., "Estuarine Water Quality Management and Forecasting." *Jour. San. Eng. Div., Proc. Amer. Soc. Civil Engr.,* 90, SA5, 9 (1964).
3. Delaware Estuary Comprehensive Study, "Preliminary Report and Findings." U. S. Dept. Interior, Federal Water Pollution Control Administration, Philadelphia, Pa. (1966).

Copyright © 1970 by the Institute of Electrical and Electronics Engineers, Inc.

Reprinted from *IEEE Trans. Sys., Sci. Cybern.*, **SSC-6**, 272–275 (Oct. 1970)

A Water Pricing and Production Model

EMMETT J. BURKE

Abstract—The purpose of the model presented is to maximize the social profitability of a governmental agency by determining prices and production levels of two different sources of water. This maximization is performed subject to both sociopolitical and physical constraints. The nonlinear stochastic model developed is reduced to a linear stochastic model of one decision variable. Also presented are the results of the application of the model to an existing situation.

I. INTRODUCTION

THIS paper is concerned with the development of optimal operating decisions for a local governmental agency which has responsibility for the acquisition, development, and conservation of water and water-related resources. Of prime interest is the problem of selecting a price and a production level of water when 1) there is more than one source of water, 2) the cost to the agency and the user depends significantly on the source, and 3) there are important interrelationships between the sources.

Manuscript received August 3, 1969; revised February 9, 1970.
The author is with the Santa Clara County Flood Control and Water District, San Jose, Calif.

The approach which has been taken to this problem is similar to the classical formulation of the theory of the firm, Hall [3] and Allen [1], i.e., a revenue function subject to a production function. In the water resources field, Dracup [2] has used linear programming in the allocation of water sources to meet a developing demand over a time horizon. Here, since the concern is with price, production level, and source, the revenue function is nonlinear, as also is one of the constraints.

II. THE ENVIRONMENT OF THE PROBLEM

The local governmental agency, the Santa Clara County Flood Control and Water District, acts essentially as a wholesaler in providing water to retailers who in turn deliver it directly to the consumer. The retailers may acquire water either from the underground through their own pumps or directly from the District's water treatment plant which treats only imported water from the South Bay aqueduct of the California State Water Plan. In either case, the retailer must pay the District a fee which depends on the source of water—a ground-water price when the source is the ground, and a treated-water price when the source is the treatment plant.

In addition, the District manages a number of small and medium size reservoirs which are connected to a system of specially constructed percolation ponds. Percolation is the mechanism by which the district seeks to replenish the underground aquifer. Water caught in the reservoirs is released to the percolation ponds and then it seeps into the underground aquifer. Imported water for which there is no treatment plant demand is also released to the percolation ponds.

The impetus for percolation is primarily to retard land subsidence caused by ground-water overdraft. It appears that a significant decline in the artesian head of the confined aquifer increases the stress which causes compaction of the underground deposits and correlative land subsidence [4], [5]. Land subsidence causes the destruction of well casings and public works, and necessitates the construction of levees to keep San Francisco Bay waters from inundating low-lying lands adjacent to the Bay.

The situation that the District found itself in immediately after completing the construction of the previously mentioned treatment plant was that the retailers were using the plant to satisfy only their peak demands, i.e., the plant was operating at ten-percent capacity. Part of the reason for this was that the price of treated water was set at slightly more than two times the price of ground water (sixty-two dollars per acre-foot as opposed to twenty-nine dollars per acre-foot). The retailers felt that they themselves could not justify the purchase of such expensive water to the California Public Utilities Commission, and the district felt that the price was necessary in order to retire the bonds (and associated interest) used to construct the plant. The model presented here was developed in order to assist the District in analyzing the problem and to provide some quantitative insight into the determination of what the optimal price and production levels should be for each source.

Thus the model sought to determine ways of improving the District's net revenue position while at the same time not increasing the costs of water to the consumer.

III. The Model

The model consists of an objective function subject to a number of constraints. The objective function is a measure of the social profitability of the District. By this it is meant that, in addition to representing revenue and operating costs, the community costs of subsidence are also incorporated in the objective function. There are two types of constraints: sociopolitical requirements and a series of mass-balance equations and flow restrictions.

The principal sociopolitical requirement aims at not increasing the consumer's cost of water. When this constraint is translated to the retailer it states that the retailer's total water acquisition costs should not exceed that which he is currently expending in acquiring water from his chosen mixture of sources. Included in the retailer's water acquisition costs is the cost of pumping ground water.

Other sociopolitical constraints are the limits on ground- and treated-water prices. These limits are set out in the enabling state legislation of the district.

The mass-balance equations indicate that water demand is satisfied from either the ground or the treatment plant, that imported water not treated and water stored in the reservoirs is percolated, and that there is some natural percolation. The flow restriction is treatment plant capacity.

Since the model seeks to maximize the profitability of the District given the existing facilities, neither additional nor sunk capital costs are considered. The model is concerned only with variable revenue and variable costs. Fixed costs, which are not entered in the model, include such items as bond redemption, interest payments, and charges levied by the State of California for the District's entitled allotment of imported water.

In order to write the objective and constraint equations we need to define the following symbols.

Decision Variables

1) 1968 values in dollars per acre-foot:

P_t price of treated water, 62

P_g price of ground water, 29.

2) 1968 values in acre-feet per year:

Q_t quantity of treated water sold, 10×10^3

Q_g quantity of ground water sold (pumped), 130×10^3.

Remaining Variables and Parameters

1) 1968 values in dollars per acre-foot:

C_n retailer's total acquisition costs now, i.e., prior to model development, 5.820×10^6

C_o operations cost of treating and distributing (power and chemicals) a unit of water from the treatment plant, 5

C_p cost per unit of water artificially percolated, 5

C_r retailer's cost of pumping per unit of ground water pumped, 11

C_s cost of subsidence per unit of overdraft of water, 13.

2) 1968 values in acre-feet per year:

Q_c quantity of water caught in reservoirs and subsequently percolated, 75×10^3

Q_d quantity of water demand (consumed), 140×10^3

Q_i quantity of water imported, 75×10^3

Q_p quantity of water percolated, 180×10^3

Q_v net quantity of water naturally percolated less agricultural uses, 40×10^3

L factor for current overdraft in same units as Q, 100×10^3.

The social objective is to maximize the difference between revenue and costs of the District. The equation is

$$P_g Q_g + P_t Q_t - [C_p(Q_p - Q_v) + C_o Q_t + C_s(L + Q_g - Q_p)].$$

The first two terms are revenue. The third term represents costs associated with percolation, which are directly proportional to the volume of artificial percolation. The fourth term gives variable costs of operating the treatment plant. The fifth term states that subsidence is directly proportional to the difference between outflow and inflow plus a factor for current overdraft for which subsidence has not yet occurred. We require that $L + Q_g - Q_p$ be always greater than zero; otherwise, subsidence costs are zero.

The first constraint requires that the new water acquisition costs of the retailer be less than or equal to what they are now. Thus we have

$$(P_g + C_r)Q_g + P_tQ_t \leq C_n.$$

The second constraint establishes that total water demand is satisfied from either ground or treated water:

$$Q_d = Q_g + Q_t.$$

The third constraint relates the total water percolated with the availability of water:

$$Q_p = Q_c + Q_v + Q_i - Q_t.$$

Other constraints such as treatment plant capacity may also be added. For example, $Q_t \leq \bar{Q}_t$, where $\bar{Q}_t$ is the upper limit on treatment plant volume.

Bringing the equations together as a system we have

$$\max \{P_gQ_g + P_tQ_t - [C_p(Q_p - Q_v)$$
$$+ C_oQ_t + C_s(L + Q_g - Q_p)]\} Q_g, Q_t, P_g, P_t \quad (1)$$

subject to

$$(P_g + C_r)Q_g + P_tQ_t \leq C_n \quad (2)$$

$$Q_d = Q_g + Q_t \quad (3)$$

$$Q_p = Q_c + Q_v + Q_i - Q_t \quad (4)$$

$$0 \leq Q \leq \bar{Q}_t \quad (5)$$

$$0 \leq P_g \leq \bar{P}_g. \quad (6)$$

The objective function is nonlinear in the decision variables and so is the first constraint. One could have recourse to nonlinear methods of solution such as the Kuhn–Tucker [6] conditions or the Lagrangian multiplier technique [6], but we will solve the system in a very straightforward fashion, i.e., by a series of substitutions.

To begin with, we will let the first constraint equation be an equality. We do not lose any generality since a parametric study can always be performed with respect to C_n. If we solve this equation for P_t and substitute the obtained representation into the objective function we have

$$\max \{C_n - Q_gC_r - C_p(Q_p - Q_v)$$
$$- C_oQ_t - C_s(L + Q_g - Q_p)\} Q_g, Q_t$$

Note that the two pricing variables have left the function. Thus we seek to maximize the function only with respect to the two sources of water. Substituting the representation of Q_p into the equation and $Q_t = Q_d - Q_g$ and reorganizing we obtain

$$\max \{C_n - LC_s + Q_i(C_s - C_p) + Q_vC_s$$
$$+ Q_d(C_p - C_o - C_s) + Q_c(C_s - C_p)$$
$$+ Q_g(C_o - C_r - C_p)\}Q_g. \quad (7)$$

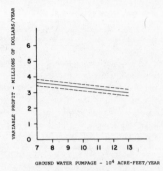

Fig. 1. District's variable profit as a function of ground-water pumpage.

This function is linear in the remaining decision variable. To maximize the District's profitability, either the treatment plant will be used to full capacity, or it will be used not at all, depending on the sign of the coefficient of Q_g, $C_o - C_r - C_p$.

Before beginning the determination of the price and production levels it is interesting to note that (7) provides additional insight into the District's operation, e.g., the return from the reservoirs is the difference between subsidence costs and percolation costs. The function also permits an analysis of the degree to which the District's profit depends on the random hydrologic variables of natural percolation Q_v and reservoir storage Q_c.

IV. APPLICATION OF THE MODEL

The reduced formulation of the variable profit of the District (7) has been plotted as a function of ground water pumpage in Fig. 1. The rate of increase of the District's profit is 11 dollars for each acre-foot of ground water not pumped. The estimates of the unit costs of treatment plant operations and ground-water percolation have been obtained from District records. The estimate of retailers' unit pumping cost has been obtained from operating reports of the retailers filed with the California Public Utilities Commission. Thus the slope of variable profit is known with significant accuracy and confidence. Since the total demand of water is 140 000 acre-feet per year, and the treatment plant has a capacity of 70 000 acre-feet per year, the minimum ground pumpage is 70 000 acre-feet per year which in turn yields the maximum variable profit to the district. With this decrease in ground-water pumping of 60 000 acre-feet per year, the District obtains a savings of 660 000 dollars per year.

While the slope of the profit line is accurately known and is deterministic, the ordinate of variable profit is a random variable due to the inclusion of the two hydrologic variables of captured runoff Q_c and water which naturally infiltrates the underground aquifer Q_v. In order to represent this randomness, a single standard deviation region has been bracketed around the expected profit line in Fig. 1. This confidence region was obtained by considering the variables

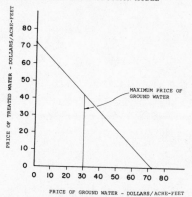

Fig. 2.   Price indifference for treated and ground water.

Q_c and Q_v as normally distributed with unequal means, equal variances, and complete linear dependence.[1]

Having determined the production levels, we return to the first constraint (2) in order to obtain the prices for the different sources of water. This constraint establishes a linear relationship between the price of ground water and treated water when the production values are known. The constraint, therefore, defines an indifference line, any point of which may be used for pricing (Fig. 2). However, since the maximum price of ground water is 30 dollars per acre-foot, as set by the state legislature, the permissible range along the line is narrowed. If this maximum price for ground water is chosen, the optimal treated water price becomes 42.14 dollars per acre-foot, as is shown in Fig. 2 by the intercept of the maximum price of ground water with the price indifference line.

In the early summer of 1969, an experimental pricing change was made. The average price for treated water was

[1] $Q_c \sim N$ (75 $\times$ 10^3 acre-feet,   11 $\times$ 10^3 acre-feet)
$Q_v \sim N$ (40 $\times$ 10^3 acre-feet,   11 $\times$ 10^3 acre-feet)
$Q_v = (\sigma Q_v / \sigma Q_c)(Q_c - \mu Q_c) + \mu Q_v$.

reduced to 40 dollars per acre-foot. There was, of course, a significant increase in treated-water usage, and associated with the decrease in ground-water usage there was, by summer's end, a significant increase in the water table level, a phenomenon which had not been observed recently in the area.

V. CONCLUDING REMARKS

The application of systems analysis to the resolution of what were thought to be two contending desires has yielded the community of Santa Clara County significant annual savings. The governmental agency now has a tool which permits it to maximize social benefit under current price levels of the state legislature and current expenditures by the water user. In addition, the model obtains the return to the community, i.e., the value of the various operations of the water supply agency. Thus the model can be used in the economic planning of the expansion of all these activities.

ACKNOWLEDGMENT

The author wishes to thank D. Clack, D. Kriege, and J. Watkins for their assistance.

REFERENCES

[1] R. G. D. Allen, *Mathematical Economics.* New York: St. Martin's Press, 1966.
[2] J. A. Dracup, "The optimum use of ground-water and surface-water system: a parametric linear programming approach," Water Resources Center, University of California, Berkeley, Rep. 107.
[3] A. D. Hall, *A Methodology for Systems Engineering.* New York: Van Nostrand Reinhold, 1962, ch. 9.
[4] J. F. Poland, "Land subsidence and aquifer system compaction, Santa Clara Valley, California, USA," presented at the Int. Symp. on Land Subsidence, Tokyo, Japan, September 1969.
[5] ——, "Land subsidence in the western United States due to ground water overdraft," presented at the Nat. Reclamation Assoc. Conv., Spokane, Wash., October 1969.
[6] T. Saaty and J. Bram, *Nonlinear Mathematics.* New York: McGraw-Hill, 1964.

V

Waste Management

Editors' Comments on Papers 18 Through 21

18 **Rose, Gibbons, and Fulkerson:** Physics Looks at Waste Management
    Phys. Today, **25**, 32–38, 41 (Feb. 1972)

19 **Drobny, Qasim, and Valentine:** Cost-Effectiveness Analysis of Waste Management
    Systems
    J. Environ. Sys., **1**, 189–209 (June 1971)

20 **Peterson and Wahi:** Interactive Computer-based Game for Decision-making in Ecology
    IBM J. Res. Develop., **16**, 154–161 (Mar. 1972)

21 **Norris, Birke, Cockburn, and Parker:** Marine Waste Disposal—A Comprehensive Environmental Approach to Planning
    J. Water Pollution Control Federation, **45**, 52–70 (Jan. 1973)

The four papers of Part V have been recycled to provide at least a basic treatment of waste management. The first selection is a general paper that provides useful perspectives on the waste problem and that suggests alternative strategies for the management of waste. The second paper, Paper 19, describes a systematic methodology for developing a measure of effectiveness for systems that are to be evaluated on the basis of several properties that may be incommensurable (i.e., lacking a common basis of comparison). Paper 20 is a discussion of a computerized Ecology Decision Game for "instructional gaming" with respect to optimizing refuse collection. The final paper of Part V considers the gathering and analysis of data to determine the effect of marine waste disposal on the waters and shores around San Francisco, California; the article concludes with waste disposal recommendations based on ecological design criteria derived from the data analysis.

Paper 18, "Physics Looks at Waste Management," was written by physicists to describe possible roles for the physicist and physical thinking in waste management. For example, several areas in which scientific training is necessary are the modeling and monitoring of environmental processes and the development of new technologies that are less damaging to the environment.

There is a natural reluctance on the part of professionals to adapt their training to research areas that are significantly different from their current pursuits. This is at least partially a consequence of the fact that the reward system (recognition, financial advancement) supports vertical rather than horizontal development. However, for those individuals who are willing to make this sacrifice, the authors examine the question of what a physicist might contribute to the resolution of environmental problems.

One interesting statement by the authors is that modern society uses energy to convert resources into junk and that temporary benefits are derived from intermediate stages in the process. The three objectives for optimizing this societal function are to control pollution, to conserve resources, and to advance material wealth. Conflicts often exist between these objectives, and it is the role of the environmental systems analyst to seek an acceptable compromise.

Specific consideration is given to alternative strategies for pollution control. These include recycling, extending product lifetime, substitution of less harmful materials in production and consumption, and dispersal of pollutants. This paper is primarily nontechnical and provides a useful perspective on the waste management problem for those with or without a background in physics. The colors used in the figures in the original publication do not show in the present reproduction, but the gray shading and labeling of flow arrows should be sufficient for a correct understanding of the authors' intent.

One approach to the problem of dealing with incommensurable variables (i.e., variables measured in different unit systems, such as cost and reliability) is presented in "Cost-Effectiveness Analysis of Waste Management Systems." Authors Drobny, Qasim, and Valentine address the following question: If one alternative action is to be selected, and the performance characteristics encompass incommensurable variables, how is a selection to be made? Of course, if a single alternative dominates all others (e.g., costs less *and* is more reliable), the selection is simple, but this is not usually the case. More commonly, higher reliability is associated with higher cost.

The multidimensional performance characteristics of an alternative may be considered to be a vector description of that alternative. A statement that one alternative is preferred is a scalar description, for the only characteristic is preference. There is no way to avoid the reduction from a vector to a scalar, despite the fact that many people object to explicit attempts to derive this scalar. At the present time, most decisions involving incommensurable variables are based upon the implicit judgments of a decision-making individual or body.

Paper 19 considers the cost effectiveness of waste water treatment systems for a military camp. Effectiveness is a scalar measure for a variety of factors, including simplicity of operation, reliability, and space requirements. A procedure is described for computing effectiveness as a weighted linear sum of performance characteristics. Dominated alternatives are discarded, and the initial array of fifteen choices is reduced to six.

Some shortcomings of this approach include the linear assumption for deriving an effectiveness measure, the aversion that many people feel toward this type of analysis, the difficulty in quantifying performance and weighting judgments, and the lack of uncertainty considerations.

It is not obvious why cost is not considered as another performance characteristic that is weighted and rated along with other measures of effectiveness. In this manner, a unique optimum alternative would emerge. The three-figure numbers in Figure 3 are probably unwarranted and suggest an accuracy that does not exist.

It appears as if the second sentence in the last paragraph on page 255 needs the insertion of ", i.e.," after the word "effectiveness" in order to be read properly. In the first paragraph on page 259, the sentence starting "One way would . . ." might more clearly indicate the suggested averaging of effectiveness measures over all participating decision makers if the phrase "all decision makers' " replaced the word "his" in the sentence.

The possibilities for using a computer to improve understanding of such problems as the management of solid waste are described in Paper 20, "Interactive Computer-based Game for Decision-making in Ecology." Authors T. I. Peterson and P. N. Wahi of IBM present a prototype Ecology Decision Game that can be used interactively and simultaneously by several "players." With such an "instructional gaming" approach, the players can experiment with decision strategies to acquire a better understanding of the model used to represent the real world and of the nature of the outcomes generated by the model.

Having stated that refuse *collection* represents 75 to 85 percent of the cost of solid-waste management, Peterson and Wahi focus on collection operations and planning in the scenario of the Ecology Decision Game. However, the flexibility of the APL (*A Programming Language*) gaming system described in this paper extends to "game authoring," the capability that allows a player to write his own scenarios for any area of ecology and revise the game according to his needs. An extensive "prompting" facility programmed into the system minimizes the need for a player to have computer expertise.

Three successively more realistic and more difficult routing problems are illustrated: (1) the shortest route problem, (2) the transportation problem with capacity constraints, and (3) the maximal flow problem. The bit of graph theory used in the third problem may be a small stumbling block; if it is, just read the section to get the gist of the formulation. References are given from which you may learn graph theory and mathematical programming solution algorithms at your leisure and inspiration.

The last article of this section on waste management considers the design of a waste disposal system that would discharge effluents into the waters around San Francisco, California. The problem is to determine the location of discharges, size of outfall, depth of discharge, and degree of treatment required. Primary concerns are costs of the system, protection of public health, avoidance of nuisance conditions (odors, unattractive visual effects), and preservation of the marine ecosystems.

Data gathering is an important aspect of a systems study, and one of the main thrusts of Paper 21, "Marine Waste Disposal—A Comprehensive Environmental Approach to Planning," is to illustrate the data needs and the methods for satisfying these needs. There are two components to the data requirements: oceanographic and ecological. The former concerns the movements and characteristics (temperature, depth) of the water, and the latter concerns the response of marine organisms to effluence and oxygen depletion in the water.

Oceanographic information was obtained by a variety of measurements. Aerial photography was useful for determining the movement of water mass, and probes and water samples provided data on temperature, salinity, depth, and dissolved oxygen. The movement of floatables may be different from the movement of water mass, and the former was observed by dropping cardboard milk bottle caps at various locations in the water. Dispersion coefficients were measured by injecting a dye and observing the three-dimensional spread of the dye.

Ecological investigations were performed both in the laboratory and in the field. Plankton, benthic (bottom) organisms, and intertidal biota were surveyed. Bioassay experiments were conducted with marine and estuarine fish and invertebrates to evaluate effects of effluent discharge. The results of the ecological studies were used to design a discharge system that would prevent detrimental changes in marine biota.

This article provides a good description of the type of data required for the analysis of a practical system and illustrates the different methods that were employed for gathering and analyzing this data. A couple of minor points, however, should be clarified. In Tables I through IV, the "dilution" is the percentage of wastewater effluent in the total amount of water available to the test organisms. Thus, if the dilution ratio is 1:5 (i.e., one part effluent to five parts dilution water), then the dilution is $[1/(1 + 5)] \times 100\% = 16.67\% \simeq 17\%$. Therefore, the columns with dilution ratio 1:10 and 1:5 in both Tables I and II should have, respectively, (9) instead of (10) and (17) instead of (20) underneath the dilution ratio. Also, in Table I, the "160" in the row for Shiner perch should undoubtedly be 100. In the column for TL_m in Table IV, there should not be a dagger (†) in the row for adult sand crabs and there should be 5–10* in the line for larval sand crabs. Finally, in the second paragraph on page 290, something was left out after the word "survive" at the top of the second column. There should probably be a period and a closing parenthesis after "survive," followed by something like "Among the species of fish tested, the skulpin" to lead into the next sentence.

Reprinted from *Phys. Today*, **25**, 32–38, 41 (Feb. 1972)

Physics looks at waste management

What are the best ways to find out what is in our waste, sort and separate it, return to service any parts that can be used again and properly dispose of the remainder?

David J. Rose, John H. Gibbons and William Fulkerson

Electromagnetic separator for light and dark glass granules salvaged from the unburned ashes of a municipal garbage incinerator. Dark glass (left) is separated from the uncolored glass (right) by differences in magnetic susceptibility. (Photo by UPI from the US Bureau of Mines Solid Waste Center.)
Figure 1

The question of what role physics plays in the garbage business invites the casual answer: "Not much." But that would be wrong, and not from any trivial viewpoint, such as the mechanics of garbage grinders. Admittedly, physics does not play the key role, and certainly what is known colloquially as "high physics" may be quite unresponsive to the problem. Nevertheless, the ethos of physics and recognizable physical principles have become peculiarly useful as part of the intellectual armamentarium of waste management. Surely we are biased—two of us are former physicists and one is a former physical metallurgist. We will try to convince you that we are serious and have a valid point of view.

First, we point out that modern high-consumption society operates by using energy to turn resources into junk; we derive comfort, pleasure, and other temporary benefits through intermediate stages of this dynamic process. Of this trio—resources, energy, and junk—the first two have received some organized attention, albeit woefully inadequate, as recent reconsideration shows. But the last one has been generally ignored until very recently (see, for example, the Solid Waste Disposal Act of 1965, as amended by the Resource Recovery Act of 1970): Throw it away, there's lots more room, and lots more resources where those came from. All this now changes.

What is waste? Our view will become obvious through what follows, but a good enough starting definition is: Waste is something that we have pro-

David Rose is a professor of nuclear engineering at Massachusetts Institute of Technology and is active in "energy and environment" work there. He recently spent a two-year leave at Oak Ridge. John Gibbons is director of the "ORNL-NSF Environmental Program" at Oak Ridge National Laboratory and William Fulkerson manages that part of the program called "Material Resources and Recycle."

duced, for which we find no satisfactory use. This avoids pejorative prejudgments about pollution and so forth; we will talk about pollution later on, but do not wish to becloud the view *ab initio*. The point is important; take, for example, the one-way beer bottle, a topic of frequent environmental opprobrium. Is it a problem of pollution, of resource conservation, or of what? Now it is a fact that if everyone on this earth drank beer to insobriety, tossed away the empties, and persevered at this forever, geologic processes would turn the empties back to sand before we ran out of it for making more bottles. The bottle is quite acceptable for sanitary landfill, and the energy required to make a new one is not very large. Thus, the glass container represents more closely a problem of convenient disposal and litter prevention (perhaps to be solved by recycle) than one of material resource conservation.

We will choose here not to emphasize the nonmaterial (aetherial) wastes such as unwanted heat, electromagnetic radiation and sound. Rather, we will emphasize material wastes, and stress how some physical principles can be constructively applied.

These wastes come in many categories that cannot be arranged in order; they range from high-volume low-cost things (mine-tailings or paper) all the way to the opposite extreme of low-volume high-cost (mercury in switches, for example). But they can also be categorized as disposal problems (mine tailings), steady-state management of renewable resources (paper), global scarcity of nonrenewable resources (mercury, tungsten, phosphorus, and so on), and hazards (untreated sewerage, most heavy metals, pesticides, for example).

In the narrowest sense, waste management involves sensing what is there, sorting, separating and/or transforming, returning to service what can be useful, and properly disposing of what is left.

In those steps lies a great deal of physical and physics understanding, and even these steps do not reach all we wish to get at.

One example of the way physics can get involved is the electromagnetic glass separator shown in figure 1.

We have observed a striking feature in all this recent environmental and socio-technological activity—waste management, resource conservation, energy, pollution and so forth; it is the disproportionately large fraction of new practitioners who come from the physical sciences, particularly from physics. We have wondered why this is so. Is some quality involved, not always recognized as valuable—in spite of what we said of physics at the very beginning? The cynic might cite money, thinking of the drying-up of physics research funds. That doesn't hold, for the new practitioners were usually quite competent in their old trade, and would have had little difficulty in pursuing it. Perhaps it is a penchant for modeling—looking for the big and little things, with orders of magnitude assigned, followed by selective analysis of what appear to be the bigger parts. Perhaps, more sharply, it lies in the difference between physics *per se* and the process of solving physical problems; some physicists delight in the former and others in the latter, which is in truth a very interdisciplinary activity. Environmentally related problems need a lot of such interdisciplinary thinking. He who neglects the connection between paper recycling and private investments in the southern pine forests is spiritual brother to he who neglects the plasma potential in his ion source.

The objective, and two principles

The objective of proper waste management can be stated easily; it is to achieve a socially optimum mass or resource flow. However, the easiness deceives, and what follows makes a sort of extended definition, with both normative and deterministic components.

By deterministic components we mean more-or-less immutable facts (such as so much sewage creates so much oxygen demand of effluent water), and our developing scientific and technological knowledge related to handling materials. We will concern ourselves here mainly with these latter parts. The normative decision of what is socially optimal depends upon the determinable facts, but only in part. Styles change from time to time; we are witnessing such a change now.

Many scientific and technological tools get used; perhaps an early nontrivial example will help. Figure 2 shows a fairly simple and almost self-explanatory flow chart of mercury through our 1970 US civilization. The need for several skills appears quite plainly; for example:

▶ The need for, and power of, overall modeling techniques. "Everything goes somewhere," as Barry Commoner says in connection with these matters; the tracing out of "where" identifies (for instance) the large use of mercury by caustic-chlorine plants (where it is used as a circulating electrode), and the (hitherto) substantial leakage of mercury into rivers and lakes. Working flow charts for most materials are much more complex than figure 2, but the power of this sort of mass-flow modeling is immediately evident.

▶ The need for much analytic detective work. The actual mass of mercury is very small compared to the mass of environment into which it is wasted: Five parts in 10^9 is a significant concentration in drinking water. Measuring such things easily, cheaply and ubiquitously demands sometimes the most artful application of mature science (mercury by neutron activation analysis, for example) and ingenious application of new techniques. In many situations it does not suffice to identify and quantify atomic components of wastes; their chemical and physical forms must be found, and also their proclivity to

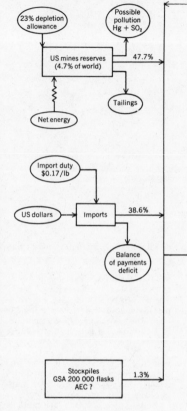

The flow of mercury through society. This is a schematic representation of mercury flow in the US economy during 1970. Ovals represent "externalities" (inputs and outputs), and colored lines show recycle paths. Figure 2

change (for example, metallic mercury into methyl-mercury). We think of the relatively new technique of photo-electron spectroscopy[1] in this vein, because it allows determining in many instances not only the element but its valence state (say, sulfur as SO_2 and as SO_3).

▶ Chemistry, physics, regional modeling, and so on do not appear very recognizably as separate activities here. They are tools, not ends in themselves. Here is a big difference, sometimes a stumbling block, in the understanding of what is to be done: We see here not physics for its own sake, but for a problem's sake. Thinking must be problem-oriented, not discipline oriented; the former is "real," in the sense that it exists apart from any artificial categorization we may make into particular disciplines.

▶ The social-scientific and normative inputs are very real. Mining subsidies in figure 2 come via societal decisions; they tend to lower the price of virgin metal and hence encourage waste and discourage recycle. Many of the pollution lawsuits in figure 2 are brought by sport fishermen and resort operators. To discover the normative content there, try placating the intrepid angler by offering him an even larger edible frozen fish in return for his inedible mercury-tainted catch.

Thus we see a general mass-flow objective, and many scientific-technological tools that may help us to accomplish the objective. Are there useful guiding principles? We see two related to our scientific side of these discussions. The first is conservation of mass, a point already mentioned and obvious, but often strangely neglected. Indeed, everything *does* go somewhere. Sometimes the fact is inconsequential, as one more bucket of water is assimilated into the ocean. But for some materials the natural capacity to assimilate wastes is remarkable small, or the wastes are reconcentrated by natural processes, or there is only so much available. We shall see examples of all these things later.

The second principle is more subtle, and we have found some difficulty even in describing it to ourselves. We call it "the principle of minimum-energy paths." By this we mean that in taking a substance from some first state to a desired second state, we prefer processes involving the least total exchange of

energy. Violating this principle generally means either the appearance of an inappropriate energy cost, or a concomitant capitalization charge, or use of additional material (hence new waste), or some combination of all these things. As a seemingly frivolous example, but a harbinger of later discussions, we separate nails from broken glass with a magnet, not with a mass spectrograph. Anyone who has explored the path cows take uphill will understand what we mean: They take the least slope, with never an unnecessary descent to be paid for later. This principle has the flavor of being restatable in terms of minimum entropy increase for the material process, but we have not explored it very much.

Continuing such an extended discussion of guiding principles leads us to figure 3, which is really a simple outline of what follows in this article.

We recognize that the idea of societally optimum mass flows can be restated in terms of three easily recognizable societal aims:

(a) Pollution control
(b) Conservation of concentrated materials
(c) Material wealth

These are to some extent in conflict, particularly the last with the other two. The question about the proper balance between them brings in normative judgments; the decision is basically political, not scientific, as almost everyone will probably agree. It is not economic either, a point less readily accepted: The aims (a) and (b) above are largely external to the market system of self-regulating management. It seems to us that the US is now shifting from (c) to (a). Note that the first two items, pollution control and conservation, are frequently related; this is not always so, as later examples will show.

Our article is about things at the left side of figure 3. Although all three general strategies abide in our societal decisions, and all three enter some of what follows, we will finally concentrate on pollution-control strategies. Thus, below that heading are the seven particular strategies that we envision and will describe in detail later.

Before turning to those matters, we will linger to discuss the interactions among material conservation, material utilization, and energy cost from one point of view. Figure 4 shows on the horizontal axis increasing degrees of

material concentration. At the left is matter in its most dispersed state, averaged over the earth's crust at what is known as the "Clarke" level;[2] depending upon the material in question, the relative abundance might be parts per million. Usually we utilize ores or other natural concentrates, at a higher level of natural organization, and process them to produce components and useful relatively pure materials. By this we have in mind iron and steel, glass, neoprene, doorknobs, and the like. This degree of "purity" is the highest in our society, and appears at the right side of the figure.

The systems we build as end products in society lie to the left of components. The intellectual organization is higher, but it represents a planned dispersion of components or refined materials: Nails, boards, glass, iron pipe and doorknobs are mixed up in carefully defined ways to build a house.

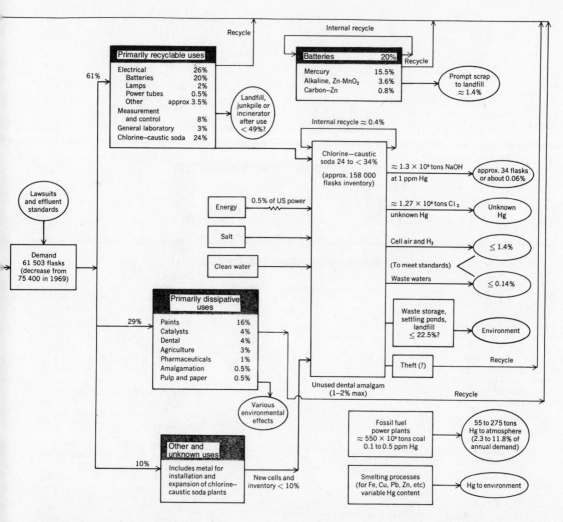

Off to the side of lower concentration, but usually higher than "ores," lie the mixed wastes that society produces: sanitary land-fill dumps, for instance. The somewhat overlapping range and nature of wastes and ores is illustrated by our occasional discussions about, and forays into, "mining our wastes" for essentially new materials.

Complicated interchanges occur among these states of concentration, and we show several of them on the figure. Start, for example, with "ores," a general term that here could include such things as forests. We mine (or cut down), smelt (or saw up) and form nails, boards, and so on. All this takes energy, and the arrow labeled "conventional mining" makes a loop of size roughly coordinated with the energy expenditure involved. This extraction from "ores" usually expends more energy (or its equivalent cost in other materials) than the alternative processes of extracting

the same thing from society's mixed wastes. Thus the loop called "mining our wastes" usually represents a smaller energy per extracted unit. But that option has not yet been much exploited, partly because ore bodies are usually much larger than waste bodies (dumps). Hence economies of scale favor mining natural ores. Other considerations enter too: depletion allowances, material homogeneity, historical custom, differential transportation rates, less pollution, and so on. Even more costly in energy, but always available as an option, are processes for extracting needed materials from the natural average abundance in the earth's crust.

So much for making components. These are assembled, usually at modest energy cost, into the systems (houses, refrigerators,...) we use and consume. These may be repaired (a similar arrow, but most thrifty of all), and sooner or later discarded as waste—an inexpen-

sive process in itself, but often expensive to recover from. Each higher stage of "organization," as distinct from mere concentration, as shown in the figure usually requires additional processing energy. Perversely, returning junked goods down the ladder (shredding cars) also usually requires energy. Wastage of goods tends generally to disperse material toward the Clarke limit, which leads to pollution if the task is not artfully accomplished, and also leads to requirements for more "ores."

From these admittedly ethereal deliberations comes the operating principle that, unless other factors supervene, we opt for the smallest loops in any generalized figure 4. That is, we elect not just individual processes that tend to be energy conserving, but search for whole options that minimize total energy costs along the entire path of ores-materials-goods-wastes. Consonant with this, we aim for relatively "shallow" recycling, as a rule, both from conservation and minimum-energy points of view. On the other hand, pollution control and material conservation are occasionally at odds, a fact we mentioned earlier. For example, the prospective pollution-abatement schemes for automobiles require more equipment and (possibly) precious platinum for catalysts, a lower engine ratio, and other changes that (in general) require higher fuel consumption and capital investment for the same horsepower output. The effluent is relatively benign, but there is more of it; thus in this case we achieve pollution control at the cost of additional resources.

Before some action decision can properly be made, the various possible strategies must be worked out so that their merits can be compared. Quite a few strategies appear, each with technological, social, scientific and normative components. Limiting ourselves even more, we will consider chiefly the technologist's approach,[3] but caution the reader to bear in mind the partial view. Here now follow seven schemes that can be used separately or in combination, to achieve optimum mass flow relative to the aim of pollution control. The first two—substitution and durability—apply "before the event" of waste production; the latter five deal more directly with waste management.

Substitution

Alternative use can be made either of systems or of materials. For a combination example, recall the use of mercury cells in the chlor-alkali process (figure 2 and accompanying discussion), and the resulting societal misery. These can be replaced by diaphragm cells (70% of the US chlorine is now supplied this way), albeit by forfeiting low quality in the sodium hydroxide product. This substitution of processes is what our Oak Ridge colleague, Hal Goeller, calls "functional substitution."

Materials themselves can be substituted, often in simple ways. Copper is a detrimental impurity in many steels because it reduces formability; thus copper wiring in automobiles makes recycling old automobiles more difficult. Aluminum presents some problems, but not as many; then why not install aluminum wiring? One trouble is that reliable and longlasting electrical connections are harder to make with it. Here, then is a place for some ingenious idea.

Again returning to the example of figure 2, we have divided the mercury uses into approximate recyclable and dissipative categories, where the latter arises from the intrinsic nature of the use. Many of these uses *depend* on the toxic nature of mercury and its derivatives (for example, for agriculture,

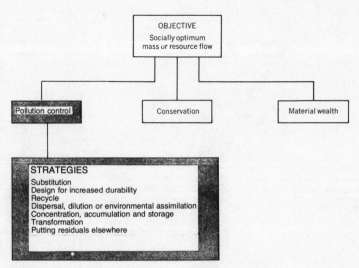

Technological strategies associated with the social objective of optimum mass or resource flow. Decisions among the three main strategies are basically political, not scientific or economic. The seven strategies listed for pollution control are discussed in detail in the text.
Figure 3

paints, and slimicides in the pulp and paper industry). Therefore here is a fertile field for substitution technology, to find more available (less scarce) materials that are less persistently toxic, thereby both abating pollution and conserving available resources.

Durability

If clever design of products can lead to longer useful lifetimes without increasing the difficulty of disposal of the object when it is finally junked, then waste volume is reduced without decreasing material affluence. For many products, the benefit from increased useful lifetime exceeds the cost of added quality. Yet normal economic incentive to the constructor is to maximize present profit, and sometimes even to design for maximum rate of obsolescence within the confines of customer acceptance. Thus our economy operates with a built-in bias towards a throw-away philosophy.

We feel that the durability strategy is very important, but it is one in which normative values enter strongly. How can the concept of durability as a "good" be promoted without stifling inventiveness, which can cause true (and often beneficial) obsolescence? What is the balance between the joy of new style and the concomitant waste? As a practical level, is it really necessary to design household appliances so that

when something goes wrong we must junk the whole thing, because fixing it requires a part that you can't buy due to design change, or needs more skill than you can conveniently hire—or perhaps the construction was basically nonreplaceable or nonrepairable? Household appliances execute large and frequent cycles on figure 4, but that need not be so. Some people comment on this point that production, *per se,* is a "good" because it creates employment. Of course employment is good, and necessary, but such comments miss the mark of societal gain that we aim at here; surely people should do better things than create objects designed to make an early transition to junk.

Private housing makes a subtler case, and one in which we recognize three time scales of concern. The first and potentially most expensive in the long run is that of the speculative builder, for whom the house must look good only until he is well out of the involvement. The second is that of the average householder with a time horizon of say 20 years, who adds insulation, storm windows and maintenance, and minimizes to some extent the total societal cost during his tenure. The third time scale is that of society as a whole, and has to do with much longer-range questions such as depletion of forests and optimum land and energy use. In general, the total pollution and wasting of resources decreases as we move from the shorter to longer time scales—sometimes at the cost of less convenient or spacious accommodations, and sometimes not, but always at the sacrifice of present savings for future rewards.

Recycle

Recycling to reduce both waste and overuse of resources is a very popular topic of conversation these days, and we spend some time talking about it. Recycling comes in many degrees of sophistication; among the simplest are such arrangements as the collection of glass and newspapers at community centers. The economic incentives being not very large at present, the work proceeds, if at all, as a labor of dedication. The potential power of it, however, is large: Something like half the paper products made can in principle be recycled, and if that were done, our forest cutting could be substantially reduced. By increasing recycle, our burgeoning demand for paper products can be met for some years without depleting the quality of our most important renewable resource. Of course, some uses for paper are inherently dissipative, and all involve some degradation of quality. Therefore demand must ultimately level off if we are to manage forests at a steady state.

Our principle of minimum energy paths is nicely illustrated here. Provided extraneous factors do not interfere (for example, the use of a gallon of gasoline to deliver an armful of newspapers to the recycle center), paper, glass, aluminum and many other things are recyclable at low-energy cost—much less than required for winning new material. This method of separation by the individual is the Maxwell Demon approach to recycling, by which the entropy of mixing is defeated by the human capacity to distinguish components of the wastestream.

Aluminum makes a noteworthy example. The total coal equivalent energy to mine, beneficiate and electrolytically refine a ton of aluminum is about 56 000 kWh, taking into account all the inefficiencies but excluding transportation.[4] This works out to 60 eV/atom, but the total energy cost of remelting is only 1–2 eV/atom. In general, recycling energies are in this order: 1–10 eV. Other examples are not so extreme as that for aluminum, but energy factors of five to ten are not unusual. Also, do not forget the resource-conservation side of the story.

At the opposite end of the recycle sophistication street lies the proposed fusion torch,[5] upon which much discussion has been bestowed. Because it has figured rather prominently in some assessments both of controlled nuclear-fusion priorities and of how physics comes to the aid of waste management, let us linger to discuss it in some detail.

In briefest terms, the idea stems from the fact that most schemes for achieving controlled nuclear fusion depend upon confining an exceedingly hot plasma—at a density of 10^{14} to $10^{15}/cm^3$, and a temperature of above 10^8 K, or 10 keV—by magnetic fields alone. The confinement is at best leaky, and it is just this leakiness that makes even the scientific demonstration of controlled fusion so

difficult.[6] Well then, the thought goes, if this superhot plasma keeps escaping (and we must keep re-injecting it, which is another story), can we not find some use for it? One scheme would be to use its heat value in a thermal cycle to make electricity; another would attempt the same thing via direct electrostatic conversion. A third scheme, and the one of interest here, is to use it as a primeval blowtorch, to ionize and reduce to the elemental state everything that is dumped into the exhaust. Unfortunately the scheme will not work, for several reasons.

The first reason involves simple materials flow: No likely fusion scheme can avoid having a substantial amount of tritium in its leakage exhaust. Thus anything introduced into the exhaust and converted to elemental states will become intimately mixed with tritium. Then for later re-use, a further separation of radioactive tritium must be made from whatever comes out of the fusion device; reducing the concentration below (say) $1:10^8$ or $1:10^9$ appears to be a worse problem than the original one.

The second reason that the fusion torch is a fundamentally poor idea involves violation of the principle of minimum-energy paths. Each ion carries with it the ionization energy, plus whatever was radiated (by excitation of incompletely stripped ions, and so on). This will be some 30–40 eV/atom at least, and probably much more, if past experience is any guide on how effective incompletely stripped atoms are in holding down fusion plasma temperatures. Remember also that this is the theoretical minimum; experience shows that complete system energy costs run several times as much. For aluminum, it is a factor of about ten. The consequences are: (a) If all this energy is not recovered with efficiency as high as if the clean plasma exhaust were dumped on a hot surface designed for maximum thermal recovery, the energy waste is huge; (b) with each recycled atom carrying so much energy on its back (so to speak), no great amount of material can be recycled. We believe it's not possible to meet requirement (a); regarding (b), we point out that with a fusion exhaust energy available of 1 kW/person (about equal to the entire US electric power generation at present), and average atomic weight of 12, the maximum amount recyclable would be 0.3 kg/person-day, if all recovery schemes were 100% efficient. This is much less than the current municipal solid waste generation of 2 to 3 kg/person-day.

The third reason why the fusion torch will not be practical has to do with recent findings about what will be the power balance in any conceivable fusion reactors. For open-ended systems (magnetic mirrors), the balance seems

so precarious that for them to work at all, energy in the copious plasma exhaust must be recovered via direct electrostatic conversion, with some 90% efficiency or more; surely we cannot do that with an energy-consuming fusion torch in the way. For closed systems (Tokamaks, stellarators and toroidal devices), the amount of exhaust will (we hope) be not very large, because of better plasma confinement than was at one time feared.

All these questions beg the very practical ones of recovering whatever materials condense from such a carbon-hydrogen-oxygen-metal-etc. plasma, in usable form at low cost. Direct mass separation appears bleak, because the multiplicity of isotopes sets us to doing one atomic mass unit at a time, and even worse. At the Oak Ridge National Laboratory, where most of the US work on electromagnetic separation of isotopes is carried out, the cheapest material (iron 56) sells for $50 000/kg.

Where does this leave us? There exists a huge middle ground of important tasks and low-energy paths to follow. For instance, the US Bureau of Mines[7] describes a scheme for separating incinerated wastes that sorts out iron, nonferrous metals, even colorless and colored glasses, at a cost of $4.00/ton, to yield products estimated to be worth $15/ton. Here are some more challenges for physicists: how can a waste stream "best" (for example, in terms of minimum energy, invested capital, and maintenance) be processed? Processing includes transport, probably shredding, and sensing or sorting into segregated streams for recycle and other fates. In the Bureau of Mines process the sensing and sorting steps exploit physical properties and principles. A density separation is followed by a magnetic one to remove ferrous materials, and a separation of colored and noncolored glasses is accomplished in a high-intensity magnetic field by way of differences in magnetic susceptibility (see figure 1). Extensions of schemes like this receive much deserved attention these days; the technical prognosis is excellent.

Further analysis of municipal waste management reveals some important problems that are probably far less tractable than the technical ones we have mentioned so far. Current waste-management costs are due to collection and transport (typically 6–10, or more, dollars per ton) and a smaller amount for disposal (about 1–2 dollars per ton for landfill or 3–7 dollars per ton for incineration). However, acceptable disposal sites for landfill near the cities are becoming extremely scarce and, because of growing environmental constraints, disposal costs (for example, for incineration) are increasing rapidly. One way to offset spiraling costs is to derive revenue from secondary-materials sales *via* a recycle operation. But a recycle system is capital-intensive, and economics dictate a relatively large-scale operation (typically 1000–2000 tons per day). Since this size corresponds roughly to a million people we are led to the "Wasteplex" concept[8] of bringing waste from a relatively large region to a central waste utilization and recycling complex. In eastern Tennessee such a potential region, which enjoys a low-cost transportation system (by navigable waterway) is twice the area of New Jersey. It encompasses several medium-sized cities and 29 counties in two states.

Thus the principal problem of municipal solid-waste management is twofold: first the development of economic recycle technologies to produce *marketable* secondary materials; second the development of long-term cooperative arrangements between political jurisdictions that historically covet independence.

Dispersal

The traditional scheme—dilution as the solution to pollution—works well enough, so long as we understand how the system works and don't overload it. Dispersal is in reality a natural recycle, and the capacity for recycling many materials—even SO_2, CO, or H_2S—is huge, because huge amounts are produced, absorbed and degraded naturally. But the dilution must be carefully done: Tall stacks for SO_2 dispersal may have worked much of the time up to now, but the effective threshold for local and even remote downwind damage has been passed in the New York City area and in much of Europe. Again, physical monitoring of the conditions and understanding the natural recycle process are things for physicists to do.

For all practical purposes, the dispersal scheme breaks down entirely for certain materials. These are particularly the toxic heavy metals, some of which can reconcentrate in the biological cycle and which are persistent and cumulative; that is, they never degrade into harmless forms. To be sure, a small amount of mercury *can* be safely dispersed; mercury occurs naturally in water at less than one part per billion, and has an average crustal abundance of 50–100 ppb. But the amount safely disposable by "dispersive" loss is so much less than the amount required to be in use by our modern society, *that no effective option exists but to recycle most of it*, or otherwise prevent it from being dispersed back into the environment.

These latter remarks touch upon a very important matter, which has limits we are not competent to judge. For millions of years, we and our evolutionary ancestors have lived in, and adjusted to, a world of certain natural concentrations, and we have little knowledge about our ability to survive (over centuries and millenia, that is) in a world of much different ones. The effects of some things we disperse now are almost wholly unknown; our ecologist and geneticist colleagues are rightly worried. This is no cry of "back to nature and damn the cost"; it is, as is the rest of this article, a plea for mature sociotechnological assessment.

Concentration, accumulation, storage

If the material cannot be disposed of any other way and is not reusable, we can collect it and store it away safely—for as long as we wish, we hope. The prime example here is radioactive wastes—they are very hard to transform into non-radioactive ones, and dangerous to leave unattended. Even though the nuclear physics of radioactivity may be well understood, the storage problem is not; physics enters here too. Should one separate the isotopes, or at least the elements, so that short-lived and long-lived residuals can be handled more expeditiously by separate schemes? The proposed program for disposing of solidified radioactive wastes in salt mines distinguished only low-level, long-lived alpha-emitters from high-level ceramicized billets of (mostly) shorter-lived materials. Whereas the main geophysical questions relating to salt beds and to the effects of the wastes on the salt deposit appear well posed and relatively well answered (in terms of geological stability, thermal effects on the salt, decay of activity in detail with time, and decay of temperature with time and position and so on), a lot more needs to be done. Even for this particular scheme, a number of further details need working out. How close together can billets be stored with safety? How close can conventional salt-mining operations be carried out? How could we recover the radioactive canisters, should something unexpected go wrong?—and so forth. Even more, a detailed assessment of the salt-mine scheme versus other storage methods (for example, elemental separation, bed rock disposal, and so on) has yet to be made. Also, while technically speaking some given method may seem preferable, we must also consider what responsibilities alternative methods may visit upon future generations.

Transformation

Almost too obvious for detailed comment here, yet important nonetheless, are the vast number of physical and chemical steps that can be taken to render waste materials harmless, or relatively so. "Sewage processing" is a generic term covering many examples. More esoteric, perhaps possible, and perhaps of particular interest to physicists is the idea of burning out selected radioactive wastes. It is possible to calculate that the idea of transforming *all* radioactive wastes to stable isotopes, *via* (n,γ) reactions, usually, seems not possible—more reactors (generating more wastes) are needed than we had in the first place. On the other hand, a number of the long-lived elements, which present particularly hazardous or long-term storage problems, might just be convertible this way. Separating these (most transuranium) elements also poses problems, the full measures of which have not yet been taken, as we said before. So the question is still somewhat open; typically, its resolution calls for the participation of many professions, including nuclear science and technology, chemical separation, long-term economics, projections of human behavioral trends, and those able to weigh the value of preserving future options.

Putting residuals elsewhere

An easy technological suggestion to make, but a hard one to carry out, is the idea of shooting particularly intransigent and harmful wastes (radioactive wastes for example) into the sun. Against such ideas are the hard facts that radioactive wastes and their shielding will weigh thousands of tons (but some isotope separation is possible) and that the launch into orbit cannot yet be made accident-free.

For some time now there has been a tendency for polluting industries to seek the wide-open spaces where the assimilative capacity of the environment is not so heavily burdened as it is in most urban centers. The present lively controversy over a large coal-burning electric power plant near Four Corners (of Colorado, New Mexico, Arizona and Utah) is an almost pure example. Provided the "export" is well managed, it can be a viable tool. However, overuse or misuse can lead to incremental irreversibilities. That is, a new area might stand one or two facilities but not a dozen; however, bit by bit, the dozen is reached. Inhabitants of the region may have little to say about the gradual deterioration that results. Indeed, an increasing fraction of the population becomes dependent for their livelihood on the industries, and there is little incentive for long-range planning. Presumably, national pollution standards will go a long way toward blunting this type of development. But also, one sees the need for careful regional planning that will depend on modeling (physical, biological, economic, demographic, and so on) of the air and watersheds involved. Hence the physicist becomes part of the regional planning team.

Another form of export of pollution occurs with regard to translocation of

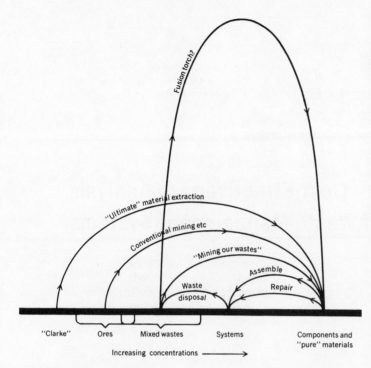

Fusion torch?

"Ultimate" material extraction

Conventional mining etc

"Mining our wastes"

Assemble

Waste
disposal

Repair

"Clarke"   Ores   Mixed wastes   Systems   Components and
                                                "pure" materials

Increasing concentrations ⟶

Energy commitments incurred in the processing of materials. The horizontal dimension represents increasing degrees of material concentration, from the most dispersed state (the "Clarke" level) on the left to pure materials on the right. The vertical dimension is related to energy expenditures for various processing paths. The "fusion torch" is probably not a realistic concept.
Figure 4

American enterprises, especially to the "underdeveloped" regions of the US and other countries where concern for the environment is, to date, very much secondary to economic development. Our own preaching about restoring and maintaining environmental quality may sound fine to us; we are a post-industrial society not only with the sociotechnological capability to effect the changes, but also with not many severe problems of other kinds. But to our friends in the tropical, less-developed countries, the four great killers are malnutrition, malaria, tuberculosis, and schistosomiasis. Relief from these horrors, in the minds of many, comes via the path of rapid generation of national wealth through industrial development—the very route the US has traveled—and smoke coming out of the chimney is a sign of progress. Dying at age 55 of lung cancer seems pretty remote compared with starving to death next year at age 15.

What is the answer to this "across-the-track" question? For the economically privileged to say that each sector (or community or nation) must decide for itself constitutes a pious hypocrisy, reminiscent of 19th-century social Darwinism. Such an attitude forces the disadvantaged sectors to opt for accepting residuals. Clearly, the advantaged sectors have to give quite a bit to the common cause, if they are to stay morally solvent. Besides that, we see a strong self-interest in aiding other nations in their problems. Our wastes wash to other shores, and *vice versa*. We all share the same air and water, and many recent global studies have shown both the connectivity and the urgency of realizing the global consequences.

These thoughts take us far from physics, to societal questions that we earlier promised to try to avoid; but see how central these issues are, how they help weave the fabric of our society, and how we cannot avoid them in any real discussion.

A role for physics

The physicist alone will not set the priorities among the three goals of figure 3. That is for society to determine. But he can be very influential in developing new options and in deciding between alternative strategies for getting to where society decides to go. The physicist will be called upon along with colleagues of other disciplines to tell how the environmental system

works or how it can work, how the system can be measured and described quantitatively, and what the limits of human activities are within the constraints of societal goals. We have tried to demonstrate this by showing that the role of physics, or at least physical thinking, is central to assessing the alternative technological strategies, for achieving one of these societal goals—pollution abatement.

Beyond all that lies a great deal more, which we have not mentioned. For example, even after the societal choices are made and the regulations are passed, a very complex activity of monitoring our previous condition—whether it was better or worse—has been a substantial physics activity, and well recognized.

Because in all these supposedly beneficial things that we hope to see come to pass we require an increasingly complex organization of society, we must be willing to pay for it. One cost is the need for much more information about our physical environment, in ways that can be usefully interpreted by society as a whole, and not just by physical scientists. We are persuaded that physical principles, physical thinking, and, yes, physicists, play an important contributory role in achieving a more socially optimum management of our waste. But the chance for a lasting contribution hangs heavily upon the practitioner's awareness of the broad sociotechnical dimensions of the problem. To proceed otherwise invites irrelevance and justifiable criticism.

* * *

This article is adapted from a talk given by John Gibbons at the American Institute of Physics Annual Meeting of Society Officers and Corporate Associates, September 1971. The research described here is sponsored by the National Science Foundation under Union Carbide Corporation's contract with the US Atomic Energy Commission.

References

1. T. A. Carlson, PHYSICS TODAY, January 1972, page 30.
2. F. W. Clarke, Washington, H. S., "The Composition of the Earth's Crust," US Geological Survey, paper No. 127 (1924).
3. R. M. Solow, Science **173**, 498 (1971).
4. J. C. Bravard, C. Portal, "Energy Expenditures Associated With the Production and Recycle of Metals" (revised by P. H. Wadia), ORNL unpublished report.
5. W. C. Gough, B. J. Eastlund, Scientific American, February 1971, page 50.
6. D. J. Rose, Science **172**, 797 (1971).
7. P. M. Sullivan, M. H. Stanczick, "Economics of Recycling Metals and Minerals from Urban Refuse," US Bureau of Mines Technical Progress Report TPR TPR 33 (April 1971).
8. "The Environment and Technology Assessment Progress Report, June-December 1970," ORNL Report ORNL NSF-EP-3, page 138 (February 1971). □

19

Copyright © 1971 by the Baywood Publishing Company, Inc.

Reprinted from *J. Environ. Sys.*, **1**, 189–209 (June 1971)

Cost-Effectiveness Analysis
of Waste Management Systems

NEIL L. DROBNY
Group Leader, Environmental Planning Group
Battelle Memorial Institute

SYED R. QASIM
Associate Professor, Dept. of Civil Engineering
Polytechnic Institute of Brooklyn

BRUCE W. VALENTINE
Technical Director
Richards of Rockford, Inc.

ABSTRACT

Cost effectiveness techniques have been used to identify the most cost effective commercially available waste-water treatment and disposal system for 500- and 1000-man military camps. Unit costs (cents per thousand gallons treated—capital plus operating) were developed using manufacturers' data. System effectiveness was determined using a decision weighting model based on paired comparisons. Results of this study have provided the Navy with guidance in terms of waste management systems that should be selected for use at advanced bases. Furthermore, this study is one example of how systems technology can be used in the solution of complex environmental quality management problems.

Introduction

A wide variety of commercially developed processes are generally available for treatment of waste water from small communities, industries, restaurants, motels, camp sites, etc. In addition, significant variations exist in both cost and performance characteristics from one manufacturer to another within the same process. Consequently, the selection of economic and effective systems from among the alternatives available is not a casual

189

exercise. An investigation was conducted to determine the most cost-effective system for the disposal of sewage at advanced military bases. The specific objective was to identify highly cost-effective commercially available systems, equipment, and related hardware which can be selected for 500- and 1000-man military camps.

Background

Remote locations (where most of the military bases are built) impose severe logistic constraints not only on the initial system delivery and installation, but also on system maintenance due to the difficulty of supplying spare parts and repair tools. Thus, advanced base sanitary facilities must be simple to build and simple to operate. Only construction skills and equipment within or available to military forces can be relied upon for construction. Similarly, operation and maintenance of the completed facilities must be within the capabilities of petty officers who may possess little prior applicable training or experience. For maximum effectiveness, construction time should be as short as possible to reduce planning time and to permit quick response to changes in operational requirements which occur rather frequently. Finally, minimal capital and operating costs are desirable in the general interest of economy. Difficulties in providing low cost facilities are compounded by the fact that advanced bases are, by their nature, temporary facilities with expected lives, and hence amortization periods, of five years or less.

During recent years, waste treatment facilities specially designed for relatively small flows from small communities, camps, motels, etc., have become commercially available. These plants, commonly referred to as "packaged waste treatment plants," have many features (simplicity, compactness, reliability, and operational flexibility) that render them potentially suitable for use at advanced military bases. The work reported in this paper was directed toward a cost-effectiveness analysis of those systems that are presently commercially available; the specific objective was to identify the most cost effective systems for use at advanced military bases in remote areas. A comprehensive survey of treatment plant and hardware manufacturers was conducted to obtain data on available systems. Technical and economic data supplied by various system manufacturers were used to perform the cost-effectiveness analysis of these systems.

Cost-Effectiveness Methodology

APPROACH

Cost-effectiveness analyses involve two types of evaluation. The first and most straightforward is cost analysis. This involves the delineation of all major system components and the development of capital and operating cost estimates for each. A second, and often the most difficult (in complex systems), is the effectiveness evaluation whereby one attempts to generate a single cardinal measure or indicator of effectiveness based upon multiple considerations. The essence of cost-effectiveness analysis *per se* then involves the trade-off of cost with effectiveness to identify the most cost-effective alternative(s).

Economic analyses of engineering systems were traditionally based strictly on cost considerations. Initially, engineers were concerned with least-cost solutions that met fixed requirements or constraints; economic efficiency was measured by cost minimization without recourse to benefits. Next, evaluations centered about net cost or net savings which represented the difference between total cost incurred and any resultant savings or benefits which could be expressed in monetary units. It has been recognized however, that combining costs and benefits into a single measure will not necessarily indicate the most economically efficient alternative. Consequently, keen interest developed in the use of cost-benefit analyses which focused attention on the cost/benefit ratio as the measure of economic efficiency.

Benefit-cost analyses are satisfactory only so long as all benefits can be expressed in dollars. This is not the case, however, in evaluating wastewater treatment and disposal systems for use at advanced military bases, since the overall effectiveness of any given alternate system depends on multiple criteria or measures of effectiveness (reliability, size, ease of operation, etc.) which cannot be expressed directly in monetary units.

Thus selection of one system from among a group of alternatives, when multiple criteria are to be considered, poses a complex problem in decision making. The difficulty arises from the multiplicity of considerations which somehow must be weighed against one another in order to reach a decision. This usually indicates a need for some type of decision-weighting model. Decision-weighting models have been criticized by several authors, and when taken in context, many criticisms are valid. One cannot escape the fact, however, that somehow the decision maker must make a final choice. Somehow he must weigh all the diverse factors so as to reach a final overriding value assessment and to arrive at a choice. A methodology is outlined below for doing this in an explicit manner.

Historically, decision makers have dealt with multiple-criteria problems largely on the basis of subjective judgment and intuition. Personal judgments have been used both to effect trade-offs among the relative importance of various effectiveness criteria and to assess the effectiveness of different levels of predicted performance. Problems of physical interaction among performance consequences and interdependencies among effectiveness criteria have similarly been handled on an intuitive basis.

If the decision problem under consideration is very simple, and if the consequences of making a poor decision are relatively inconsequential, the subjective, intuitive approach may be the best way to proceed. Additional gains that might be realized from formalizing and systematizing the decision process probably would not justify the extra time, cost, and effort required. However, when the decision problem becomes complex and the consequences become important, strict reliance on subjective judgment is not satisfactory. In these cases, a need exists for a systematic procedure to develop an explicit, logically consistent, and replicable procedure to aid in the assessment of effectiveness.

It should be emphasized that explicitness, logical consistency, and replicability do not preclude the use of judgment. One can, however, attempt to make his judgments in an objective manner rather than in a subjective one. Judgments must be used both in trading off relative importance among effectiveness criteria and in assessing measures of effectiveness for various performance levels. When judgment is used, it should be made explicit, should be thoroughly scrutinized for logical consistency, and should be elicited by a uniformly applicable procedure.

The advantages of such a procedure are several. The explicit statement of assumptions will help ensure that personal judgments are not based on false information. In addition, a decision process thoroughly scrutinized for logical consistency will help to remove random elements and inconsistencies from the decision process.

The decision-weighting model employed in the effectiveness analysis of advanced-base wastewater treatment and disposal systems may be outlined with the aid of Figure 1. Assume that one is faced with a series of alternative concepts $(A_1, A_2, A_3, \ldots, A_m)$ which must be evaluated in terms of several measures of effectiveness $(M_1, M_2, M_3, \ldots, M_n)$. The general procedure is to first assign relative weights (w) to each of the n measures of effectiveness. These weights merely reflect the relative importance of each of the measures of effectiveness. A convenient ground rule to follow in deriving these relative weights is that they should add up to one; when this is done the resulting overall effectiveness ratings (computed as the sum of weighted individual effectiveness ratings) may be subjected to the same interpretation as the effectiveness ratings or scores (discussed below) assigned to individual performance levels. This renders far more manageable the task of checking assigned weights for intuitive reasonableness and consensual validation.

A score or rating is then assigned which reflects the degree to which each alternative satisfies one of the effectiveness measures. Referring to Figure 1, r_{ij} is the rating of alternative A_i with respect to effectiveness M_j. The overall effectiveness of any given alternative is then equal to the sum

Alternate System Concepts	Measures of Effectiveness (M) and Relative Weights (w)				
	M_1 (w_1)	M_2 (w_2)	...	M_j (w_j)	... M_n (w_n)
A_1				.	
A_2				.	
.				.	
.				.	
.				.	
A_i	. . . .			r_{ij}	
.					
.					
.					
A_m					

Effectiveness of alternative $i \equiv E[A_i] = \sum_{j=1} w_j r_{ij}$

where

$$\sum_{j=1}^{n} w_j = 1$$

$r_{ij} \equiv$ Rating of alternative i with respect to measure j

Figure 1. Methodology for System-Effectiveness Analysis

of the product of each rating multiplied by its relative weight; the calculation procedures are outlined in Figure 1.

To implement the above decision procedure, judgments are required at two critical steps: first, in the assignment of weights; second, in the assignment of ratings. The need for judgment on these elements cannot be avoided since they must, by definition, reflect the decision-maker's opinion (as opposed to an absolute measure) on the relative importance of the measures and on the relative merit of alternatives being considered. One can, however, use a procedure to render these judgments objective, explicit, consistent, and replicable. Such a procedure, based on the technique of paired comparisons, is outlined below.

Technique of Paired Comparisons

The decision model employed in the effectiveness analysis is based on a technique that forces the decision maker to make a series of paired comparisons. In considering each pair of items a decision (judgment) is made as to which item is more important or scores higher. As will be shown in the following discussion and example, the decision maker then relates all decisions to a common reference and makes the thought processes followed in assigning his judgments explicit.

The paired comparison technique also has the advantage that the decisions required of the decision maker are simple. He need compare only two items at a time, as contrasted to the dozens of items that would need to be considered simultaneously (at least implicitly) if one attempted a direct assignment of weights and ratings. Decision weighting models and the paired comparison technique are discussed extensively by Eckenrode[1] and Miller.[2] The procedure following is a condensation of the basic technique outlined by Miller. Other less sophisticated applications of the paired comparison technique have also been proposed but their simplicity negates their utility.[3,4]

The technique is best illustrated by an example. Throughout this discussion, the reader may find it helpful to refer to Figure 1: Consider that three alternative systems, A_1, A_2, and A_3 are to be evaluated; It is further determined that only three criteria or measures of effectiveness are relevant to the evaluation. For illustration purposes, let these be M_1 —reliability; M_2 —weight; and M_3 —size. In selecting measures of effectiveness one must ensure that they are:

1. *complete* (i.e., all criteria which the decision maker is able and willing to formulate and consider must be included)
2. *mutually exclusive* (i.e., individual criteria must neither encompass nor be encompassed by other criteria on the list)
3. *free of effectiveness interdependence* (the effectiveness of a given alternative with respect to any one measure should be independent of its effectiveness with respect to any other measure)

Returning to the example, it has been determined that only three measures of effectiveness—reliability, weight, and size are to be considered. The next step is to determine the relative weights to be assigned to each measure of effectiveness the relative importance the decision maker wishes each to have in determining his decision. The effectiveness measures are then tabulated in a list. It is often convenient to structure this list by ranking the items in order of decreasing importance. Starting at the bottom of the list, successive paired comparisons are made between contiguous measures; for each comparison the decision maker indicates, in terms of a ratio, the perceived relative importance of the two items being considered. Stated alternatively, the decision maker indicates the rate at which he would be willing to accept reduced satisfaction with one criterion in return or increased satisfaction with another.

255

Effectiveness measures	Relative Importance		Normalized relative weights
	With respect to next item on list	With respect to last item on list	
M_1 Reliability	3	6	0.67
M_2 Weight	2	2	0.22
M_3 Size	–	1	0.11
		$\Sigma = 9$	$\Sigma = 1.00$

Figure 2. Assignment of Weight to Measures of Effectiveness

For illustration purposes, assume that weight is twice as important as size, and reliability is three times as important as weight. The relative importance of each item, with respect to the one immediately bedow it on the list, is then indicated (see Figure 2). One then calculates by simple multiplication the importance of each item relative to the last one on the list. In the example being considered, weight is twice as important as size, and reliability is six times (2×3) as important as size. These values are indicated in the third column of Figure 2; in some cases, this column can be set down directly. The best approach depends in part on the type of information one has to work with and also on the psychological preferences of the decision maker. By normalizing this third column (so that the sum of the individual values is 1), one can derive the relative weights (w) for each of the measures of effectiveness; these values are set down in column 4*. Although this procedure guarantees that the resultant weights will possess certain desirable logical properties, it must be emphasized that their validity depends upon the decision maker's ability to provide accurate judgments in making the paired comparisons.

Having derived the relative weights in a consistent and explicit manner, one then uses a similar procedure to derive the respective rating scores (r_{ij}). Consider first the assignment of ratings to the three alternative systems with respect to effectiveness measure M_1, reliability. The reader is referred to Figure 3 for the following discussion. The alternatives may be tabulated in any order. Successive paired comparisons are then made between contiguous alternatives on the list, starting at the bottom and working up. For each comparison the decision maker makes a judgment about the relative degree to which each of the alternatives satisfies the effectiveness measure under consideration. As in the case of assigning weights to the effectiveness measures, the decision maker is asked to indicate, in terms of

* The third column of numbers contains all the information needed to establish relative weights, but for reasons mentioned earlier, it is desirable to have the weights sum to 1; this is the only reason for computing the averaged weights listed in the fourth column of Figure 2.

Effectiveness measures	Alternative systems	With respect to next alternative	With respect to last alternative	Ratings, r_{ij}
M_1 Reliability	A_1	3.4	8.50	$r_{11} = 0.708$
	A_2	2.5	2.50	$r_{21} = 0.208$
	A_3	—	1.00	$r_{31} = 0.084$

r_{ij} = Effectiveness of alternative i with respect to measure j.

Figure 3. Derivation of Ratings, r_{ij} with Respect to Effectiveness Measure, M_1

a ratio, the degree to which one alternative is superior to another in terms of its effectiveness with respect to the measure being considered.

For example, assume in the comparison being considered, that in terms of reliability, system A_2 is judged to be 2.5 times more effective than alternative A_3 and alternative A_1 is judged to be 3.4 times more effective than alternative A_2. These judgments are entered into the table as shown in Figure 3. By appropriate multiplication of these values, the relative effectiveness of each alternative with respect to the last alternative on the list (A_3) is calculated; each of these numbers is then averaged by dividing by their sum. The averaged ratings are then listed in the fifth column as shown in Figure 3, and are the respective ratings of the three alternative systems in terms of reliability. The motivation for averaging these ratings will become obvious shortly.

The above procedure is then repeated for the other two measures of effectiveness (weight and size); the reader is now referred to Figure 4. For example, system A_2 in terms of weight is considered to be twice as effective as system A_3, and A_1 is considered to be only 9/10 as effective as system A_2. Similarly with respect to size, system A_2 is considered to be only 1/2 as effective as system A_3; system A_1 is considered to be only 1/2 as effective as system A_2. Respective ratings are calculated according to the procedure outlined previously and are summarized in Figure 4. It should now be obvious that the reason for averaging the ratings is to put all of the ratings derived for the various measures of effectiveness on the same basis, i.e., in the range of zero to one. (Any other common range could be used just as well.)

Having derived the respective weights for the three measures of effectiveness and the ratings for each of the alternative systems with respect to the three measures of effectiveness, one has all of the data required to enter the calculation matrix outlined in Figure 1. The data and calculations of overall effectiveness are outlined in Figure 5. It is determined that alternative A_1 has an effectiveness of 0.573, alternative A_2

Effectiveness measures	Alternative systems	Relative Effectiveness		Ratings, r_{ij}
		With respect to next alternative	With respect to last alternative	
M_1 Reliability	A_1	3.4	8.50	$r_{11} = 0.708$
	A_2	2.5	2.50	$r_{21} = 0.208$
	A_3	—	1.00	$r_{31} = 0.084$
			$\Sigma = 12.00$	$\Sigma = 1.000$
M_2 Weight	A_1	0.9	1.80	$r_{12} = 0.375$
	A_2	2.0	2.00	$r_{22} = 0.416$
	A_3	—	1.00	$r_{32} = 0.209$
			$\Sigma = 4.80$	$\Sigma = 1.000$
M_3 Size	A_1	0.5	0.25	$r_{13} = 0.143$
	A_2	0.5	0.50	$r_{23} = 0.285$
	A_3	—	1.00	$r_{33} = 0.572$
			1.75	$\Sigma = 1.000$

Figure 4. Summary of Ratings, r_{ij}

has an effectiveness of 0.262, and alternative A_3, an effectiveness of 0.165. It should be noted that the alternatives turn out to have been ranked in the order of decreasing effectiveness purely as a matter of coincidence.

One shortcoming of the procedure is that it does not provide any means for handling risk and/or uncertainty considerations. In assessing the relative importance of a given system or effectiveness measure, it is assumed that an outcome will definitely occur. In reality, however, any given outcome or occurrence would properly be described by a probability distribution function. The above procedure will not reflect explicitly the aversion which a decision maker may feel toward either the risk or uncertainty regarding the likelihood of the actual occurrence of a given outcome. Furthermore, the process provides no mechanism for reflecting perceived tradeoffs between the effectiveness of a given outcome conditional upon actual occurrence and the variable risk or uncertainty surrounding its occurrence. This is, in some respects, an academic question since in most cases the decision maker has no information regarding the risks or uncertainty, and so could not incorporate it even if he had a procedure that, in principle, would permit him to do so. An excellent discussion of these limitations and others, along with their implications, and the general methodology of cost effectiveness is contained in a recently published textbook.[5]

258

Measures of effectiveness, (M)
and Relative Weights, (w)

Alternative system concepts	M_1 reliability $w_1 = 0.67$	M_2 weight $w_2 = 0.22$	M_3 size $w_3 = 0.11$	Total effectiveness $E\,[A_1]$
A_1	0.708	0.375	0.143	0.573
A_2	0.208	0.416	0.285	0.262
A_3	0.084	0.209	0.572	0.165
				$\Sigma = 1.000$

$$E[A_1] = \sum_{j=1}^{3} w_j r_{1j} = (0.67) \times (0.708) + (0.22) \times (0.375) + (0.11) \times (0.143)$$
$$= 0.573$$

$$E[A_2] = \sum_{j=1}^{3} w_j r_{2j} = (0.67) \times (0.208) + (0.22) \times (0.416) + (0.11) \times (0.285)$$
$$= 0.262$$

$$E[A_3] = \sum_{j=1}^{3} w_j r_{3j} = (0.67) \times (0.084) + (0.22) \times (0.209) + (0.11) \times (0.572)$$
$$= 0.165$$

Figure 5. Summary of effectiveness analyses

A few more comments regarding the implementation of the procedure outlined are in order. In actual practice, it is generally desirable to get the judgments of several decision makers as inputs to the procedure. This can be done in several ways. One way would have each decision maker go through the process independently and then average his final effectiveness measures for the respective alternative. Another way would use average values for each of the paired comparisons, and then conduct the evaluation using these average values.

Having arrived at a final measure of effectiveness for each of the alternatives under consideration, one can employ sensitivity analyses to explore the effect of changes in the value judgments assigned at any step in the paired comparison procedure on the final decision. The purpose of a sensitivity analysis is to identify areas where emphasis should be placed in system improvements, since one desires to concentrate on areas wherein a given level of improvement will produce the greatest improvements in overall system effectiveness.

Another dimension which can readily be incorporated into the procedure is to use several levels of the effectiveness measures. For example, within the measure of size, one may wish to consider explicitly length, width, and height. Such considerations can be factored into the procedure rather easily, and Miller[2] demonstrates a technique for doing so. Similarly, when evaluating the effectiveness of alternative systems it may on occasion be desirable to break the systems down into respective components so that explicit judgments can be made on each of the important items comprising the system. This simply adds another dimension to the decision matrix but poses no conceptual problems. This degree of sophistication was not warranted for the analyses reported in this paper.

Process Alternatives

A complete technical description of the process alternatives (biological and chemical) available in various packaged waste treatment plants is not germane to the cost-effectivenss methodology described in this paper; readers are referred to an earlier publication[6] for a general discussion of different treatment processes, along with a description of the manufacturers' variations available within each category. For descriptions of an oxidation pond, aerated lagoon and oxidation ditch, four additional sources may be consulted.[7,8,9,10]

Evaluation of Commercially Available Systems

TECHNICAL REQUIREMENTS

Technical requirements used to select various biological and chemical systems for evaluation are summarized below. These requirements were treated as minimum performance criteria; commercially available systems failing to meet all of these standards were excluded from the intensive cost-effectiveness analysis.

- Environment: All but polar
- Camp Population: 500 and 1000 men
- Waste Loading:
    Hydraulic: 65 gal/man/day
        Total: 500-man camp—32,500 gal/day
        Total: 100-man camp—65,000 gal/day
    Organic (BOD:) 400 milligrams/liter
    Suspended Solids: 400 milligrams/liter
    Sludge Quantities: Dry weight—500-man camp—92 lb/day
                       1000-man camp—184 lb/day
            Volume—500-man camp—400 gal/day
                   1000-man camp—800 gal/day

260

- Degree of Treatment Required: Secondary treatment or equivalent (80 percent or more BOD and suspended solids removal)
- Equipment Life: 5 years
- Maximum Size of Shipment Packages: 8 X 8 X 20 ft
- Maximum Weight of Shipment Packages: 25,000 lb

Based upon a survey of 52 manufacturers 15 different process systems were found to meet the minimum performance criteria outlined above.

COST EVALUATIONS

Summary of Cost Criteria

Specific cost parameters and assumptions are outlined below:
- Capital cost amortization parameters
  - —Interest rate: 5 percent
  - —Equipment life: 5 years
- Manpower: $10.25/hr (total cost to support Second Class Petty Officer in the field)
- Power: $0.05/kwh (includes amortized capital cost of equipment, shipping of equipment and fuel to Southeast Asia, the cost of shelter or housing for equipment and allowance for power—transmission system)
- Shipping: $1.00/cu ft (U.S.A. to Southeast Asia—assumed to be independent of weight)
- Earthwork:
  - —Site preparation: $80/1,000 sq yd
  - —General excavation: $550/1,000 cu yd

Two sets of cost estimates were developed for all systems investigated; one for the capital investment required to install the complete system at the job site, and the other for operating the plant in the manner specified by the manufacturer. For evaluation purposes, both of these costs were reduced to equivalent annual costs and unit costs (dollars/1000 gallons).

Unit Capital Cost Data. The unit capital cost ($/1000 gallons) was obtained from the annual capital cost data and the average quantity of waste water treated annually. For the annual capital cost, the total capital cost is amortized over 5 years at an interest rate of 5 percent.

Unit Operating Cost Data. The unit operating cost ($/1000 gallons) was estimated from the annual operating cost data and the average plant flow rate. The annual operating cost includes the cost of chemicals, equipment, repairs, power, and general operating maintenance labor. Allowances for supervisory staff and overhead items are not included since these cost items are difficult to define for a military combat situation.

261

EFFECTIVENESS EVALUATION

Measures of Effectiveness

As indicated, the first step in applying the decision model outlined above is to develop the effectiveness criteria and to derive their relative weights. Eight effectiveness measures were identified. Using the technique of paired comparisons outlined earlier, relative weights for each were calculated; these are summarized in Table 1. Some discussion of what is meant by each measure of effectiveness will also be instructive.

Table 1. Effectiveness Criteria and Adjusted Relative Weights

	Effectiveness Measures	Adjusted Relative Weights
M_1	Simplicity of Operation	0.37
M_2	Simplicity of Installation	0.25
M_3	Operational Flexibility and Reliability	0.13
M_4	Environmental Quality Control	0.10
M_5	Manpower Requirements	0.06
M_6	Space Requirements	0.04
M_7	Power Requirements	0.03
M_8	Relocatability	0.02
		1.00

M_1–**Simplicity of Operation**. To minimize routine daily demands on manpower and other resources, operational simplicity is of paramount importance. Specific items to be considered include the degree of skill needed to operate and maintain the system, the frequency of laboratory tests and/or other monitoring activities required, the extent of automatic *versus* manual controls, and general housekeeping requirements.

For automatic controls, the following position was taken. In situations where automatic controls are auxiliary to manual controls, they are considered to be an advantage. In situations, however, where automatic controls are a substitute for manual controls, a low ranking is assigned.

M_2–**Simplicity of Installation/Construction**. After simplicity of operation, simplicity of installation and construction is considered to be the next most important feature. Here one is concerned with such things as the extent of field construction required (concrete base, excavation, heavy equipment needed, etc.) and the degree of field assembly required in terms of the number and ease of required field connections (welded versus nut and bolt). One factor which contributes significantly to field assembly considerations, is that in order to meet the 8 X 8 X 20 ft shipment package size constraint, several manufacturers indicated a need to cut their equipment into pieces which would have to be refabricated on site. The extent of this varies considerably from one manufacturer to another and from one system to another.

M₃ –Operational Flexibility and Reliability. Because of the changing activities in military operations and the sporadic availability of manpower and spare parts, operational flexibility and reliability is an important factor to be considered in the selection of advanced-base waste water treatment and disposal systems. Items of concern here are the ability of a system to perform satisfactorily under varying hydraulic and organic loading and the extent of process controls available for the control of air rate, return sludge, etc.

It should also be noted that significant differences also exist within a specific group of packaged plants in that the operational flexibility and reliability of two smaller systems, each designed to treat one-half the total flow, is considered to be greater than that for a single large system.

M₄ –Environmental Quality Control. Environmental quality control is important for health and aesthetic reasons which in turn directly affect troop morale and effectiveness. All systems evaluated meet the minimum basic technical requirement of equivalent secondary treatment established previously. Environmental control factors over and above the minimum requirements are given due consideration in the overall system evaluations. Items of concern here are degree of treatment over and above secondary treatment, adequate handling and disposal of sludge, expected odor problems, the potential of surface and subsurface water pollution, etc.

M₅ –Manpower Requirements. Manpower requirements refer to the manpower required to construct and operate a given system. Although it is true that differences in manpower requirements will also be reflected in the cost analysis, it is felt that because manpower is a limited resource, manpower needs contain dimensions over and above cost that relate purely to effectiveness.

Consider, for example two hypothetical systems, the total combined operating and capital costs of which are equal, but one system requires less manpower than the other. Since manpower is a prime resource at advanced bases, the system requiring less labor should receive more favorable overall evaluation (in spite of the fact that the two systems compare equally in terms of cost and other performance characteristics) besides what would be reflected in the labor-dollar-cost analysis. Considerations such as these are incorporated in the effectiveness portion of the analysis.

For purposes of comparing individual systems, primary importance was attached to low operating labor requirements. There are two reasons for this. First, operating labor is a continual requirement whereas construction labor is a "one shot" demand. Additionally, military forces find it more difficult to provide personnel with operational skills than those with construction skills. Thus in the system evaluations the systems which received the highest ratings were those with the lowest operating labor requirements. Differences in construction labor requirements were used only to break ties between separate systems with identical operating-labor requirements.

M_6 –Space Requirements. The land area required for a given system determines the space required. For security reasons it is desirable to maintain military systems and supporting facilities in as confined an area as possible.

M_7 –Power Requirements. Electrical energy and/or fuel needed to operate a given system are included in Power Requirements. As in the case of manpower needs, low power requirements will reflect favorably in the cost analysis, but since power is a limited resource, low power requirements also contain dimensions which relate purely to effectiveness. Generally speaking, the relative effectiveness of the various systems with respect to power needs was considered inversely proportional to the absolute power required. For example, a system requiring 7 kw to operate was considered (in terms of power requirements) to be twice as good as one requiring 14 kw.

M_8 –Relocatability. Due to the constantly changing demands on military forces, especially under combat conditions, there is some merit in having military hardware and supporting systems (such as waste water treatment plants) that are easily transportable. To reflect this factor in the evaluations conducted, systems which would require extensive disassembly, refabrication, and reinstallation to be moved were rated lower than those which could be transported essentially intact.

RESULTS

A total of 15 different treatment systems met the minimum performance criteria set forth earlier. These systems are given in Table 2. All were included in cost-effectiveness evaluation.

Table 2. Summary of Treatment Systems Used in
Cost-Effectiveness Evaluation

Treatment Process	Number of Systems
Biological Process	
Aerated lagoon	1
Completely mixed activated sludge	2
Contact stabilization	2
Extended aeration	3
Oxidation ditch	1
Oxidation pond	1
Rotating biological contactor	1
Trickling filter	1
Ultrafiltration	1
Chemical Processes	
Chemical precipitation	1
Electrochemical flotation	1
Total	15

Results of the cost and effectiveness evaluations for 500- and 1000-man systems are summarized in Tables 3 and 4. Since there are 15 potentially feasible alternative systems available for both 500- and 1000-man camps, perception of the cost effectiveness tradeoffs, and identification of the most cost effective systems from these tabular data is rather difficult.

To facilitate the identification of the most cost-effective systems, Figures 6 and 7 were prepared. These figures represent graphically the relationship between total unit costs (capital plus operating) and total effectiveness for the various systems. It can be ascertained visually, from these figures, that waste treatment processes designated $A_8, A_9, A_{10}, A_{11}, A_{12}$, and A_{14} are the most cost-effective systems in the entire group. This is true for both 500- and 1000-man systems. This may be determined by using the following decision rule. If any system, e.g., A_i, has a greater cost but lesser or equal effectiveness than any other system, i.e., A_j, A_i is eliminated. Repeated applications of this elimination procedure leaves systems $A_8, A_9, A_{10}, A_{11}, A_{12}$, and A_{14}.

A critical inspection of Figures 6 and 7 indicates that A_{10} is the least costly of all systems but is also the least effective; system A_{12}, although having the highest effectiveness, is also most expensive. The curve possesses two distinct ranges, one below and the other above system A_8. The range

Table 3. Summary of Cost-Effectiveness Analysis—Alternative Treatment Processes for 500-Man Camps

Alternative Systems A_i	Measure of Effectiveness, (M_{ij}), and Relative Weight, (w_{ij})[a]								Total Effectiveness, $E(A_i)$	Total Unit Treatment Cost, $/1000 gal
	M_1 (0.37)	M_2 (0.25)	M_3 (0.13)	M_4 (0.10)	M_5 (0.06)	M_6 (0.04)	M_7 (0.03)	M_8 (0.02)		
A_1	0.0303[b]	0.0133	0.0109	0.0010	0.0034	0.0004	0.0040	0.0010	0.0643	0.53
A_2	0.0303	0.0133	0.0109	0.0010	0.0034	0.0004	0.0024	0.0010	0.0627	0.60
A_3	0.0228	0.0140	0.0077	0.0097	0.0026	0.0035	0.0013	0.0011	0.0627	1.05
A_4	0.0228	0.0140	0.0077	0.0097	0.0023	0.0035	0.0007	0.0011	0.0618	1.70
A_5	0.0228	0.0140	0.0077	0.0097	0.0033	0.0035	0.0011	0.0011	0.0632	1.05
A_6	0.0228	0.0140	0.0077	0.0097	0.0030	0.0035	0.0016	0.0012	0.0635	1.03
A_7	0.0251	0.0140	0.0090	0.0097	0.0047	0.0035	0.0016	0.0012	0.0687	0.85
A_8	0.0251	0.0182	0.0090	0.0065	0.0037	0.0035	0.0031	0.0018	0.0709	0.60
A_9	0.0313	0.0111	0.0105	0.0020	0.0055	0.0069	0.0024	0.0009	0.0706	0.70
A_{10}	0.0331	0.0127	0.0114	0.0003	0.0035	0.0001	0.0050	0.0009	0.0670	0.39
A_{11}	0.0289	0.0191	0.0077	0.0015	0.0064	0.0035	0.0046	0.0014	0.0731	1.09
A_{12}	0.0207	0.0241	0.0081	0.0097	0.0049	0.0035	0.0004	0.0020	0.0734	1.30
A_{13}	0.0178	0.0202	0.0067	0.0097	0.0040	0.0035	0.0003	0.0014	0.0636	2.90
A_{14}	0.0187	0.0241	0.0070	0.0097	0.0058	0.0035	0.0013	0.0020	0.0721	0.93
A_{15}	0.0178	0.0241	0.0067	0.0097	0.0040	0.0035	0.0003	0.0020	0.0681	2.48

[a] Relative weights are given in parentheses and are obtained from Table 1.
[b] All matrix entries are products of respective ratings and averaged relative weights.

Table 4. Summary of Cost-Effectiveness Analysis—Alternative Treatment Processes for 1000-Man Camp

Alternative System (A_i)	Measure of Effectiveness, (M_j), and Relative Weight, $(w_j)^a$								Total Effectiveness, $E(A_i)$	Total Unit Treatment Cost, $/1000 gal
	M_1 (0.37)	M_2 (0.25)	M_3 (0.13)	M_4 (0.10)	M_5 (0.06)	M_6 (0.04)	M_7 (0.03)	M_8 (0.02)		
A_1	0.0303[b]	0.0132	0.0109	0.0010	0.0032	0.0004	0.0039	0.0010	0.0639	0.47
A_2	0.0303	0.0132	0.0109	0.0010	0.0032	0.0004	0.0021	0.0010	0.0621	0.51
A_3	0.0228	0.0140	0.0077	0.0097	0.0032	0.0035	0.0014	0.0011	0.0634	0.83
A_4	0.0228	0.0140	0.0077	0.0097	0.0022	0.0035	0.0007	0.0011	0.0627	1.47
A_5	0.0228	0.0140	0.0077	0.0097	0.0037	0.0035	0.0012	0.0011	0.0637	0.75
A_6	0.0228	0.0140	0.0077	0.0097	0.0035	0.0035	0.0017	0.0011	0.0640	0.77
A_7	0.0252	0.0140	0.0090	0.0097	0.0050	0.0035	0.0022	0.0012	0.0698	0.70
A_8	0.0252	0.0182	0.0090	0.0065	0.0033	0.0035	0.0032	0.0018	0.0707	0.46
A_9	0.0314	0.0110	0.0105	0.0020	0.0038	0.0069	0.0017	0.0009	0.0682	0.58
A_{10}	0.0331	0.0128	0.0114	0.0003	0.0029	0.0001	0.0047	0.0009	0.0661	0.31
A_{11}	0.0288	0.0190	0.0077	0.0015	0.0067	0.0035	0.0043	0.0014	0.0729	0.85
A_{12}	0.0206	0.0241	0.0081	0.0097	0.0053	0.0035	0.0005	0.0020	0.0738	1.04
A_{13}	0.0178	0.0202	0.0065	0.0097	0.0046	0.0035	0.0003	0.0014	0.0640	2.22
A_{14}	0.0187	0.0242	0.0070	0.0097	0.0061	0.0035	0.0014	0.0020	0.0726	0.80
A_{15}	0.0178	0.0292	0.0067	0.0097	0.0024	0.0035	0.0003	0.0020	0.0716	2.48

[a] Relative weights are given in parentheses and are obtained from Table 1.
[b] All matrix entries are product of respective rating and normalized relative weights.

267

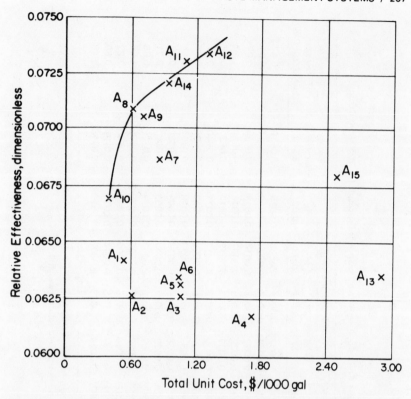

Figure 6. Relationship between total unit cost and total relative effectiveness of various alternative treatment systems suitable for installation at 500-man camps

below system A_8 has higher slope than that above, indicating that systems in the lower range yield a greater gain in effectiveness for any given incremental cost than do systems in the range above system A_8. The sharp change in the slope at system A_8 is a break point, indicating that the system is highly cost effective as compared to the others. Other alternative systems such as A_{14}, A_{11}, and A_{12} exhibit a nearly linear relationship between cost and effectiveness. Selection of any of these systems would depend upon how much one can afford to pay for the additional effectiveness gained.

Results of this study have provided the Navy with guidance in terms of waste management systems that should be selected for use at advanced bases. Furthermore, this study is one example of how systems technology can be adapted to the solution of complex environmental quality management problems.

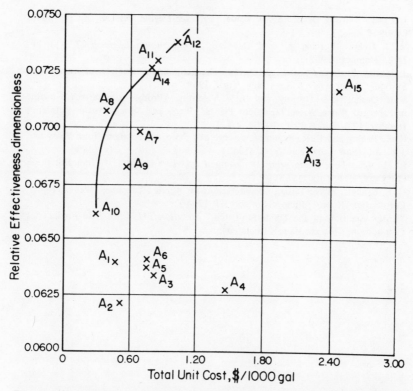

Figure 7. Relationship between total unit cost and total relative effectiveness of various alternative treatment systems suitable for installation at 1000-man camps

ACKNOWLEDGMENT

Work reported in this paper was developed in conjunction with, and supported by the U. S. Naval Civil Engineering Laboratory under Contract No. N-62399-69-C-0036, as part of Work Unit YF38,534.003.01.003, "Advanced Base Sewage Treatment Study."

REFERENCES

1. R. T. Eckenrode, "Weighting Multiple Objectives," *Management Science* 12 (3), pp. 180-192 (1965).
2. J. R. Miller, "A Systematic Procedure for Assessing the Worth of Complex Alternatives," Defense Documentation Center, AD 662001 (November, 1967), 189 pages.

3. J. Fasal, "Forced Decisions for Value," *Produce Engineering,* 36 (8), pp. 84-86 (April 12, 1965).
4. M. J. Gelpi, "Forcing a Good Decision," *Westinghouse Engineer,* 27 (1), pp. 24-25 (January, 1967).
5. J. Morley English, "Cost-Effectiveness—The Economic Evaluation of Engineering Systems," John Wiley & Sons, New York, 1968.
6. S. R. Qasim, N. L. Drobny, and B. W. Valentine, "Process Alternatives Available in Packaged Waste Water Treatment Plants," *Water and Sewage Wastes* Reference Number, 1970, pp. R-99 to R-108.
7. H. E. Babbit and E. R. Baumann, *Sewage and Sewage Treatment,* John Wiley and Sons, Inc. New York, New York (1952).
8. P. H. McGauhey, *Engineering Management of Water Quality,* McGraw-Hill Book Company, New York (1968).
9. C. E. Bannett, "Basic Design Considerations for Aerated Lagoons," Welles Projects Corporation, Roscoe, Illinois, Bulletin WP 1204-1.
10. "Rotor Aeration in the Oxidation-Ditch," *Pollution 141,* Lake Side Engineering Corporation, Chicago, Illinois (undated).

Reprinted from *IBM J. Res. Develop.*, **16**, 154–161 (Mar. 1972)

T. I. Peterson
P. N. Wahi

Interactive Computer-based Game for Decision-making in Ecology

Abstract: This paper describes a prototype Ecology Decision Game which has been developed for experimental use within IBM. The paper is directed to those in ecology desiring to use similar techniques in developing programs that interrelate computing, management science, mathematics, and APL for training and educational purposes. The game is implemented in two modes: an *author* mode, which permits an author to write his own scenario; and a *player* mode, which enables a person to play the game. Features of the game exploit interactive capabilities for both modes. The particular scenario written for the game treats decision-making in the environmental area of solid-waste management. Three submodules explore progressively more complicated situations that give rise to management science problems: shortest route, transportation, and maximal flow. By active and passive role-playing, and controlled and uncontrolled learning, the player is given the opportunity to use quantitative tools to refine his subjective judgments.

Introduction

Administrators, planners, and those generally involved in ecology face admittedly complex problems, and are avidly seeking tools that can aid in solving these problems. This paper is concerned in part with solutions, particularly of the management science type. More central, however, is the elucidation of an approach, hopefully an easy, interesting, and challenging approach, for training and education through the device of an interactive computer-based game, designated the Ecology Decision Game.

• *Gaming*

Games are a natural adjunct to our way of life. Huizinga[1] has even argued that, Thomas Aquinas not withstanding, man is not only a "reasoning animal" but a "playing animal." Games of the serious class can also be interesting, as evidenced by the unusual popularity of management games. The American Management Association first made available their widely accepted game in 1956, and since that time the number of such games has grown prodigiously[2]. This is not to say that games cannot be made challenging, particularly for the player who wants to have a greater understanding of the areas that the game treats. In fact, Coleman[3] has suggested that the challenge lies not only with the player, but also with the observer of the player.

• *Ecology*

In this paper the particular area of ecology selected for gaming is one that is attracting increased attention, the

more so because it absorbs the largest expenditures in pollution control: solid-waste management. Nevertheless, although federal legislation regulating water pollution originated in 1899 and laws governing air pollution date back to 1955, the first Solid Waste Disposal Act did not come into existence until 1965. In recognition of the burgeoning problem, the federal government has moved to enact laws and establish agencies in an area that traditionally has been the function of local government[4].

A few statistics indicate the dimensions of the problem[5]. In 1966 the cost of refuse collection and disposal was reported to exceed three billion dollars annually, third among public services only to expenditures for schools and roads. By 1970 the cost had reached at least four billion dollars, and this outlay is expected to triple by 1980. In 1966 refuse production averaged about 4.5 pounds daily per person, a rate which by 1970 had risen to more than five, and which is projected to increase possibly to eight by 1980. Collection, as contrasted with disposal, accounts for 75 to 85 percent of the total cost, which is continuing to rise rapidly, primarily as a result of increased labor charges.

• *Management science*

From a management science standpoint, routing-problem solutions have been identified as the possible key to improved planning and more economic collection operations[6]. In fact, Klee[7] has described a simulation game developed by the Bureau of Solid Waste Manage-

154

ment in which players attempt to optimize solid-waste management costs by assigning different routings.

Gaming

A "scenario" has been written that involves players in routing problems associated with collection operations and planning in solid-waste management. Before examining the game in detail, an outline of its salient features is presented.

• Interactive gaming

The game is, first of all, a prototype intended for experimental use as a demonstration of capabilities not usually found in a single game. It is interactive with provision for multi-accessing. One consequence is that each player at a terminal has the impression that he has his own computer at all times. This capability is provided by the APL system[8]. A further implication of interaction is that there is a "dialogue" established between the player and the computer via the terminal. The player sees natural language text and answers in constructed responses, stylized where necessary to avoid the pitfalls of full natural language processing. He can, of course, command the full capabilities of the APL system and is given the opportunity to do so at various points in the game.

• Instructional gaming

Another facet of the game attempts to bridge the gap that often exists when a player sees only input and output, and wonders about the intervening process. The "black-box" approach is appropriate in simulation games in which the attempt is to create the real world through an approximation model. One disadvantage, however, is that the player usually has no knowledge of the model except by inference. As a complement to such games, instructional gaming[9] can lead the player to an understanding of the model that constitutes the black box and to problem solutions that can be obtained with the model.

The Ecology Decision Game is primarily concerned with instructional gaming. Teaching is based on the Socratic method and the use of tutorial dialogue. The player is given learner control over his individualized instruction in that he can branch around parts of the game. He can also cause the computer to perform tedious chores for him. He has at his service, for example, a desk calculator, a plotter, a message sender and receiver, and a note taker and report generator, in addition to a problem solver for well posed problems.

In this type of instructional gaming the player may want to take a passive role and compete, say, only against nature as represented by the computer. On the other hand, he might enjoy vying with other players in a more active role. He might at some points in the game

want to be controlled as to the method of learning. At other times, he might want to explore various possibilities that suggest themselves in an uncontrolled setting. The Ecology Decision Game exhibits these features. It leaves open considerations that are often closed in simulation gaming, such as the number of players and the roles that the players adopt.

• Game authoring

No matter how well conceived and executed a game may be, players and authors find deficiencies, especially during the validation and evaluation phase of game development. The literature reflects the diverse approaches taken to gaming[10]. For this reason, and for others such as currency of content, a game should be readily modifiable, even to the point of extensive rewriting.

The Ecology Decision Game, as described thus far, is a particular scenario treating a given area of ecology. The game, however, also provides an author with the capability of writing his own scenarios to treat other areas of ecology. To do so he uses an interactive and conversational mode, coupled with prompting, so that his need for computing expertise is minimized. In this sense the game resembles computer-aided instruction systems such as COURSEWRITER and PLANIT[11].

Frames are constructed by means of a restricted natural language. Dialogue is established between the computer and the author, who provides his script as input. The author can also link to pre-established APL functions to augment features of the game. As in conventional text processing, he is given editing and composition capabilities so that he can revise and format the game to his requirements.

Ecology Decision Game

• Objectives

The Ecology Decision Game has been developed to assist both the author in writing scenarios for a game and the player who wants to learn by playing the game. More particularly, the game is directed to authors with limited knowledge of computing and to players largely unfamiliar with computing, mathematics, management science, and APL. With this target population in mind, the objectives of the game are summarized in Table 1.

• Strategy

The game is intended to be used either alone or as a complement to a total gaming environment. Alone, it is a terminal-oriented, self-contained package requiring minimal instructor interaction. As a complement it can serve, for example, to reinforce lectures and demonstrations in any of the areas outlined under objectives, or to introduce other topics in ecology. Emphasis is placed on

155

Table 1 Overall objectives of the Ecology Decision Game.

Enable authors to write scenarios in
    Natural language insofar as possible
    Prompting method to minimize need for computing expertise

Expose players to the application of computing in ecology through
    Management science techniques
    Decisions resting in part on these techniques
    Mathematical concepts
    APL system capabilities

Table 3 System features in the *player* mode.

    Linkage to game
        Entry to beginning, intermediate points
        End linkage, resume linkage

    Game functions
        Access full use of APL system
        Form tables
        Plot graphs
        Right-justify text
        Provide "help"
        Deliver "mail"

Table 2 System features in the *author* mode.

    Questions
        Merge questions
        Right-justify text

    Answers
        Single or multiple
        Expected or unexpected
        Full or partial
        Ordered or unordered
        Scored or unscored

    Branches
        To questions:
            Global
            Local
        To APL functions:
            Numeric
            Non-numeric

    Scenario
        Edit
        Print
        Logic map

Table 4 Submodule objectives.

Sub-module	Management science technique	Decision-making area	Mathematical concepts	APL capabilities
MOD1A	Shortest route problem	Number of sites	Scalars Vectors	Numeric processing
MOD1B	Transportation problem (also linear regression)	Cost-planning	Vectors Matrices	Numeric processing Tables, graphs, reports Messages
MOD1C	Maximal flow problem	Truck fleet size	Matrices Heuristics Simulation	Numeric processing Non-numeric processing

developing cognitive skills and strategies rather than on creating win-lose situations.

• *Author mode*

Features of the *author* mode are shown in Table 2. The author is prompted to write the scenario as a question followed by an answer followed by a branch to another question. This basic sequence is elaborated upon to give added flexibility for modifying paradigms. A "question" can in fact be a question, but it can also be a statement or even a blank line. Similarly, an "answer" can be an expected answer in which the whole character string is matched, or only part of the character string, or it can be an unexpected answer. In fact, a sequence of answers can be formed, but the sequence ends when the author indicates that he has made provision for an unexpected answer. The branch to each answer must be satisfied as the answer is constructed.

A global request can be permitted by the author so that the player may at any time ask, for example, to quit, restart, or use the full capabilities of the APL system. Also, branching can lead to an APL function for numeric or non-numeric processing. Additionally, the author can merge successive questions into one question, score answers, accept unordered answers, right-justify text, edit any portion of the scenario (which includes both text and logic), and print the scenario or an abbreviated "logic map."

• *Player mode*

The *player* mode, features of which are listed in Table 3, is initiated by the player or anyone acting that role. The player can start at the beginning or at some intermediate point in the game. He can use the APL system to send and receive messages, or use global game functions such as those for plotting graphs, forming tables, and composing reports. At specific points in the game he can also call for local game commands such as HELP or MAIL. Of course, he always has available the APL system should he want to construct his own functions or use any of the library functions.

• *Scenario features*

The scenario for the game is divided into modules. Each module in turn consists of submodules, which progressively increase in complexity with respect to the stated objectives. In general, the player is presented with a situation requiring attention and action and he uses management science tools as aids in the decision-making process.

The major module, MOD1 Collection, is directed to the environmental area of solid-waste collection operations and planning. This module consists of three submodules, each of which relates to the objectives outlined for the player as indicated in Table 4.

MOD1A Shortest route problem

The player is confronted with a situation involving a shortest route solution [12, 13]. A dialogue is established between him and the computer as illustrated in Fig. 1. The case stems from a request by a municipal agency to improve environmental conditions and at the same time upgrade operating efficiency. A simplified map is provided at the terminal which shows collection points and disposal sites, distances between these points, and allowable routings. The problem is to find the shortest allowable route between each collection point and a disposal site so that the sum of these distances is a minimum. All routes are assumed to be two-way for simplicity of exposition.

From a management science standpoint, the problem reduces to that of finding shortest paths between pairs of nodes of the Fig. 1 network. We are given a set of $N = 8$ nodes and the $N \times N$ matrix D, whose elements d_{ij} represent the distances between nodes i and j (d_{ij} can differ from d_{ji} in real problems). We assume $d_{ii} = 0$ and $d_{ij} > 0$ for all i and j ($i, j = 1, 2, \cdots, N$). If there is no route (arc) from node i to node j, then $d_{ij} = \infty$ or, for computational purposes, d_{ij} can be assumed to be arbitrarily large.

Mathematically, the problem can be stated as

Minimize d_{ij} for all i and j,

$$d_{ij} \geq 0 \text{ and } i, j = 1, 2, \cdots, N. \tag{1}$$

To solve this problem, any of several algorithms can be used; in this game, that of Hu[13] has been adopted. Figures 2(a) and 2(b) illustrate how the player is introduced to the correct answer.

The number of points, eight in all, is not unlike the construct described by Klee[7]. This number is inten-

Figure 1   Initial problem statement and presentation of map.

THE MAYOR'S ENVIRONMENTAL PROTECTION AGENCY (EPA) HAS ASKED FOR A REVIEW OF REFUSE COLLECTION. IT PARTICULARLY WANTS TO KNOW IF THE CITY CAN BE MADE CLEANER BY BETTER USE OF COLLECTION TRUCKS AND TEAMS.
TO GET SOME FEELING FOR THE PROBLEM, YOU HAVE DECIDED TO CONSIDER A VERY SIMPLIFIED SITUATION: REFUSE PICKED UP BY COLLECTION TRUCKS AT SPECIFIC POINTS AND TRANSPORTED TO DUMP SITES.

⋮

LOOK AT THE FOLLOWING MAP. WE SHOW COLLECTION POINTS BY o AND DISPOSAL POINTS BY *. WE ALSO SHOW A TYPICAL DISTANCE (=6) FROM POINT [3] TO POINT [4] WHICH WE LABEL, WITH SOME FORESIGHT, D[3;4] ← 6.

$$D[3;4] \leftarrow 6$$
$$\bullet[3]\text{---------}o[4]$$

o[2]

o[6]

o[1]

o[8]

•[5]

•[7]

HERE ARE THE ALLOWED ROUTES AND DISTANCES SO YOU CAN FILL IN THE REST OF THE MAP BY HAND. HIT CR.

D[1; 2] ← 3	D[2; 4] ← 4	D[6; 7] ← 13
D[1; 5] ← 1	D[3; 6] ← 5	D[6; 8] ← 2
D[2; 3] ← 3	D[5; 6] ← 10	D[7; 8] ← 14

tionally small so that the player can solve the shortest route problem by inspection. In a tutorial dialogue he verifies that he has analyzed the problem correctly. He then asks for the computer solution. The player is encouraged to analyze a similar situation of his own choosing and is guided to provide input data [Fig. 2(c)]. Following this opportunity for individualized exploration he is introduced to costs associated with the various routes.

Solution of this shortest route problem suggests abandoning one of the disposal sites. The player is alerted through newspaper articles to consider other factors:

157

274

Figure 2 Instructional gaming: (a) tutorial dialogue; (b) function evaluation; (c) opportunity to initiate data.

Figure 3 Decision-making: (a) request for decision; (b) advice on factors in decision.

YOU NOW HAVE ENOUGH INFORMATION TO FIND THE SHORTEST ROUTE. FOR EXAMPLE, CONSIDER COLLECTION POINT [1]. ITS CLOSEST DISPOSAL POINT IS [5]. YOU'D THEN ROUTE [1] TO [5], NOT TO [3] OR [7].

SO, LET'S SEE. [1] GOES TO ? (THE ANSWER IS 5 – JUST PUT IN THAT NUMBER).

⋮

FINALLY, [8] WOULD GO TO ? (CAREFUL)

7

THE CORRECT ROUTING IS TO [3]. WE TOLD YOU TO BE CAREFUL. HIT CR AND WE'LL GO ON.

(a)

WE CAN ALSO ASK THE COMPUTER TO TYPE OUT THE SHORTEST PATH FROM [8] TO [3]. TO DO SO, TYPE IN '8 3', LEAVING AT LEAST ONE BLANK BETWEEN THE NUMBERS TO SEPARATE THEM; OTHERWISE HIT CR.

8 3

8   6   3

AGREED THAT THE PATH FROM [8] TO [3] IS [8] TO [6] TO [3]? OK, HIT CR.

(b)

BEFORE WE LOOK AT THE SITUATION MORE CRITICALLY, MAYBE YOU'D LIKE TO CONSIDER THE SAME TYPE OF PROBLEM IN YOUR OWN CITY OR REGION. IF SO, YOU'LL HAVE TO GIVE DISTANCES BETWEEN POINTS THAT CAN BE CONNECTED, JUST AS WE'VE DONE BEFORE. DO YOU WANT TO ENTER YOUR OWN DATA? TYPE Y OR N.

(c)

BRR .. INN .. GGG. MAIL FOR YOU. TYPE 'MAIL'.

MAIL

THE EPA ADMINISTRATOR HAS JUST CALLED. WANTS YOUR PRELIMINARY THOUGHTS ABOUT IMPROVING COLLECTION OPERATIONS. WHAT ARE YOU GOING TO TELL HIM? EVEN THOUGH THIS IS A PRELIMINARY REPORT, YOU SHOULD TAKE INTO CONSIDERATION OTHER POSSIBLE FACTORS. REMEMBER ALSO THAT AS A PLAYER YOU HAVE ADOPTED A CERTAIN ROLE, SO YOU SHOULD PRESENT THAT VIEWPOINT.

(a)

BEFORE YOU DECIDE ON WHETHER TO REPORT NOW AND WHETHER TO ABANDON DUMP SITE [7], YOU MIGHT LIKE TO REFER TO A PARTICULAR SITUATION REPORTED IN THE NEWSPAPER THAT ATTRACTED YOUR ATTENTION. TYPE 'MAIL' IF YOU WANT TO SEE EXCERPTS THAT YOUR SECRETARY RECORDED FOR YOU. OTHERWISE HIT CR TO GO ON.

MAIL

UNHAPPY SAUGUS SEEKS TRASH DISPOSAL SOLUTION

THE TRUCKS KEEP ROLLING INTO SAUGUS AND NO ONE KNOWS HOW TO STOP THEM.

THE TRUCKS – 300 A DAY – ALL HAVE SLOGANS ON THEIR SIDES: 'KEEP EVERETT CLEAN.' 'HELP KEEP OUR CITY CLEAN' ON MELROSE TRUCKS, AND SIMILAR MESSAGES FROM 14 OTHER NORTH SHORE CITIES AND TOWNS.

(b)

diminishing available space, rising costs, and the political and social implications associated with solid-waste collection and disposal (Fig. 3). The decision that confronts the player, who adopts a particular role, is whether he should surrender the disposal site.

In this portion of the scenario, the player is exposed to scalars and vectors through the desk calculator capability of the game. He uses black boxes, which are later shown to be APL functions, and other game commands when they are made available to him, as in printing the contents of his "mailbox." Further, he uses APL system commands that permit him to perform ancillary operations, for example, loading various parts of the scenario.

MOD1B Transportation problem

In the second submodule the situation previously presented is indicated to be an over-simplification, in that there are incinerators at two of the disposal locations. Capacity constraints are thus placed on nodes 3 and 5, while node 7, which is designated a sanitary landfill, has no capacity constraint.

In the game the player learns that collection points 1, 2, 4, 6, and 8 in Fig. 1 generate 10, 15, 5, 10, and 15 tons of refuse daily, respectively; that the refuse handling capacities of the incinerators at disposal sites 3 and 5 are 25 and 20 tons per day, respectively; and that the cost of transportation is assumed to be proportional to the distance traveled.

158

The total refuse generated is 55 tons per day, whereas the incinerators can handle only 45 tons per day. Since the incinerator sites are preferred by the shortest route solution, only the excess ten tons of refuse should be routed to site 7. The question arises, Which ten tons and at what extra cost? The player is encouraged to try a hand solution to convince himself that even in this restricted and small-size problem, an optimal solution is found only after substantial effort.

Formally, the problem can be regarded as a transportation problem[14] in which commodities are transported from sources to destinations with associated costs, subject to constraints on the capacities of the sources and the demands of the destinations. Let x_{ij} denote the amount transported from source i ($i = 1, 2, \cdots, m$) to destination j ($j = 1, 2, \cdots, n$) with associated unit cost c_{ij}. Let a_i denote the capacity of source i and b_j denote the demand of destination j. In the game, sources are collection points, destinations are disposal sites, costs are proportional to distances (found by the shortest route algorithm), and the sum of all a_i equals the sum of all b_j. We are interested in the values of x_{ij} that correspond to minimum cost.

Mathematically, the problem can be stated as

$$\text{Minimize} \sum_{i=1}^{m} \sum_{j=1}^{n} c_{ij} x_{ij}$$

such that $\sum_{j=1}^{n} x_{ij} = a_i$ for $i = 1, 2, \cdots, m$;

$\qquad\quad \sum_{i=1}^{m} x_{ij} = b_j$ for $j = 1, 2, \cdots, n$;

$\qquad\qquad x_{ij} \geq 0$ for all i and j.   (2)

A discussion of transportation algorithms can be found in Ref. 14.

The matrix for this transportation problem is shown in Fig. 4. The transportation algorithm is readily available to the player at the terminal and he can obtain such a solution on request. The total cost in this example is 320, whereas the cost for an unconstrained shortest route would have been 240. The increase in cost, 80, is due to the capacity constraints on the incinerators. Clark and Helms[15] deal with a similar but more difficult (nonlinear) routing problem.

In the course of working toward the solution, the player is introduced to further manipulations with vectors and matrices, as well as to the construction of tables and graphs. He is also exposed to non-numeric processing in preparing composed reports. The decisions that he makes require that he use the message sending capability. He is encouraged to challenge other players to find the minimum cost of operations for reduced capacity of the incinerators. For example, the game decrees that a lower level of operation will be necessary because air

Figure 4 Transportation problem matrix. The first number of each pair is the assigned unit cost c_{ij}; the second, in boldface, is the solution x_{ij} for the amount of material transported.

	DESTINATION			
SOURCE	3	5	7	SUPPLY
1	6, **0**	1, **10**	24, **0**	10
2	3, **5**	4, **10**	21, **0**	15
4	6, **5**	8, **0**	24, **0**	5
6	5, **10**	10, **0**	13, **0**	10
8	7, **5**	12, **0**	14, **10**	15
DEMAND	25	20	10	55
MINIMUM COST ← 320				

pollution standards are projected as becoming more stringent. At this point the player is given the opportunity to use linear regression (curve fitting) analysis in the projection of quality standards.

MODIC Maximal flow problem

In the third submodule an even more realistic situation is studied. Not only are the previously inroduced constraints present, but there is also a schedule that must be met and trucks that can handle only a fixed-size load. (The problem of determining a schedule is assumed to have been solved separately.) There are 22 loads (55 tons) of refuse to be handled and all trucks are assumed to be the same size, namely, 2.5 tons capacity. Other conditions are that truck travel time is assumed to be proportional to distance and that after a loaded trip to a disposal site, the truck is available for routing to any of the collection sites, subject to pickup time constraints.

The objective is to meet the schedule with the minimum number of trucks. The player readily observes that a trial and error solution to this problem is very time-consuming. Clearly, the maximum number of trucks required would be 22, one assigned to each pickup. To formulate the problem mathematically, we consider a directed graph $G(N,A)$, having N nodes (collection and disposal points) and A directed arcs (collection → disposal and disposal → collection routes), to which the following theorem of graph theory may be applied[12, 16].

Theorem: Let G' (N,A') be a chain-decomposed acyclic graph of $G(N,A)$ (an equivalent bipartite graph) with C being the number of chains and D the set of arcs that are parts of the chains. Then $C + |D| = N$.

In this formulation, C is the number of trucks. We can minimize C by maximizing $|D|$. If we assign a flow ca-

159

Figure 5 Example of a maximal flow/routing solution of the collection-disposal problem. Such solutions are dependent on a predetermined schedule map for pickups.

COLLECTION POINTS ← 5 (1, 2, 4, 6, 8)
LOADS ← 22 (55 TONS)
DISPOSAL SITES ← 3 (3, 5, 7)
NODES N ← 36
MAXIMAL FLOW |D| ← 28
MINIMUM FLEET SIZE C ← 8

TRUCK	ROUTE ASSIGNED
1	$1 \rightarrow 5 \rightarrow 1 \rightarrow 5 \rightarrow 1 \rightarrow 5 \rightarrow 1 \rightarrow 5 \rightarrow 2 \rightarrow 5$
2	$2 \rightarrow 5 \rightarrow 2 \rightarrow 3 \rightarrow 6 \rightarrow 3$
3	$2 \rightarrow 3 \rightarrow 8 \rightarrow 7$
4	$8 \rightarrow 3 \rightarrow 2 \rightarrow 5 \rightarrow 4 \rightarrow 3$
5	$4 \rightarrow 3 \rightarrow 8 \rightarrow 7$
6	$6 \rightarrow 3 \rightarrow 6 \rightarrow 3 \rightarrow 2 \rightarrow 5$
7	$6 \rightarrow 3 \rightarrow 8 \rightarrow 7$
8	$8 \rightarrow 3 \rightarrow 8 \rightarrow 7$

pacity d_{ij} of one unit to each arc and define $x_{ij} = 1$ if nodes i and j belong to a chain, but $x_{ij} = 0$ otherwise, the maximum value of $|D|$ is equal to the maximal flow through the graph G' (N, A'). Thus, the problem may be restated as

$$\text{Maximize} \sum_{i=1}^{m} \sum_{j=1}^{n} x_{ij}, \quad x_{ij} = 0 \text{ or } 1, \quad 0 \le x_{ij} \le d_{ij}. \quad (3)$$

for all i, j and flow conservation constraints.

The optimal fleet size is $N - |D|$. Hence, the solution may be obtained by determining the maximal flow through the network, for which the Ford-Fulkerson algorithm[16] is used in the game. The solution is printed at the terminal, along with the truck routings (Fig. 5). Effects on fleet size and routings (sensitivity analysis) are observed by the player, who is exposed to changes in the schedule map. Further work is fostered to familiarize the player with capabilities for both numeric and non-numeric processing.

The player subsequently learns that the analytic techniques thus far presented are not the only management science methods available. Heuristic methods may be useful for more complex and larger problems. Clark and Helms[15], for example, have used heuristic procedures to find a solution to a solid-waste disposal problem formulated as a fixed-charge transportation problem. Simulation methods are similarly referenced in the game to identify their role in still more complex problems[17].

Summary

Systems analysis is introduced through modeling a situation in which the player participates from the outset. Where a hand solution is possible, his analysis and that of the computer are shown to be coincident. With progressive sophistication, he appeals to the computer when he cannot expect to find an optimal solution by hand. The strength of the computer is emphasized as being its ability to assist in the decision-making by implementing management science algorithms and to relieve the decision maker of burdensome tasks.

The management science solution, however, is identified as only part of the solution to the overall problem. The ultimate decision rests with the player. In this way, the player can accommodate the imprecision that the problem always poses.

The game has been developed to assist both an author, with a limited knowledge of computing, to write scenarios for a particular topical area, and a player, largely unfamiliar with computing, management science, and the APL system, to learn by playing the game.

Effectiveness of the Ecology Decision Game with respect to the stated objectives has been partially evaluated. Further evaluation as an adjunct to other methods of instruction is being developed as various groups use the game.

Acknowledgments

The authors express their appreciation to the Cambridge Scientific Center staff, K. M. Chandy (now at the University of Texas at Austin) for his consultation, and D. Coppersmith for his programming assistance.

References and notes

1. J. Huizinga, *Homo Ludens*, Beacon Press, Boston 1950.
2. See, e.g., J. R. Raser, *Simulation and Society*, Allyn and Bacon, Inc., Boston 1969, Chap. 6.
3. J. C. Coleman, "In Defense of Games," *American Behavioral Scientist* X(2), 3 (1966).
4. See, e.g., H. F. Sherrod, Jr., *Environmental Law Review—1970*, Sage Hill Publishers, Inc., Albany, N.Y., and Clark Boardman Co., Ltd., New York, N.Y., 1970; W. E. Small, *Third Pollution*, Praeger Publishers, New York, N.Y., 1971.
5. For data cited, see "Waste Management and Control," report to the Federal Council for Science and Technology, *NAS-NRC Publication 1400*, National Academy of Sciences-National Research Council, Washington, D.C., 1966; R. J. Black and C. Clemons, "The National Solid Wastes Survey: An Interim Report," U.S. Dept. of Health, Education, and Welfare, Cincinnati, Ohio, 1968; W. E. Small, *loc. cit.*
6. R. M. Bodner, E. A. Cassell, and P. J. Andros, "Optimal Routing of Refuse Collection Vehicles," *Proc. A.S.C.E., J. Sanitary Eng. Div.* 96(SA4), 893 (1970); A. J. Klee, "Tactics, Strategy: The Solid Waste Battle," *Environmental Science & Technology* 3, 898 (1969).
7. A. J. Klee, "DISCUS—A Solid-Waste Management Game," *IEEE Trans. Geoscience Electronics* GE-8(3), 125 (1970).

160

8. K. E. Iverson, *A Programming Language*, John Wiley and Sons, Inc., New York 1962; *APL/360 User's Manual*. Report GH20-0683, IBM Corporation, White Plains, N.Y., 1968.

9. See, e.g., *Computer-Assisted Instruction*, edited by R. C. Atkinson and H. A. Wilson, Academic Press, Inc., New York 1969, p. 3.

10. In addition to Ref. 2, see, e.g., S. S. Boocock and E. O. Schild, *Simulation Games in Learning*, Sage Publications, Inc., Beverly Hills, Calif., 1968; *Simulation and Games*, edited by M. Inbar, *ibid.*, 1968.

11. C. H. Frye, "CAI Languages: Capabilities and Applications," in *Computer-Assisted Instruction, loc. cit.*, p. 317; K. L. Zinn, "Programming Conversational Uses of Computers for Instruction," *ibid.*, p. 253.

12. C. Berge, *The Theory of Graphs and Their Applications*, John Wiley and Sons, Inc., New York 1966.

13. T. C. Hu, *Integer Programming and Network Flows*, Addison-Wesley Publishing Co., Inc., Reading, Mass., 1969.

14. S. I. Gass, *Linear Programming*, McGraw-Hill Book Co., Inc., New York 1969.

15. R. M. Clark and B. P. Helms, "Decentralized Solid Waste Collection Facilities," *Proc. A.S.C.E., J. Sanitary Eng. Div.* **96**(SA5), 1035 (1970).

16. L. R. Ford and D. R. Fulkerson, *Flows in Networks*, Princeton University Press, Princeton, N.J., 1962.

17. In addition to the simulation described in Refs. 6 and 7, see, e.g., J. M. Betz and R. P. Stearns, "Simulation and Analysis of a Refuse Collection System," *Proc. A.S.C.E., J. Sanitary Eng. Div.* **92**(SA3), 17 (1966); M. M. Truitt, J. C. Liebman, and C. W. Kruse, "Simulation Model of Urban Refuse Collection," *ibid.* **95**(SA2), 289 (1969); J. E. Quon, M. Tanaka, and S. J. Wersan, "Simulation Model of Refuse Collection Policies," *ibid.* **95**(SA3), 575 (1969); J. A. Green, N. S. Lister, and B. Whitworth, "Refuse Disposal in the Tyneside/Wearside Area—The Evidence for Co-operation," Royal Institute of Public Administration, Local Government Operational Research Unit, Reading, England, February 1967.

Received July 7, 1971; revised November 11, 1971

The authors are located at the IBM Data Processing Division Scientific Center, 545 Technology Square, Cambridge, Massachusetts 02139.

21

Reprinted from *J. Water Pollution Control Federation*, **45**, 52–70 (Jan. 1973)

Marine waste disposal—a comprehensive environmental approach to planning

DAN P. NORRIS, LAWRENCE E. BIRKE, JR., ROBERT T. COCKBURN, AND DENNY S. PARKER

THE CITY of San Francisco, Calif., is presently confronting water pollution with a total liquid waste management program based on a comprehensive evaluation of the water pollution abatement problems in the areas of collection, industrial wastes control, treatment, and ultimate disposal of waterborne wastes. While this paper is devoted to the problem of marine disposal of the liquid waste fraction, a brief history and description of the other facets of the total program will place the marine disposal aspect in proper perspective.

San Francisco has over 900 miles (1,450 km) of totally combined sewers tributary to three primary treatment plants (Figure 1). The daily average dry-weather flow is 100 mgd (378,500 cu m/day), and there are 43 combined sewer overflow points along the shoreline. Under dry-weather conditions all sanitary wastes flow to the plants for treatment and disposal. Each plant can accommodate about three times the average dry-weather flow rate during periods of light rainfall [0.02 in./hr (0.51 mm/hr) or less]. Higher intensity precipitation results in an overflow of untreated combined wastewater. Thus, the city is faced with two different but interrelated sets of water pollution problems, one in dry weather and the other in wet weather.

Present dry-weather pollution problems include inadequate settleable and floatable solids control and aesthetic degradation of the receiving water in the discharge zones. The latter condition is aggravated by surface discharges at two of the three plants. More stringent discharge requirements are anticipated in the near future, including more effective control of grease, heavy metals, and acute and chronic toxicity.

All wet-weather pollution problems are complicated by the random, intermittent nature of the overflow occurrences and by the fact that the range of wet-weather flows is up to 50 times the available treatment capacity. The present uncontrolled overflows result in bacteriological degradation of the receiving waters and unpleasant nuisance conditions along the shoreline.

Since the development of San Francisco's first sewerage master plan in the 1890's, the city has had a continuous program of improving its sewerage facilities. The first treatment plant was completed in 1938 and the remaining two in 1951. For the past 3 yr, the city has worked on the development of a new master plan for wastewater management. This plan will present an integrated solution to problems of sewerage system improvements, industrial waste control, dry-weather treatment and disposal, and the storage, treatment, and disposal of wet-weather overflows.

The master plan seeks to achieve a balanced approach to all of the city's sewerage problems by evaluating the following:

1. The cost of treatment necessary to achieve various levels of effluent quality in the dry-weather flow;

2. The feasibility of treating wet-weather overflows and the optimum balance between treatment and storage capacity; and

3. The location and manner in which it is feasible to dispose of both dry- and wet-weather effluents to the adjacent waters of San Francisco Bay or the Pacific Ocean (this paper addresses itself to this facet of the problem).

In the past, most studies of marine waste disposal have been concerned with the ob-

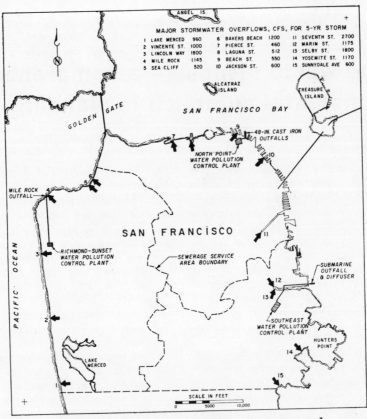

FIGURE 1.—Existing wastewater treatment and disposal facilities.

vious effects of the discharge. Design efforts have been directed primarily toward avoiding nuisance conditions and maintaining bacterial standards acceptable to public health authorities. Effects on the marine environment itself, if assessed at all, have generally been established by after-the-fact monitoring. The city of San Francisco was determined to reverse this procedure and include the overall environmental effect of marine waste discharges as part of the feasibility study. The stress that a waste discharge might exert on the total marine ecosystem could then be used as another design input, together with conventional water quality parameters and construction costs. To define the overall environmental effect, a study program was designed that would describe the oceanographic characteristics of waters near shore,

measure the effect of wastewater effluent on marine biota, and give recommendations for outfall locations and discharge quantities that would maintain acceptable water quality, meet discharge requirements, and preserve a healthy marine ecosystem.

PHYSICAL OCEANOGRAPHIC INVESTIGATIONS

The city of San Francisco is surrounded on three sides by the waters of San Francisco Bay and the Pacific Ocean. An understanding of the oceanographic characteristics of these waters is essential to the prediction of the fate of a marine waste discharge. Studies of mass water movement, drift of particulate matter on the water surface, and dispersion characteristics of both the bay and ocean were conducted during representative oceanographic seasons for a full year. Special investigations were also conducted to measure the performance of the city's two existing outfalls and to define tidal exchange of bay-ocean waters through the Golden Gate.

Water characteristics and movement. Aerial photography was found to be the single most valuable tool for defining mass water movement in the Gulf of the Farallones. At all seasons of the year, the water flowing out of San Francisco Bay is more turbid than the water of the Gulf of the Farallones. This phenomenon permits the water mass to be tracked by aerial observation. In the winter and spring, when outflowing bay water has the highest turbidity, it is usually possible to see the color lines created by the two most recent ebb tides.

Several different photographic techniques were investigated, including vertical shots and oblique perspectives using black and white, color, and infrared film. The best definition of water masses was obtained by using panchromatic film with a red filter. The most useful information was obtained by taking obliques from a light plane. The advantage of this technique lay in having an engineer-photographer who was familiar with the goals of the program and capable of coupling the photographic effort with sketches and qualitative judgments of the observed phenomena.

The aerial photography was very successful in recording the surface manifestations of mass water movement. Information on changes in the water characteristics with depth was provided by vertical profiles of temperature, dissolved oxygen (DO), and salinity measured from a boat equipped for oceanographic research. Instrumentation systems were kept as simple and portable as was consistent with good results. Temperature, salinity, depth, and DO were measured by submersible probes that were lowered to the desired depth by a small electric winch. The sensing devices were connected by multiconductor underwater cable to indicating instruments on the deck of the boat. Water clarity was measured by Secchi disk, and wind direction and velocity were measured periodically with a hand-held anemometer and compass. These simple instrument systems permitted the use on short notice of any available boat that had sufficient work space and was adequately constructed and equipped for operation in the unfavorable sea and weather conditions of the Gulf of the Farallones. Both commercial fishing boats and chartered research vessels were used.

A considerable accumulation of historical current data on the study area was obtained from the U. S. Coast and Geodetic Survey. Additional measurements were taken in the course of this study using both captive and *in situ* current meters. Because of the complicated current regime that exists around the mouth of the Golden Gate, the periodic vertical current profiles obtained with the captive meter were found to be more useful than a continuous record at a single depth.

Current measurements were translated into displacement vectors and plotted as advective flow diagrams showing magnitude and direction of displacement over a 25-hr tidal cycle at a single station. A

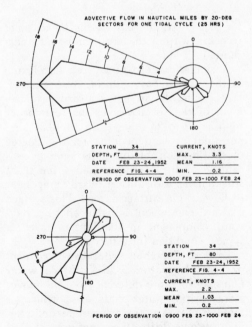

ADVECTIVE FLOW IN NAUTICAL MILES BY 20-DEG
SECTORS FOR ONE TIDAL CYCLE (25 HRS)

STATION	34	CURRENT, KNOTS	
DEPTH, FT	8	MAX.	3.3
DATE	FEB 23-24,1952	MEAN	1.16
REFERENCE	FIG. 4-4	MIN.	0.2
PERIOD OF OBSERVATION	0900 FEB 23-1000 FEB 24		

STATION	34
DEPTH, FT	80
DATE	FEB 23-24, 1952
REFERENCE FIG. 4-4	

CURRENT, KNOTS
MAX.	2.2
MEAN	1.03
MIN.	0.2

PERIOD OF OBSERVATION 0900 FEB 23-1000 FEB 24

FIGURE 2.—Typical advective flow diagram.

FIGURE 3.—Surface advection on the Gulf of the Farallones.

typical example of the advective flow diagrams is shown in Figure 2. Where sufficient data were available, diagrams were prepared for both the surface and bottom layers.

The analysis of aerial photography, current data, and temperature and salinity of the water mass gave a clear picture of the pattern of mass water movement in the central bay, through the Golden Gate, and in the Gulf of the Farallones. The movement of surface water in the Gulf of the Farallones is shown schematically in Figure 3. The ebbing tide from San Francisco Bay moves strongly westward and southward. During months of high freshwater outflow, the water mass is strongly layered with a much lower salinity in the surface layer than in the bottom layer. During this period, the length of the surface ebb increases, and at maximum freshwater outflow the surface flow through the Golden Gate may ebb continuously. Water that enters the bay on the flood comes principally from the north and south near shore.

A similar pattern of layered flow exists in the central bay. During wet weather, layering results in net seaward displacement of the surface layer that can be as high as 10 to 25 miles (16.1 to 40.25 km) per tidal cycle depending on the degree of stratification. During the summer season the strong layering subsides, but there still remains a distinct seaward component to flow in the surface layer. This surface seaward component is balanced by a bayward component in the bottom layer, a typical condition for estuaries.

Drift of floatables. Because movement of floatable matter in the surface water film may be quite different from the movement of the water mass as a whole, a number of measurements were made to define surface drift. The method selected involved the release of old-fashioned, cardboard milk bottle caps at various locations and tidal conditions. After the caps had been given time to drift ashore, approximately 20 miles (32 km) of beaches and shorelines were walked by pickup crews who collected the caps and recorded the location where they drifted onto the shoreline. Further information on surface drift was obtained from the movement of oil released in the major oil spill at the Golden Gate in January 1971.

Analysis of the results of the surface drift studies produced several important conclusions:

1. Floatables discharged in the central bay in the vicinity of Alcatraz will not accumulate on the bay shoreline in significant quantities.

2. Floatables released outside the Golden Gate will not enter the bay.

3. The heaviest accumulation of floatable material from outfall sites reasonably available to the city of San Francisco will occur on beaches south of the Golden Gate.

4. Floatables released at or near any of the available sites may be expected to accumulate on the shoreline at a rate of 1.6 percent/mile (1.0 percent/km) south of the Golden Gate from Point Lobos to Point San Pedro and 0.4 percent/mile (0.25 percent/km) north of the Golden Gate from Point Bonita to the Bolinas Peninsula.

On January 18, 1971, two oil tankers collided under the Golden Gate Bridge, resulting in damage that released several hundred thousand gallons of oil into the waters of the central bay. The ships were towed to a site near Angel Island, where the oil release continued for several hours. The spill resulted in severe oil contamination of the ocean beaches from Duxbury Point to Point San Pedro, the same area where the majority of the cardboard floats were recovered the preceding summer. Within the bay, however, oil contamination of the shoreline was confined to a small area seaward of the point of release. The oil spill thus confirmed in dramatic fashion the results of the surface drift studies, which indicated that floatable material released within the central bay moves rapidly seaward without significant effect on the shoreline of the bay.

Dispersion studies. The major factor controlling the subsequent dilution of an effluent field after it reaches stability with the surrounding water mass is dispersion resulting from turbulent diffusion. A computer program was written to calculate subsequent dilution using dispersion coefficients determined by field measurement. Nearly 100 individual measurements of dispersion coefficients were made in San Francisco Bay and the Gulf of the Farallones.

A procedure was developed for determining dispersion coefficients from measurements of the rate of dispersion of rhodamine WT fluorescent dye. For each series of measurements, 5 lb (2.3 kg) of dye was dropped in the area of concern. A research vessel then made periodic traverses of the patch in alternating longitudinal and transverse directions, taking a continuous record of dye concentration along each transect. Vertical profiles were run periodically to record the depth of dye penetration. A submersible pump positioned by a power winch delivered seawater continuously to a recording fluorometer, which produced a continuous record of dye concentration. The fluorometer could detect dye concentrations as low as 0.05 μg/l corresponding to the normal background level of natural fluorescing substances in ocean water.

During the first cruise, it was found that the spreading dye patch was far more visible from the air than from the surface, and thereafter, the research vessel was directed from a light plane. The dye patch was visible from the air at dye concentrations of about 2 μg/l. With this limit of visual detectability, the dye patch could be followed for up to 6 hr, depending on field conditions such as sunlight, water clarity, and rate of dispersion. The scale of the dye patch was determined by relating controlled boat speed to time of travel and then checked by sextant shots and reference to colored aerial photographs using the length of the research vessel for scale.

A definition of the turbulent dispersion coefficient given by Brooks[1] and Fischer[2] states that

$$k = \frac{1}{2}\frac{d\sigma^2}{dt} \qquad (1)$$

where

σ = standard deviation of the concentration distribution,
t = time, and
k = dispersion coefficient, sq ft/hr.

The scale of the dye patch, L, is defined as

$$L = 2\,\sigma$$

Therefore, Equation 1 can be integrated and simplified to

$$k = \frac{L_2{}^2 - L_1{}^2}{8\,(T_2 - T_1)} \qquad (2)$$

where

L_1 = scale at time T_1, and
L_2 = scale at time T_2.

Equation 2 is the basic equation used to develop values for k from the field data. The standard deviation of the concentration distribution, σ, was determined directly from the analysis of the fluorometer chart records.

Results for dye dispersion studies for the Gulf of the Farallones are indicated in Figure 4, with dispersion coefficients plotted against scale. The plot presents the best-fit curves as computed by least squares and as plotted in accordance with the so-called "four-thirds law" for the values of k_x and k_y only. For comparison, Pearson's[3] four-thirds law correlation of dispersion coefficients for other ocean measurements is also shown. The Gulf of the Farallones data points are mostly well above Pearson's line, suggesting that dispersion takes place at a faster rate than in most coastal areas. This may be postulated to result from a higher level of turbulence created by the massive tidal flows through the Golden Gate. The four-thirds law relationship shown in Figure 4 ($k = 20\,L^{1.33}$) was used for all Gulf of the Farallones applications because it has a theo-

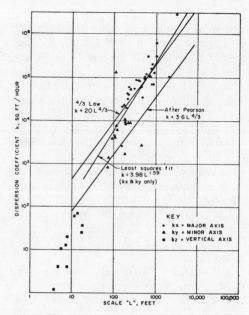

FIGURE 4.—Dispersion coefficients in the Gulf of the Farallones.

retical basis founded in fundamental turbulence theory.[4] The comparable value determined for central San Francisco Bay was $k = 33\,L^{1.33}$.

Outfall performance studies. The city of San Francisco has two submarine outfalls into San Francisco Bay, one at its North Point water pollution control plant and one at the Southeast plant (Figure 1). Field tests were conducted on each outfall to obtain a precise quantitative measure of outfall performance. The results provided a check on the ability of outfall theory to predict accurately what the performance of an outfall will be. The tests were directed toward measurement of three parameters: initial dilution, subsequent dilution of the effluent field, and effluent field visibility.

The same basic procedure was used at both outfalls. Rhodamine WT dye was metered into the effluent stream at a rate precisely proportional to flow, producing a constant concentration of dye in the effluent. The effluent field was then monitored by a boat equipped with a submersible pump and a recording fluorometer

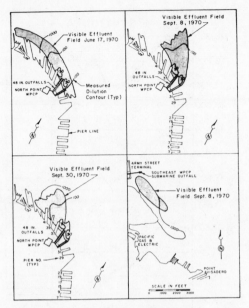

FIGURE 5.—Probable dilution of visible effluent fields.

using the same techniques utilized in measurement of dispersion coefficients. The measured dilution contours for the effluent fields were compared with aerial photographs and Secchi disk profiles taken at comparable tidal conditions, but with no dye in the effluent, to check probable dilution at which the effluent fields were visible (Figure 5).

The Southeast outfall discharges an average dry-weather flow of 20 mgd (75,700 cu m/day) through a 300-ft (91-m) diffuser located in 40 ft (12.2 m) of water. A computer program, which describes buoyant jets discharging to a stagnant ambient field based on work by Ditmars [5] and Brooks,[6] predicted an initial dilution of 40:1. Minimum dilution by field measurement was 53:1 at slack water and 140:1 at maximum tidal current of about 1.5 knots. Subsequent dilution at a fixed point in time was measured at 7:1 and computed at 4:1 using the dispersion coefficients derived in the course of the study. However, the computer program ignored vertical dispersion in arriving at the value of 4:1. The field check was considered

good verification of the computer programs for both initial dilution and subsequent dispersion of the effluent field.

The North Point outfall was the principal source of information on effluent field visibility. The data demonstrated that a field of primary effluent in central San Francisco Bay will at times be visible in dilutions of 1,000:1 or greater because of the high color distinction capability of the human eye.

Tidal exchange. Tidal flushing is a major factor in removing pollutants from central San Francisco Bay. Because the central bay is a potential deep water disposal site, a definition of the exchange of ocean and bay waters through the Golden Gate is of primary importance in projecting future water quality conditions. Not all of the water that ebbs from the bay is returned to it on the flood tide. Some is transported away by ocean currents and replaced by new ocean water. Tidal exchange ratio, simply stated, is that portion of the flooding tide that is ocean water entering the bay for the first time.

The tidal exchange measurement was an excellent example of interagency cooperation. The California Department of Water Resources, which needed the tidal exchange ratio for use with their mathematical modeling studies of the total waste assimilative capacity of San Francisco Bay, joined in the study. As the work program developed, a number of other agencies expressed interest in the program and offered to participate. Each of the two oceanographic cruises involved from 25 to 30 individuals representing 11 agencies.

All water that passes through the Golden Gate on a flood tide consists of water that has been in the bay before (old water) and water that has not (new water). The tracer level in old water (chlorides in this case) can be defined by measuring the level on an ebb tide. The tracer level in new water can be determined by measuring a background condition in the ocean. Measurement of tracer level on the following flood tide will then serve to establish the fractions of old and new water in the flooding tide. A detailed derivation presented elsewhere [7] yields the basic expression used in data analysis:

$$\bar{r} = \frac{\bar{C}_F - \bar{C}_E}{C_0 - \bar{C}_E} \quad (3)$$

where

$\bar{r}$ = tidal exchange ratio,
$\bar{C}_F$ = average tracer level in flood phase,
$\bar{C}_E$ = average tracer level in ebb phase, and
C_0 = tracer level of new ocean water entering the bay during the flood phase.

The basic data were collected for two different tidal conditions on September 2 and 3, and September 16, 1970, representing tides of different amplitudes. Each measurement was made during a period without diurnal inequality, which considerably reduced the work load by permitting the measurements to be made over only one ebb and one flood instead of a full 25-hr cycle. One research vessel made continuous measurements at four stations across the mouth of the Golden Gate over an ebb and the succeeding flood tide. A second research vessel simultaneously measured background conditions in the ocean beyond the influence of bay outflow.

Samples at the Golden Gate were taken with a submersible pump, which was drawn up through the water column at a constant rate to obtain a composite sample. Each sample was titrated for chlorides. Measurements made during the September 16 cruise are shown in Figure 6. Evidence of exchange is obvious in the figure, as flooding waters contain higher chloride levels after the time of current reversal (approximately 8:30 AM).

Tidal exchange was found to increase with tidal range. Because one of the factors affecting tidal exchange is the volume of water drawn into the bay on a flood tide, it is logical to expect a larger fraction of new water as the volume passed increases. The relationship is shown graphically in Figure 7, where measured tidal exchange ratio is plotted against flood-tide range. The origin of the curve

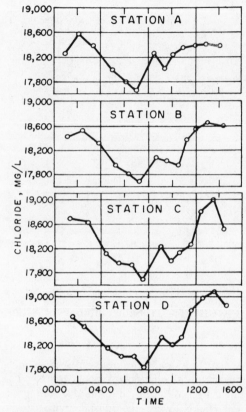

FIGURE 6.—Chlorides measured at the mouth of the Golden Gate on September 16, 1970.

is treated as a data point because the exchange ratio must approach zero as the tidal amplitude approaches zero. From

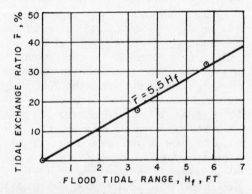

FIGURE 7.—Tidal exchange versus flood-tide range for dry weather conditions.

Figure 7, the tidal exchange ratio at the Golden Gate, expressed as a percentage of the water passing the gate on a flood tide, was determined to be 5.5 times the flood-tide range. Applying the tidal exchange equation to the mean tidal amplitude of 4.1 ft [8] (1.75 m) gives a mean tidal exchange ratio of 23 percent. Recognizing that the tidal volume transferred in a tidal phase (the tidal prism) also increases with tidal stage, increasing the volume of exchanged waters,[9] the mean tidal exchange at the Golden Gate is conservatively estimated to be 24 percent during dry weather.

ECOLOGICAL INVESTIGATIONS

Ecology is defined as the study of the total relationship of an organism to its environment. In recent years, however, the definition has been modified by general usage to encompass all studies of conservation and environmental biology. The fact that modern technology can influence the environment to a greater extent today than at any time in the past emphasizes the need for an ecological approach to many engineering studies. The environment is a regenerating resource, and like other resources it can either be utilized to its fullest potential or overexploited. The disposal of large volumes of wastewater effluent to a marine environment creates a potential for large-scale environmental change, and ecological investigations are essential to insure that detrimental changes will be avoided.

The ecological research program was designed to integrate a field investigation of the marine environment surrounding San Francisco with a laboratory investigation of the responses of selected marine organisms to dilutions of San Francisco wastewater effluent. The integrated data were then used to predict the effects of discharging San Francisco's liquid wastes to the marine environment. All basic data collected in the course of the investigations were published with the report to the city and county of San Francisco.[10]

As the first step in the study, the literature on marine resources of San Francisco Bay and the Gulf of the Farallones was carefully reviewed. The literature review gave special attention to the Dungeness crab, *Cancer magister*. Historically an important commercial fishery of the San Francisco area, the Dungeness crab suffered a natural decline all along the Pacific Coast in 1958–62, from which it has never recovered. The California State Department of Fish and Game emphasized the importance of protecting the Dungeness crab and suggested that larvae of the crab would be a suitable indicator organism for determining the effects of pollution.

Field investigations. Biological field investigations were conducted in San Francisco Bay and the Gulf of the Farallones for a full year during 1969–70. The investigations were divided into five phases: plankton studies, benthic studies, diving studies, intertidal studies, and *in situ* cage experiments.

Plankton studies. A cooperative program of plankton sampling and sample identification was carried out in co-operation with the California State Department of Fish and Game. In the course of this program, 33 points were sampled regularly at near-shore stations. Plankton samples were analyzed for diversity and population size, and the results were used to interpret the role of the plankters in the food chain and the productivity of the waters in different areas. Fluctuations in plankton numbers and diversity could be correlated with the upwelling of deep ocean waters along the coast and with high freshwater runoff from the San Francisco Bay system.

Benthic studies. Benthic organisms are those life forms that grow on or in the bottom materials in an aquatic environment. Because their lack of mobility makes them subject to the environmental characteristics at a single location, they are generally regarded as effective indicators of environmental stress. The objective of the benthic studies was the establishment of present background conditions for benthic organisms in the study area.

The study defined four benthic communities, identified by habitat as the Near Reef, Continental Shelf, Bar, and Bay communities. The Near Reef community, in the vicinity of the Duxbury Reef north of the Golden Gate had the highest benthic diversity and productivity, while the Continental Shelf community had the lowest. Both communities were dominated by forams and mollusks. The Bar community showed variable conditions of currents and wave action, with corresponding changes in productivity and predominant organisms. Sand dollars were the most abundant organism in the Bar community. The Bay community had a low diversity with high numbers of polychaetes and phoronids (segmented worms and tube worms).

Diving studies. Four near-shore locations were sampled by trained diver-biologists in scuba gear. Two of the surveys were conducted in the area influenced by the existing outfalls from the North Point and Richmond-Sunset water pollution control plants, and the remaining two were conducted for comparative purposes at Point Bonita and Muir Beach, which is on the Marin coast north of the Golden Gate. The North Point discharge, which averages about 60 mgd (227.000 cu m/day), reduced the species diversity within 200 ft (61 m) of the discharge point. The Richmond-Sunset outfall of 18 mgd (68.000 cu m/day) showed little or no reduction in subtidal species diversity.

Intertidal studies. Intertidal biota live between extreme low water and the upper splash zone. The wide range of environmental conditions results in a great diversity of organisms. As with the diving studies, the intertidal surveys were designed to provide background information and to assess the effects of existing discharges. A total of seven intertidal areas were investigated along the bay and ocean shoreline, including again the area influenced by the Richmond-Sunset outfall. Within 50 ft (15.2 m) of the outfall the intertidal rocks were devoid of macroscopic algae, and the flora was noticeably reduced for a distance of 400 ft (122 m). Beyond that point, recovery was rapid. The number and diversity of faunal organisms was reduced in the vicinity of the outfall, but at a distance of 400 ft (122 m)

to the east and 100 ft (30.5 m) to the west of the discharge, faunal composition had returned to normal.

In situ bioassays. To complement the controlled laboratory experiments on toxicity, an attempt was made to assess the actual toxicity of the North Point effluent field by suspending cages of fish in the effluent field at various distances from the point of discharge. Seven experiments were conducted using three-spined stickleback as the test organism. Test results fluctuated widely, with observed survivals ranging from 0 to 100 percent after 96 hr. Survival was apparently unrelated to distance from the point of effluent discharge, and the only conclusion that could be drawn was that some factor other than North Point effluent affected survival of the test fish. There was some evidence to indicate that the unidentified factor might be ship traffic in the vicinity of the piers where the cages were suspended.

Laboratory investigations. Laboratory investigations were conducted at the San Francisco Bay Marine Research Center in a laboratory specifically equipped for the study at Fort Baker on Horseshoe Bay, near the northern end of the Golden Gate Bridge. The purpose of the laboratory research was the development, through controlled experiments, of quantitative information on the toxicity of San Francisco wastewater effluent to marine organisms. The experiments modeled the marine environments of the Gulf of the Farallones and central San Francisco Bay as defined by the field investigations. Each experiment held all variables constant except the addition of wastewater, and each experiment included a control replicate with no wastewater added.

Effluent used in the experiments was collected as a 24-hr flow proportional composite sample at each of the city's three treatment plants and then composited in proportion to the average daily flow through each plant. The result was an effluent that was a representative composite of the total wastewater flow from the city of San Francisco. The basic reason behind the decision to composite the wastes from the city's three plants was that testing the effluent of each plant separately would have tripled the number of tests required for each organism, thereby reducing the scope of the study that could be conducted with the available funds. It was felt that more comprehensive results could be achieved by compositing. The following factors led to the conclusion that compositing the sample would not represent a significant compromise of the results.

1. At the time of the study, plans were being formulated to combine all three of the effluent streams into a single ocean outfall. The tests on composited effluent more accurately model the condition of combined outfalls.

2. While recognizing that there might be variances in the vicinity of individual outfalls, it was reasoned that for the estuary and coastline as a whole, the effect of all discharges, whether single or combined, would be cumulative.

3. Effluent from each of the plants constituted a large fraction of each composited sample, and any unusual toxicity would have shown up in the bioassays. No unusual toxicities were observed.

Dilution water was supplied continuously to the laboratory by a submersible pump installed near the mouth of Horseshoe Bay.

Two basic criteria governed the selection of test organisms. First, the organism had to be obtainable from the environment in quantities large enough for statistically significant results; and second, the natural habitat of the organism had to be suitable for laboratory modeling. Many thousands of organisms representing 22 species were tested. The tests included eggs, larvae, and juveniles as well as adult organisms. Experiments were designed to evaluate both acute and chronic toxicity as well as other possible effects of effluent discharge. Test methods included static and continuous flow bioassays, environmental chamber studies, microcosm studies, and special studies of biostimulation and of stickleback blood response to dilutions of wastewater effluent.

TABLE I.—Static Bioassays of Marine and Estuarine Fish

Organism	Average Survival after 96 hr (%) (dilution ratio, effluent: dilution water)											
	Control (0)*	1:100 (1)	1:50 (2)	1:20 (5)	1:10 (10)	1:5 (20)	1:2† (33)	1:2‡ (33)	1:1† (50)	1:1‡ (50)	1:0† (100)	1:0‡ (100)
Shiner perch *Cymatogaster aggregata*	79		160	80	73	75	18	23				
Walleye surf perch *Hyperprosodon argenteum*	96	100	95	83	83	67						
Mottled sand dab *Citharichthys sordidus*	100			95	84	95	100	100	88	75		
Ling cod *Ophiodon elongatus*	100	100	100									
Skulpin *Scorpaena guttata*	94	100	100	80	80	80						
Rock bass *Sebastodes* sp.	100	100	100	100	100	100	100	100	100	90	0	
Three-spined stickle-back *Gasterosteus aculeatus*	90	95	90	100	100	83	87	98	70	68	3	53

* Figures in parentheses are the percentages of wastewater effluent.
† Salinity adjusted to the level of the diluting water by addition of NaCl.
‡ Salinity not adjusted.

TABLE II.—Static Bioassays of Marine and Estuarine Invertebrates

Organism	Average Survival after 96 hr (%) (dilution ratio, effluent: dilution water)											
	Control (0)*	1:100 (1)	1:50 (2)	1:20 (5)	1:10 (10)	1:5 (20)	1:2† (33)	1:2‡ (33)	1:1† (50)	1:1‡ (50)	1:0† (100)	1:0‡ (100)
Hermit crab												
Pagurus samuelis	96	100	95	100	75	100						
Shore crab[a]												
Hemigrapsus oregonensis	97	95	98	97	97							
Sand crab												
Emerita analoga (adult)	93	76	71	73	36.5	—	0	0				
Emerita analoga (larvae)	82	76	59	71	35		0	0				
Bay shrimp												
Crago sp.	60	53	45	47	45	42	17	23	7	7	0	0
Pink ghost shrimp												
Callianassa californiensis	100	83	83									
Turban snail												
Tegula funebralis	100	100	100	100	94	100						
Sea urchin												
Strongylocentrotus purpuratus	100	100	100									
Bent-nosed clam												
Macoma nasuta	100	100	100	100	100	91	91	100	100	82	0	0
Ribbed horse mussel												
Volsella demissus	100	100	100	100	100	100	100	100	100	100		
Broken-back shrimp												
Spirontocaris paludicola (larvae)	100	90	90							70		0

* Figures in parentheses are the percentage of wastewater effluent.
† Salinity adjusted to the level of the diluting water by addition of NaCl.
‡ Salinity not adjusted.

Static bioassays. Static bioassays were used to determine the acute toxicity of the effluent to adult fish and invertebrates by means of mortality in short-term tests of 96 hr each. Procedures in these tests[5] were modeled after those described in "Standard Methods".[11] Effluent dilutions tested ranged from one part wastewater in 200 parts dilution water (1:200) to pure effluent (1:0).

The average percent survivals for all species studied are presented in Tables I and II, and the corresponding TL$_m$ and LD$_{10}$ values are presented in Tables III and IV. (The TL$_m$ represents median tolerance limit, or the dilution at which half of the test organisms die during the test period, while LD$_{10}$ represents the dilution at

which 90 percent of the organisms survive and the walleye surf perch required the highest dilutions (1:20 to 1:50) to assure 90 percent survival. The bay shrimp, sand crab, and pink ghost shrimp were the most sensitive organisms tested, requiring dilutions of about 1:100 for 90 percent survival. All studies indicated that no significant toxic effect could be demonstrated on any organism tested after 96 hr of exposure in dilutions of primary effluent greater than 1:100.

Environmental chamber experiments. Toxicity tests of microorganisms were conducted in the rigorously controlled climate of an environmental chamber. The environmental chamber was used to test the effect of San Francisco wastewater effluent

TABLE III.—Bioassay Results for Static Fish

Organism	TLm (%)	Estimated Minimum Dilution for 90 Percent Survival (LD-10)
Shiner perch		
Cymatogaster aggregata	20–33*	1:10-20
Walleye surf perch		
Hyperprosodon argenteum	†	1:20-50
Mottled sand dab		
Citharichthys sordidus	†	1:1-2
Ling cod		
Ophiodon elongatus	†	‡
Skulpin		
Scorpaena guttata	†	1:20-50
Rock bass		
Sebastodes sp.	50–100*	1:1
Three-spined stickleback		
Basterosteus aculeatus	50–100*	1:2

* Estimated values from Table I.
† Greater than 50 percent survival in all dilutions tested.
‡ Greater than 90 percent survival in all dilutions tested.

on the hatching of Dungeness crab eggs and on the survival of larval stages of Dungeness crab, sand crab, bay shrimp, and broken-back shrimp.

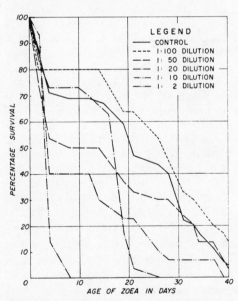

LEGEND
— CONTROL
---- 1:100 DILUTION
— 1:50 DILUTION
— 1:20 DILUTION
—·— 1:10 DILUTION
—··— 1:2 DILUTION

FIGURE 8.—Survival of sand crab zoea in San Francisco composite wastewater effluent.

TABLE IV.—Bioassay Results for Static Invertebrate

Organism	TLm (%)	Estimated Minimum Dilution for 90 Percent Survival (LD-10)
Hermit crab		
Pagurus samuelis	†	‡
Shore Crab		
Hemigrapsis oregonensis	†	‡
Sand crab		
Emerita analoga (adult)	5–10*·†	1:100
Emerita analoga (larvae)		1:100
Bay shrimp		
Crago sp.	20–33	1:100
Pink ghost shrimp		
Callianassa californiensis	†	1:100
Broken-back shrimp (larvae)		
Spirontocaris paludicola	†	1:50
Turban snail		
Tegula funebralis	†	‡
Purple sea urchin		
Strongylocentrotus purpuratus	†	‡
Bent-nose clam		
Macoma nasuta	50	1:1-2
Ribbed horse mussel		
Volcella demissa	†	‡

* Estimated values from Table II.
† Greater than 50 percent survival in all dilutions tested.
‡ Greater than 90 percent survival in all dilutions tested.

The Dungeness crab egg hatching bioassays showed a significant toxic effect of effluent at a dilution of 1:50. On the other hand, a dilution of 1:100 not only exhibited no toxic effect, but produced a slightly higher hatch than the control.

Bioassays on crab larvae suggested a beneficial effect on the first two larval stages of sand crab and Dungeness crab at dilutions greater than 1:50. Typical survival curves for sand crab larvae, illustrating the apparent beneficial effect of 1:100 dilution, are shown in Figure 8. The bay shrimp larvae and the third stage Dungeness crab larvae, however, showed some negative response even at dilutions of 1:100.

Continuous-flow bioassays. The continuous-flow bioassay more accurately models the environment than does the static bioassay and thereby permits longer

291

tests to determine chronic effects of the test medium. The continuous flow apparatus unfortunately was not completed in time for extensive application on this study, but bioassays were conducted by this method on Dungeness crab zoea and three-spined sticklebacks. The results of all of the continuous-flow experiments showed no toxic effect of San Francisco wastewater effluent in dilutions equal to or greater than 1:50.

Microcosm studies. A microcosm is a miniature world that is designed to model an existing ecosystem. A long-term microcosm experiment with Dungeness crab larvae was conducted by George Schumann of Marine Associates, San Diego, Calif., who grew Dungeness crabs from newly hatched larvae to adult form in three 1,000-gal (3,785-l) aquaria. One aquarium served as a control, and the other two received measured amounts of San Francisco effluent to produce dilutions of 1:100 and 1:200. Equal amounts of food in the form of plankton and brine shrimp were added daily to each aquarium. The

Other marine organisms that were inadvertently introduced in larval form into the microcosms with the dilution water or with the food organisms and grew to adulthood in the diluted effluent included top smelt, gobys, barnacles, shrimp, mussels, crabs, tunicates, annelids, isopods, and bryozoans. In a separate test, American lobsters grown from larvae to adulthood in a 1:100 dilution of San Francisco effluent exhibited a survival rate virtually identical to the control.

Stickleback blood study. An experimental study of the effects of San Francisco wastewater effluent on blood composition of three-spined sticklebacks was conducted by Gretchen Kaplan at San Francisco State College under the direction of Curtis L. Newcombe in 1970. The hematological responses of the stickleback to various dilutions of effluent were investigated over a 7-day exposure period. The results showed that dilutions less than 1:10 exerted a stress on sticklebacks as documented by a change in the number of white blood cells and

thrombocytes. The initial response seemed to be a reaction of protection against large numbers of bacteria. Hematological response seems to be a very sensitive index of environmental stress and should be investigated further.

Biostimulation study. Surface seawater was collected from a point just north of the Farallon Islands in September 1970 and tested for increase in chlorophyll *a* 4 days after the addition of various amounts of wastewater effluent. Two controls and two effluent dilutions each of 1:100, 1:50, and 1:20 were tested. The results showed no significant change in chlorophyll concentration in the controls during the test period and no deviation from the controls in the dilutions tested. While limited in scope, the fact remains that the work did not document any stimulatory effect of San Francisco wastewater on phytoplankton even in dilutions as low as 1:20.

Ecological design criteria. The information obtained from the field and laboratory investigations in the course of this study was evaluated together with the literature on ecological investigations in the vicinity of existing submarine outfalls, and ecological design criteria were formulated for waste discharges to the waters of San Francisco Bay and the Gulf of the Farallones adjacent to San Francisco. These criteria formed the basis for recommendations concerning the location, size of outfall, depth of discharge, and degree of treatment required for various disposal alternatives investigated. It is the conclusion of this study that these criteria, if followed, will prevent new waste discharges from affecting the marine environment detrimentally. The ecological design criteria may be summarized as follows:

1. The ecological design criteria for discharge to the Gulf of the Farallones requires a location south of the Golden Gate. The shoreline in that area is a sandy beach, which is one of the least sensitive habitats.

2. Studies of both large and small discharges in southern California have indicated that it would be desirable to limit

the dry-weather flow from a single outfall to about 100 mgd (378.5 cu m/day) unless a degree of treatment beyond primary is provided.[12]

3. The marine biota tested in the laboratory may be divided by habitat into three groups: the intertidal community, the benthic community, and the community within the water column. For each community, the recommended effluent dilution was set at 10 times the value that first showed adverse reaction in the most sensitive organism. The recommended dilutions thus determined were 1,000:1 for the intertidal community, 500:1 for the benthic community, and 200:1 for the inhabitants of the water column. These recommendations are for long-term exposure to an effluent equivalent to the present effluent from San Francisco's three primary plants.

4. For large intermittent wet-weather discharges, it is ecologically desirable to have a surface field because there is a strong surface movement out of the central bay and away from the shoreline in the Gulf of the Farallones.

5. The diffuser and pipeline should be constructed in a manner that will not impede the shoreward migration of benthic animals.

Monitoring program. In order to document any environmental change that a new waste discharge might create, a comprehensive monitoring program is recommended for any new outfall. The program should start at least 1 yr in advance of outfall construction and should be intensive during the first few years of discharge. The necessity for a monitoring program of considerable scope stems from the fact that the changes which the program seeks to document will be subtle ones, if they occur at all. Data analysis should begin immediately, both to permit early detection of possible changes and to permit modifications in the monitoring program where necessary.

WASTE DISPOSAL RECOMMENDATIONS

After the oceanographic and ecological studies had been completed and the re-

sults reduced and analyzed, a series of recommendations that were consistent with the findings of the study was developed. Specific effluent flow conditions posed by the Department of Public Works were then tested within this framework, and methods of discharge were recommended that will meet all of the ecological design criteria.

It is important to note that the recommendations of this study were based primarily on protection of public health, avoidance of nuisance conditions, and protection of the marine environment. At the time this study was made, the requirements of the regulatory agencies were in a state of flux. Some discharge requirements have been more severe and some less severe than the criteria found necessary for protection of the marine environment. Discharge requirements above and beyond what is needed for environmental protection may be enforced. For example, nothing in this study indicated a need for 85 percent biochemical oxygen demand removal as required by present federal standards.

The basic findings and recommendations derived from the study are summarized as follows:

1. The ecological design criteria set forth in this study can be met while discharging dry- and wet-weather flows from the North Point zone to the central bay.

2. The ecological design criteria can be met while discharging the total dry- and wet-weather flows from the entire city to the Gulf of the Farallones at specific points.

3. A degree of treatment equivalent to efficient primary treatment, plus disinfection, will be adequate for dry-weather flows, and treatment equivalent to substantially complete removal of gross floatable and settleable material, plus disinfection, will be adequate for intermittent wet-weather overflows.

4. Any effluent field that rises to the surface will at times be visible in the receiving waters regardless of degree of treatment, and a visible discharge should

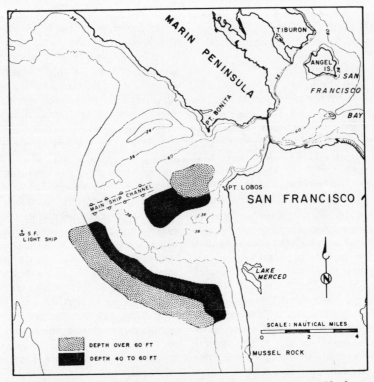

FIGURE 9.—Suitable waste disposal areas in the Gulf of the Farallones.

not necessarily be considered an unsatisfactory discharge.

5. It is a basic finding of this study that environmental protection of the central bay and Gulf of the Farallones will be best achieved by discharging intermittent wet-weather flows in a manner that will create a surface effluent field. Surface advection will then transport the field seaward and dissipate it in the shortest possible time.

6. The total dry-weather flow from the entire city can be disposed of either inside or outside of the bar which surrounds the mouth of the Golden Gate (Figure 9). Over the bar itself the water is too shallow for a major outfall.

7. The total dry- and wet-weather flow from the entire city can be adequately disposed of outside the bar. Some combinations of total dry- and wet-weather

discharge inside the bar would violate the ecological design criteria.

Working within the framework of the findings set forth above, preliminary designs were prepared for a number of outfall schemes representing different options available to the city in its master planning study. A brief summary of the design calculations for an outfall from the North Point water pollution control plant into central San Francisco Bay will illustrate the manner in which the design criteria developed by the study were used.

Under one alternative of the city's master plan study, the North Point plant would ultimately discharge an average dry-weather flow of 83 mgd (314,000 cu m/day) to the central bay. Two sites were investigated (Figure 10). Site A was selected because an analysis of hydrographic charts for the last 100 yr showed

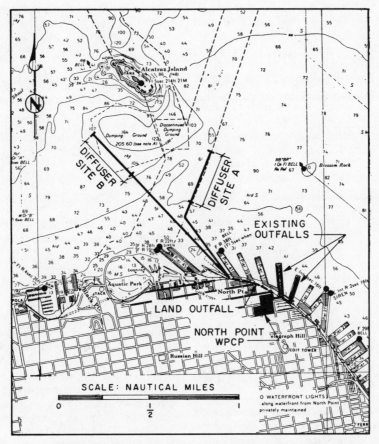

FIGURE 10.—North Point outfall alternative sites.

Site B to lie in an area subject to large shifts in bottom elevation. For Site A, the outfall will consist of approximately 5,000 ft (1,524 m) of 102-in. (258-cm) pipe, including a 1,850-ft (564-m) diffuser section. A series of computer analyses of outfall hydraulics and initial dilution indicated that with the head available, the optimum diffuser design would contain 462 ports, each 3 in. (7.6 cm) in diam, spaced at 4-ft (1.2-m) intervals, and staggered on opposite sides of the pipe. Head loss and initial dilution characteristics are shown in Figure 11. Initial dilution at slack water condition will be 200:1 or greater, and dilution at the average current of 1.5 knots will be about three times that value. Assuming 200:1 initial dilution, the effluent field depth at average current will be

about 4 ft (1.2 m). To affect the benthos, the field must reach the bottom by vertical dispersion. Because the minimum depth in areas likely to be affected by the effluent field is 40 ft (12.2 m), the field will be diluted at least 10:1 by the time it reaches the bottom. Minimum dilution at the bottom will thus be 2,000:1, a value well in excess of the ecological design parameter of 500:1. This analysis ignores horizontal dispersion, which will increase the total dilution.

On the ebb tide, the effluent field will move directly through the Golden Gate entrained in the main water mass and will not contact the shoreline. On the flood tide the field will move toward ecologically sensitive shallow areas in the south bay, dispersing horizontally in accordance with the curves in Figure 12. After 3 hr, the field dilution resulting from horizontal dispersion will be 5:1, giving an overall dilution of 1,000:1, and the field will still not have reached the shallow areas of the south bay. This analysis ignores vertical dispersion, which occurs simultaneously and multiplies the total dilution.

Average initial dilution of 200:1, which will be achieved in less than 2 min after discharge, will be adequate to protect the organisms dwelling in the water mass.

A similar series of calculations, applied to each discharge considered in the city's master plan, demonstrated that marine disposal of the entire liquid waste flow from the city of San Francisco can be accomplished in the waters adjacent to the city without adversely affecting the marine environment.

Summary

The city of San Francisco undertook a comprehensive study of the marine environment to determine where and in what quantities it is feasible to dispose of the city's dry- and wet-weather wastewater effluents. The investigation was divided into two phases. The oceanographic phase defined the characteristics of potential discharge sites and determined what will happen to a discharge at any site. The ecological phase provided the basic data for determining what will happen to the

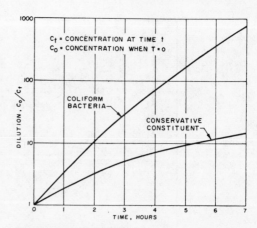

FIGURE 12.—Horizontal dispersion of the effluent field recommended for the North Point outfall.

marine organisms that may be exposed to the discharge. Considered together, the results of the two phases indicated that marine disposal is feasible and that the marine environment can be adequately protected by discharging chlorinated primary effluent through one or more submarine outfalls with properly designed diffuser systems. The requirements of regulatory agencies may in some cases necessitate additional treatment, but the study did not indicate that additional treatment is necessary to protect the marine ecosystem.

References

1. Brooks, N. H., "Diffusion of a Sewage Effluent in an Ocean Current." *Proc. 1st Intl. Conf. on Waste Disposal in the Environ.*, Pergamon Press, New York, N. Y. (1960).
2. Fischer, H. B., "Methods for Predicting Dispersion Coefficients in Natural Streams with Applications to the Lower Reaches of the Green and Duwamish Rivers, Washington." Geol. Surv. Professional Paper 582-A, U. S. Govt. Printing Office, Washington, D. C. (1968).
3. Pearson, E. A., "An Investigation of the Efficacy of Submarine Outfall Disposal of Sewage and Digested Sludge." California State Water Poll. Control Bd. Publ. No. 14 (1956).
4. Batchelor, G. K., "The Theory of Homogeneous Turbulence." Cambridge Univ. Press, Eng. (1953).

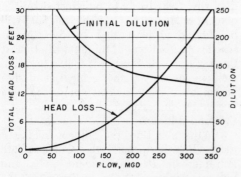

FIGURE 11.—Head loss and initial dilution characteristics recommended for the North Point outfall.

5. Ditmars, J. D., "Computer Program for Round Buoyant Jets into Stratified Ambient Environments." W. M. Keck Laboratory of Hydraulics and Water Resources, Technical Memorandum 69–1 (Mar. 1969).

6. Brooks, N. H., and Koh, R. C. Y., "Discharge of Sewage Effluent from a Line Source into a Stratified Ocean." Intl. Assn. for Hydraulic Res., Proc. XI Congress, Leningrad, U.S.S.R. (Sept. 1965).

7. Parker, D. S., *et al.*, "Tidal Exchange at the Golden Gate." *Jour. San. Eng. Div., Proc. Amer. Soc. Civil Eng.*, 98, 2 (1972).

8. "Tidal Current Tables—1970, Pacific Coast of North America and Asia." U. S. Coast and Geodetic Survey, Washington, D. C. (June 1969).

9. "Salinity Incursion and Water Resources." State of California, Dept. of Water Resources, Appendix to Bull. No. 76 (Apr. 1962).

10. "A Predesign Report on Marine Waste Disposal, Vol. II, Data Supplement." Brown and Caldwell Consulting Engineers for the city of San Francisco, Calif. (Sept. 1971).

11. "Standard Methods for the Examination of Water and Wastewater." 12th Ed., Amer. Pub. Health Assn., New York, N. Y. (1965).

12. North, W. J. "Review of Biological Literature on Pacific Coast Marine Waste Disposal as a Guide to Prediction of Ecological Effects of a Submarine Outfall in the Gulf of Farallones." Rept. to the city of San Francisco, Calif. (Dec. 1970).

VI
Coupled-Systems Analysis

Editors' Comments on Papers 22 Through 26

22 **Vanyo:** Law, Operations Research, and the Environment
J. Environ. Sys., **1**, 213–236 (Sept. 1971)

23 **Koenig, Cooper, and Falvey:** Engineering for Ecological, Sociological, and Economic
Compatibility
IEEE Trans. Sys., Man Cybern., **SMC-2**, 319–331 (July 1972)

24 **Young:** Optimal Pollution Regulation—A Data-based Study
Original manuscript

25 **Leontief:** Environmental Repercussions and the Economic Structure: An Input–Output
Approach
Rev. Econ. Statist., **52**, 262–271 (Aug. 1970)

26 **Ayres and Kneese:** Production, Consumption, and Externalities
Amer. Econ. Rev., **59**, 282–297 (June 1969)

The human environment includes not just the physical world; it has social, economic, and political components as well. We have our groups and organizations with their respective rules and methods of sanction and their procedures for procuring and allocating goods and services. This means that the modeling of the physical environment cannot lead to significant changes if pursued without regard to the links between natural ecosystems and a human community's social, legal, political, and economic systems. Paper 6, by Koenig and Tummala, has already provided a glimpse of coupled-systems analysis, principally with respect to the relationship between the ecological system and the economic system. Part IV contains five papers that purposefully develop the interrelationships among various systems. Paper 22 relates the legal structure to environmental concerns. The remaining four papers (Papers 23–26) concentrate primarily on the relationship of the economic system to the ecosystem, although Paper 23 also includes considerations of social and political factors. Interestingly, Papers 23, 25, and 26 are cited in Paper 6.

The first paper of Part VI is "Law, Operations Research, and the Environment" by James P. Vanyo. Vanyo, who is both a lawyer and an engineer, summarizes in Part I of his article the major subdivisions of law, pointing out those areas that have a bearing on legal redress of environmental problems. Taken together, these areas have become the basis for a new field of legal practice called "environmental law."

In Part II of this paper, "Environmental Law as Part of a Systems Approach," Vanyo first gives a brief history of the birth and childhood of operations research and systems analysis (OR/SA). Then he describes, as nonmathematically as possible, the potential application of several OR/SA techniques in the practice of environmental law. Vignettes on linear programming, network analysis (PERT), game theory, systems (feedback) analysis, and simulation are included. Except for simulation, the gist of each technique is generally made clear. Two main points need to be addressed with regard to simulation. Vanyo states:

> Unlike the other techniques a formal mathematical model is not required.
> This says that one need not formulate the functional relationships of a

problem. The solution is characteristically not only accurate but it also yields an estimate of what the accuracy is. The disadvantage is that while a person may learn the answer to a given problem he might never learn why he got that answer.

For the reader with a knowledge of probability or statistics it is apparent that enough data now exists to compute standard deviations (a measure of the accuracy) for the averaged results.

Because there are no equations to formulate or solve, the simulation may be made as complex and realistic and as accurate as desired.

The first point is that Vanyo's comments in the "Simulation" section can apply to simulations of only one kind. Such Monte Carlo (stochastic or probabilistic) simulations based solely on empirical frequency distributions represent only a tiny fraction of all possible types of simulations. In general, quantified and appropriately shaped functional relationships are necessary with simulation. Rules and relations for the processing of the data input and the updating of variables must be provided. Even with most Monte Carlo simulations, not all required probability distributions can be obtained empirically, and there is a need to choose a functional form for probability distributions to be used in "coin-flipping." Additionally, the alleged disadvantage of simulation "that while a person may learn the answer . . . he might never learn why he got the answer" is simply not experienced very often. Most simulations allow a tracing back to causes of simulation results through examination of intermediate calculations, the running of special test inputs, the substitution of alternative models, and/or the changing of parameters. In sum, the reader should consider the field of simulation to be much broader than indicated in this paper.

The second point to be made is that in the quotations above and in other statements in the "Simulation" section, Vanyo uses the word "accuracy" where standard practice would require the word "precision." The standard deviation of a variable is a measure of the spread or dispersion of its values. A small standard deviation corresponds to high precision, a large standard deviation to low precision. But standard deviation has nothing to do with "accuracy" as this term is used in statistics. Accuracy refers to the closeness of a result to the true or target value. If a man shoots at a tin can sitting on a fencepost and always hits the fencepost in a 1-inch diameter circular area below the can, then the shooter has shot with precision but not with accuracy. He would have to hit the can itself before one would say that he had made an accurate shot. The important thing to note is that accuracy can be determined only when one knows what the target is. One cannot determine the accuracy of a simulation by looking only at the output of simulation runs; accuracy is determined by the closeness of the output of simulation runs to real-world values or preselected target values.

An engineer, a zoologist, and an economist collaborated to write Paper 23, "Engineering for Ecological, Sociological, and Economic Compatability." Developmentally, this paper is a direct predecessor of the mathematically formalized Paper 6, with Herman Koenig being the authorial link between the two papers. Paper 23 qualifies for this section because it considers the interrelationships between biological, geophysical, and human systems. A simplified but useful characterization of the ecology of an industrialized nation is presented in verbal and diagrammatic form. Several instructive examples of far-reaching socioecological interactions are provided. Outlined briefly are four avenues for the improvement of "the industrialized ecosystem": spatial distribution, material recycling, longer-life products, and environmental design.

This paper raises a number of important sociopolitical points. For example, recognition of an important problem confronting humanity is shown in the authors' statement that "technology provides a mechanism for concentrating and storing as potential energy the effort of many men in the hands of one or a relatively small number of individuals." Later in the paper, they note that

> with the power of modern technology, the industrialized nations might develop national and international ecosystems that are not technically, economically, and politically controllable. . . . Is it possible that technological specialization in certain areas should be bounded for reasons of controllability as well as mass–energy compatibility with the environment?

An important question indeed!

Attention is given to describing why today's price system, by underpricing public goods (what Ayres and Kneese in Paper 26 call externalities), prevents the market economy from adequately controlling developments affecting environmental quality. The authors consider ecosystem design and conclude with a section describing the scientific, technological, and managerial developments that must precede the successful design and management of life-supporting ecosystems.

Perhaps the greatest difficulty one might have with this paper is a practical and/or philosophical rejection of the authors' thesis and expressions of technological optimism:

> The fundamental thesis of this paper is that as a dominating species on the surface of the earth, man must learn how to engineer the developments in industry, agriculture, and human habitats as components of an industrialized ecosystem.
>
> . . . seminatural ecosystems can be designed to fluctuate within the region of stability required to ensure the desired system characteristics.

To some, the possibilities of man's designing stable ecological systems appear remote for several reasons. First, current knowledge requisite to rational ecosystem design is almost entirely lacking. Procurement of the required information and understanding would cost fantastic amounts and take decades. Second, the nature of life is change, and even natural communities are not permanently stable (e.g., biological succession). Man-shaped ecosystems are likely to be even less stable because their design will be based on incomplete information, the inclusion of fewer species (less complex ecosystems are generally less stable), the introduction of species not fully adapted to the new environment (this means a built-in pressure for change), or the alteration of energy inputs, nutrients, or habitats in the target area (again, these manipulations create a pressure for change). Third, meteorological variation and geologic events tend to operate against this goal, often in rather powerful ways (heat, drought, flood, earthquake, volcanic eruption, tidal wave).

Many people, particularly ecologists and Buddhists, do not conceptualize man as a dominant species. Seeing an interrelatedness in all things, these people see each life, whether it be human, animal, or plant, as equally important. From this philosophical vantage point, the development of an ecological ethic ("spaceship earth ethic") is more desirable than the attempt to engineer ecosystems.

Of the several environment-related papers presented at the 41st National Meeting of the Operations Research Society of America in New Orleans, April 26–28, 1972, Donovan Young's "Optimal Pollution Regulation—A Data-based Study" was almost the only work that treated a specific application. Pollution abatement cost data was gathered from ten industrial firms that discharged particulate matter into the air around Houston, Texas. The purpose of the research was to evaluate different air pollution control strategies using real-life assumptions and data in a dynamic programming simulation.

This paper effectively couples economics and engineering through consideration of rational economic decision-making by industrial firms and the technology-defined curves of cost (equivalent uniform annual cost) versus emissions. The overall analysis approach is akin to cost–benefit analysis, although benefits are not quantified in money terms and no benefit/cost ratio is formed. The best alternative is the one with the minimum cost (the sum of environmental damage and abatement control costs) to meet or exceed a given level of benefit (i.e., emissions not exceeding a specified number of tons per year).

In his preconvention abstract, Young states: "This paper presents the first extensive data-based quantitative argument in favor of effluent fees."* He also mentions the inclusion of a "definitive bibliography." The paper is short and easy to read, although familiarity with basic economics and engineering economy is certainly beneficial. It may be helpful in relating Young's study to other work to recognize that the phrase "residuals charges" is often used synonomously with "effluent fees" (see Freeman and Haveman, "Residuals Charges for Pollution Control: A Policy Evaluation," cited in the subsection "Coupled-Systems Analysis" in the Bibliography).

Wassily Leontief is the father of input–output analysis, a technique that relates the amount of goods or services produced by each production or service sector (e.g., production sectors for agriculture or manufacturing and service sectors for banking or health care) to the final consumption demands (e.g., household consumption or consumption by the federal government). It is generally assumed that the coefficients relating final demands to production are slowly varying functions of time, whereupon

ORSA Bulletin, **20**, Suppl. 1, B-174 (Spring 1972).

changes in production may be calculated solely from changes in consumption. Input–output analysis involves a linear model, for it is usually assumed that the coefficients are not functions of the amount of either production or consumption. Linearity is a reasonable assumption for small percentage changes, but it may not be appropriate for larger changes.

In Paper 25, "Environmental Repercussions and the Economic Structure: An Input–Output Approach," Leontief incorporates the generation of pollution into the input–output concept by defining coefficients relating the amount of emission of each type of pollutant to the level of production and final demands. This approach can be used to predict the change in pollution emission as final demands change. A cost for pollution control is calculated as the sum of the price paid by each production sector to other production sectors for the use of their products in the control effort and the cost of pollution-controlling services (labor cost). A practical limitation of this input–output method is the significant amount of data required for its effective use.

Paper 25 is valuable for its introduction to input–output analysis and its conceptualization of the dependence of pollutant emissions on production and final demands, but, unfortunately, a disconcerting number of errors appear in the text. On page 361, four lines below Equation (2), Y_1 should be set equal to 55, not 50. A few lines later in the same paragraph, (50 bushels) should be (100 bushels) and (100 yards) should be (50 yards). There is a mix-up on page 362 in the third and fourth paragraphs of Section IV. In the third paragraph, the coefficients for manufacturing and agriculture are reversed; the manufacture sector should have a pollution coefficient of 0.20 grams of pollutant per yard of cloth produced, and the agriculture sector should produce 0.50 grams of pollutant per bushel of wheat. Similarly, in the fourth paragraph, the discussion of Equations (5) should associate agriculture with X_1 and manufacture with X_2, not the reverse. From Table 4 onward in the paper, it appears as if labor inputs should consistently be described in units of man-hours, not man-years, in keeping with Table 5 and the magnitudes of labor and value-added entries in Tables 4 through 8. In Equations (10), the coefficient of Y_2 in the X_3 equation should be 0.655, not 9.655, and the coefficient of Y_4 in the L equation should be 1.000, not 0.000. Three lines below Equations (14) on page 367, the value of v_3 should be \$2.00 instead of \$2.60. In Table 8, the entries (\$11.70), (\$222.04), and (\$3.93) in the Labor row should simply be 11.70, 222.04, and 3.93. Finally, in Equation (15) of the Mathematical Appendix, an equals sign should be placed between the **X** and **Y** matrices.

In "Production, Consumption, and Externalities," a paper that is rapidly becoming a classic, Robert Ayres and Allen Kneese are concerned about externalities (i.e., by-products of production and consumption that are not taken into account by the marketplace because no price or cost is assigned to them). The lack of any cost associated with pollution, for example, qualifies it as an externality. In order for pollutors to take steps to reduce their effluents, pollution must either be brought under the market pressures of supply and demand (thus removing it from the externality category) or directly regulated by government. Acting as a broker for the public, a government can set a pollution tax or effluent fee to bring pollution into the market system (indirect regulation); alternatively, a government can limit the permissible amount of effluents (direct regulation).

Ayres and Kneese model the production of residuals (pollutants) by writing mass conservation equations. They state that their "main objective is to make some progress toward defining a system in which flows of services and materials are simultaneously accounted for and related to welfare." Three sectors are considered: the environment, a production sector, and the final demands or consumption sector. Figure 1 illustrates the mass flows between these sectors. Direct contributions from the environment to consumption are neglected. From this diagram, we see that in the steady state the residual mass returning to the environment equals the mass of raw materials extracted.

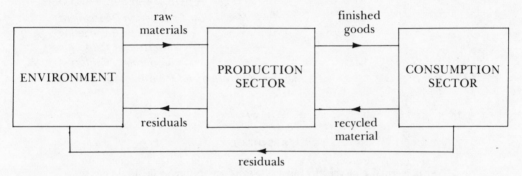

Figure 1

Three main conclusions of this article are as follows:

1. The residuals problem, which is an inherent part of the production and consumption processes, becomes progressively more significant with population and economic growth because the assimilative capacity of the environment is then exceeded.

2. Environmental media (e.g., air, water) should not be considered independently.

3. Ad hoc restrictions for pollution control are not adequate; public investment programs must be planned in relation to the amounts and effects of residuals.

Features of Paper 26 that may be of particular interest are the mass conservation equations and the inclusion of a cost for pollutants in the price of a commodity. The analysis appears to be more of theoretical and attitudinal importance than of practical value, for it would be extremely difficult to obtain the data, to estimate the residual costs, to establish a management system, and to perform the computation necessary to apply the model. The input–output notation introduced by Leontief is used in this article, and if the notation is unfamiliar, it may be helpful to read Paper 25 before proceeding with Paper 26. It is also worth mentioning that the type of analysis suggested by Ayres and Kneese in their last paragraph, namely, the look at a residuals-producing industry in terms of its costs versus emission levels, is precisely what Young describes in Paper 24.

22

Reprinted from *J. Environ. Sys.*, **1**, 213–236 (Sept. 1971)

Law, Operations Research, and the Environment

PROF. JAMES P. VANYO

Mechanical Engineering Department
College of Engineering
University of California
Santa Barbara, California
Member, California and Ohio Bar Associations

ABSTRACT

This paper discusses the legal component of a systems approach to environmental problems. In Part I, the scope and nature of the legal process is reviewed for the nonattorney reader. Aspects of the legal process that have a major impact on environmental areas are considered in more detail. In Part II, a systems approach integrating legal and technical solutions to environmental problems by operational research techniques is outlined.

Introduction

An initial attempt has been made to formulate a general legal-technical systems solution to environmental problems. Because the paper bridges several disciplines, summaries of each of the disciplines have been included. Part I, on the legal process, is intended as a survey of the relevant features of environmental law for the nonattorney reader. Part II includes surveys of the major techniques of operations research and system analysis for the reader not familiar with these developments. They are only surveys and as such lose the validity of more rigorous expositions.

In each instance the material is selected with a prior expectation that it interacts well with the other discipline. The reader familiar with operations research will, as he reads Part I, see suggestions of several operations research techniques that have been used for centuries by the legal profession—although not as part of a mathematical formulation. Likewise,

213

the attorney reader will, when reading Part II, see balancing the equities in tradeoff and optimization studies, and courtroom strategy in game theory.

Part I. THE LEGAL PROCESS AND ENVIRONMENTAL LAW

This section consists of a review of the scope and nature of the legal process in preparation for the remainder of the paper. To investigate the legal process we might profitably investigate the educational process by which a person becomes familiar with the law and its daily implementation.

The list in Table 1 is typical of the legal topics that make up an average law school curriculum. An estimate of the topics most relevant to the solution of environmental problems is shown in the last column. An assessment of the nature of the topics listed in Table 1 is next made.

Substantive Law

Contracts, torts, property, and remedies constitute the basic substantive areas of law. Here, *substantive* is used to distinguish from *procedural*. It does not imply "of more importance."

The law of contracts concerns agreements between two or more people (or organizations) to do or not to do a particular thing. Once an express or implied agreement is reached, obligations normally ensue flowing from each party to the other. Failure to meet an obligation may result in damages to the aggrieved party. The law of torts imposes obligations on persons as a matter of social policy. As with contracts, failure to meet an obligation may result in damages.

The word *tort* is a term applied to a miscellaneous and more or less unconnected group of civil wrongs, other than breach of contract, for which a court may afford a remedy in the form of an action for damages or other relief.[1] Which civil wrong will be classified as a tort for which damages may ensue is a matter of custom, previous legal precedent, and statutory law. Included as torts are assault, trespass, nuisance, negligence, defamation, misrepresentation, and several types of interference.

The degree of severity of a tort and the nature of the liability of a tort feasor is, again, a matter of custom, legal precedent, and statute. The severity of a tort ranges from intentional behavior (all crimes against the person are also torts), to willful and reckless conduct, and to negligence.

Negligence is further delineated as ordinary negligence (failure to act as a prudent person—here the plaintiff must prove negligence), negligence under the heading of *res ipsa loquitor* ("the thing speaks for itself"—here the defendant must prove he was not negligent), and negligence, *per se* (if the act was committed it is defined to be negligence by a specific statute). Certain types of conduct result in strict liability for which the concept of negligence or fault does not even enter. These types of conduct have been held to include keeping dangerous animals, the use of explosives and, generally, any "ultrahazardous activity."

The substantive law of torts plays a major role in the solution of environmental problems by legal process. The law of contracts does not. Neither does the law of property—except perhaps in the use of property,

Table 1. Typical Law School Curriculum

General nature of course	Course	Most relevant to environmental solutions
Introduction	Legal methods	
	Bibliography	
Basic substantive	Contracts	
law courses	Torts	Yes
	Property	
	Remedies	Yes
General law	Criminal law	
	Domestic relations	
	Wills	
	Trusts	
	Municipal law	Yes
Business law	Accounting	
	Taxation	
	Negotiable instruments	
	Business associations	Yes
Procedure	Code pleading	Yes
	Trial practice	
	Evidence	Yes
	Creditors' rights	
	Conflict of laws	
Origin of laws	Administrative law	Yes
	Legislative procedure	Yes
	Constitutional law	
Law practice	Moot court	
	Practice & Ethics	Yes

and here one gets more involved in other areas of law, for example, remedies.

The study of remedies derives from the common law concept of "equity" as opposed to "law," which in turn derives from the distinction between the King's Bench and the Courts of Chancery operated by the ecclesiastics in early England. The Anglo-Norman kings, beginning in 1066, developed a system of law which had become hardened into a rigid system (the King's Bench) by the middle of the thirteenth century. Only a very limited number of specific actions (writs) were permitted, and relief consisted only of money damages. Recognizing that money damages were sometimes insufficient, the Courts of Chancery began to grant injunctions (an order to do or not to do a specific act) and other relief.

The system gradually evolved into the Courts of Law and Courts of Equity that persisted until the nineteenth and twentieth centuries. New York first abolished the distinction by statute in 1848. In 1938 the U.S. Supreme Court, by Federal Rules of Civil Procedure, abolished the distinction between actions at law and suits in equity and fused their administration into one procedural system.[2] However, the language, forms of action, and relief tend to persist.

The study of remedies includes injunction, declaratory judgment, specific performance of a contract (as opposed to award of money damages), mandamus (a court order compelling a public official to perform his duty), and restitution. In granting an injunction the issues usually raised are adequacy of the remedy, relative hardship, and delay in bringing the suit. The defenses of mistake, misrepresentation, undue influence, and hardship are generally included in the study of remedies. The material included in the topic of remedies plays a major role in many environmental actions.

Criminal law concerns behavior that is considered to be so negative that its occurrence is detrimental to and punishable by society as a whole. Although an individual may affirm that a crime has occurred, only the public through its officers may initiate a criminal action in the courts. The common law of crimes has been abolished throughout the United States; for a behavior to be criminal, it must be declared to be so by specific statute. Very limited use has been made of criminal law in the area of environmental problems. Such matters generally arise in the way of fines against organizations for pollution resulting from violation of regulatory rules and statutes.

The law of trusts concerns situations whereby a person, as trustee, holds property according to the terms of a trust for the benefit of a beneficiary. Trusts have been used as a device whereby a large number of persons pool their resources to create a trust for the purpose of promoting some environmental cause.

The law of municipalities varies from state to state. Included are the rights to tax, to regulate. and to exercise police power. "Police power" includes much more than the patrolman on his beat. It includes providing utilities, sewage facilities, garbage collection and, in general, any matters relating to the public health, welfare, and safety not preempted by state or county governments. Municipal law obviously plays a major role in urban environmental problems.

Taxation by various taxing authorities has been proposed as a technique for regulating pollution and for transferring the cost of polluting to the polluter.[3] Many practical problems arise in attempting to do so, including the difficulty of equitably assessing the potential damage and the difficulty of obtaining the necessary legislation.

Business associations, especially coporations, have participated directly or indirectly in almost all environmental problems and actions. The reasons are several but do not necessarily relate to corporate law, as such.

In early England favorites of the crown were given licenses, often exclusive in nature, to engage collectively in some business operation. In the eighteen hundreds in the United States, this evolved into permission for groups of businessmen to collectively finance and operate large projects such as railroads. Included was the idea of liability limited to the amount of the capital devoted to the venture. The conventional business of the period was a sole proprietorship or partnership whereby the participants were all liable for the total debts of the business. Corporations were sometimes looked upon by the conservative businessman of that period as a questionable way of doing business. In the early nineteen hundreds, the states successively relaxed their laws governing incorporation. The corporation, as a common form of business operation, came into being. This permitted large accretions of capital, large industries, mass production; and along with the wealth and benefits, came mass pollution.

With the initial uncontrolled use of the corporate form of business came financial instability. This, at least to some extent, led to the crash of 1929. Changes in corporate and securities laws have attempted to reduce this financial instability. However, the corporate objective function remains—to maximize the equitable interest of the stockholders. The principal obligation of a board of directors, by statute and reinforced by court decisions, is to maximize profit and equitable growth of the corporate assets for the benefit of the stockholders. At least in theory, if a director considered public interests over the interests of the stockholders, he would be guilty of a breach of his office and, potentially, would be personally liable.

The Importance of Procedural Law

Statutory rules of pleading (code pleading) have replaced the common law types of pleading. Pleading here refers to the preparation and submission of written documents to a court in the process of instigating a legal action. It relates to who may successfully present what types of pleading to what courts and for what purpose. The purpose of pleading is to define precisely the parties to an action and the issues of law and fact that are to be resolved. Successful pleading for the plaintiff results in either bringing the case to trial, with the hope that the outcome will be favorable, or to reach a position that appears to the defendant to be so formidable that he settles out of court. Successful pleading for the defendant results in a decision by a judge that no justifiable action exists, whereby the case is dismissed before trial, or to reach such a formidable position that the plaintiff decides to drop the matter.

Terminology varies; the first pleading filed by the plaintiff is a *complaint* or *petition.* The defendant then files either:

1. a motion to dismiss,
2. a demurrer, which says in effect, "Even if everything the plaintiff alleges is true, it still does not amount to a valid cause of action,"
3. an answer which denies some or all of the complaint, or
4. an answer plus a cross complaint.

Depending on the facts, some of the parties (if there are more than two) may be dropped or additional parties may be joined. Also, the judge, type of court, or physical location of the action may change during the pleadings.

Discovery techniques will certainly be used by one or all parties including filing of interogatories (lists of questions which the opponent is obligated to answer under oath before trial) and depositions (examination and cross examination of witnesses under oath before trial). When and if the case comes to trial, both parties will have attempted to maximize their knowledge of the opponent's case. "Surprises" in the modern courtroom are very rare.

Code pleading moves directly into the topic of trial practice, which relates to the conduct of the trial itself. Trial practice involves selecting a jury, strategy in presentation of evidence to support allegations in the pleadings, and the conduct of the trial itself. Trial practice relies heavily on a knowledge of *the rules of evidence.*

The rules of evidence attempt to restrict the trial to those evidentiary matters needed to resolve the issues met in the pleadings. The rationale in the use of rules of evidence is that an orderly and expeditious trial, restricted to matters material and relevant to the issues, will optimize the concept "justice." However, the complexity of the common law and statutory rules of evidence—and their exceptions—is such that both parties have considerable room for maneuvering.

Trial practice generally includes an introduction to appellate procedures. Normally appeals may be made only on questions of law—not on questions of fact. In certain cases, however, an appeal is made *de novo* during which appeal all questions, including evidentiary matters, may be reexamined.

Conflict of laws concerns the situation where a question exists as to which law to apply during pleading or in trial. Potential conflicts exist in many cases as to whether federal or state law should apply and, if state law, the law of which state. Applications to environmental cases usually arise before the pleadings are filed or during the early stages of pleading when the parties are attempting to get the pleadings and trial heard in the jurisdiction with the most favorable rules.

The procedural aspects of law have great importance in the field of environmental law, as indeed they do in all types of legal practice. To say "For every wrong, there is a remedy" is, at least, misleading. Certain matters have been held by courts to be "not justiciable." A typical case might be disagreements between members of a religious sect over matters that could have only a theological resolution. Also, courts, relying on the doctrine of separation of powers between the legislative, executive, and judicial branches of government have declined jurisdiction over matters that should be left to the voters, for example, to resolve.

As for matters in which jurisdiction exists, a substantive legal right exists only to the extent that a procedural method exists for its implementation. If the procedural method is simple and routine, then substantive rights will be obtained easily and routinely. If the procedural method is complex and difficult to achieve, then substantive rights, however beautifully expressed, will tend to disappear.

The Origin of Rules and Laws

Applying effort relative to the origin of laws opens the possibility of adding to, deleting, or revising any or all of the previously discussed substantive or procedural matters.

The people vote for a constitution and for legislators. The legislators pass laws and sometimes, in doing so, create administrative agencies. The agencies make rules which, if promulgated properly, have the force and effect of law. The idealistic separation of governmental powers—legislative, executive, and judicial—is almost nonexistent in many of the administrative agencies that exist on the federal, state, and municipal levels. It has been stated that "the administrative agency is a distinct organ of government unlike any of the other three in that it refers exclusively to an agency which combines all three powers."[4]

Questions raised in connection with practice before administrative agencies include delegation of power (the National Industrial Recovery Act of 1933 was held unconstitutional on the basis of an improper delegation of power), rule-making power, power to investigate, the licensing function, jurisdiction, right to hearing and notice, enforcement of decisions, exhaustion of administrative remedies, right to judicial review, and the scope of judicial review. Acts such as the Federal Register Act[5] and the Administrative Procedure Act[6] were passed to correct early difficulties with the operation of federal administrative agencies.

Although administrative agencies were often used as mechanisms for reform consistent with the changing nature of society, their growth and power have sometimes resulted in rigidity and loss of procedural justice. In 1954, Jaffe said, "It does not follow from what has been said that today or forever the administrative process is the only instrument of law reform. . . . The judicial process has not stood still. It has reformed its procedure and reoriented many of its approaches to doctrine."[7] In some contemporary areas of environmental law, the agencies are the defendants and recourse is made to the courts and the pressure of public opinion[8] the influence the administrative agencies toward a more flexible position.

Legislative procedure on the federal level is well documented elsewhere.[9] Municipal procedures are too specialized to discuss in a general article. The legislative procedure for the State of California is outlined in Table 2 as an example of how to use the legislative process in seeking environmental legislation.[10]

Constitutional law begins by studying the wording of the actual constitutions and their amendments. The philosophy of the original writers is examined, and the history of court decisions, relative to social and economic conditions, is carefully analyzed. Constitutional law applies generally to whether actions of the executive, legislative, and sometimes judicial branches of government are appropriate in terms of the wording and/or intent of the applicable constitution.

Proper use of administrative law and procedures and of the legislative process are of major importance in environmental actions. Only in unusual, but perhaps significant, situations does the matter of constitutionality arise.

The Daily Exigencies of Law Practice

One of the greatest constraints on the use of the legal process in environmental matters has been the inability of attorneys to earn a living while specializing in the practice of environmental law. A number of law firms, acting in the public interest, are currently supporting young attorneys who engage solely in environmental areas. However this has an obvious limitation; members of the firm doing the more mundane but

Table 2. Use of the Legislative Process: Summary for California

---

Legislature convenes at beginning of January.

All bills are introduced during first ninety days of session—except emergencies by 2/3 vote and amendments. No carryovers from previous year.

Only member of legislature can introduce bill—persons seeking legislative change should establish a contact—consider a local author or co-author for a local problem.

Draft the bill—seek assistance from experts.

Keep track of introduced bills—very difficult due to the large quantity—seek help from groups like the Planning Conservation League.

Follow the bill—keep in touch with author and legislative publications—daily journal, weekly history, legislative index, etc.

Bill to committee—most important part of process—Speaker of Assembly or Senate Rules Committee can make or break—try to influence selection of committee—committee cannot act for thirty days.

Bill must go through a policy committee, a fiscal committee, and both houses.

Review membership of selected committees—try to influence favorably by letters, friends, groups, government agencies and reports, other legislators, and personal contacts.

Committee hearing—give author your support—make sure friendly members are in attendance—have people attend hearing to show interest.

Strategy—have amendments ready for author's use—if lack votes, put it over to work out amendments or for interim study rather than defeat.

Floor—three readings before vote—if in doubt, keep up campaign.

Other house—same procedure—amendments may result in selection of a special conference committee.

Passage—both houses plus Governor—keep up campaign—no veto overrides since 1946.

---

profitable work to support the environmental attorneys will eventually become unhappy and seek to stop the arrangement.

The use of trusts, as mentioned earlier, is a possible solution as are class actions (actions taken on behalf of a group of persons having a common complaint). Both of these have procedural disadvantages and also place the attorney in a situation bordering on advertising. Groups such as the Sierra Club offer a solution but with the difficulty of proving a "standing to sue." This is a procedural difficulty that asks whether the party involved has a claim or injury sufficient to justify *his* bringing the suit. An additional solution is to award court costs and reasonable attorney fees to a litigant who successfully prosecutes an environmental suit and thereby protects the public interest. Taxpayers' suits (a suit to prevent the improper expenditure of public funds) usually award costs and attorney's fees,[11] as do a number of other actions.

The following example shows how law is gradually changed and concludes this introduction to the legal process.

The tort of nuisance relative to environmental matters has an extended history. For example, in 1705 a defendant was found liable when his sewage percolated into the cellar of an adjoining house.[12] In 1899 pollution of a stream which inconvenienced downstream property owners was held to be a nuisance.[13] Both of these were private nuisances—a term applied to an unreasonable interference with the interest of an individual in the use or enjoyment of land.[14] When pollution killed fish, there was an interference with a public interest; it became a public nuisance; and it was prosecutable by the state.[15] Note that a private party may act on a public nuisance only if he suffers special damage different from the damage suffered by the public, in general. Also a nuisance requires a substantial interference; an occasional unpleasant odor, for example, is not sufficient.[16]

However, a tresspass may consist of a mere technical invasion and, furthermore, is occasionally the basis for punitive damages in addition to the actual damages incurred. The majority of decisions hold that a trespass has to be by a tactilely perceivable object—a rock, a person, a stream of water, etc. Recently, Oregon and California have recognized trespass by fumes, dust, and noise.[17,18] In a 1970 review article[19] Evans and Kratter presented arguments for the existence of "A New Tort: Mass Trespass by Air Pollution," in which they would permit use of a class action. Use of the trespass concept would avoid the need to show special damages and would permit the granting of punitive damages as a deterrent. A legislative approach[20] was presented by Senator Moscone of the California Senate who authored a bill to expand the definition of nuisance to include anything potentially injurious to health, or which constitutes air, water, or noise pollution. It would permit a private person to maintain an action for a public nuisance in certain instances and would permit the person to collect costs and attorney's fees, if successful. All three of the above approaches are probably indicative of a coming trend.

This discussion of nuisance and trespass illustrates the type of legal thinking and development that makes up much of the legal process. General rules of law are formulated from a series of court decisions based on isolated fact situations. A continuing series of legislative enactments parallel the court decisions. They are complemented by changes in governmental structure.

As the social scene changes further, continued application of the rules and laws may yield results that will again appear unattractive. In response to this stimulus, other law review articles will be written; judges will again experiment with departures from precedent; and legislators will again author new bills. This dynamic continuum—too fast for some—too slow for others—is discussed further in the second part of this paper where environmental law is viewed as part of a systems approach.

Part II. ENVIRONMENTAL LAW AS PART OF A SYSTEMS APPROACH

Operations research and systems analysis, like torts, are a more or less unconnected group of concepts. Which concept will be classified as a part of operations research or systems analysis, again, like torts, is a matter of precedent. However, the combined set of concepts that make up these analyses, unlike torts and the common law, have developed very recently. Almost all of the concepts have originated during and since World War II. There are very sparse references to use of the techniques based on these concepts prior to 1940.

The need to satisfy the operational problems of World War II led Great Britain and, soon afterwards, the United States into extensive *research* into the area of *operations* as a general concept. This included optimizing the allocation of resources, the analysis of cooperative and competitive strategies, and the analysis of technical and nontechnical systems. After the war some of the techniques were picked up by industry and government as a means for analyzing inventory problems, waiting line problems, assignment of people and resources, traffic scheduling, and even portfolio analysis.

In all cases the numerical procedures required extensive computations. The growth of the computer industry has been influenced very directly by financial and other gains resulting from the proper formulation and solution of operational research and systems analysis type problems.

In the late 1950's, a concept called *Program Evaluation and Review Technique* (PERT) was developed and implemented as a means for controlling a Navy contract to Lockheed to design and build the Polaris submarine and missile system. In general, defense contracts of the 1950's had already begun to treat the technical design phases of planned projects in terms of a system to achieve an objective. This is to be contrasted to a much earlier approach where in the case of an aircraft, for example, the fuselage would be procured by the government from one manufacturer almost as an isolated object, the engines from another, communication equipment from a third, etc. Efficiency of the overall system was rarely analyzed—at least in part because of the lack of mathematical techniques and computers.

With the advent of NASA and Apollo, the use of operations research and systems analysis reached a zenith. Without extensive use of these techniques, the operational problems in implementing the required resources and the problems associated with solving the enormously complex technical systems would never have been achieved—not in the required ten years or in a hundred.

One of the few significant benefits to mankind that can be attributed to the earth's collected military and space spending since 1940 are these techniques. However, there remains the task of using these techniques in the solution of the problems of people and their societies. One such application is the use of these techniques to combine legal and technical approaches in the solution of environmental problems. It is to that goal that this paper is addressed.

Application of Techniques to Environmental Law

At this point it will be useful to survey several of the important techniques of operations research and systems analysis and their potential application to environmental law. Although the concepts will be outlined for the nonmathematically oriented reader, it is impossible to completely avoid the language of mathematics.

An application to a simplified environmental problem might be the following description (model) for air *purity*. Let's say that air *pollution* depends only on automobile exhaust emissions which can be reduced by 1) smog control devices, and 2) use of rapid transit systems. The objective is to minimize pollution (i.e., maximize purity). In quantizing pollution we might use, for example, parts of certain pollutants per million parts of air (ppm), or even use some arbitrary but *quantifiably definable* scale of pollution. This could be a scale running from 0 to 1000, where 0 represents pure air and 1000 represents a definable upper limit to allowable pollution. Obviously, air purity would be just the reverse scale—when pollution is 1000, purity is 0, and when pollution is 0 purity is 1000.

Let's further stipulate limits on the amount politicians will require of their constituents to pay for smog control devices and as taxes for rapid transit systems. We will use in our model (description) a maximum cost for a smog control device of $100 ($x \leqslant 100$), and a maximum tax, over an equivalent period, for rapid transit of $60 ($y \leqslant 60$). To satisfy the opposition that both may be imposed, we can stipulate that $x + y \leqslant 120$ (i.e., the sum of the cost for a smog control device plus the tax for rapid transit shall not exceed $120).

317

Completion of the problem formulation requires a definable relationship between 1) the use of smog control devices and rapid transit, and 2) a reduction in pollution (increase in purity). Suppose that a technical analysis results in the conclusion that z, representing air purity, can be quantifiably related to dollars spent for smog control devices (x) and rapid transit (y) by the equation $z = 4x + 2y$ (the objective function), where z is air purity on the scale 0 to 1000. The complete analytical formulation of the model is shown in Figure 1a.

The problem resolves into the simple mathematical question, "What values of x and y should be chosen so that z is maximized without violating the restrictions imposed on x and y?"

A graphical formulation exists for this simple linear problem involving only two variables (x and y). It is shown in Figure 1b. The constraints show up on the graph as the straight lines labeled with the equations $x = 100$, $y = 60$, and $x + y = 120$. Any combination of values of x and y inside the shaded area are within the constraints placed on the money to be spent for smog control devices, rapid transit, or both. Figure 1c shows the solution using a graphical approach plus some logic. First, select an arbitrary point A within the shaded area to get $x = 40$ and $y = 40$ leading to a value of $z = 240$. The line in Figure 1c passing through point A represents the equation $240 = 4x + 2y$. It now is apparent that as x and y increase, z will increase but the slope of the line representing the equation $z = 4x + 2y$ will remain a constant. The maximum value of z is, therefore, the value of $z = 4x + 2y$ for the line that goes through the corner labeled B. Values of x and y for corner B are easily obtained as $x = 100$ and $y = 20$. Substituting those into the equation for z yields $z_{max} = 440$.

What is the useful conclusion? The conclusion is, if laws are passed with the given constraints, that the maximum air purity will be 440 on a scale of 0 to 1000 and that this purity will be obtained when $100 is spent on smog control devices and $20 is spent on rapid transit. For this model any other possible combination of expenditures (that meet the constraints) will result in a lower air purity. If this maximum air purity is still unacceptable then some change must be made in the constraints.

In this problem the constraints were built into laws or regulations governing pollution. Other constraints can exist because of physical laws or because of the expected social reaction to stimuli. In any event, the sample problem shown here is a very simple model (mathematical description) of a real world situation. A realistic model would contain many variables, not just two, and many constraints of differing kinds on the variables. A graphical solution would no longer be feasible, but an analytical technique, the simplex method,[21] is available for linear problems of any size. Use of a computer becomes essential.

This first and very basic technique of operations research is referred to

318

The objective function: $z = 4x + 2y$

the constraints: $x \leqslant 100$
$y \leqslant 60$
$x + y \leqslant 120$

plus the conditions: $x \geqslant 0$ and $y \geqslant 0$

Figure 1a Analytical Formulation of the Air Purity Model

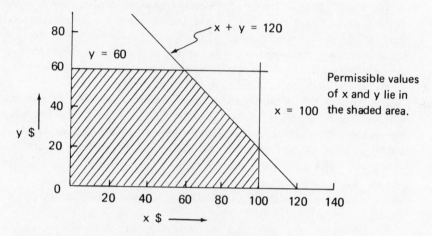

Figure 1b. Graphical Formulation of the Model

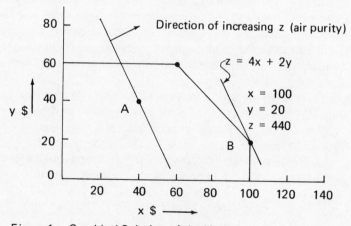

Figure 1c. Graphical Solution of the Model

319

as *linear programming* and has been used to illustrate an application whereby the impact of a law relating to the environment can be assessed. By predicting such an impact, the law could have been revised so that with proper tradeoffs the degree of air purity might have been much better with no additional increase in expenditures.

It should not be inferred that these techniques are not being used in environmental areas. They are, but only rarely do they include the legal process as a critical element.

Network Analysis

The next technique to be discussed is network analysis upon which PERT (Program Evaluation and Review Technique) is based. This has been used widely as a technique for planning large, complex operations, in scheduling time, costs, and other resources, in monitoring the results, in assessing tradeoffs, and in guiding the overall operation. An application to a legal-technical-environmental problem is illustrated next.

Los Angeles has an environment, very beautiful on several days of the year, very unacceptable on many days. The prevailing direction of the wind makes the difference—the other factors remain almost the same. These factors include all the environmental factors of population density, air pollutants, water pollutants, disposal of solid waste, use of land, etc. The factors also include the set of rules and laws that directly or indirectly relate to, or fail to relate to, the technical and physical characteristics of the environment.

In developing a master plan for improving the environment of Los Angeles, or California, or the United States, or the Earth, the total system must be considered. As with the aircraft example, where the engines affect the structure of the fuselage and the fuselage affects the performance of the engines, here the rules and laws affect the technical solutions and the technical solutions affect the rules and laws. Some more than others, admittedly, but all to so some degree interact. A PERT chart is one way of including legal matters and their interaction with other parts of the system into overall planning.

A sample PERT chart showing a very simplified plan for changing Los Angeles, say, from its present state to some more desirable environment is shown in Figure 2. The numbered circles represent events that can be precisely defined to have occurred or not to have occurred. The lines connecting the events represent the required effort to get from a prior event to a subsequent event. For example, the line connecting event 8 with event 12 stipulates that event 8 must occur (a condition precedent) before the effort required to get to event 12 can be initiated. The connecting line

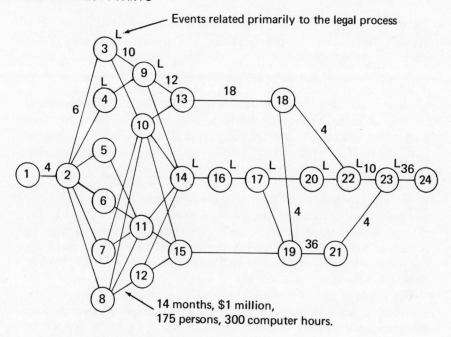

Figure 2. Simplified PERT Plan for Achieving a Desired Environment.

1. Initiate conference.
2. Set up study groups for events 3 through 8.
3. Complete study of deficiencies in agency structure.
4. Complete study of deficiencies in legal procedure.
5. Complete study of technical problems—air.
6. Complete study of technical problems—water.
7. Complete study of technical problems—solid waste.
8. Complete study of transportation problems.
9. Complete restructuring plan for regulatory agencies.
10. Complete land use plan.
11. Complete restructuring plan for monitoring technical systems.
12. Complete transportation plan.
13. Make land transfer offers for open spaces and transit system.
14. Formulate proposed laws.
15. Submit requests for proposals for rapid transit systems.
16. Complete public relations program for proposed laws.
17. Pass required laws.
18. Complete land transfers for open spaces and transit system.
19. Let contracts for rapid transit system.
20. Implement changes in agency structure.
21. Complete rapid transit system.
22. Promulgate agency rules relative to the environment.
23. Complete procedures for monitoring and regulating environment.
24. Achievement of desired environment.

can be used to represent the estimated time to complete effort 8-12 (14 months), the cost of the effort ($1 million), the persons required (175), etc.

A scheduling use is shown along the top line where the required time to complete the series of events 1-2-3-9-13-18-22-23-24 is estimated. Adding up these times gives a total of 4+6+10+12+18+4+10+36 = 100 months as the time to completion along *this* path. Along another path it might be greater. For example, instead of going from 18-22-23 a path exists going from 18 -19-21-23 which takes 30 additional months for completion. The *longest* such path from beginning to end is called the *critical* path and represents the *shortest* time in which the completion date (event 24) can be reached. To reduce the time to completion, one examines the events along the critical path and applies more resources in these areas at the expense of less critical areas. As events are completed, new estimates for remaining events are made with the result that new critical paths are discovered. Many texts discuss PERT techniques; the reader is referred elsewhere[22] for more detail.

The emphasis of this example is to show a technique for including the legal process and its interaction with technical matters into a systems plan. By following the circles with a superscript "L" in Figure 2 one sees the central path of the legal portion of an environmental plan. The interaction with nonlegal matters becomes apparent. The need to estimate the time required to formulate and pass the necessary laws—along with estimates of the required cost—is also apparent.

Development of a realistic PERT chart for restructuring the environment of a major urban area would lead to both discouraging and encouraging results. The discouraging result would be the realization of the time and effort that would be required to change an existing major situation. The encouraging result would be the realization that it actually could be done at all.

Game Theory

A third important technique that has relevance to the legal process and environmental problems is game theory. In 1928, John von Neumann did some early research in constructing mathematical models of poker and other games. It resulted ultimately in a corroboration with Oskar Morgenstern (1944) and the rigorous and now classic "Theory of Games and Economic Behavior."[23] A light and often amusing book for the nonmathematical reader is McDonald's "Strategy in Business, Poker, and War."[24]

Game theory presupposes a "game" with two or more players (investors, poker players, generals, litigants, etc.). The theory also includes the concept that the opponent is intelligent and rational and takes into account that the opponent will capitalize on your mistakes and indiscretions. In the very limited number of "games" that can be mathematically solved the conclusion has been that the optimum strategy to follow is to seek to minimize your losses rather than to maximize your gains. Attempting to maximize gain often results in a situation where the opponent can take a later action and ultimately inflict a serious loss. By minimizing loss you protect yourself from later actions of an opponent—the process of winning involves waiting for *him* to make a mistake. In the case of a "fair," two-player rational game in which one wins what the other loses (two person—zero sum game) the outcome is always a tie.

Games with more than two players and games where all can win or lose (i.e., the stockmarket) are difficult to handle with any present techniques. However the present techniques do enable one to systematically evaluate a complex situation in a rational manner. In the words of Rapoport, "As much as anything else, then, the achievement (of game theory) was in focusing attention on the nature of reasoning involved in the logic of events where conflict . . . of interest enters . . . perhaps for the first time the difficulties of reasoning about typically human affairs has been pointed out and made explicit."[25]

Cooperative games are even more difficult to formulate adequately than competitive games. In fact, in games with a large number of players, the concepts of cooperation and competition tend to merge and sometimes it is uncertain whether a cooperative or competitive situation exists. In the battle of "conservationalists vs. polluters" the sides sometimes become so intermingled one suspects that, like Walt Kelly's comic strip character Pogo said, "We have met the enemy and they are us."

Some potential applications of game theory, relevant to the legal process, are in the pleading and trial of a legal action and in the legislative process. Because of the complexity of even simple examples, the reader is referred to the current literature for numerical applications.

Systems Analysis

A fourth technique, sometimes referred to as "servo theory," "control theory," or, loosely, as "systems analysis," deals with the understanding and control of dynamic processes involving "feedback." Feedback is that characteristic of a system whereby the result (output) affects the cause (input). Such a system must be carefully balanced to avoid erratic system response.

In a physical system, too much feedback (overcontrol) results in oscillations of increasing amplitude which reach a steady state only if sufficient damping is present to absorb the excess energy. If too much damping exists the system fails to respond internally to any stimuli. The optimum physical system is one in which feedback and damping are balanced so that the system responds as quickly as possible without going into destructive oscillations.

The mathematical techniques of control theory are very well established and have become very sophisticated.[26] It is essentially impossible to mathematically formulate even a simple problem without the use of at least calculus. In applying the theory to the legal process, the difficulty is again in collecting the necessary data and mathematically formulating the problem—not in solving the problem once it is stated.

This technique has tremendous potential application in understanding and influencing the dynamic continuum of the legal process mentioned at the end of Part I of this paper. The block diagram of a simplified model is shown in Figure 3. Responses (eg., motion or energy for a technical system—information or social action for a nontechnical system) flow in the direction of the arrows. The rectangles denote the condition or state of

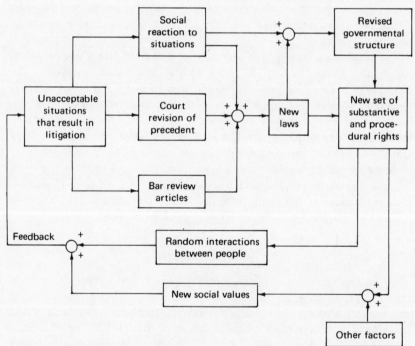

Figure 3. Simplified Block Diagram of the Dynamics of the Legal Process.

different portions of the system. The circles with the + symbols denote that the sums of the output of two or more portions of the system are computed in some definable way to produce the total response or input to the subsequent portion of the system. The essential characteristic showing that feedback exists is the closed loop structure of the block diagram. A system in which all responses flow in one direction is not a closed loop and does not have feedback.

Simulation

The last technique to be discussed is called *simulation*.[27] It is probably the most useful and most general of all the operations research and systems analysis concepts. As used today, simulation is a technique for reproducing inside a computer situations that occur in the real world. Changes are made inside the computer resulting in variations in the computed result. These results are used to infer that similar results would occur in the real world if the same variations were made. The utility of the results depends, of course, on the accuracy of the data.

Unlike the other techniques a formal mathematical model is not required. This says that one need not formulate the functional relationships of a problem. The solution is characteristically not only accurate but it also yields an estimate of what the accuracy is. The disadvantage is that while a person may learn the answer to a given problem he might never learn why he got that answer. However, one can vary the input data and observe changes in the output in order to get an idea of causative relationships. Included in the disadvantage is the large amount of computer time often required to analyze a problem fully.

Almost any kind of a problem, technical or nontechnical, can be solved. The essential ingredients are a large computer and imagination. As an example, suppose one had a wilderness area whose physical characteristics can be well defined—or if not precisely definable, then at least an estimate of the error can be made (i.e., 1000 trees plus or minus 25). Assume also that there exists data on the probability that a hiker will arrive on a given type day and data with an estimatable accuracy on what an average hiker will do during different parts of a day. This data is all put into the computer and the computer is started.

On the first day (in computer time) the computer "flips a coin" that has been "mathematically adjusted" to come up yes or no based on the probabilities of the input data. On the first day the coin may come up yes once and one person comes into the area. On the second day perhaps no one; on the third day maybe eight. Also, on some probability basis, the hikers leave. The computer keeps track of all people in the area and what they have done. The number of campfires is accumulated, the number of tin cans left in the area is accumulated, etc.

After a year has "elapsed" in the computer the impact of the hikers during that year can be assessed. But that is only one of all the infinite ways that that year could have happened. The computer records the results of that year, goes back to day one and starts all over again. Perhaps the second time four people arrive the first day and the whole series of events and their impact on the area are different. The computer records the results of the year that happened to occur that way. Many years are repeated in quick succession and the average of all the years is then taken. For the reader with a knowledge of probability or statistics it is apparent that enough data now exists to compute standard deviations (a measure of the accuracy) for the averaged results. The entire process can occur very quickly in a modern computer even for a very, very complicated situation. A years worth of events in the wilderness area could conceivably occur in seconds in real time. The whole computation could occur in minutes.

Suppose one wished to assess the impact of a regulation relative to campfires. The assessment can easily be made by including the regulation into the computer's data bank with the instruction that none of the hikers may violate it—or perhaps that a certain percentage will. The entire computation is then repeated; the expected impact of the hikers on the area is again summed up; and this result is compared with the earlier result without the regulation to assess the effectiveness of the regulation.

Because there are no equations to formulate or solve, the simulation may be made as complex and realistic and as accurate as desired. The limitation is only in collecting accurate and useful data and the cost of computer time.

The entire spectrum of legal-technical-environmental problems is susceptible to this technique. For example, one may assess:

1. Changing the structure of administrative agencies,
2. The effect of imposing an obligation on corporate directors to evaluate the impact of their operations on the environment.
3. Variations in court procedure,
4. The impact of awarding costs and fees in successful environmental actions, and
5. The effect in general of revisions in laws.

The real utility in all of the techniques of operations research and systems analysis lies in the ability of rationally quantizing and then mathematically manipulating quantities that to many persons would appear to be "nonmeasurable." Measurements may be based on physical laws or they may be based on an arbitrary but definable scale. Most physical dimensions in science originated as arbitrary but definable scales. Only after a phenomena was well understood were the arbitrary scales related to general physical laws.

Measurable may also be categorized as exact or approximate. An assumption of exact values and exact laws leads to a deterministic world. The real world is probabilistic and includes the fact that even physical laws do not always hold and that all data and measurements are inexact. A sophisticated analysis treats a problem in a probabilistic way and yields an expected answer plus some measure of its accuracy. For simplicity, the examples included above were given as deterministic models. However, in the usual solution to such problems the computation techniques used yield probabilistic results.

One can make an estimate of any physical, emotional, or intellectual concept with at least some estimatable degree of accuracy and therefore obtain a mathematical solution in a probabilistic sense. That the error in the answer might be so large as to limit its immediate usefulness does not argue against attempting the computation.

Conclusion

The paper has presented some initial formulations of models that combine the legal and technical aspects of solutions to environmental problems using the mathematical techniques of operations research and system analysis. The sample applications were kept very simple and were intended only as indicative of how operations research techniques might be used. The intended emphasis was to show that the sample applications could, with modest resources, be extended to realistic and useful predictive models.

Earlier in the paper a comparison was made between the length of time that operations research has been in existence and the length of time during which the common law has evolved. The one has a history of some 30 years, the other some ten centuries. The comparison is only relative. With respect to the time it takes a tree to fall both are very long. With respect to the lifetime of mankind both are very short. Before the common law there were some ten centuries of evolving Christian law. The Judaic law, back to the time of Moses, predates the Christian era by another twelve centuries. Before this (3000 years ago) the history of man becomes very uncertain. Prior to 5000 years ago the nature of human civilization is almost as much speculation as fact.

There is a tree living high in the White Mountains between California and Nevada that started life 5000 years ago. When Moses walked on Mt. Sinai, the tree was 2000 years old. Relative to the life of this tree the development of modern law, technology, and environmental problems are very very recent. Yet these factors have collectively and significantly deteriorated the environment on earth and have, in recent years, even caused the disappearance of several species of life form and endangered many others.

Long before the tree existed was the age of the dinosaurs. This prehistoric life form is said to have ended his transient stay on earth due to an inability to adapt to the changing needs of his existence. The age of the dinosaurs lasted some 140 million years. The age of man in his present human form appears to be on the order of 0.2 million years.

Informed persons have made estimates ranging from several centuries to as low as several decades for the continued existence of human society. Given the present state of the earth, the rate at which it is being degraded, and man's inertia for change, some of the estimates may be well founded. Systems techniques offer a method for understanding the complex interactions of society and, in conjunction with other knowledge, provide a useful approach for solving its problems.

REFERENCES

1. William L. Prosser, "Handbook of the Law of Torts," 2nd. Ed., West Publishing Co., 1955, p. 1.
2. Walter Wheeler Cook, "Cases and Materials on Equity," 4th. Ed. by M. T. Van Hecke, West Publishing Co., 1948, p. 11.
3. James E. Krier, "Air Pollution and Legal Institutions," Project Clean Air, Task Force Assessments, vol. 3, sec. 5, University of California, September, 1970.
4. Louis L. Jaffe, "Administrative Law, Cases and Materials," quoting James M. Landis, Prentice-Hall, New York, 1954, p. 1.
5. 49 Stat. 500 (1935), 44 U.S.C. § 301 (1940).
6. 60 Stat. 237 (1946), 5 U.S.C. § 1001 *et seq* (1946).
7. Louis L. Jaffe, above. p. 6.
8. Joseph L. Brecher and Manuel E. Nestle, "Environmental Law Handbook," California Continuing Education of the Bar series sponsored by The State Bar of California and University of California Extension, Berkeley, 1970, § 5.5.
9. William J. Keefe and Morris S. Ogul, "The American Legislative Process: Congress and the States," Prentice-Hall, 1964.
10. Nicholas C. Yost, "The Legislative Process," Environmental Law Institute, University of California, Los Angeles, March 6, 1971.
11. Ohio Revised Code § 309.13 (as an example).
12. Tenant vs. Goldwin, 1 Salk. 360 (1705).
13. Smith vs. City of Sedalia, 152 Mo. 283, 53 S.W. 907 (1899).
14. William L. Prosser, above, p. 389.
15. State *ex rel.* Wear vs. Springfield Gas & Electric Co., 204 S. W. 942 (1918).
16. Jones vs. Adler, 183 Ala. 435, 62 So. 777 (1913).
17. Roberts vs. Permanente Corporation, 188 Ca. App. 2nd 526, 10 Cal. Rptr. 519 (1961).

18. Davis vs. Georgia-Pacific Corp., 445 P. 2nd 481 (1968).
19. Marjorie W. Evans and L. M. Kratter, "A New Tort: Mass Trespass by Air Pollution," *California Trial Lawyers Journal,* IX(4), Fall 1970.
20. George R. Moscone (Senator), California Senate Bill 490 (1971).
21. G. Dantzig, "Linear Programming and Extensions," Princeton University Press, 1963.
22. R. W. Miller, "Schedule, Cost, and Profit Control with PERT," McGraw-Hill, 1963.
23. John von Neumann and Oskar Morgenstern, "Theory of Games and Economic Behavior," Princeton University Press, 1944.
24. John McDonald, "Strategy in Poker, Business, and War," Norton and Co., New York, 1950.
25. A. Rapoport, "Critiques of Game Theory," *Behavioral Science,* 4, 52, 1959.
26. Katsuhiko Ogata, "Modern Control Engineering," Prentice-Hall, 1970.
27. G. W. Morgenthaler, "The Theory and Application of Simulation in Operations Research," in "Progress in Operations Research," Vol. I, (R. L. Ackoff, ed.), John Wiley and Sons, 1961, pp. 363-419.

23

Reprinted from *IEEE Trans. Sys. Man Cybern.*, **SMC-2**, 319–331 (July 1972)

Engineering for Ecological, Sociological, and Economic Compatibility

HERMAN E. KOENIG, SENIOR MEMBER, IEEE, WILLIAM E. COOPER, AND JAMES M. FALVEY

Abstract—Industrial societies have evolved under the influence of a political–economic reward system and a technological revolution that literally views the environment as infinite in its waste absorbing capabilities, if not in resource producing capability as well. As such they are not ecologically feasible in the long term. It is precisely their inconsistencies with the laws of material and energy balance under the pressures of an expanding human population that are generating the environmental crisis. The thesis of this paper is that man, as a dominating species on the surface of the earth, must learn to engineer the developments of industry, agriculture, and human habitats, giving explicit consideration to the effects of these developments on the environment. Environmental components are viewed as conceptually similar to industrial production processes, having many possible alternative uses. Since alternative uses are often mutually exclusive, the choice of use is critical. Some of the ecological and sociological considerations implicit in the problems of choice are demonstrated herein along with some of the problems of economic regulation.

INTRODUCTION

MAN'S HISTORY is a series of self-inflicted divorces from his dependence on natural ecosystems to supply his material needs. During the agricultural revolution, man was divorced from a dependence on natural ecosystems for food. The industrial revolution, tapping ancient ecosystems as a source of energy, separated man from a dependence on living ecosystems for energy. Our growing independence from finite reserves of fossil fuels has already progressed through hydroelectric power toward harnessing nuclear fuels and less toward direct conversion from solar energy. The metallurgical and chemical revolutions have largely freed man from dependence on natural ecosystems for structured materials, and managed forests have accentuated this.

However, with this independence from natural ecosystems for food and energy has not come an independence from them for waste processing. Until now physical dilution capacity and "slack" in the natural ecosystems has been able to accommodate man with only localized changes. Cycling of materials on earth is probably limited by the more "steady-state" capacity of natural ecosystems to process the products of man and by the finite reserves of raw materials. Through sheer weight of numbers and his technological power, industrialized man is affecting most if not all of the components of the biosphere. He no longer merely

Manuscript received February 21, 1972. This work was supported by the Office of Interdisciplinary Research, NSF, under Grant GI-20.
H. E. Koenig is with the Department of Electrical Engineering and Systems Science, Michigan State University, East Lansing, Mich.
W. E. Cooper is with the Department of Zoology, Michigan State University, East Lansing, Mich.
J. M. Falvey is with the Department of Economics, Michigan State University, East Lansing, Mich.

responds to ecological patterns, but is instrumental in the processes that are reshaping them; if not for the present generation, then certainly for generations to come. The problem of determining what these patterns shall be and how to regulate them may be the most crucial test ever faced by industrialized society. It is not known to what degree independence from natural ecosystems for waste processing must be established. The degree to which we establish this independence will eventually depend to a great extent upon the confidence and values man places on natural ecosystems in relation to the ecological patterns he generates for himself.

The fundamental thesis of this paper is that as a dominating species on the surface of the earth, man must learn how to engineer[1] the developments in industry, agriculture, and human habitats as components of an industrialized ecosystem. In this greatest and most challenging of all engineering efforts we must be concerned with feasible alternative ecosystem goals and how to direct landscape development toward these goals, rather than with projections of present trends. To limit one's attention to the class of alternatives that can be investigated as perturbations or extrapolations in simulation models of existing systems restricts the alternatives to a prohibitively small class. Industrial societies evolved under the influence of a political–economic reward system and a technological revolution that literally viewed the environment as infinite in its waste processing capabilities, if not in resource producing capability as well. As such, they are not ecologically feasible in the long term. It is precisely their inconsistencies with the laws of material and energy balance under the pressures of an expanding human population that are generating the "environmental crisis."

In this paper the mass–energy features of the industrialized ecosystem are conceptualized as a system of material transformation, transportation, and storage processes driven by solar and synthetic forms of energy—literally a "material processing machine." In principle this material processing machine must obey the laws of material and energy balance, and as such it is subject to fundamentally the same rational design perspective that stands as the hallmark of engineering.

The class of ecosystem structures that satisfies the laws of material and energy balance over a given planning horizon are viewed as 1) a class C_t of ecologically feasible target structures (equilibrium states) within which tech-

[1] *Engineering:* A science by which the properties of matter and the sources of energy in nature are made useful to man. *Webster's Seventh New International Dictionary.*

nological and economic developments must be confined and 2) the class of pilot structures to which service industries, human habitat, and social activities adapt.

In principle, the class C_t of ecologically feasible structures will include some that are not economically and politically viable in the sense that they cannot be regulated by realizable social instruments. For example, some of these systems may be so critically structured that inherent delays in the political feedback system may lead to both technical and social instability. Thus, within the set C_t there is in principle a subset C_e that is economically and politically feasible. The size of the subset C_e depends directly upon the dynamic features of the economic and political regulating system.

The economic system is viewed as one of the major instruments that can be used to direct technological and economic development toward selected target states. But the economy as now structured ignores the limited capacity of local and regional environmental components to process wastes. Several alternative strategies for inserting forces into the economy to direct the trajectory of future economic growth toward ecologically feasible target states are analyzed from the point of view of their effectiveness and practical implementation.

The design perspective upon which this paper is based recognizes that within the class C_e of technically and politically feasible life support systems, some are socially acceptable, others are not. And within the class of socially acceptable structures some are more acceptable than others. As one progresses from the identification of ecologically feasible alternatives toward the identification of socially preferred structures, one moves from science and rational analysis toward art forms, subjective judgment, and social and individual human values. But the class of alternatives open to society should be no larger than the set of ecologically, economically, and politically feasible alternatives. In this sense it is through the design of the mass–energy features of the ecosystem and the associated social regulating instrument that potential for humanitarian values is either created or lost. Some of the important structural features of an industrialized ecosystem are discussed in light of their potential for improvements in environmental and social conditions.

Finally, it is suggested that social and humanistic value and long-term stability of the ecological patterns man sets for himself will depend critically upon the development of an ecosystem design and management science and technology. Some of the important components of this science are discussed in the light of both long-term and short-term objectives.

ECOLOGY OF AN INDUSTRIALIZED NATION

The macro features of the material processing system through which industrialized man relates to the physical resources and biotic structures of the aquatic and terrestrial regions of the earth is shown in Fig. 1. In concept, this system is divided into two major subsystems: 1) a material processing substructure and 2) a physical energy conversion and transmission substructure.

The biological sector represents the collection of all natural ecosystems on the landscape—the biological communities in aquatic and terrestrial regions resulting from ages of evolution without man's influence. These biological processes structure the elemental materials into biological tissue (fats, proteins, carbohydrates) with water as a primary vehicle of material transport. These structured compounds are the physical means for transmitting energy derived from solar radiation (in the form of structural bonding) through the natural biological system. Green plants, the primary producers, provide the bulk of the biological energy production in the form of plant tissue through the process of photosynthesis. Consumers convert energy held in biological tissue into dissipated energy. Chemicals released as elemental forms by the transformation are available for recycling within the biological sector.

Natural ecosystems have a limited capacity to adapt to new levels of chemicals introduced into these closed cycles. These limits depend both upon the genetic characteristics of the biological organisms and the structural characteristics of the communities of organisms as determined by the flow rates and concentrations of chemicals experienced within recent evolutionary history. When the chemical input rates and concentrations are changed, the ecosystems respond through adaptive processes at both levels. Responses of these types take varying lengths of time and are restricted by the richness of the genetic information available within the biological arrays. These changes are not always considered socially and economically desirable by man and are not always reversible. Changes of an irreversible nature are most often associated with extinction of species populations. Also, the state of the ecosystem may be driven beyond the region of stability of the desired equilibrium point into the region of stability of new and undesirable equilibrium points. Such is the ecologist's fear of changes occurring in the Great Lakes, particularly Lake Erie. It is in this sense that natural ecosystems have a limited capacity to process biodegradable wastes and still maintain their structural integrity.

The material transformation processes within the agricultural sectors are fundamentally biological, and at this level of detail are similar to the processes for the natural sector except in three important respects: 1) material inputs are required from the industrial sectors in the form of chemicals, fertilizers, pesticides and machines; 2) auxiliary man-made sources of energy are required to drive the machines, and human effort in the form of physical labor and management is required to culture the biological material transformation processes; and 3) in current agricultural practices recycling within the sector has been virtually eliminated.

The industrial sector represents the collection of physical, chemical, and technological processes developed by man for the restructuring of materials. The basic material inputs come either from the agricultural sector as biomass or from the raw physical reserves of ore. The material transformation processes are driven entirely by auxiliary man-made energy sources under human control. The materials

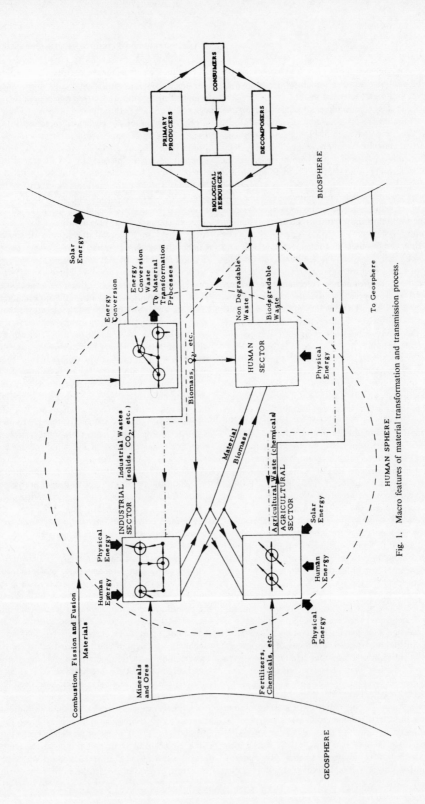

Fig. 1. Macro features of material transformation and transmission process.

332

are transported between sectors by energy-driven transportation systems. The outputs are restructured material products—chemical fertilizers, pesticides, and machinery—which serve as inputs to both the agricultural and human sectors.

The human sector refers to the material processing features of the aggregate human population. It receives material inputs from the agricultural and industrial sectors as food (structured biological products), shelter, transport devices, recreational devices, etc. The human sector depends on the natural environment for materials such as oxygen and water that are essential to maintaining the metabolic functions of life and for providing a wide variety of important aesthetic and recreational opportunities.

The energy transformation sector represents the set of all auxiliary physical energy sources, including the processes of mining, refining and conversion by combustion, fission or fusion (hopefully in the future), and transmission in solid, gaseous, liquid or electrical forms. Physical energy derived from this sector (measurable in megawatt hours) is used to supplement solar energy (measurable in hectare-days) and human energy (measurable in man-hours) in driving the material processing and transportation processes. Increasing the available energy will in itself only result in the augmentation of existing flow rates and concentrations of materials. It will not solve the existing environmental problems.

The wastes generated within the energy transformation sector and material processing substructure are regarded as "unwanted" materials. Some of these wastes (including organic materials, some inorganic materials, and limited classes of synthetic materials) are processable by the natural environment and therefore could actually be recycled at little energy cost to man. Other synthetic materials (persistent insecticides), compounds of heavy metals (lead, mercury, cadmium), and radioactive wastes are toxic to biological organisms, and therefore actually threaten the recycling capacity of the natural environment. For the nondegradable forms of materials the environment becomes a repository. From a material balance point of view the materials that cannot be transformed by the natural environment should be recycled within the man-made structure.

MASS–ENERGY CHARACTERISTICS

Each component within the industrial, agricultural, and human sectors of the ecosystem illustrated in Fig. 1 can be viewed as performing one or a combination of three basic functions: material transformation, transportation, and storage. Each of these functions is carried out at a cost to society in human energy e_1 (labor), land e_2 (solar energy), and physical energy e_3. They may be viewed at a variety of functional and spatial levels of organization depending upon the class of question and policy decisions under consideration. If, for example, the concern is with policy regarding the plant and human settlement size and geographic location in relationship to waste assimilation capacity of the local biosphere, then specific industrial plants and processes appear as functional objects. However, if regional environmental policies are of concern, specific industrial plants need not appear explicitly as functional objects—regional aggregates may be used.

The action of the natural environment can likewise be viewed as a complex of interconnected material processes having many properties analogous to the production processes carried out by man. It can be subdivided into three obvious classes of processes: terrestrial, aquatic, and atmospheric. These can be further subdivided and particularized to specific geographic regions to whatever degree of resolution is desired. The environmental components at any such level of detail can, in principle, be characterized in terms of their material and energy inputs and output rates and internal states. The internal states in general characterize the "quality" of the environmental unit as measured by such variables as the aesthetic characteristics of the biological species, stability of the community array, net production of desirable organic matter, and processing capability of natural material inputs. In general, these states will depend upon both the biophysical material processing rates and the physical storage. A stream, for example, may process both organic solids and dissolved inorganic material by the mechanical action of the stream and the activities of the biotic communities. In principle, it should be possible to determine how much material per day of known chemical composition the stream can process without changing its internal state beyond a preassigned tolerance limit, given its physical and ecological structure.

The industrialized ecosystem can be viewed literally as a material processing "machine" driven by solar, physical, and human forms of energy. The mass (material)–energy characteristics of the system are determined by the physical laws governing the mass–energy transformation and transport characteristics of the components and the laws of material and energy balance governing their interaction. The energy vector $x = (e_1, e_2, e_3)$ represents the cost of nonrenewable resources used to alter the spatial location and molecular, chemical, and physical forms of the renewable resources (materials)—renewable in the sense that they can, in principle, be recycled.

The factors ultimately limiting the volume of material return to society are the nonrenewable resources (energy) and the overall energy conversion efficiency of the ecosystem, including both the man-made and natural sectors. From this perspective the question is not whether to discharge materials and energy from man-made processes into the natural environment, but rather to design the technological and spatial features of the man-made processes to take advantage of the inherent mass–energy processing capacities of natural systems—the only component of the ecosystem involving no energy cost to man.

The inherent mass–energy processing capacities of natural systems are both material and regionally specific and they depend upon the "quality" of the environmental component to be maintained. A river in Minnesota, for example, and a seemingly similar river in Florida will have quite different capacities for material processing if for no other reason than their normal differences in water temperature. Discrete levels of quality can be based on additional uses to which the environmental component can be allocated. In principle, zoning procedures at the national, regional, and local levels can encompass spillover effects between regions and they can exclude inter-regional competition for economic expansion by imposing uniform environmental quality standards or they can provide for controlled variation in standards consistent with the total needs of the community. Use of the term "mass–energy capacity" of environmental components is always in this context.

The components of the industrial, agricultural, and energy generating sectors are certainly designed by man. But until this point in history they have not been designed as integral parts of the total ecosystem with any particular ecosystem design objective or mass–energy equilibrium state in mind. Indeed, the physical and technological designs, and the economic rewards that motivate them are based literally on a concept of the environment as an infinite source and as an infinite sink. The overall system that has thus evolved is basically a "once-through" system whose mass–energy equilibrium state is at best socially undesirable and at worst ecologically nonexistent. Since materials cannot be destroyed anywhere in the system, the amount of waste entering the natural biosphere must approximately equal the weight of basic fuels and raw materials extracted from the geosphere [2]. The material and regionally specific waste assimilation capacity of the environment must now be taken into consideration as specifications to which the man-made structures are designed. Serious environmental "hot spots" already exist on our landscape as irrevocable evidence of this need.

Within the general framework of existing technology, there are four basic and essentially independent avenues through which the mass–energy characteristics (the ecological characteristics) of our life-support system can be improved.

1) Spatial Distribution: Systematically distribute certain classes of man-made material processing activities and the human habitat over the physical dimension of the landscape according to the type and volume of wastes discharged and the potential assimilation capacity of the local environment.

2) Material Recycling: Systematically restructure the man-made sectors of the ecosystem so that it becomes more efficient in its use of renewable resources by recycling the bulk of the material within the man-made structures, the natural environment being used primarily for final low-grade cleanup processes (low contamination) that are technically difficult and expensive to handle by man-made processes.

3) Obsolescence Rate: Minimize the flow rates within the total ecosystem by systematically extending the expected life of the durable goods and products demanded by the human sector, such as dwellings, household equipment, automobiles, etc.

4) Environmental Design: Within the limits imposed by the existing array of biologic components (species populations) stable ecosystems can be designed to provide tradeoffs between optimal states, processing capability, and external maintenance cost to society. If properly designed, these seminatural ecosystems can be designed to fluctuate within

the region of stability required to ensure the desired system characteristics.

Avenue 1) clearly leads to maximum utilization of the natural environment for processing biodegradable wastes. Avenue 2) clearly leads to the preservation of scarce resources and minimizes the potential for international conflict arising from disputes over geological resources. A life-support machine based on recycling is the only structure that can operate over an indefinite period of time on the surface of a finite object. Avenue 3) leads to a reduction of the number of "service packages" that are processed in a given period of time, in both recycling and open-loop structures. In general, the extent to which improved performance can be achieved through this avenue depends on the obsolescence rate of the particular product or durable goods. Avenue 4) admits of the possibility of allocating given lakes and streams or terrestrial regions to waste processing and managing them so a; to maximize their capability to carry out this function. Design and management of multispecies communities is certainly a feasible goal of modern society.

Socio-Ecological Considerations

As important as the mass–energy characteristics of the life-support system are in an ecological and environmental sense, they are by no means sufficient. The physical features of the material processing system (type, size, and geographic location of the processing units) also have both a direct and indirect influence on the social amenities and political and economic stability of society. They represent the pilot structure to which human populations and associated service functions (including the vocational aspects of education, for example) must adapt and adjust. Since social institutions are inherently limited in their capabilities to adjust and adapt, rapid changes in the physical and technical structure of the material processing system may impose excessive stresses on society in a variety of social, political, and economic forms. The transient stresses of adjustment along with potential loss in long-term environmental quality, social amenities, and cultural stability are central to the problem of technological assessment as defined by Daddario [1]. Long-range planning efforts for industrial and technological developments must therefore also take into consideration the relative merits of alternative material processing structures against the social and cultural values of the society it is to service. The following examples illustrate some of these ecological and sociological considerations.

The development of a mechanized agriculture in conjunction with new genetic strains of domesticated organisms to be sure has provided critical increases in the yield per acre of our agricultural lands. In this sense the solar energy conversion of the photosynthetic processes has been significantly increased. The improved energy conversion efficiency, however, has required a disproportionate increase in the volume of chemicals that must be processed and broken down by the environment. In the United States, for example, the per-hectare rate of application of chemicals is approximately 11 times greater than in Africa with twice the yield. In Japan the application rate is ten times greater than in the United States with another two-fold increase in yield [3].

Up until the past decade or so, other components on the landscape had the capacity to dilute and to assimilate the pesticide and fertilizer runoff with no apparent environmental degradation. But it is clear that these chemicals along with other production processes are beginning to impose reductions in the real outputs of the natural environment and its aesthetic value. No one is yet able to evaluate costs (externalities) of other landscape components in producing the biomaterials that are important to the recreational, aesthetic, and metabolic needs and wants of man. In fact, it cannot be stated with assurance that the effects are always detrimental. Empirical evidence would suggest that limited levels of wastes of the proper form may actually increase the aesthetic value of certain types of landscape components. But there is reason to question how far present agricultural practices can or should be extrapolated in the United States for increased yields and to what extent they should be exported in their present form to developing nations.

The use of machines driven from auxiliary man-made energy sources has relieved back-breaking labor operations and improved the human labor efficiency of man in carrying out the cultural practices. The optimal size of these machines is specific to the size of the agricultural unit in acres, but they can and have been designed to carry out the cultural practices over a wide range of sizes of production units. In the interest of endlessly increasing the labor efficiency in carrying out the cultural operations, there has been a steady movement to larger and larger machines. If a small version of a particular machine is good, a larger version of the same machine has to be better! It is the same philosophy that has led to the supersonic transport. But since in agricultural processes the size of machines is specific to the size of land holdings on which they operate, the drive for labor efficiency through the use of larger machines has a particularly high social and environmental impact. Indeed, it has all but eliminated the "family farm" and with it the relatively small (and some not so small) urban villages and cities distributed rather uniformly over the landscape where their wastes were reasonably well matched to the carrying capacity of the landscape components. It has been said that the greatest migration in the history of man has been from the rural to urban United States.

The biomass resulting from the photosynthetic operations must also be gathered and geographically concentrated for human consumption. The urban sewage systems collect these biomass materials after they are processed by the human body and ultimately discharge them into the nearest lake or stream. From a material balance point of view this material should be redistributed over the thousands of acres from which it was collected.

Increased labor efficiency (but not necessarily photosynthesis efficiency) through the use of increasingly large machines also leads to increased specialization and spatial concentration of operations such as poultry, beef, and milk

production, and virtually eliminates crop rotation and diversification—all of which aggravate the material recycling and material distribution problems. The animal wastes from specialized high-density cattle feeding operations, for example, amount to approximately ten pounds for every pound of beef consumed by man. The potential magnitude of these wastes and their concentration levels compared to human wastes can be appreciated by noting that for each pound of beef consumed by the city dweller, ten pounds of material must be absorbed by the natural environment, much of which is concentrated by cattle populations as high as 100 cattle per acre. On one hand it is easy to see why the subject of solid animal waste disposal has become an important subject of applied research, but on the other hand, we must also in a more fundamental ecological sense question the concentrated feed operation as an acceptable alternative to the more diversified beef production practices. In the diversified production practice the potential runoff from inorganic fertilizer is considerably reduced since it need be used only as a supplement to animal waste. Agricultural practices developed before the advent of cheap inorganic fertilizers suggest that diversified operations and crop rotation can be used to further reduce both the fertilizer and the pesticide requirements. It is clear that adequate agricultural design coupled with biological control techniques, now in the very early stages of development, are a potential replacement or supplement to many of the broad-spectrum high-volume chemical applications.

The drive toward increased labor efficiency in agriculture through advances in machine size also loads the environment in another and more subtle way that is not always appreciated. It is evident from the demands of the city dwellers for lake cottages, camping, hunting, hiking, and other outdoor activities that an opportunity for social interaction with other forms of life is an integral part of man's needs and wants—a need that is at least as important as many other material goods and which seems to increase when his supply of material needs have reached a certain level. This need, which at one time apparently was satisfied to a great extent by the diversified farm operation (for the family and friends and relatives in nearby communities), must now be "mass produced" for the city dweller from natural resources that at one time were not used, but are now in short supply. However, on the other hand, an urban community of a critical size provides cultural and social amenities (human) that are not obtainable from the smaller rural communities. What, more specifically, are the trade-offs and how are they to be evaluated? To what extent is the "move to the city" attributed to social factors in contrast to economic conditions generated by economic policy and technological development?

Finally, still in reference to the agricultural sector, many of the environmental and social impacts occur in forms that are very difficult to identify and evaluate. To cite one of the more obvious, the plight of the migrant worker who must provide the seasonable labor to pick grapes, lettuce, and other vegetable crops on giant land holdings is both a social and economic tragedy. The land sizes of the holdings are structured to accommodate the largest machine designed to carry out the other cultural operations, but they are badly matched to the labor intensive operations. Would it be possible to obtain basically the same yield per acre in grapes and lettuce production, for example, from smaller units owned and operated by tightly knit family groups in which harvest is a festive, delightful time of the year? Certainly the machines can be and have been designed for operations on smaller units. How does one evaluate the optimum size land holdings for production processes such as these, where sequential operations range from essentially zero mechanization level to nearly 100-percent mechanization. From a standard economic point of view, where is the break-even point between the taxes and other forms of social costs to alleviate some of the misery of migrant workers on one hand and the additional cost of a glass of wine or a head of lettuce on the other that is attributable to the use of smaller machines for the mechanized operations?

In the material and energy transformation industries, distinction must also be made between the process technologies and the size of the machine designed to carry out these processes. Certainly no one will debate the value to society of knowing how to make fibers from coal and how to generate electrical energy through hydrogen fusion, for example. There is clearly a lower bound on the physical size of the machines or processing plants that can be designed to carry out these processes. The upper bound, however, is not so clear. Indeed, the costs in labor, land, and capital per unit of output decrease monotonically for rather extended ranges of production capacity (physical size) of the processing units. Transport energy costs (both physical and human) generally decrease with spatial proximity of one process to another. It is clear that machine implementations of material and energy processing technologies motivated by improvements in manpower, physical energy, and efficiency lead to geographic concentration of industrial processes with associated concentration of wastes. From a social and political point of view these economies of scale provide a form of interfirm competition that ultimately leads to monopolies.[2] For a constant or nearly constant output demand the competition is more in the nature of a baseball series. In many industries there is ultimately only one winner unless political action is taken to interrupt the "rules of the game."

The problems of waste concentrations associated with large-scale processes are compounded by the fact that the nonbiodegradable materials and toxic compounds are only transported by the environment. The processing rate through natural physical and chemical reactions may be very low. Consequently, the apparent capacity of the environment to tolerate these classes of wastes drastically decreases as toxic levels of concentrations are approached. The technical need for recycling of nonbiodegradable wastes in the material processing sector of the ecosystem

is self-evident. It is also clear that the labor and energy costs in recycling will in general decrease with increased concentration of processes.

Recycling in itself will not and perhaps should not promote a reduction in concentration. But 100-percent recycling of materials is not a realistic goal even for nondegradable materials. Consequently, the physical concentrations of these processes must eventually be regulated to match the limited tolerance capacity of the components of the landscape to absorb the "leakage." What this implies in regard to technically acceptable size and geographic concentration of material processing units relative to present concentration levels cannot be assessed at this point. Recycling may permit and actually p omote future increases in geographic concentration.

Future increases in geographic concentration of industrial processes to reduce recycling costs also implies corresponding increases in concentrations of manpower to develop and to operate the processes. The increase in human population is compounded by expansion in countless service industries (and disservice industries) that adapt directly or indirectly to serve the added manpower employed in the material processing system. Remember, in a specialized industrial society no man is anywhere near sufficient unto himself! Thus, a small percentage increase in concentration of industrially employed manpower will in general produce a greatly amplified increase in human wastes and wastes from associated service industries. What are the amplification factors relating population levels and service industries to the employment levels in the basic materials processing industries? Such factors are as critical to evaluating the technical admissable concentration levels of material processing operations as are the technical production coefficients (including externalities) of the industries. To what extent will recycling of industrial wastes really reduce total pollution levels of a given area, and to what extent will it simply shift the waste from industrial to human forms?

It is clear that in a closed recycling structure the waste discharge to the environment decreases with the decrease in the flow rate of material within the loop. Thus, from a material balance point of view and from the point of view of maximum total efficiency of the ecosystem, life expectancy of household appliances, automobiles, and other service products (machines) should be maximized. Since the technology exists now for doubling the life expectancy of American-made automobiles, for example, wastes generated from all processes associated with their production could be reduced to roughly one-half of its present level. But this would also imply a drastic reduction of the labor force in the automotive industry. Could we survive the economic shock? At what stage of economic development is such a shift necessary or desirable?

From an ecological point of view an automobile is a bundle of organized materials designed to provide man the services of transportation. The bundle of materials is renewed periodically. In concept, dwelling facilities in our urban communities are fundamentally no different, except that the biodegradable material content (wood and wood

---

[2] Many firms have attempted to hedge against this type of competition through diversification of processes under their jurisdiction.

fibers) is higher and the recycling time is longer. Since the biodegradable materials for the most part recycle through the natural environment, the environmental load decreases with increased life expectancy. Plans for urban development should not be considered complete without explicit plans for continuous and/or periodic urban renewal. Many of the proposed plans for low-cost housing should be looked at carefully in terms of these considerations.

An interesting political implication of the social impact of technology results when one views physical energy driven machines (structured materials) as stored forms of physical and human energy (man-hours of labor and inventive "genius"). This "stored" energy is available for potential use at the command of the user of the machine. From this point of view technology provides a mechanism for concentrating and storing as potential energy the efforts of many men in the hands of one or a relatively small number of individuals. This interpretation is particularly vivid in the case of a technological defense machine. The larger and more sophisticated the machine, the greater is the level of human effort available at the command of the user. The longer the technical life of the machine, the greater is the extent of time over which these human efforts are held in storage for use. What are the tradeoffs between production efficiency on one hand and political stability on the other?

In general it can be said that highly specialized ecosystems require sophisticated social instruments of control with a high degree of centralized and reliable decision making. In contrast, a highly diversified ecosystem with parallel or redundant mass–energy pathways may have a lower overall processing efficiency but will also require a less sophisticated centralized control. The current lack of job opportunities for highly specialized scientists and engineers illustrates both the profound difficulty of planning for the man-power requirements of a highly specialized industry with highly specialized educational programs, and the social costs that can be incurred. The problems (and results) faced by the U.S. Department of Agriculture in managing corn production in the midwest in 1970 and 1971 illustrate these same principles for a highly specialized agriculture [4]. These and other examples would suggest that with the power of modern technology, the industrialized nations might develop national and international ecosystems that are not technically, economically, and politically controllable. The time delays in economic and political feedback loops, for example, may be too long in comparison to the response time of the mass–energy system and the "noise level" may mask the essential "signal." Is it possible that technological specialization in certain areas should be bounded for reasons of controllability as well as mass–energy compatibility with the environment?

DESIGN–MANAGEMENT PERSPECTIVE

Ecosystem *design* is used here to refer to the procedures through which the technical, spatial, and temporal features of the material transformation, transportation, and storage processes in industry, agriculture, and human habitats are brought into long-term dynamic equilibrium with the counterpart processes of the natural environment as a closed industrialized ecosystem. Ecosystem *management* is used to refer to the procedures through which society selects, implements, and maintains the socio-ecological options implicit in the design process on preferential, humanitarian, and other normative scales.

It is the specific objective of ecosystem design to articulate the ecological options available at the various levels of technological, spatial, and social organization so that normative decisions are well informed. One of the specific objectives of the management system is to provide reliable and effective social instruments through which well-informed normative decisions can be negotiated and implemented at and between levels of organization.

The design and management processes are intimately related and enormously complex. They must provide for evolutionary development of the landscape, and they must account for geographic differences in natural environmental and geophysical features, temporal changes in technology, temporal and regional variations in social and cultural values, and many other socio-ecological factors. Within the historical frame of reference in which it was developed, the pricing system in our economy has served essentially as a "computing device" for allocating resources, expressing individual and group preferences—in general, managing many aspects of regional developments as an evolutionary process.

Referring to Fig. 1, the monetary cost $\$_i$ of a renewable resource $y_i(t)$ is among other things a scalar function of the energy cost vector $x_i = (e_1, e_2, e_3)_i$ required to put material y_i in its corresponding technological, spatial, and temporal state, i.e.,

$$\$_i = f_i(e_1, e_2, e_3, \cdots, \text{other factors}).$$

To this extent the monetary cost depends upon the overall energy efficiency of the ecosystem as determined by its technological structure. In principle, the weightings placed on e_1, e_2, e_3 by the function f_i depends upon the relative availability of the three forms of energy, the preference order of society for renewable resource y_i in relation to other renewable resources, and the monetary values society places on the mass–energy exchange rates of materials with the components of the natural environment.

The economic characteristics of the system are determined by the same basic laws of processing and interaction that determine the mass–energy characteristics, except that energy costs are measured in monetary value and aggregated with capital and other costs. Consequently, the economic equilibrium state of the ecosystem will not correspond to desirable mass–energy equilibrium states unless special efforts are made to structure the economy so that it will. Herein is the fundamental problem: the equilibrium states of a free economy designed to operate in a finite environment must correspond to mass–energy equilibrium states that are ecologically feasible.

To the extent that the mass–energy exchange rate with the natural environment is underpriced or that society has no way of expressing its valuations of environmental resources, the price system is transmitting incorrect information to the market economy as a control mechanism. Since common property resources of the environment, such as air and water, for example, are not bought and sold, there is no way to register the social preference for these commodities in relation to other material products. As long as these resources can be assumed infinite in supply, there is no need to register a preference order. But when the ecosystem production rates reach a level where the demand for material produced and processed by natural environmental components exceeds their capacity, preference orders must be expressed. A zero price implies that society is indifferent to whether or not it has clean air and water and it implies that all production processes and social uses carry equal priorities.

Underpriced environmental resources also tend to skew most industrial research and development efforts toward overuse of the natural environment. The technological processes of our present industrialized ecosystem have evolved with little emphasis on conserving common property resources. A firm selecting production procedures will choose from the set of all available technological alternatives those practices that conserve nonrenewable resources and costly materials at the expense of increased use of common property resources. Finally, it is a basic principle of economics that the use rate of products by society varies inversely with the price. Consequently, structured materials whose production involves heavy air and water pollution are underpriced in direct relation to the amount of pollution resulting from their manufacture—the demand for high pollution products is artificially high. The bias generated by a zero price in all such decisions is additive, if not compounding, in overall ecosystem effects.

These inconsistencies represent fundamental malfunctions in our present market economy as a control mechanism for an ecosystem developing and operating in a finite environment. If malfunctions cannot be or are not corrected by adjustments in the economy, then many fundamental and very powerful and pervasive economic forces must be overridden by other instruments of management such as authoritative planning, legislation, social resistance, ecological ethics, religious taboos, etc. Under any condition economic forces should complement rather than counter other forms of social control wherever possible.

All economic and other social instruments designed to regulate the technological development of the landscape toward ecologically feasible states require as a prerequisite knowledge of the material and energy processing capacities of the natural environmental components. As indicated in earlier sections of this paper, these capacities are based on ecological considerations and they vary from one type of natural landscape component to another, with class or type of material and the quality state to be maintained. Given these material processing capacities, there are basically two ways in which the mass–energy discharge rates from manmade processes can be regulated: 1) by planning and direct regulation of material rates and 2) by economic regulation. Where the mass–energy exchange rates with the natural

environment are well defined and undisputably associated with specific processes, as is the case with many industrial firms and municipalities, costs in the form of taxes or other pricing mechanisms might be used to adjust the economic equilibrium state to correspond to desirable ecological states. In the field of agriculture, urban habitats, and many other areas involving diffused or nonstationary processes, planning and regulation may be the only feasible approach. Although many, if not most, situations involve a combination of both forms of management, a brief perspective on each is perhaps helpful.

In the context of regional planning, the temporal scale can be conceptually broken into discrete planning intervals or horizons over which the design and management processes are considered to be essentially sequential. In this context it is possible to consider many aspects of the problem from a fairly conventional and well-established multilevel engineering design perspective. For example, the tradeoffs in the mass–energy and economic characteristics of alternate regional developments along with the implied environmental quality, distribution of labor, material transport requirements, diversification of labor skills and industrial activities, etc., can provide an effective information base for assessing the relative social value of alternative regional developments as targets. Clearly, the choice of ecological target states or classes of target states is a subjective judgment involving tradeoffs between quantitative measures and judgments of individual values and social utility to the socio-political areas affected by the regional target. The planner must rely on social utility indicators or social welfare indicators and functions to help provide a ranking of ecological feasible alternatives. One of the specific objectives of such a planning effort is to allocate the limited processing capacities of the environment to the competing material processing operations.

In the context of economic regulation there are a variety of pricing mechanisms that might be used. We shall consider three very briefly: tax, subsidy, and property rights.

Theoretically, there exists a tax rate which will equate the discharge rate to the ecological limits. A tax has some advantage over direct regulation: it motivates those with cheap alternatives to discharging to utilize those alternatives, and new firms and municipalities are on equal footing with existing entities—they would merely buy their way in by paying the per unit tax. The tax rate must increase, however, with the entry of new users to keep the total discharge within ecological limits. Uncertainty in the effect of tax rate on discharge rates makes it necessary to hedge on the conservative side, thus not utilizing fully the assimilative capacity allowable. From the point of view of the firm or municipality paying the tax, it introduces a new dimension of uncertainty in their planning and budgeting procedures —their tax bill would depend not only on how much they discharge, but also on what their neighbors discharge.

Subsidy is essentially the reverse of the taxing scheme— instead of taxing discharge activity, would-be dischargers are paid not to discharge. In practice, subsidies often take the form of contributions for specific waste processing equipment, but in theory they could be straight cash payments to "bribe" the processor to curtail his activities in whatever way he sees fit, so long as he decreases his discharge rate. Such a scheme has all the limitations of a tax plus problems of financing.

Although many environmental economists point to the existence of common property as the reason for pollution problems, Dales [6] is the only one that seriously puts forward an argument for breaking up that common property into private holdings as a solution to the problem. This approach depends critically on the concept of property rights. Before exploring this alternative this concept merits some amplification.

The term "property rights" refers to that set of legally sanctioned activities open to the property owner emanating solely from the condition of ownership. Although conceptually separable, the "property" and the "rights" are analytically indistinguishable. Ownership of the property implies ownership of the rights; a change in the property rights (sanctioning additional activities or withdrawing previous sanctions) changes the property itself, even though the property is unchanged physically. The basis of the rights, the property itself, may be something as concrete as a tract of land or as abstract as an idea (e.g., patents).

The essential element common to all private property is that the rights are solely dependent on ownership and not who owns it. It therefore represents what might be called a "store of value" to the individual property owner that can be traded. Other types of property and other types of rights undoubtedly have value, but their value is only in their use. They cannot be "cashed in" through sale. A license to practice medicine for example, is not a piece of private property, since the right to practice medicine cannot be transferred to a new owner. In contrast, a license to sell liquor is true private property in most states since the right to sell liquor is transferred from one owner to the other when the license is purchased. The distinction can be made another way. The store of value is essentially a stock concept. Conversely, use is a flow concept. Private property is viewed by the owner as both a stock and a flow—he has domain over both. The medical license is an example of common property having only the flow dimension to the individual user. The solitary fisherman among many that fish from a common property fishery looks only at the fish he catches (flow from the fishery). The state of the fishery (stock) is largely beyond his control.

Using a private property rights arrangement to "save" our environmental resources involves a major institutional change. Two comments are necessary at the outset: 1) a private property arrangement does not mean that common property such as a lake, stream, airshed, etc., is privately owned—only specifically designated mass–energy processing capacities, and 2) although the capacity is not in itself divisible the energy and material inputs utilizing that capacity are divisible.

For example, given that a stream has already been zoned and the ecological admissable mass–energy exchange rates established, the rates are subdivided into conveniently sized denominations. Discharge certificates are then issued and sold as private property. It is illegal under legislation for anyone to discharge material in excess of the rate stipulated by the certificates he holds. There are many possible ways to sell the discharge certificates the first time around—fixed fees, public auction, etc., but once issued and sold, a free market might be allowed to function essentially on its own between buyers and sellers of discharge rights.

Where implemental, such a system has the advantage over other forms of regulation in that it both sets the total use rate (by setting the number of discharge certificates) and allocates that total between competing users. Users are motivated to minimize the number of rights they have to hold by implementing recycling procedures and other alternatives. New firms and municipalities are required to buy (or rent) discharge certificates, just as they must purchase or rent land. The certificates could be purchased by an individual or group of individuals who wish to retain the quality of the stream above the level for which it was originally zoned. From the point of view of the firms and municipalities, future needs for discharge rates are secured by simply purchasing and holding requisite discharge certificates.

Enforcement is crucial to all forms of regulation. Effective and inexpensive methods for monitoring mass–energy discharge rates from man-made processes remains as an area for further technological development. Neither taxes, subsidies, nor regulatory planning will work without enforcement.

As a footnote to the design-management perspective presented here, it should be emphasized that the mass–energy exchange rates with the environment as presented here are based on ecological considerations (mass–energy characteristics of the ecosystem) rather than on estimates of the "cost of pollution." The problem of measuring the costs of pollution is at best a very difficult one. It has two aspects: 1) the problem of determining the actual physical, chemical, and biological effects of a given pollutant on the environmental component, and 2) the problem of putting a monetary figure on these effects. The first problem is a technical ecological question, related to the mass–energy characteristics of ecosystems and should be answerable through research. Conceptually, the social cost of pollution is merely the sum of the costs borne by the individuals affected. To go from this concept to an operational definition of individual costs is very difficult in principle, however.

Many of the effects of pollution on individuals concern things that are not bought and sold and consequently have no price. To the extent that pollution reduces the yield of marketable outputs of environmental components, part of the cost of pollution can be evaluated as a loss in real goods and services [5]. Thus, the cost borne by the Great Lakes commercial fishing industry attributable to the accumulation of DDT might be fairly accurately estimated.[3] Likewise, work days lost due to increased respiratory ailments

---

[3] It should be noted that this estimate is possible only because of a federal ban on the sale of fish containing high concentrations of DDT. Without the ban the DDT costs would presumably be borne by the consumer, which would be impossible to calculate. Some might even say these costs are not attributable to DDT but to the ban itself.

can be estimated as part of the cost of air pollution. In neither of these cases is the full social cost evaluated, only the effect on marketed items is captured.

Pollution may induce essentially equal physical effects on all members of a local population or it may affect individuals in varying degrees. But irrespective of the distribution of the observable physical and biological effects, one can safely say that the valuation of the pollution effects will vary markedly from one individual to another. The social cost of the inconvenience, discomfort, and frustrations associated with poor health, an abrasive environment, and reduced life expectancy brought on by pollution are not measurable within the context of the present economy. They are very personal and subjective values of the individuals that collectively make up a society. For marketplace items, and only for marketplace items, does the economic system serve as an instrument for expressing preference orders. Within the context of a given technology and distribution of wealth, the price of a marketable item serves as a measure of the relative value the individual associates with an item. Since common property resources are not bought and sold there is absolutely no way for the individual to register the relative economic value he places on these commodities in relation to other items. Since the social value is the sum of the values expressed by its individual members, there is as yet no conceptual or operational way for measuring the social value of clean air or the social cost of pollution.

REQUISITE SCIENTIFIC AND TECHNOLOGICAL DEVELOPMENTS

A comprehensive solution to the environmental problems of the industrialized nations requires the development and implementation of scientific, technological, and management capability in at least the following four major areas:

1) Mass–Energy Rate Capacities of Environmental Components: This is the capability to assess the material and energy processing capacities of major components of the natural environment as a function of their physical and ecological features and in relationship to projected changes in their quality as measured by their aesthetic, humanitarian, recreational, and health characteristics.

2) Design and Management of Multispecies Communities: This means the capability to manipulate the ecological communities of restricted portions of the natural environment to increase the mass–energy assimilation capacity and/or other designated characteristics.

3) Ecosystem Design: This requires the capability to structure the technological and spatial features of the industrial, agricultural, and urban sectors so as to retain ecological compatibility with the environment.

4) Social Instruments of Control: This refers to the capability to design and implement comprehensive economic, political, and other social instruments of control to manage technological and economic development to be ecologically compatible with the environment and socially desirable.

The capability to assess the mass–energy processing capacities of natural landscape components—lakes, streams, coastal regions, airshed, terrestrial regions, etc.—is an absolute prerequisite to all other areas. In spite of all the scientific advances, precious little is known about the dynamics of natural ecosystems from which the material and energy carrying capacities of landscape components can be evaluated in relation to their recreational, aesthetic, and other potential uses.

Complete knowledge of the effects of all possible inputs to the wide variety of natural landscape components is not likely to be available in the near future. Research required to provide reliable answers to these questions is sure to be extremely expensive, and a relatively long gestation period will be required before complete understanding is achieved. Consequently, decisions as to what are rational regional environmental standards for material and energy injection rates into the environmental components will have to be made in the context of incomplete information, possibly for several decades. This implies an element of risk, the magnitude of which depends upon the physical size and ecological uniqueness of the environmental components and the degree of uncertainty. The risks may be very high for major lakes and streams. The question as to who shall bear the risk represents a very fundamental social concern. If the specifications are too restrictive, society stands to lose major material benefits. If they are set too liberally, at best there is a risk of temporarily destroying specific ecological features, and at worst, permanently upsetting the biotic stability of geographic regions. In the absence of specifications, the present pricing mechanisms heavily bias the risk in the direction of destroying environmental components.

This situation underscores the need for both a short-term and long-term research strategy in this area—a short-term pragmatic approach to establish general guidelines (with large safety factors) and a long-term more basic approach to advance the scientific understanding of the dynamic response of biological communities to changes in their environment.

The capability to restructure ecological communities intelligently so as to alter their characteristics as multispecies cultures is closely related to and depends upon the understanding of the community dynamics. In addition, it requires the development of technologies for managing and maintaining these multispecies cultures as stable communities.

A capability to structure the technological and spatial features of industrial, agricultural, and urban communities to retain ecological compatibility with the environment as an industrialized ecosystem has already been defined as a problem in the design of energy driven material processing "machines" to operate within the constraints of the environment. In an operational sense, each component of the natural environment—each lake, stream, airshed, and terrestrial region—has a limited capacity for processing materials and energy, given that a minimum quality of the environmental component is to be preserved. The quality of the regional environment will depend in a very real sense on the extent to which man-made industrial and agricultural and urban processes can be designed or redesigned to operate in harmony with the natural regional

IEEE TRANSACTIONS ON SYSTEMS, MAN, AND CYBERNETICS, JULY 1972

environmental components as a stable closed life-support system. The structure and behavior of these life-support systems depend upon the tradeoffs society is willing to make between material products of the industrial and agricultural processes, the fruits of natural environmental components, the level of human population to be supported, and the level of material standard of living.

The tradeoffs society is willing to make between the material returns and the aesthetic and recreational features of the emerging ecosystem must be reflected in the material and energy exchange rates regulating agents assign to environmental components. In principle, social instruments of control and management must provide a mechanism for setting and modifying regionally specific standards on exchange rates. For any given set of processing rates assigned to the environmental components of a given region, there also exists the problem of allocating the exchange rate capacities to the alternative industrial, agricultural, and human habitat processes.

Social instruments must provide a mechanism for making these allocations and exchange rate assignments as a function of the dynamic changes in social needs. The economic system as a social instrument for making some of these tradeoffs has been briefly discussed. However, it is not likely that the economic system alone can be used to regulate all aspects. What is needed is a comprehensive combination of coordinated economic, political, and other social instruments.

The range of alternatives that are available, and hence the potential of the evolving ecosystem for satisfying human needs, will surely depend upon the extent to which individuals and corporate groups are free to exercise their creative talents and powers. Such talent and power should be bounded only by the class of mass–energy behaviors that are ecologically consistent with the environment.

Virtually no scientific base exists at present for dealing with the design of the ecosystem or its social control. Research strategy in these areas must therefore also contain a long-term effort directed at establishing a sound theoretical base and a short-term pragmatic approach to deal with urgent aspects.

Practical application of the design–management perspective to industrialized ecosystems requires that scientific and technological developments (and their associated implementations) in each of the four major areas previously defined must relate in a specific way to virtually all functional and spatial levels of organization in the ecosystem—the natural environment, the agricultural and industrial sectors, the human habitat, etc., at local, regional, national, and international levels.

Coordination between these research and implementation efforts depends critically upon the emergence of a logically consistent theoretical structure. This structure must provide a language of documentation and communication between the community of people who in one way or another are concerned with some aspects of the system. It must provide a framework for analysis and tradeoff studies at all functional and spatial levels of organization. A capability to move systematically from one level of organization to another in tradeoff evaluations is absolutely essential.

The technical and social tradeoffs, negotiations, and iterative decisions that take place within this array of activities are extremely complex. However, they are not unlike the complex array of computations and iterations that take place among the community of engineers responsible for the design of a great aircraft or the generation and distribution network that supplies society with reliable electrical energy. To ensure coordination and compatibility between the various functional and spatial levels of organization of the emerging system, all participants are required to demonstrate a degree of proficiency in essentially the same theoretical principles and computational procedures. So also in the design of the emerging industrialized ecosystem, the objective is not to simulate these complex processes of negotiations, iterative decisions, and technical and social tradeoffs, but to provide a network of theoretical and computational structures and social instruments through which these processes can operate effectively.

In this context, conceptual and computational models are tools for abstracting the essential features of the systems at a particular level of organization for the purpose of evaluating technical and management feasibility within the context of a logically consistent structure. The systems-theoretic model is induced by the translation of the system elements from the conceptual and qualitative realm into a well-defined mathematical space. Hence, the complete system model is isomorphic to the actual system in structure and behavior.

Most of the ideas out of which new and important system features originate are likely to come largely from a qualitative understanding of the structural features of system models. Numerical solutions of particular structures may be used to sharpen the intuition or improve the understanding of particular tradeoffs between alternative candidate structures.

Models of the mass–energy system are strictly tools for presenting expected performance characteristics to the social regulating process. Models of the social regulating process are tools for documenting the design of the social decision-making process as an orderly procedure and for evaluating response times, reliabilities, and other procedural characteristics. While a social regulating model may generate the expected behavior of a statistically large community of individuals, it does not model decision makers. Nor can a model make decisions.

Conclusions

The fundamental thesis of this paper is that as a technologically sophisticated and dominating species on the surface of the earth, man must learn how to engineer the developments in industry, agriculture, and human habitats as components of a life-support system (industrialized ecosystem) that obeys the laws of material and energy balance and operates within the limited mass–energy rate capacities of regionally specific natural environmental components.

331

In this context we are concerned with feasible alternative ecosystem goals and how to structure pricing mechanisms, regulating policies, and other social instruments to direct landscape development toward these goals, rather than with projections of present trends or minor perturbations in present trends.

Since it imputes zero monetary value to mass–energy exchange rates with the natural environment, our present market economy cannot in principle function as a control mechanism for managing the development and operation of our life-support system in the context of a limiting environment. To the extent that these fundamental malfunctions cannot be or are not corrected, very powerful and pervasive economic forces must be overridden by other instruments of social management.

The fundamental hypothesis is that the environmental and social amenities and cultural stability of society are directly or indirectly related to the structural features of the emerging life-support system through discernible parameters. On the basis of this hypothesis, solutions to many of the contemporary environmental and social problems are found in the rational design of the physical and technological features of the life-support system as a closed industrialized ecosystem. From this design perspective, equitability of resource utilization, distribution of benefits and risks between current versus future generations, maintenance of individual freedoms, and incentives and opportunities for achieving a quality life style remain as normative decisions within a set of goals that are ecologically feasible in a technical sense.

REFERENCES

[1] E. Q. Daddario, "Priorities for the future: guidelines for technology assessment," in *Britannica Yearbook of Science and the Future*, 1971.
[2] R. V. Ayres and A. V. Kneese, "Production, consumption, and externalities," *Amer. Econ. Rev.*, vol. 59, June 1969.
[3] *Man's Impact on the Global Environment—Assessments and Recommendation for Action.* Cambridge, Mass.: M.I.T. Press.
[4] "The farmer's bursting cornucopia," *Time*, p. 85, Sept. 27, 1971. "Growing unrest on the farm," *Time*, pp. 20–21, Nov. 22, 1971.
[5] R. A. Musgrave, "Cost benefit analysis and the theory of public finance," *J. Econ. Liter.*, vol. 7, 3, pp. 797–806, Sept. 1969.
[6] J. H. Dales, *Pollution, Property, and Prices.* Toronto, Ont., Canada: Univ. Toronto Press, 1968.

24

Optimal Pollution Regulation—
A Data-based Study

DONOVAN YOUNG

Georgia Institute of Technology

ABSTRACT

Practical political considerations impose severe constraints on policies for public regulation of the actions of air pollutors, water pollutors or others whose actions tend to impose external diseconomies on their surrounding communities.

Present regulatory practice consists almost entirely of the enforcement of effluent-limit regulations, but there has been increasing discussion of effluent fees, open bidding and other structures.

This research considers the question of whether there exists a feasible regulation structure which *dominates* present practice and all other feasible structures in the sense of providing lower or equal levels of pollution while achieving lower economic costs excluding costs due to pollution reception. Further, an attempt is made to *quantify* total benefits to a degree sufficient to ascertain whether a dominant structure is really a significant improvement over the structures it dominates.

Proposed structures which strongly violate American egalitarian notions, or which assume unprecedented regulatory power or scrutiny, were eliminated from consideration as infeasible.

The advantages of one regulatory structure over another depend critically upon factors which cannot be resolved without appeal to hard data. For example, an effluent-fee system has well-known theoretical advantages over an effluent-limit system, but one can easily construct reasonable situations where the advantages are arbitrarily small (e.g., shallow cost curves or cost curves similar from pollutor to pollutor) or where the theoretical development breaks down (e.g., non-convex cost curves). The specific goal of this project was to use real cost data from an actual group of pollutors to provide what amounts to a small set of simulation experiments in the operation of various control structures.

For the ten particulate sources studied, it was discovered that present practice is doing better than the *optimal* ton-per-year-per-pollutor limit, but far worse (5 times the pollution *and* 30 per cent more abatement cost) than a well-enforced (but not necessarily accurately-designed) dollar-per-ton-per-year effluent fee.

This paper gives the results of a data-based study directed by the author to test the assumptions behind theoretical arguments about the relative efficiencies of various pollution control strategies which can be undertaken by regulatory agencies. The economic theory is discussed in the author's dissertation (1970) and elsewhere, while complete details of the data collection and analysis are given in a thesis by David K. Kumpf (The University of Texas at Austin, August, 1972).

The experimental work described herein is exploratory rather than definitive, as it is based on a very small sample.

The reader is presumed familiar with the current discussions about the relative merits of various structures of public regulatory systems for the control of pollution -- the "carrot" (economic incentives) vs. the "stick" (direct regulation). Direct regulation has traditionally been used; surveys show that existing Federal, state and local policies for both air and water pollution abatement are based primarily on the prohibition of effluent rates in excess of specified effluent-limit rates, with the limit rates set by law or administrative regulation having the force of law. Kumpf classifies regulations into such devices as licenses, permits, compulsory standards, zoning and registration procedures. By contrast, economic incentives include direct payments, subsidies, tax exemptions, accelerated depreciation and fees for the discharge of specified amounts of specified pollutants (or taxes in lieu of fees). An accelerated amortization of non-productive pollution control equipment was included in the Federal Tax Reform Act of 1969, and other tax incentives such as higher investment tax credits and even a sulfur-oxide effluent fee are being discussed, but economic incentives are not significant in today's control picture.

There is a growing literature on the supposed superiority of effluent fees over other incentives and particularly over all types of direct regulation, but the

344

theoretical arguments are based on abstractions so extreme that further discussion is fruitless without appeal to experiment. The work reported here was a small experiment of the kind that should help to clarify matters. The concept is simple. Ten industrial sources of particulate emissions reported their most favorable combinations of equipment, operating policies, process changes, etc., to meet various emission levels, along with the associated costs. The costs were as of each pollutor's most recent major capital decision about particulate emissions (when all costs and opportunities were best known). Given these costs, and assuming rational self-interest decision making (an assumption confirmed, incidentally, by the actual past actions of the ten pollutors in response to past changes in pollution requirements), the resulting pollution levels and abatement costs were predicted for each of several regulatory strategies.

Overview of Results

Here are some facts which hold true for the group of pollutors studied. Generalizations and extensions, however inescapable they may seem, should be tested on a much larger sample in the future.

    1. The ten pollutors studied were rational decision makers.

Most discussion about control systems assumes that the pollutors act so as to maximize profits (minimize costs) on a discounted cash flow basis. These ten pollutors had demonstrably acted with the intent of minimizing costs in the past, including rational consideration of the time value of money in balancing investments against continuing annual expenditures, and their predicted future actions were of the same nature and intent.

    2. The cost and marginal cost of particulate abatement, as a
           function of the amount of abatement, varied widely from
           pollutor to pollutor.

Wide cost variations among pollutors cause rigid control structures to waste money. At the point in time considered, our ten pollutors could, for example, have operated at a total 40,000 ton/yr emission level with emission control costs of $420,000/yr if a control scheme were used which selected the lowest-cost abatements, whereas the best across-the-board rules that did not allow cost tradeoffs among pollutors gave a $1,000,000/yr cost at the same pollution level.

3. Pollution abatement costs were not convex for any pollutor.

All pollution abatement cost curves had the general character of Figure 1. A few of the actual curves are given on the following pages. The lack of convexity actually was found to *increase* the extent of possible cost-saving tradeoffs among pollutors, although it made the tradeoffs more difficult to find computationally.

4. Present practice is *better* than any uniform ton/yr limit for these pollutors.

Present practice in particulate control is almost universally based on regulations which specify an upper limit, say in units of ton/yr, which is the same for all sources, or for all sources in a given area or zone. But such limits are only part of a complex actual practice matrix in which pollutors hardest hit by rigid regulations *do* have some recourse and influence. These ten pollutors presently emit about 58,000 ton/yr of particulates at a total abatement cost which is millions of dollars per year less than any strict ton/yr limit could achieve.

5. A uniform *abatement efficiency* limit, strictly enforced, could have given results significantly better than present practice.

Removal-efficiency requirements have not been widely discussed in the literature. But if each of the ten pollutors studied had been required to operate at, for example, at least a 90% particulate removal efficiency based on his natural raw emission rate, then costs would have been $200,000/yr less than now while emissions would have been 40,000 ton/yr instead of 58,000 ton/yr.

6. An effluent-fee system, without an accurate fee, would have allowed huge savings. Further, no other scheme, even omniscient central control, could have done significantly better.

Figure 2 shows how various schemes fared.

346

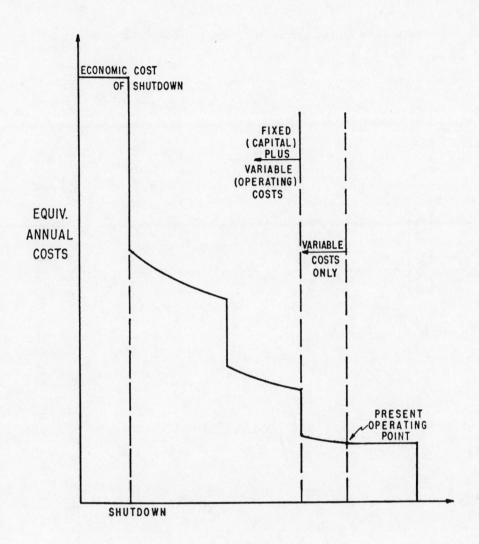

Figure 1: The General Form of Pollution Abatement Cost Curves

Addendum to Figure 1: One of the Actual Pollution Abatement Cost Curves

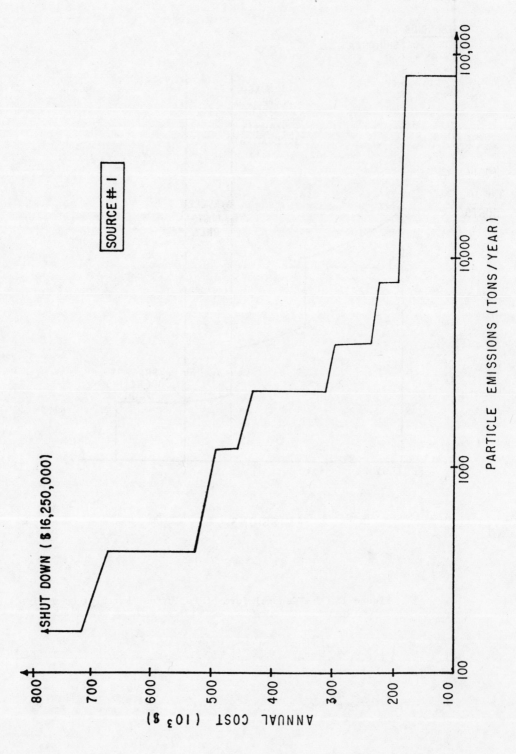

Addendum to Figure 1: One of the Actual Pollution Abatement Cost Curves

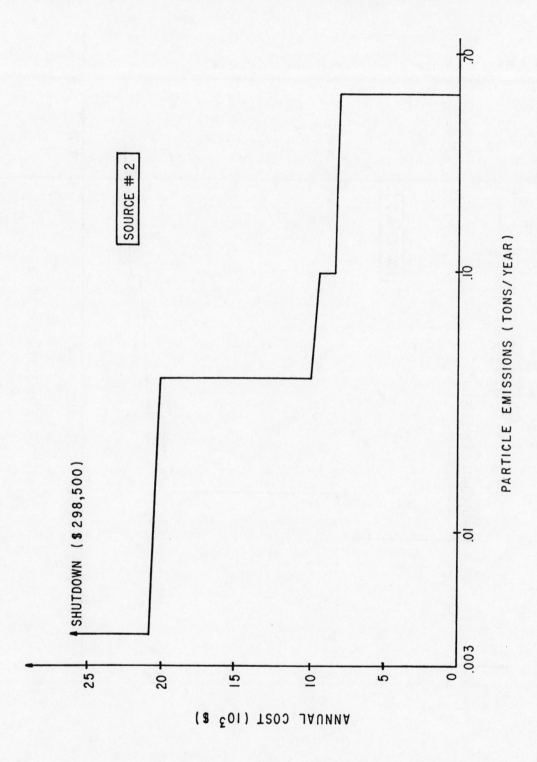

Addendum to Figure 1: One of the Actual Pollution Abatement Cost Curves

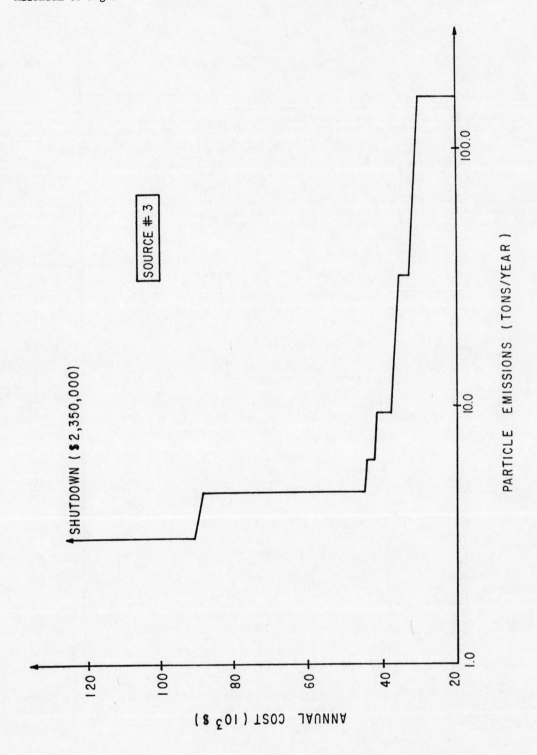

		OMNISCIENT CENTRAL CONTROL	EFFLUENT FEE CONTROL	EFFICIENCY LIMIT CONTROL	PRESENT CONTROL PRACTICE
LENIENT DAMAGE VALUATION	EMISSIONS, TON/YR	40,000	37,000	40,000	58,000
	CONTROL COST, $/YR	420,000	430,000	1,000,000	1,300,000
	PARAMETER*		$6/TON	96.75%	
MEDIUM DAMAGE VALUATION	EMISSIONS, TON/YR	9,500	9,200	23,000	58,000
	CONTROL COST, $/YR	1,100,000	1,200,000	1,600,000	1,300,000
	PARAMETER*		$51/TON	98.20%	
STRICT DAMAGE VALUATION	EMISSIONS, TON/YR	7,300	7,300	5,000	58,000
	CONTROL COST, $/YR	1,300,000	1,300,000	5,500,000	1,300,000
	PARAMETER*		$144/TON	98.60%	

*THE EMISSIONS AND CONTROL COSTS ARE CALCULATED USING THE VALUE OF THE CONTROL PARAMETER THAT MINIMIZES THE SUM OF THE DAMAGE AND CONTROL COST, GIVEN THE DAMAGE VALUATION CURVE ASSUMED.

Figure 2: Optimal Control Costs and Emission Levels for Various Control Strategies Under Various Damage Valuations

Abatement and Administration Costs

Excluding the damage caused by the pollution itself, the costs that vary with
levels of pollution and with various control strategies are

    abatement costs
        plant shutdown or relocation
        capital expenditures
        operation and maintenance

    administration costs
        scrutiny costs
            agency
            pollutor
        enforcement costs
        redistribution costs

The pollutor faced with a change in regulations makes decisions among alternatives in which he weighs differences in capital expenditures, differences in
continuing annual operation and maintenance costs, and possibly the costs of a
complete change such as relocation, shutdown or major process change. It is
these costs on which detailed data were gathered. A time value of money of 11
per cent per year was used to balance lump sums against annual costs, and the cost
of shutdown was taken as net profits plus a given fraction (.10) of gross re-
ceipts. Except for these two items (which were found to be better estimated
by policy capture techniques than by asking the pollutors), the abatement costs
were taken as given by the pollutors – subject of course to some "auditing" which
consisted mainly of checking equipment manufacturers' prices and cross-checking
installation, operation and maintenance costs. For these ten pollutors, abatement costs for all levels of pollution may be regarded as known to the pollutors
and known to us.

Administration cost differences are not possible to estimate as accurately, and
it was decided not to attempt to gather data, but rather to treat these costs
separately. The costs in Figure 2 do *not* include differences in administration
costs, and the savings of one control strategy over another are to be read as
sums from which any increased administration costs must be subtracted to calculate
net savings. It should be noted that all of the control strategies considered
feasible and hence included in the analysis required far less scrutiny than the
present administration of income tax regulations for the same pollutors, so that
scrutiny and enforcement costs are not unbounded.

Redistribution costs for control structures involving economic transfers are the largest unknown item. It may be feasible for effluent fees simply to go to public coffers, but most discussion in the literature assumes not. A discussion of this problem is given in Kumpf's thesis, Chapter 2.

Damage Costs and Dominant Strategies

Every control structure contains at least one parameter, the setting of which will determine the total particulate emissions of the ten pollutors. The various structures balance emissions among the various pollutors in different ways, but if the damage to receptors is considered to be a function of the total emissions, then abatement and administration costs *only* are changed from one control structure to another. Hence one way to compare structures is to compare costs at the same level of pollution. Alternatively, one could compare pollution levels at the same level of costs.

The overall research project, however, was concerned with finding and comparing *best* versions of each strategy, rather than with comparing strategies artificially adjusted for easy comparison. Hence a policy capture technique was used to establish an approximate upper and lower bound on the curves of damage costs vs. emission rates (see Figure 3), and the operating points chosen for comparison were those which minimized the sum of abatement costs and assumed damage costs. Damage Function #2 in Figure 3 may be interpreted as a "lenient" damage function in the sense that the pollutors, the control agencies and the public all act as if the harm caused by various pollution levels is at least as great as that given by reading the curve. Conversely, Damage Function #3 may be interpreted as a "strict" damage function -- pollution levels resulting from this function would be less than the public is willing to pay for (in increased prices of products manufactured by the pollutors, for example).

Analysis

A dynamic programming technique was used to find pollution levels and abatement costs. Each pollutor was assumed to minimize the sum of his abatement costs plus any additional costs imposed by the control structure, subject to limitations also imposed by the control structure and by his process. The control agency was assumed to set control parameters so as to minimize the sum of the damage cost plus the pollutors' abatement costs (administration costs were assumed not

to vary as the parameters of a given control structure were varied). The calculations were made for ten pollutors in the same situation as the actual pollutors
were *before* their last major capital expenditure decisions, because the data for
analyzing such decisions was better than the data for analyzing such decisions
made now. (If the decisions were made now, the sunk cost of the last major equipment change, the one most accurately known, would need to be ignored).

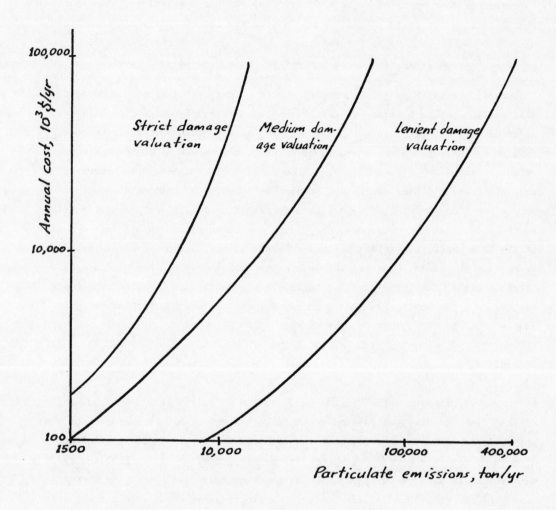

Figure 3: Assumed Pollution Damage Costs

354

Conclusions

Any conclusions must be preliminary, based on assuming the present results are not atypical.

For ten pollutors it has been found, regardless of what the desirable levels of pollution are seen to be, that control by *effluent fees* can save the pollutors an average of about $50,000 per year per pollutor in abatement costs, as compared to the best feasible direct regulation plan. This is nearly the highest theoretically-possible savings, so there appears to be little justification for considering more elaborate schemes.

Administering groups of pollutors as groups would have essentially the same results as effluent fees if the pollutors were allowed to establish among themselves the equivalent of an effluent fee plan. *Open bidding* is a version of effluent fee control, and would have essentially the same results as effluent fees. Its advocacy is based on the idea that bidding would be a better way to set and readjust the fees than the usual process of public hearings, court actions, etc. However...

The *amount* of an effluent fee tends to set the amount of pollution, but the sensitivity is far less (for these ten pollutors, at least) than has been assumed in the literature. Any fee from about $10 per ton to $90 per ton would set the pollution at about 10,000 tons per year, which seems to be approximately the "best" rate. (Present pollution for these ten is 58,000 tons per year).

It appears reasonable to assume that the savings of going to an effluent fee or similar system are substantial enough to cover the added administrative costs, if distribution of the fee monies can be handled in an acceptable manner.

BIBLIOGRAPHY

Aoki, Masahiko, "Marshallian External Economies and Optimal Tax-Subsidy Structures", _Econometrica 39_, No. 1, January 1971

Arrow, Kenneth J., "Criteria for Social Investment", _Water Resources Research 1_, No. 1, 1965, p1

Auld, D. A., _Economic Thinking and Pollution Problems_, Toronto, The University Press, 1971

Ayres, R. U., "Air Pollution in Cities", _Natural Resources Journal 9_, No. 1, 1971

Bohm, P., "Pollution, Purification and the Theory of External Effects", _Swedish Journal of Economics_, June 1970

Bonem, G. W., "On the Marginal Cost Pricing of Municipal Water", _Water Resources Research 4_, No. 1, February 1968

Boyd, J. Hayden, "Pollution Charges Income and the Costs of Water Quality Management", _Water Resources Research 7_, No. 4, August 1971, pp. 759-769

Breton, Albert A., "A Theory of the Demand for Public Goods", _Canadian Journal of Economics and Political Science 32_, November 1966

Brown, Gardner Jr., and McGuire, C. B., "A Socially Optimum Pricing Policy for a Public Water Agency", _Water Resources Research 3_, No. 1, p. 35

Brown, Gardner Jr. and Brian, Mary, "Dynamic Efficiency of Water Quality Standards or Charges", _Water Resources Research 4_, No. 6, December 1968, p. 1153

Buchanan, J. M. and Kafoglis, M., "A Note on Public Goods Supply", _American Economic Review 53_, June 1963

Buchanan, J. M., "External Diseconomies, Corrective Taxes and the Market Structure", American Economic Review (Communications), March 1969

Burrows, P., "On External Cost and the Visible Arm of the Law", _Oxford Economic Papers 22_, No. 1, March 1970, pp. 1-17

Ciriacy-Wantrup, S. V., "The Economics of Environmental Policy", _Land Economics_, February 1971

Crocker, T. D., "Some Economics of Air Pollution Control", _Natural Resources Journal 8_, No. 2, 1968, p. 236

Crocker, T. D., "On Air Pollution Control Instruments", to be published in _Social Science Quarterly_, 1972

Crocker, T. D., _Urban Air Pollution Damage Functions: Theory and Measurement_, U. S. Department of Commerce, 1970

Crocker, T. D., "The Structuring of Atmospheric Pollution Control Systems", in _The Economics of Air Pollution_, Harold Wolozin (ed.), W. W. Norton, New York, 1966

Currie, D. V., "Land and Water Resources Management for Environmental Improvement: A Comment", Water Resources Bulletin 6, No. 5, 1970

Dales, J. H., Pollution, Property and Prices, Toronto, University of Toronto Press, 1968

David, Elizabeth L., "Public Perceptions of Water Quality", Water Resources Research 7, No. 3, June 1971

Davis, O. A. and Whinston, A. B., "Some Notes on Equating Private and Social Cost", Southern Economic Journal 32, No. 2, October 1965

Eckstein, Otto, "Investment Criteria for Economic Development and the Theory of Intertemporal Welfare Economics", Quarterly Journal of Economics 5, February 1967

Fisher, Anthony C., "Population and Environmental Quality", Public Policy 19, No. 1, Winter 1971

Fredrickson, H. George, and Magnes, Howard, "Comparing Attitudes Toward Water Pollution in Syracuse", Water Resources Research 3, No. 1, p. 57

Gerhardt, Paul H., "Incentives to Air Pollution Control", in Law and Contemporary Problems, 33, No. 2, School of Law, Duke University, 1968

Ginsburg, A. L., "Public Provision, Jointness and Efficiency", Public Finance 4, 1970 (The Hague), p. 465

Goetz, Charles and Buchanan, James M., "External Diseconomies in Competitive Supply", American Economic Review 56, No. 5, December 1971, p. 883

Goldman, Marchall, Controlling Pollution - The Economics of A Cleaner America, Prentice-Hall, Inc., Englewood Cliffs, New Jersey, 1967

Grad, F. B., Rathgens, G. W. and Rosenthall, A. J., Environmental Controls: Priorities, Policies and the Law, Columbia University Press, New York, 1971

Gramm, W. P., "A Theoretical Note on the Capacity of the Market System to Abate Pollution", Land Economics 45, August 1969

Hagevik, George, "Legislating for Air Quality Management: Reducing Theory to Practice", in Law and Contemporary Problems 33, No. 2, School of Law, Duke University, 1968

Hags, Jerome C., "Optimal Taxing for the Abatement of Water Pollution", Water Resources Research 6, No. 2, April 1970

Kade, Gehard, "Introduction: The Economics of Pollution and the Interdisciplinary Approach to Environmental Planning", Public Policy, Winter 1971

Karber, James W., "Competition and the Regulatory Process", The Quarterly Review of Economics and Business 9, No. 3, Autumn 1968, pp. 57-64

357

Kapp, S. William, The Social Costs of Private Enterprise, Harvard University Press, 1950

Kneese, A. V. and Ayres, R. U., "Environmental Pollution", in Federal Programs for the Development of Human Resources, Subcommittee on Economic Progress of the Joint Economic Committee, U. S. Congress, Volume 2, GPO, 1968

Kneese, Allen V., "Air Pollution - General Background and Some Economic Aspects", in The Economics of Air Pollution, Harold Wolozin (ed.), W. W. Norton, New York, 1966

Kumpf, David K., "A Data-based Investigation of Effluent Fees for Air Pollution Control", unpublished master's thesis, The University of Texas at Austin, August 1972

Marchand, Maurice, "A Note on Optimal Tolls in an Imperfect Environment", Econometrica 36, No. 3,4, July-October 1968

Macauley, Hugh H., Use of Taxes, Subsidies and Regulations for Pollution Abatement, Report No. 16, Water Resources Research Institute, Clemson University, Clemson, S. C., June 1970

Meyer, Robert A., "Private Costs of Using Public Goods", Southern Economic Journal 37, April 1971

Michelson, I., and Tourin, B., "Comparative Method for Studying Costs of Air Pollution", Public Health Reports 81, No. 6, June 1966

Mills, Edwin S., "Economic Incentives in Air Pollution Control", in The Economics of Air Pollution, Harold Wolozin (ed.), W. W. Norton, New York, 1966

Mills, Edwin S., "Federal Fiscal Policy in Air Pollution Control", in Proceedings: The Third National Conference on Air Pollution, U. S. Department of Health, Education and Welfare, Washington, D. C., 1966

Mishan, E. J., The Costs of Economic Growth, Fredrick A. Praeger, New York, 1967

National Association of Manufacturers, "Pollution Abatement Incentives for Industry", (A Directory of Federal and State Tax Laws), New York, 1970

Nourse, H. O., "The Effects of Air Pollution on House Values", Land Economics, May 1967

Plott, C. R., "Externalities and Corrective Taxes", Economica 33, No. 129, February 1966

Pollack, Lawrence W., "Legal Boundaries of Air Pollution Control - State and Local Legislative Purpose and Techniques" in Law and Contemporary Problems 33, No. 2, School of Law, Duke University, 1968

Reiter, S., Sherman, G., "Allocating Indivisible Resources Affording External Economies or Diseconomies", International Economic Review 3, No. 1, 1962

Ridker, Ronald and Henning, J., "The Determinants of Residential Property Values with Special Reference to Air Pollution", Review of Economics and Statistics 49, No. 2, May 1967

Ridker, Ronald, Economic Costs of Air Pollution, Praeger, New York, 1967

Rogers, A. J. and Crocker, T. D., Environmental Economics, The Dryden Press, Inc., Hinsdale, Illinois, 1971

Schall, Lawrence, "Technological Externalities and Resource Allocation", Journal of Political Economy 79, No. 5, October 1971, p. 983

Scherer, Joseph, "Pollution and Environmental Control", Federal Reserve Bank of New York, Monthly Review 53, No. 6, p. 132

Sharp, C., "Congestion and Welfare - an Examination of the Case for a Congestion Tax", The Economic Journal 76, December 1966

Sobel, Matthew, "Water Quality Improvement Programming Problems", Water Resources Research 1, No. 4, 1965, p. 477

Solow, R. M., "The Economists Approach to Pollution and its Control", Science 163, 1971

Starr, Chauncey, "Social Benefit vs. Technological Risk", Science 165, September 19, 1969

Stiglitz, J. E., and Dasguptaz, P., "Differential Taxation, Public Goods and Economic Efficiency", Review of Economics Studies, Vol. 88, No. 117, April 1971

Thomas, Harold A., "The Animal Farm: A Mathematical Model for the Discussion of Social Standards for the Control of the Environment", Quarterly Journal of Economics 1963

Turrey, Ralph, "On Divergences Between Social Cost and Private Cost", Economica 30, August 1963

Upton, Charles, "Optimal Taxing of Water Pollution", Water Resources Research 4, No. 5, October 1968

Upton, Charles, "Application of User Charges to Water Quality Management", Water Resources Research 6, No. 3, June 1970

Weisbrod, B. A., "Collective Consumption Services of Individual Consumption Goods", Quarterly Journal of Economics, August 1965, p. 71

Wilson, Richard D. and Minnotte, David W., "A Cost-Benefit Approach to Air Pollution Control", Journal of the Air Pollution C A 19, No. 5, May 1969

Wolozin, Harold (ed.), The Economics of Air Pollution, W. W. Norton, New York, 1966

Young, Donovan, Benefit Profile Analysis in Environmental Decision Making, Doctoral thesis, The University of Texas at Austin, August 1970

25

Copyright © 1970 by the President and Fellows of Harvard College

Reprinted from *Rev. Econ. Statist.*, **52**, 262–271 (Aug. 1970)

ENVIRONMENTAL REPERCUSSIONS AND THE ECONOMIC STRUCTURE: AN INPUT-OUTPUT APPROACH

Wassily Leontief *

I

POLLUTION is a by-product of regular economic activities. In each of its many forms it is related in a measurable way to some particular consumption or production process: The quantity of carbon monoxide released in the air bears, for example, a definite relationship to the amount of fuel burned by various types of automotive engines; the discharge of polluted water into our streams and lakes is linked directly to the level of output of the steel, the paper, the textile and all the other water-using industries and its amount depends, in each instance, on the technological characteristics of the particular industry.

Input-output analysis describes and explains the level of output of each sector of a given national economy in terms of its relationships to the corresponding levels of activities in all the other sectors. In its more complicated multi-regional and dynamic versions the input-output approach permits us to explain the spatial distribution of output and consumption of various goods and services and of their growth or decline — as the case may be — over time.

Frequently unnoticed and too often disregarded, undesirable by-products (as well as certain valuable, but unpaid-for natural inputs) are linked directly to the network of physical relationships that govern the day-to-day operations of our economic system. The technical interdependence between the levels of desirable and undesirable outputs can be described in terms of structural coefficients similar to those used to trace the structural interdependence between all the regular branches of production and consumption. As a matter of fact, it can

be described and analyzed as an integral part of that network.

It is the purpose of this report first to explain how such "externalities" can be incorporated into the conventional input-output picture of a national economy and, second, to demonstrate that — once this has been done — conventional input-output computations can yield concrete replies to some of the fundamental factual questions that should be asked and answered before a practical solution can be found to problems raised by the undesirable environmental effects of modern technology and uncontrolled economic growth.

II

Proceeding on the assumption that the basic conceptual framework of a static input-output analysis is familiar to the reader, I will link up the following exposition to the numerical examples and elementary equations presented in chapter 7 of my book entitled *"Input Output Economics"* (Oxford University Press, N.Y. 1966).

Consider a simple economy consisting of two producing sectors, say, Agriculture and Manufacture, and Households. Each one of the two industries absorbs some of its annual output itself, supplies some to the other industry and delivers the rest to final consumers — in this case represented by the Households. These inter-sectoral flows can be conveniently entered in an input-output table. For example:

* This paper was presented in Tokyo, Japan, March 1970 at the International Symposium on Environmental Disruption in the Modern World held under the auspices of the International Social Science Council, Standing Committee on Environmental Disruption.

Peter Petri and Ed Wolff, both members of the research staff of the Harvard Economic Research Project, have programmed and carried out the computations described in this paper. For their invaluable assistance I owe my sincerest thanks.

TABLE 1. — INPUT-OUTPUT TABLE OF A NATIONAL ECONOMY (IN PHYSICAL UNITS)

From \ Into	Sector 1 Agriculture	Sector 2 Manufacture	Final Demand Households	Total Output
Sector 1 Agriculture	25	20	55	100 bushels of wheat
Sector 2 Manufacture	14	6	30	50 yards of cloth

The magnitude of the total outputs of the two industries and of the two different kinds of inputs absorbed in each of them depends on, (1) the amounts of agricultural and manufactured goods that had to be delivered to the final consumers, i.e., the Households and, (2) the input requirements of the two industries determined by their specific technological structures. In this particular instance Agriculture is assumed to require 0.25 ($= 25/100$) units of agricultural and 0.14 ($= 14/100$) units of manufactured inputs to produce a bushel of wheat, while the manufacturing sector needs 0.40 ($= 20/50$) units of agricultural and 0.12 ($= 6/50$) units of manufactured product to make a yard of cloth.

The "cooking recipes" of the two producing sectors can also be presented in a compact tabular form:

TABLE 2. — INPUT REQUIREMENTS PER UNIT OF OUTPUT

From \ Into	Sector 1 Agriculture	Sector 2 Manufacture
Sector 1 Agriculture	0.25	0.40
Sector 2 Manufacture	0.14	0.12

This is the "structural matrix" of the economy. The numbers entered in the first column are the technical input coefficients of the Agriculture sector and those shown in the second are the input coefficients of the Manufacture sector.

III

The technical coefficients determine how large the total annual outputs of agricultural and of manufactured goods must be if they are to satisfy not only the given direct demand (for each of the two kinds of goods) by the final users, i.e., the Households, but also the intermediate demand depending in its turn on the total level of output in each of the two productive sectors.

These somewhat circular relationships are described concisely by the following two equations:

$$X_1 - 0.25X_1 - 0.40X_2 = Y_1$$
$$X_2 - 0.12X_2 - 0.14X_1 = Y_2$$

or in a rearranged form,

$$0.75X_1 - 0.40X_2 = Y_1$$
$$-0.14X_1 + 0.88X_2 = Y_2 \qquad (1)$$

X_1 and X_2 represent the unknown total outputs of agricultural and manufactured commodities respectively; Y_1 and Y_2 the given amounts of agricultural and manufactured products to be delivered to the final consumers.

These two linear equations with two unknowns can obviously be solved, for X_1 and X_2 in terms of any given Y_1 and Y_2.

Their "general" solution can be written in form of the following two equations:

$$X_1 = 1.457Y_1 + 0.662Y_2$$
$$X_2 = 0.232Y_1 + 1.242Y_2. \qquad (2)$$

By inserting on the right-hand side the given magnitudes of Y_1 and Y_2 we can compute the magnitudes of X_1 and X_2. In the particular case described in table 1, $Y_1 = 50$ and $Y_2 = 30$. Performing the necessary multiplications and additions one finds the corresponding magnitudes of X_1 and X_2 to be, indeed, equal to the total outputs of agricultural (50 bushels) and manufactured (100 yards) goods, as shown in table 1.

The matrix, i.e., the square set table of numbers appearing on the right-hand side of (2),

$$\begin{bmatrix} 1.457 & 0.662 \\ 0.232 & 1.242 \end{bmatrix} \qquad (3)$$

is called the "inverse" of matrix,

$$\begin{bmatrix} 0.75 & -0.40 \\ -0.14 & 0.88 \end{bmatrix} \qquad (4)$$

describing the set constants appearing on the left-hand side of the original equations in (1).

Any change in the technology of either Manufacture or Agriculture, i.e., in any one of the four input coefficients entered in table 2, would entail a corresponding change in the structural matrix (4) and, consequently, of its inverse (3). Even if the final demand for agricultural (Y_1) and manufactured (Y_2) goods remained the same, their total outputs, X_1 and X_2, would have to change, if the balance between the total outputs and inputs of both kinds of goods were to be maintained. On the other hand, if the level of the final demands Y_1 and Y_2 had changed, but the technology remained the same, the corresponding changes in the total outputs X_1 and X_2 could be determined from the same general solution (2).

In dealing with real economic problems one takes, of course, into account simultaneously the effect both of technological changes and of anticipated shifts in the levels of final deliveries. The structural matrices used in such computations contain not two but several hundred sectors, but the analytical approach remains the same. In order to keep the following verbal argument and the numerical examples illustrating it quite simple, pollution produced directly by Households and other final users is not considered in it. A concise description of the way in which pollution generated by the final demand sectors can be introduced — along with pollution originating in the producing sectors — into the quantitative description and numerical solution of the input-output system is relegated to the Mathematical Appendix.

IV

As has been said before, pollution and other undesirable — or desirable — external effects of productive or consumptive activities should for all practical purposes be considered part of the economic system.

The quantitative dependence of each kind of external output (or input) on the level of one or more conventional economic activities to which it is known to be related must be described by an appropriate technical coefficient and all these coefficients have to be incorporated in the structural matrix of economy in question.

Let it be assumed, for example, that the technology employed by the Manufacture sector leads to a release into the air of 0.50 grams of a solid pollutant per yard of cloth produced by it, while agricultural technology adds 0.20 grams per unit (i.e., each bushel of wheat) of its total output.

Using $\bar{X}_3$ to represent the yet unknown total quantity of this external output, we can add to the two original equations of output system (1) a third,

$$
\begin{aligned}
0.75X_1 - 0.40X_2 &= Y_1 \\
-0.14X_1 + 0.88X_2 &= Y_2 \\
0.50X_1 + 0.20X_2 - \bar{X}_3 &= 0
\end{aligned}
\qquad (5)
$$

In the last equation the first term describes the amount of pollution produced by Manufacture as depending on that sector's total output, X_1, while the second represents, in the same way, the pollution originating in Agriculture as a function of X_2; the equation as a whole simply

states that X_3, i.e., the total amount of that particular type pollution generated by the economic system as a whole, equals the sum total of the amounts produced by all its separate sectors.

Given the final demands Y_1 and Y_2 for agricultural and manufactured products, this set of three equations can be solved not only for their total outputs X_1 and X_2 but also for the unknown total output $\bar{X}_3$ of the undesirable pollutant.

The coefficients of the left-hand side of augmented input-output system (5) form the matrix,

$$
\begin{Bmatrix}
0.75 & -0.40 & 0 \\
-0.14 & 0.88 & 0 \\
0.50 & 0.20 & -1
\end{Bmatrix}
\qquad (5a)
$$

A "general solution" of system (5) would in its form be similar to the general solution (2) of system (1); only it would consist of three rather than two equations and the "inverse" of the structural matrix (4) appearing on the right-hand side would have three rows and columns.

Instead of inverting the enlarged structural matrix one can obtain the same result in two steps. First, use the inverse (4) of the original smaller matrix to derive, from the two-equation system (2), the outputs of agricultural (X_1) and manufactured (X_2) goods required to satisfy any given combination of final demands Y_1 and Y_2. Second, determine the corresponding "output" of pollutants, i.e., $\bar{X}_3$, by entering the values of X_1 and X_2 thus obtained in the last equation of set (5).

Let $Y_1 = 55$ and $Y_2 = 30$; these are the levels of the final demand for agricultural and manufactured products as shown on the input-output table 1. Inserting these numbers on the right-hand side of (5), we find — using the general solution (2) of the first two equations — that $X_1 = 100$ and $X_2 = 50$. As should have been expected they are identical with the corresponding total output figures in table 1. Using the third equation in (5) we find, $X_3 = 60$. This is the total amount of the pollutant generated by both industries.

By performing a similar computation for $Y_1 = 55$ and $Y_2 = 0$ and then for $Y_1 = 0$ and $Y_2 = 30$, we could find out that 42.62 of these

60 grams of pollution are associated with agricultural and manufactured activities contributing directly and indirectly to the delivery to Households of 55 bushels of wheat, while the remaining 17.38 grams can be imputed to productive activities contributing directly and indirectly to final delivery of the 30 yards of cloth.

Had the final demand for cloth fallen from 30 yards to 15, the amount of pollution traceable in it would be reduced from 17.38 to 8.69 grams.

V

Before proceeding with further analytical exploration, it seems to be appropriate to introduce the pollution-flows explicitly in the original table 1:

TABLE 3. — INPUT-OUTPUT TABLE OF THE NATIONAL ECONOMY WITH POLLUTANTS INCLUDED (IN PHYSICAL UNITS)

Into From	Sector 1 Agriculture	Sector 2 Manufacture	Households	Total Output
Sector 1 Agriculture	25	20	55	100 bushels of wheat
Sector 2 Manufacture	14	6	30	50 yards of cloth
Sector 3 Air pollution	50	10		60 grams of pollutant

The entry at the bottom of final column in table 3 indicates that Agriculture produced 50 grams of pollutant and 0.50 grams per bushel of wheat. Multiplying the pollutant-output-coefficient of the manufacturing sector with its total output we find that it has contributed 10 to the grand total of 60 grams of pollution.

Conventional economic statistics concern themselves with production and consumption of goods and services that are supposed to have in our competitive private enterprise economy some positive market value. This explains why the production and consumption of DDT is, for example, entered in conventional input-output tables while the production and the consumption of carbon-monoxide generated by internal combustion engines is not. Since private and public bookkeeping, that constitutes the ultimate source of the most conventional economic statistics, does not concern itself with

such "non-market" transactions, their magnitude has to be estimated indirectly through detailed analysis of the underlying technical relationships.

Problems of costing and of pricing are bound, however, to arise as soon as we go beyond explaining and measuring pollution toward doing something about it.

VI

A conventional national or regional input-output table contains a "value-added" row. It shows, in dollar figures, the wages, depreciation charges, profits, taxes and other costs incurred by each producing sector in addition to payments for inputs purchased from other producing sectors. Most of that "value-added" represents the cost of labor, capital, and other so-called primary factors of production, and depends on the physical amounts of such inputs and their prices. The wage bill of an industry equals, for example, the total number of man-years times the wage rate per man-year.

In table 4 the original national input-output table is extended to include labor input or total employment row.

TABLE 4. — INPUT-OUTPUT TABLE WITH LABOR INPUTS INCLUDED (IN PHYSICAL AND IN MONEY UNITS)

Into From	Sector 1 Agriculture	Sector 2 Manufacture	Households	Total Output
Sector 1 Agriculture	25	20	55	100 bushels of wheat
Sector 2 Manufacture	14	6	30	50 yards of cloth
Labor inputs (value-added)	80 ($80)	180 ($180)		260 man-years ($260)

The "cooking recipes" as shown on table 2 can be accordingly extended to include the labor input coefficients of both industries expressed in man-hours as well as in money units.

In section III it was shown how the general solution of the original input-output system (2) can be used to determine the total outputs of agricultural and manufactured products (X_1 and X_2) required to satisfy any given combination of deliveries of these goods (Y_1 and Y_2) to final Households. The corresponding

TABLE 5. — INPUT REQUIREMENTS PER UNIT OF OUTPUT
(INCLUDING LABOR OR VALUE-ADDED)

From \ Into	Sector 1 Agriculture	Sector 2 Manufacture
Sector 1 Agriculture	0.25	0.40
Sector 2 Manufacture	0.14	0.12
Primary input-labor in man-hours (at $1 per hour)	0.80 ($0.80)	3.60 ($3.60)

total labor inputs can be derived by multiplying the appropriate labor coefficients (k_1 and k_2) with each sector's total output. The sum of both products yields the labor input L of the economy as a whole.

$$L = k_1 X_1 + k_2 X_2. \qquad (6)$$

Assuming a wage rate of $1 per hour we find (see table 5) the payment for primary inputs per unit of the total output to be $0.80 in Agriculture and $3.60 in Manufacture. That implies that the prices of one bushel of wheat (p_1) and of a yard of cloth (p_2) must be just high enough to permit Agriculture to yield a "value-added" of v_1 ($= 0.80$) and Manufacture v_2 ($= 3.60$) per unit of their respective outputs after having paid for all the other inputs specified by their respective "cooking recipes."

$$p_1 - 0.25p_1 - 0.14p_2 = v_1$$
$$p_2 - 0.12p_2 - 0.40p_1 = v_2$$

or in a rearranged form,

$$0.75p_1 - 0.14p_2 = v_1$$
$$-0.40p_1 + 0.88p_2 = v_2 \qquad (7)$$

The "general solution" of these two equations permitting to compute p_1 and p_2 from any given combination of values-added, v_1 and v_2 is,

$$p_1 = 1.457v_1 + 0.232v_2$$
$$p_2 = 0.662v_1 + 1.242v_2 \qquad (8)$$

with $v_1 = \$0.80$ and $v_2 = \$3.60$ we have, $p_1 = \$2.00$ and $p_2 = \$5.00$. Multiplying the physical quantities of wheat and cloth entered in the first and second rows of table 4 with appropriate prices, we can transform it into a familiar input-output table in which all transactions are shown in dollars.

VII

Within the framework of the open input-output system described above any reduction or increase in the output level of pollutants can be traced either to changes in the final demand for specific goods and services, changes in the technical structure of one or more sectors of the economy, or to some combination of the two.

The economist cannot devise new technology, but, as has been demonstrated above, he can explain or even anticipate the effect of any given technological change on the output of pollutants (as well as of all the other goods and services). He can determine the effects of such a change on sectoral, and, consequently, also the total demand for the "primary factor of production." With given "values-added" coefficients he can, moreover, estimate the effect of such a change on prices of various goods and services.

After the explanations given above, a single example should suffice to show how any of these questions can be formulated and answered in input-output terms.

Consider the simple two-sector economy whose original state and structure were described in tables 3, 4, 5 and 6. Assume that a

TABLE 6. — STRUCTURAL MATRIX OF A NATIONAL
ECONOMY WITH POLLUTION OUTPUT AND
ANTI-POLLUTION INPUT COEFFICIENTS INCLUDED

Inputs and Pollutants' Output \ Output Sectors	Sector 1 Agriculture	Sector 2 Manufacture	Elimination of Pollutant
Sector 1 Agriculture	0.25	0.40	0
Sector 2 Manufacture	0.14	0.12	0.20
Pollutant (output)	0.50	0.20	
Labor (value-added)	0.80 ($0.80)	3.60 ($3.60)	2.00 ($2.00)

process has been introduced permitting elimination (or prevention) of pollution and that the input requirements of that process amount to two man-years of labor (or $2.00 of value-added) and 0.20 yards of cloth per gram of pollutant prevented from being discharged — either by Agriculture or Manufacture — into the air.

Combined with the previously introduced sets of technical coefficients this additional information yields the following complex structural matrix of the national economy.

The input-output balance of the entire economy can be described by the following set of four equations:

$$0.75X_1 - 0.40X_2 = Y_1 \quad \text{(wheat)}$$
$$-0.14X_1 + 0.88X_2 - 0.20X_3 = Y_2 \quad \text{(cotton cloth)}$$
$$0.50X_1 + 0.20X_2 - X_3 = Y_3 \quad \text{(pollutant)}$$
$$-0.80X_1 - 3.60X_2 - 2.00X_3 + L = Y_4 \quad \text{(labor)}$$
(9)

Variables:

X_1 : total output of agricultural products
X_2 : total output of manufactured products
X_3 : total amount of eliminated pollutant
L : employment
Y_1 : final demand for agricultural products
Y_2 : final demand for manufactured products
Y_3 : total uneliminated amount of pollutant
Y_4 : total amount of labor employed by Household and other "final demand" sectors.[1]

Instead of describing complete elimination of all pollution, the third equation contains on its right-hand side Y_3, the amount of uneliminated pollutant. Unlike all other elements of the given vector of final deliveries it is not "demanded" but, rather, tolerated.[2]

The general solution of that system, for the unknown X's in terms of any given set of Y's is written out in full below

$$X_1 = 1.573Y_1 + 0.749Y_2 - 0.149Y_3 + 0.000Y_4 \quad \text{Agriculture}$$
$$X_2 = 0.449Y_1 + 1.404Y_2 - 0.280Y_3 + 0.000Y_4 \quad \text{Manufacture}$$
$$X_3 = 0.876Y_1 + 9.655Y_2 - 1.131Y_3 + 0.000Y_4 \quad \text{Pollutant}$$
$$L = 4.628Y_1 + 6.965Y_2 - 3.393Y_3 + 0.000Y_4 \quad \text{Labor}$$
(10)

The square set of coefficients (each multiplied with the appropriate Y) on the right-hand side of (10) is the inverse of the matrix of constants appearing on the left-hand side of (9). The

inversion was, of course, performed on a computer.

The first equation shows that each additional bushel of agricultural product delivered to final consumers (i.e., Households) would require (directly and indirectly) an increase of the total output of agricultural sector (X_1) by 1.573 bushels, while the final delivery of an additional yard of cloth would imply a rise of total agricultural outputs by 0.749 bushels.

The next term in the same equation measures the (direct and indirect) relationship between the total output of agricultural products (X_1) and the "delivery" to final users of Y_3 grams of uneliminated pollutants.

The constant -0.149 associated with it in this final equation indicates that a reduction in the total amount of pollutant delivered to final consumers by one gram would require an increase of agricultural output by 0.149 bushels.

Tracing down the column of coefficients associated with Y_3 in the second, third and fourth equations we can see what effect a reduction in the amount of pollutant delivered to the final users would have on the total output levels of all other industries. Manufacture would have to produce additional yards of cloth. Sector 3, the anti-pollution industry itself, would be required to eliminate 1.131 grams of pollutant to make possible the reduction of its final delivery by 1 gram, the reason for this being that economic activities required (directly and indirectly) for elimination of pollution do, in fact, generate some of it themselves.

The coefficients of the first two terms on the right-hand side of the third equation show how the level of operation of the anti-pollution industry (X_3) would have to vary with changes in the amounts of agricultural and manufactured goods purchased by final consumers, if the amount of uneliminated pollutant (Y_3) were kept constant. The last equation shows that the total, i.e., direct and indirect, labor input required to reduce Y_3 by 1 gram amounts to 3.393 man-years. This can be compared with 4.628 man-years required for delivery to the final users of an additional bushel of wheat and 6.965 man-years needed to let them have one more yard of cloth.

Starting with the assumption that Households, i.e., the final users, consume 55 bushels

[1] In all numerical examples presented in this paper Y_4 is assumed to be equal zero.
[2] In (6) that describes a system that generates pollution, but does not contain any activity combating it, the variable X_3 stands for the total amount of uneliminated pollution that is in system (8) represented by Y_3.

of wheat and 30 yards of cloth and also are ready to tolerate 30 grams of uneliminated pollution, the general solution (10) was used to determine the physical magnitudes of the intersectoral input-output flows shown in table 7.

The entries in the third row show that the agricultural and manufactured sectors generate 63.93 (= 52.25 + 11.68) grams of pollution of which 33.93 are eliminated by anti-industry pollution and the remaining 30 are delivered to Households.

VIII

The dollar figures entered in parentheses are based on prices the derivation of which is explained below.

The original equation, system (7), describing the price-cost relationships within the agricultural and manufacturing sectors has now to be expanded through inclusion of a third equation stating that the price of "eliminating one gram of pollution" (i.e., p_3) should be just high enough to cover — after payment for inputs purchased from other industries has been

met — the value-added, v_3, i.e., the payments to labor and other primary factors employed directly by the anti-pollution industry.

$$p_1 - 0.25p_1 - 0.14p_2 = v_1$$
$$p_2 - 0.12p_2 - 0.40p_1 \doteq v_2$$
$$p_3 \qquad - 0.20p_2 = v_3$$

or in rearranged form,

$$0.75p_1 - 0.14p_2 \qquad = v_1$$
$$-0.40p_1 + 0.88p_2 \qquad = v_2$$
$$\qquad - 0.20p_2 + p_3 = v_3. \qquad (11)$$

The general solution of these equations — analogous to (8) is

$$p_1 = 1.457v_1 + 0.232v_2$$
$$p_2 = 0.662v_1 + 1.242v_2$$
$$p_3 = 0.132v_1 + 0.248v_2 + v_3. \qquad (12)$$

Assuming as before, $v_1 = 0.80$, $v_2 = 3.60$ and $v_3 = 2.00$, we find,

$$p_1 = \$2.00$$
$$p_2 = \$5.00$$
$$p_3 = \$3.00$$

The price (= cost per unit) of eliminating pollution turns out to be $3.00 per gram. The prices of agricultural and manufactured products remain the same as they were before.

TABLE 7. — INPUT-OUTPUT TABLE OF THE NATIONAL ECONOMY
(SURPLUS POLLUTION IS ELIMINATED BY THE ANTI-POLLUTION INDUSTRY)

Inputs and Pollutants' Output	Sector 1 Agriculture	Sector 2 Manufacture	Anti-Pollution	Final Deliveries to Households	Totals
Sector I Agriculture (bushels)	26.12	23.37	0	55	104.50
	($52.24)	($46.74)		($110.00)	($208.99)
Sector 2 Manufacture (yards)	14.63	7.01	6.79	30	58.43
	($73.15)	($35.05)	($33.94)	($150.00)	($292.13)
Pollutant (grams)	52.25	11.68	−33.93	30 ($101.80 paid for elimination of 33.93 grams of pollutant)	
Labor (man-years)	83.60	210.34	67.86	0	361.80
	($83.60)	($210.34)	($67.86)		($361.80)
Column Totals	$208.99	$292.13	$101.80	$361.80	

$p_1 = \$2.00$, $p_2 = \$5.00$, $p_3 = \$3.00$, $p_k = \$1.00$ (wage rate).

Putting corresponding dollar values on all the physical transactions shown on the input-output table 7 we find that the labor employed by the three sectors add up to $361.80. The

wheat and cloth delivered to final consumers cost $260.00. The remaining $101.80 of the value-added earned by the Households will just suffice to pay the price, i.e., to defray the costs of eliminating 33.93 of the total of 63.93 grams of pollution generated by the system. These payments could be made directly or they might be collected in form of taxes imposed on the Households and used by the Government to cover the costs of the privately or publicly operated anti-pollution industry.

The price system would be different, if through voluntary action or to obey a special law, each industry undertook to eliminate, at its own expense, all or at least some specified fraction of the pollution generated by it. The added costs would, of course, be included in the price of its marketable product.

Let, for example, the agricultural and manufacturing sectors bear the costs of eliminating, say, 50 per cent of the pollution that, under prevailing technical conditions, would be generated by each one of them. They may either engage in anti-pollution operations on their own account or pay an appropriately prorated tax.

In either case the first two equations in (11) have to be modified by inclusion of additional terms: the outlay for eliminating 0.25 grams and 0.10 grams of pollutant per unit of agricultural and industrial output respectively.

$$0.75p_1 - 0.14p_2 - 0.25p_3 = v_3$$
$$-0.40p_1 + 0.88p_2 - 0.10p_3 = v_2$$
$$- 0.20p_2 + \quad p_3 = v_3. \quad (13)$$

The "inversion" of the modified matrix of structural coefficients appearing on the left-hand side yields the following general solution of the price system:

$$p_1 = 1.511v_1 + 0.334v_2 + 0.411v_3$$
$$p_2 = 0.703v_1 + 1.318v_2 + 0.308v_3$$
$$p_3 = 0.141v_2 + 0.264v_2 + 1.062v_3. \quad (14)$$

With "values-added" in all the three sectors remaining the same as they were before (i.e., $v_1 = \$.80$, $v_2 = \$3.60$, $v_3 = \$2.60$) these new sets of prices are as follows:

$$p_1 = \$3.234$$
$$p_2 = \$5.923$$
$$p_3 = \$3.185$$

While purchasing a bushel of wheat or a yard of cloth the purchaser now pays for elimination of some of the pollution generated in production of that good. The prices are now higher than they were before. From the point of view of Households, i.e., of the final consumer, the relationship between real costs and real benefits remain, nevertheless, the same; having paid for some anti-pollution activities indirectly he will have to spend less on them directly.

IX

The final table 8 shows the flows of goods and services between all the sectors of the national economy analyzed above. The structural characteristics of the system — presented in the form of a complete set of technical input-output coefficients — were assumed to be given; so was the vector of final demand, i.e., quantities of products of each industry delivered to Households (and other final users) as well as the uneliminated amount of pollutant that, for one reason or another, they are prepared to "tolerate." Each industry is assumed to be responsible for elimination of 50 per cent of pollution that would have been generated in the absence of such counter measures. The Households defray — directly or through tax contributions — the cost of reducing the net output of pollution still further to the amount that they do, in fact, accept.

On the basis of this structural information we can compute the outputs and the inputs of all sectors of the economy, including the anti-pollution industries, corresponding to any given "bill of final demand." With information on "value-added," i.e., the income paid out by each sector per unit of its total output, we can, furthermore, determine the prices of all outputs, the total income received by the final consumer and the breakdown of their total expenditures by types of goods consumed.

The 30 grams of pollutant entered in the "bill of final demand" are delivered free of charge. The $6.26 entered in the same box represent the costs of that part of anti-pollution activities that were covered by Households directly, rather than through payment of higher prices for agricultural and manufactured goods.

The input requirements of anti-pollution activities paid for by the agricultural and manufacturing sectors and all the other input requirements are shown separately and then combined in the total input columns. The figures entered in the pollution row show ac-

TABLE 8. — INPUT-OUTPUT TABLE OF A NATIONAL ECONOMY
WITH POLLUTION-RELATED ACTIVITIES PRESENTED SEPARATELY

	Agriculture			Manufacture			Anti-pollution	Final Deliveries to Households	National Totals
	Wheat	Anti-pollution	Total	Cloth	Anti-pollution	Total			
Agriculture	26.12 ($84.47)	0	26.12 ($84.47)	23.37 ($75.58)	0	23.37 ($75.58)	0	55 ($177.87)	105.50 ($337.96)
Manufacture	14.63 ($86.65)	5.23 ($30.98)	19.86 ($117.63)	7.01 ($41.52)	1.17 ($6.93)	8.18 ($48.45)	.39 ($2.33)	30 ($117.69)	58.43 ($346.07)
Pollutant	52.25	−26.13	26.12	11.69	−5.85	5.84	−1.97	30 ($6.26 paid for elimination of 1.97 grams of pollutant)	
Labor (value-added)	83.60 ($83.60)	52.26 ($52.26)	135.86 ($135.86)	210.34 ($210.34)	($11.70)	($222.04)	($3.93)		361.8 ($361.80)
Totals Costs	($254.72)	($83.24)	($337.96)	($327.44)	($18.63)	($346.07)	($6.26)	($361.80)	

$p_1 = \$3.23, \quad p_2 = \$5.92, \quad p_3 = \$3.19.$
$v_1 = \$0.80, \quad v_2 = \$3.60, \quad v_3 = \$2.00.$

cordingly the amount of pollution that would be generated by the principal production process, the amount eliminated (entered with a minus sign), and finally the amount actually released by the industry in question. The amount (1.97) eliminated by anti-pollution activities not controlled by other sectors is entered in a separate column that shows also the corresponding inputs.

From a purely formal point of view the only difference between table 8 and table 7 is that in the latter all input requirements of Agriculture and Manufacture and the amount of pollutant released by each of them are shown in a single column, while in the former the productive and anti-pollution activities are described also separately. If such subdivision proves to be impossible and if, furthermore, no separate anti-pollution industry can be identified, we have to rely on the still simpler analytical approach that led up to the construction of table 3.

X

Once appropriate sets of technical input and output coefficients have been compiled, generation and elimination of all the various kinds of pollutants can be analyzed as what they actually are — integral parts of the economic process.

Studies of regional and multi-regional systems, multi-sectoral projections of economic growth and, in particular, the effects of anticipated technological changes, as well as all other special types of input-output analysis can, thus, be extended so as to cover the production and elimination of pollution as well.

The compilation and organization of additional quantitative information required for such extension could be accelerated by systematic utilization of practical experience gained by public and private research organizations already actively engaged in compilation of various types of input-output tables.

MATHEMATICAL APPENDIX

Static-Open Input-Output System with Pollution-Related Activities Built In

Notation

Commodities and Services

$1, 2, 3, \ldots i \ldots j \ldots m, \quad m + 1, m + 2, \ldots g \ldots k \ldots n$
   useful goods   pollutants

Technical Coefficients

a_{ij} — input of good i per unit of output of good j (produced by sector j)

a_{ig} — input of good i per unit of eliminated pollutant g (eliminated by sector g)

a_{gi} — output of pollutant g per unit of output of good i (produced by sector i)

a_{gk} — output of pollutant g per unit of eliminated pollutant k (eliminated by sector k)

$r_{gi}. r_{gk}$ — proportion of pollutant g generated by industry i or k eliminated at the expense of that industry.

Variables

x_i — total output of good i

x_g — total amount of pollutant g eliminated

y_i — final delivery of good i (to Households)

y_g — final delivery of pollutant g (to Households)

p_i — price of good

p_g — the "price" of eliminating one unit of pollutant g

v_i — "value-added" in industry i per unit of good i produced by it

v_g — "value-added" in anti-pollution sector g per unit of pollutant g eliminated by it.

Vectors and Matrices

$A_{11} = [a_{ij}]$ $i, j = 1, 2, 3, \ldots, m$

$\left.\begin{array}{l} A_{21} = [a_{gi}] \\ A_{12} = [a_{ig}] \end{array}\right\}$ $\begin{array}{l} i = 1, 2, 3, \ldots m \\ g = m+1, \ m+2, \ m+3, \ldots, n \end{array}$

$A_{22} = [a_{gk}]$ $g, k = m+1, \ m+2, \ m+3, \ldots, n$

$Q_{21} = [q_{gi}]$ $\begin{array}{l} i = 1, 2, \ldots m \\ g = m+1, \ m+2, \ldots n \end{array}$

$Q_{22} = [q_{gk}]$ $g, k = m+1, \ m+2, \ldots, n$

where $q_{gi} = r_{gi} a_{gi}$

$\qquad q_{gk} = r_{gk} a_{gk}$

$$X_1 = \left\{\begin{array}{c} x_1 \\ x_2 \\ \cdot \\ \cdot \\ x_m \end{array}\right\} \quad Y_1 = \left\{\begin{array}{c} y_1 \\ y_2 \\ \cdot \\ \cdot \\ y_m \end{array}\right\} \quad V_1 = \left\{\begin{array}{c} v_1 \\ v_2 \\ \cdot \\ \cdot \\ v_m \end{array}\right\}$$

$$X_2 = \left\{\begin{array}{c} x_{m+1} \\ x_{m+2} \\ \cdot \\ \cdot \\ x_n \end{array}\right\} \quad Y_2 = \left\{\begin{array}{c} y_{m+1} \\ y_{m+2} \\ \cdot \\ \cdot \\ y_n \end{array}\right\} \quad V_2 = \left\{\begin{array}{c} v_{m+1} \\ v_{m+2} \\ \cdot \\ \cdot \\ v_n \end{array}\right\}$$

PHYSICAL INPUT-OUTPUT BALANCE

$$\left[\begin{array}{c|c} I - A_{11} & -A_{12} \\ \hline A_{21} & -I + A_{22} \end{array}\right] \left[\begin{array}{c} X_1 \\ \hline X_2 \end{array}\right] \left[\begin{array}{c} Y_1 \\ \hline Y_2 \end{array}\right] \quad (15)$$

$$\left[\begin{array}{c} X_1 \\ \hline X_2 \end{array}\right] = \left[\begin{array}{c|c} I - A_{11} & -A_{12} \\ \hline A_{21} & -I + A_{22} \end{array}\right]^{-1} \left[\begin{array}{c} Y_1 \\ \hline Y_2 \end{array}\right] \quad (16)$$

INPUT-OUTPUT BALANCE BETWEEN PRICES AND VALUES-ADDED

$$\left[\begin{array}{c|c} I - A'_{11} & - Q'_{21} \\ \hline - A'_{12} & I - Q'_{22} \end{array}\right] \left[\begin{array}{c} P_1 \\ \hline P_2 \end{array}\right] = \left[\begin{array}{c} V_1 \\ \hline V_2 \end{array}\right] \quad (17)$$

$$\left[\begin{array}{c} P_1 \\ \hline P_2 \end{array}\right] = \left[\begin{array}{c|c} I - A'_{11} & - Q'_{21} \\ \hline - A'_{12} & I - Q'_{22} \end{array}\right]^{-1} \left[\begin{array}{c} V_1 \\ \hline V_2 \end{array}\right] \quad (18)$$

Supplementary Notation and Equations Accounting for Pollution Generated Directly by Final Consumption

Notation

Technical Coefficients

$a_{gy,(i)}$ — output of pollutant generated by consumption of one unit of commodity i delivered to final demand.

Variables

y_g^* — sum total of pollutant g "delivered" from all industries to and generated within the final demand factor,

x_g^* — total gross output of pollutant g generated by all industries and in the final demand sector.

$$A_y = \left\{\begin{array}{cccc} a_{m+1, y(1)} & a_{m+1, y(1)} & \cdots & a_{m+1, y(m)} \\ a_{m+2, y(2)} & a_{m+2, y(2)} & \cdots & a_{m+2, y(m)} \\ \cdot & \cdot & & \\ \cdot & \cdot & & \\ a_{n} \ y_1 & a_n \ y_2 & \cdots & a_n \ y_m \end{array}\right\}$$

$$Y_2^* = \left\{\begin{array}{c} y^*_{m+1} \\ y^*_{m+2} \\ \cdot \\ \cdot \\ y_n^* \end{array}\right\} \qquad x_g^* = \left\{\begin{array}{c} x^*_{m+1} \\ x^*_{m+2} \\ \cdot \\ \cdot \\ x_n^* \end{array}\right\}$$

In case some pollution is generated within the final demand sector itself, the vector Y_2 appearing on the right-hand side of (15) and (16) has to be replaced by vector $Y_2 - Y_2^*$, where

$$Y_2^* = A_y Y_1. \quad (19)$$

The price-values added equations (17), (18) do not have to be modified.

Total gross output of pollutants generated by all industries and the final demand sector does not enter explicitly in any of the equations presented above; it can, however, be computed on the basis of the following equation,

$$X^* = [A_{21} : A_{22}] \left[\begin{array}{c} X_1 \\ \hline X_2 \end{array}\right] + Y_2^*. \quad (20)$$

Reprinted from *Amer. Econ. Rev.*, **59**, 282–297 (June 1969)

Production, Consumption, and Externalities

By ROBERT U. AYRES AND ALLEN V. KNEESE*

"For all that, welfare economics can no more reach conclusions applicable to the real world without some knowledge of the real world than can positive economics" [21].

Despite tremendous public and governmental concern with problems such as environmental pollution, there has been a tendency in the economics literature to view externalities as exceptional cases. They may distort the allocation of resources but can be dealt with adequately through simple *ad hoc* arrangements. To quote Pigou:

When it was urged above, that in certain industries a wrong amount of resources is being invested because the value of the marginal social net product there differs from the value of the marginal private net product, it was tacitly assumed that in the main body of industries these two values are equal [22][1].

And Scitovsky, after having described his cases two and four which deal with technological externalities affecting consumers and producers respectively, says:

The second case seems exceptional, because most instances of it can be and usually are eliminated by zoning ord-nances and industrial regulations concerned with public health and safety. The fourth case seems unimportant, simply because examples of it seem to be few and exceptional [25].

We believe that at least one class of externalities—those associated with the disposal of residuals resulting from the consumption and production process—must be viewed quite differently.[2] They are a normal, indeed, inevitable part of these processes. Their economic significance tends to increase as economic development proceeds, and the ability of the ambient environment to receive and assimilate them is an important natural resource of increasing value.[3] We will argue below that

[2] We by no means wish to imply that this is the only important class of externalities associated with production and consumption. Also, we do not wish to imply that there has been a lack of theoretical attention to the externalities problem. In fact, the past few years have seen the publication of several excellent articles which have gone far toward systematizing definitions and illuminating certain policy issues. Of special note are Coase [9], Davis and Whinston [12], Buchanan and Stubblebine [6], and Turvey [27]. However, all these contributions deal with externality as a comparatively minor aberration from Pareto optimality in competitive markets and focus upon externalities between two parties. Mishan, after a careful review of the literature, has commented on this as follows: "The form in which external effects have been presented in the literature is that of partial equilibrium analysis; a situation in which a single industry produces an equilibrium output, usually under conditions of perfect competition, some form of intervention being required in order to induce the industry to produce an "ideal" or "optimal" output. If the point is not made explicitly, it is tacitly understood that unless the rest of the economy remains organized in conformity with optimum conditions, one runs smack into Second Best problems" [21].

[3] That external diseconomies are integrally related to economic development and increasing congestion has been noted in passing in the literature. Mishan has commented: "The attention given to external effects in

* The authors are respectively visiting scholar and director, Quality of the Environment Program, Resources for the Future, Inc. We are indebted to our colleagues Blair Bower, Orris Herfindahl, Charles Howe, John Krutilla, and Robert Steinberg for comments on an earlier draft. We have also benefited from comments by James Buchanan, Paul Davidson, Robert Dorfman, Otto Eckstein, Myrick Freeman, Mason Gaffney, Lester Lave, Herbert Mohring, and Gordon Tullock.

[1] Even Baumol who saw externalities as a rather pervasive feature of the economy tends to discuss external diseconomies like "smoke nuisance" entirely in terms of particular examples [3]. A perspective more like that of the present paper is found in Kapp [16].

the common failure to recognize these facts may result from viewing the production and consumption processes in a manner that is somewhat at variance with the fundamental law of conservation of mass.

Modern welfare economics concludes that if (1) preference orderings of consumers and production functions of producers are independent and their shapes appropriately constrained, (2) consumers maximize utility subject to given income and price parameters, and (3) producers maximize profits subject to the price parameters; a set of prices exists such that no individual can be made better off without making some other individual worse off. For a given distribution of income this is an efficient state. Given certain further assumptions concerning the structure of markets, this "Pareto optimum" can be achieved via a pricing mechanism and voluntary decentralized exchange.

If waste assimilative capacity of the environment is scarce, the decentralized voluntary exchange process cannot be free of uncompensated technological external diseconomies unless (1) all inputs are fully converted into outputs, with no unwanted material residuals along the way,[4] and all final outputs are utterly destroyed in the process of consumption, or (2) property rights are so arranged that all relevant environmental attributes are in private ownership and these rights are exchanged in competitive markets. Neither of these conditions can be expected to hold in an actual economy and they do not.

Nature does not permit the destruction of matter except by annihilation with anti-matter, and the means of disposal of unwanted residuals which maximizes the internal return of decentralized decision units is by discharge to the environment, principally, watercourses and the atmosphere. Water and air are traditionally examples of free goods in economics. But in reality, in developed economies they are common property resources of great and increasing value presenting society with important and difficult allocation problems which exchange in private markets cannot resolve. These problems loom larger as increased population and industrial production put more pressure on the environment's ability to dilute and chemically degrade waste products. Only the crudest estimates of present external costs associated with residuals discharge exist but it would not be surprising if these costs were in the tens of billions of dollars annually.[5] Moreover, as we shall emphasize again, technological means for processing or purifying one or another type of waste discharge do not destroy the residuals but only alter their form. Thus, given the level, patterns, and technology of production and consumption, recycle of materials into productive uses or discharge into an alternative medium are the only general options for protecting a particular environmental medium such as water. Residual problems must be seen in a broad regional or economy-wide context rather

---

the recent literature is, I think, fully justified by the unfortunate, albeit inescapable, fact that as societies grow in material wealth, the incidence of these effects grows rapidly . . . " [21]; and Buchanan and Tullock have stated that as economic development proceeds, "congestion" tends to replace "co-operation" as the underlying motive force behind collective action, i.e., controlling external diseconomies tends to become more important than cooperation to realize external economies [7].

[4] Or any residuals which occur must be stored on the producer's premises.

[5] It is interesting to compare this with estimates of the cost of another well known misallocation of resources that has occupied a central place in economic theory and research. In 1954, Harberger published an estimate of the welfare cost of monopoly which indicated that it amounted to about .07 percent of GNP [15]. In a later study, Schwartzman calculated the allocative cost at only .01 percent of GNP [24]. Leibenstein generalized studies such as these to the statement that " . . . in a great many instances the amount to be gained by increasing allocative efficiency is trivial . . . " [19]. But Leibenstein did not consider the allocative costs associated with environmental pollution.

than as separate and isolated problems of disposal of gas, liquid, and solid wastes.

Frank Knight perhaps provides a key to why these elementary facts have played so small a role in economic theorizing and empirical research.

> The next heading to be mentioned ties up with the question of dimensions from another angle, and relates to the second main error mentioned earlier as connected with taking food and eating as the type of economic activity. The basic economic magnitude (value or utility) is service, not good. It is inherently a stream or flow in time . . . [18].[6]

Almost all of standard economic theory is in reality concerned with services. Material objects are merely the vehicles which carry some of these services, and they are exchanged because of consumer preferences for the services associated with their use or because they can help to add value in the manufacturing process. Yet we persist in referring to the "final consumption" of goods as though material objects such as fuels, materials, and finished goods somehow disappeared into the void—a practice which was comparatively harmless so long as air and water were almost literally free goods.[7] Of course, residuals from both the production and consumption processes remain and they usually render disservices (like killing fish, increasing the difficulty of water treatment, reducing public health, soiling and deteriorating buildings, etc.) rather than services. Control efforts are aimed at eliminating or reducing those disservices which flow to consumers and pro-

ducers whether they want them or not and which, except in unusual cases, they cannot control by engaging in individual exchanges.[8]

I. *The Flow of Materials*

To elaborate on these points, we find it useful initially to view environmental pollution and its control as a materials balance problem for the entire economy.[9] The inputs to the system are fuels, foods, and raw materials which are partly converted into final goods and partly become waste residuals. Except for increases in inventory, final goods also ultimately enter the waste stream. Thus goods which are "consumed" really only render certain services. Their material substance remains in existence and must either be reused or discharged to the ambient environment.

In an economy which is closed (no imports or exports) and where there is no net accumulation of stocks (plant, equipment, inventories, consumer durables, or residential buildings), the amount of residuals inserted into the natural environment must be approximately equal to the weight of basic fuels, food, and raw materials entering the processing and production system, plus oxygen taken from the atmosphere.[10] This result, while obvious

---

[6] The point was also clearly made by Fisher: "The only true method, in our view, is to regard uniformly as income the *service* of a dwelling to its owner (shelter or money rental), the *service* of a piano (music), and the *service* of food (nourishment) . . . " (emphasis in original) [14].

[7] We are tempted to suggest that the word consumption be dropped entirely from the economist's vocabulary as being basically deceptive. It is difficult to think of a suitable substitute, however. At least, the word consumption should not be used in connection with goods, but only with regard to services or flows of "utility."

[8] There is a substantial literature dealing with the question of under what conditions individual exchanges can optimally control technological external diseconomies. A discussion of this literature, as it relates to waterborne residuals, is found in Kneese and Bower [17].

[9] As far as we know, the idea of applying materials balance concepts to waste disposal problems was first expressed by Smith [26]. We also benefitted from an unpublished paper by Joseph Headley in which a pollution "matrix" is suggested. We have also found references by Boulding to a "spaceship economy" suggestive [4]. One of the authors has previously used a similar approach in ecological studies of nutrient interchange among plants and animals; see [1].

[10] To simplify our language, we will not repeat this essential qualification at each opportunity, but assume it applies throughout the following discussion. In addition, we must include residuals such as NO and NO_2 arising from reactions between components of the air itself but occurring as combustion by-products.

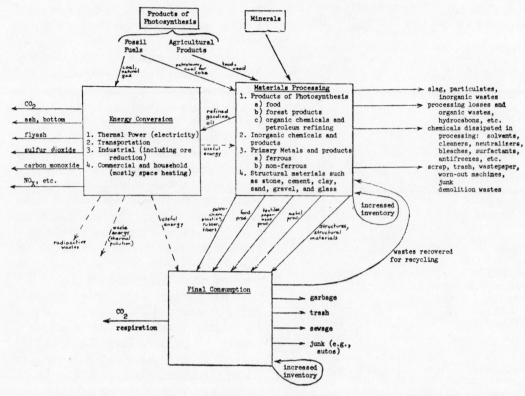

FIGURE 1.—MATERIALS FLOW

upon reflection, leads to the, at first rather surprising, corollary that residuals disposal involves a greater tonnage of materials than basic materials processing, although many of the residuals, being gaseous, require no physical "handling."

Figure 1 shows a materials flow of the type we have in mind in greater detail and relates it to a broad classification of economic sectors for convenience in our later discussion, and for general consistency with the Standard Industrial Classification. In an open (regional or national) economy, it would be necessary to add flows representing imports and exports. In an economy undergoing stock or capital accumulation, the production of residuals in any given year would be less by that amount than the basic inputs. In the entire U.S. economy, accumulation accounts for about 10–15 percent of basic annual inputs, mostly in the form of

construction materials, and there is some net importation of raw and partially processed materials amounting to 4 or 5 percent of domestic production. Table 1 shows estimates of the weight of raw materials produced in the United States in several recent years, plus net imports of raw and partially processed materials.

Of the active inputs,[11] perhaps three-quarters of the overall weight is eventually discharged to the atmosphere as carbon (combined with atmospheric oxygen in the form of CO or CO_2) and hydrogen (combined with atmospheric oxygen as H_2O) under current conditions. This results from combustion of fossil fuels and from animal respiration. Discharge of carbon dioxide can be considered harmless in the short run. There are large "sinks" (in the form of vegetation and large water bodies,

[11] See footnote to Table 1.

TABLE 1—WEIGHT OF BASIC MATERIALS PRODUCTION
IN THE UNITED STATES PLUS NET IMPORTS,
1963 (10^6 tons)

	1963	1964	1965
Agricultural (incl. fishery and wildlife and forest) products			
Food { Crops (excl. livestock feed)	125	128	130
Food { Livestock	100	103	102
Other products	5	6	6
Fishery	3	3	3
Forestry products (85 per cent dry wt. basis)			
Sawlogs	53	55	56
Pulpwood	107	116	120
Other	41	41	42
Total	434	452	459
Mineral fuels	1,337	1,399	1,448
Other minerals			
Iron ore	204	237	245
Other metal ores	161	171	191
Other nonmetals	125	133	149
Total	490	541	585
Grand total[a]	2,261	2,392	2,492

[a] Excluding construction materials, stone, sand, gravel, and other minerals used for structural purposes, ballast, fillers, insulation, etc. Gangue and mine tailings are also excluded from this total. These materials account for enormous tonnages but undergo essentially no chemical change. Hence, their use is more or less tantamount to physically moving them from one location to another. If this were to be included, there is no logical reason to exclude material shifted in highway cut and fill operations, harbor dredging, land-fill, plowing, and even silt moved by rivers. Since a line must be drawn somewhere, we chose to draw it as indicated above.

Source: R. U. Ayres and A. V. Kneese [2, p. 630].

mainly the oceans) which reabsorb this gas, although there is evidence of net accumulation of CO_2 in the atmosphere. Some experts believe that the latter is likely to show a large relative increase, as much as 50 per cent by the end of the century, possibly giving rise to significant —and probably, on balance, adverse— weather changes.[12] Thus continued com-

bustion of fossil fuels at a high rate could produce externalities affecting the entire world. The effects associated with most residuals will normally be more confined, however, usually limited to regional air and water sheds.

The remaining residuals are either gases (like carbon monoxide, nitrogen dioxide, and sulfur dioxide—all potentially harmful even in the short run), dry solids (like rubbish and scrap), or wet solids (like garbage, sewage, and industrial wastes suspended or dissolved in water). In a sense, the dry solids are an irreducible, limiting form of waste. By the application of appropriate equipment and energy, most undesirable substances can, in principle, be removed from water and air streams[13]— but what is left must be disposed of in solid form, transformed, or reused. Looking at the matter in this way clearly reveals a primary interdependence between the various waste streams which casts into doubt the traditional classification of air, water, and land pollution as individual categories for purposes of planning and control policy.

Residuals do not necessarily have to be discharged to the environment. In many instances, it is possible to recycle them back into the productive system. The materials balance view underlines the fact that the throughput of new materials necessary to maintain a given level of production and consumption decreases as the technical efficiency of energy conversion and materials utilization increases. Similarly, other things being equal, the longer that cars, buildings, machinery, and other durables remain in service, the fewer new materials are required to compensate for loss, wear, and obsolescence— although the use of old or worn machinery (e.g., automobiles) tends to increase other residuals problems. Technically efficient combustion of (desulfurized) fossil fuels

[12] See [30]. There is strong evidence that discharge of residuals has already affected the climate of individual cities; see Lowry [20].

[13] Except CO_2, which may be harmful in the long run, as noted.

would leave only water, ash, and carbon dioxide as residuals, while nuclear energy conversion need leave only negligible quantities of material residuals (although thermal pollution and radiation hazards cannot be dismissed by any means).

Given the population, industrial production, and transport service in an economy (a regional rather than a national economy would normally be the relevant unit), it is possible to visualize combinations of social policy which could lead to quite different relative burdens placed on the various residuals-receiving environmental media; or, given the possibilities for recycle and less residual-generating production processes, the overall burden to be placed upon the environment as a whole. To take one extreme, a region which went in heavily for electric space heating and wet scrubbing of stack gases (from steam plants and industries), which ground up its garbage and delivered it to the sewers and then discharged the raw sewage to watercourses, would protect its air resources to an exceptional degree. But this would come at the sacrifice of placing a heavy residuals load upon water resources. On the other hand, a region which treated municipal and industrial waste water streams to a high level and relied heavily on the incineration of sludges and solid wastes would protect its water and land resources at the expense of discharging waste residuals predominantly to the air. Finally, a region which practiced high level recovery and recycle of waste materials and fostered low residual production processes to a far reaching extent in each of the economic sectors might discharge very little residual waste to any of the environmental media.

Further complexities are added by the fact that sometimes it is possible to modify an environmental medium through investment in control facilities so as to improve its assimilative capacity. The clearest, but far from only, example is with respect to watercourses where reservoir storage can be used to augment low river flows that ordinarily are associated with critical pollution (high external cost situations).[14] Thus internalization of external costs associated with particular discharges, by means of taxes or other restrictions, even if done perfectly, cannot guarantee Pareto optimality. Investments involving public good aspects must enter into an optimal solution.[15]

To recapitulate our main points briefly: (1) Technological external diseconomies are not freakish anomalies in the processes of production and consumption but an inherent and normal part of them. (2) These external diseconomies are quantitatively negligible in a low-population or economically undeveloped setting, but they become progressively (nonlinearly) more important as the population rises and the level of output increases (i.e., as the natural reservoirs of dilution and assimilative capacity become exhausted).[16] (3) They cannot be properly dealt with by considering environmental media such as air and water in isolation. (4) Isolated and *ad hoc* taxes and other restrictions are not sufficient for their optimum control, although they are essential elements in a more systematic and coherent program of environmental quality management. (5) Public investment programs, particularly including transportation systems, sewage disposal, and river flow regulation, are intimately related to the amounts and

[14] Careful empirical work has shown that this technique can fit efficiently into water quality management systems. See Davis [11].

[15] A discussion of the theory of such public investments with respect to water quality management is found in Boyd [5].

[16] Externalities associated with residuals discharge may appear only at certain threshold values which are relevant only at some stage of economic development and industrial and population concentrations. This may account for their general treatment as "exceptional" cases in the economics literature. These threshold values truly were exceptional cases for less developed economies.

effects of residuals and must be planned in light of them.

It is important to develop not only improved measures of the external costs resulting from differing concentrations and duration of residuals in the environment but more systematic methods for forecasting emissions of external-cost-producing residuals, technical and economic trade-offs between them, and the effects of recycle on environmental quality.

In the hope of contributing to this effort and of revealing more clearly the types of information which would be needed to implement such a program, we set forth a more formal model of the materials balance approach in the following sections and relate it to some conventional economic models of production and consumption. The main objective is to make some progress toward defining a system in which flows of services and materials are simultaneously accounted for and related to welfare.

II. Basic Model

The take off point for our discussion is the Walras-Cassel general equilibrium model,[17] extended to include intermediate consumption, which involve the following quantities:

resources and services

$$r_1, \cdots\cdots\cdots, r_M$$

products or commodities

$$X_1, \cdots\cdots\cdots, X_N$$

resource prices

$$v_1, \cdots\cdots\cdots, v_M$$

product or commodity prices

$$p_1, \cdots\cdots\cdots, p_N$$

final demands

$$Y_1, \cdots\cdots\cdots, Y_N$$

[17] The original references are Walras [28] and Cassel [8]. Our own treatment is largely based on Dorfman *et al.* [13].

The M basic resources are allocated among the N sectors as follows:

$$r_1 = a_{11}X_1 + a_{12}X_2 + \cdots + a_{1N}X_N$$
$$r_2 = a_{21}X_1 + a_{22}X_2 + \cdots + a_{2N}X_N$$
$$\vdots$$
$$r_M = a_{M1}X_1 + a_{M2}X_2 + \cdots + a_{MN}X_N$$

(1a) or

$$r_j = \sum_{k=1}^{N} a_{jk}X_k \qquad j = 1, \cdots. M$$

In (1a) we have implicitly assumed that there is no possibility of factor or process substitution and no joint production. These conditions will be discussed later. In matrix notation we can write:

$$(1b) \qquad [r_{j1}]_{M,1} = [a_{jk}]_{M,N} \cdot [X_{k1}]_{N,1}$$

where $[a]$ is an $M \times N$ matrix.

A similar set of equations describes the relations between commodity production and final demand:

$$(2a) \qquad X_k = \sum_{l=1}^{N} A_{kl}Y_l \qquad k = 1, \cdots, N$$

$$(2b) \qquad [X_{k1}]_{N,1} = [A_{kl}]_{N,N} \cdot [Y_{l1}]_{N,1}$$

and the matrix $[A]$ is given by

$$(3) \qquad [A] = [I - C]^{-1}$$

where $[I]$ is the unit diagonal matrix and the elements C_{ij} of the matrix $[C]$ are essentially the well known Leontief input coefficients. In principle these are functions of the existing technology and, therefore, are fixed for any given situation.

By combining (1) and (2), we obtain a set of equations relating resource inputs directly to final demand, viz.,

$$r_j = \sum_{k=1}^{N} a_{jk} \sum_{l=1}^{N} A_{kl}Y_l = \sum_{k,l=1}^{N} a_{jk}A_{kl}Y_l$$

(4a)

$$= \sum_{l=1}^{N} b_{jl}Y_l \qquad j = 1, \cdots, M$$

or, of course, in matrix notation (4b).

(4b)
$$[r_{j1}]_{M,1} = [a_{jk}]_{M,N} \cdot [A_{kl}]_{N,N} \cdot [Y_{l1}]_{N,1}$$
$$= |b_{jl}|_{M,N} \cdot [Y_{l1}]_{N,1}$$

We can also impute the prices of N intermediate goods and commodities to the prices of the M basic resources, as follows:

(5a) $\quad p_k = \sum_{j=1}^{M} v_j b_{jk} \qquad k = 1, \cdots, N$

(5b) $\quad [p_{1k}]_{1,N} = [v_{1j}]_{1,M} \cdot [b_{jk}]_{M,N}$

To complete the system, it may be supposed that demand and supply relationships are given, a priori, by Pareto-type preference functions:

(6) Demand: $\quad Y_k = F_k(p_1, \cdots, p_N)$
$$k = 1, \cdots, N$$

(7) Supply: $\quad r_k = G_k(v_1, \cdots, v_M)$
$$k = 1, \cdots, M$$

where, of course, the p_j are functions of the v_j as in (5b).

In order to interpret the X's as physical production, it is necessary for the sake of consistency to arrange that outputs and inputs always balance, which implies that the C_{ij} must comprise *all* materials exchanges including residuals. To complete the system so that there is no net gain or loss of physical substances, it is also convenient to introduce two additional sectors, viz., an "environmental" sector whose (physical) output is X_0 and a "final consumption" sector whose output is denoted X_f. The system is then easily balanced by explicitly including flows both to and from these sectors.

To implement this further modification of the Walras-Cassel model, it is convenient to subdivide and relabel the resource category into tangible raw materials $\{r^m\}$ and services $\{r^s\}$:

$$\begin{bmatrix} r_1 \\ r_2 \\ \cdot \\ \cdot \\ \cdot \\ r_L \end{bmatrix} \qquad \begin{bmatrix} r_1^m \\ r_2^m \\ \cdot \\ \cdot \\ \cdot \\ r_L^m \end{bmatrix} \Big\} \begin{matrix} \text{raw materials} \\ \text{(units)} \end{matrix}$$

becomes

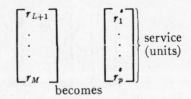

$$\begin{bmatrix} r_{L+1} \\ \cdot \\ \cdot \\ \cdot \\ r_M \end{bmatrix} \qquad \begin{bmatrix} r_1^s \\ \cdot \\ \cdot \\ \cdot \\ r_p^s \end{bmatrix} \Big\} \begin{matrix} \text{service} \\ \text{(units)} \end{matrix}$$

becomes

where, of course,

(8) $\qquad L + P = M$

It is understood that services, while not counted in tons, can be measured in meaningful units, such as man-days, with well defined prices. Thus, we similarly relabel the price variables as follows:

$$\begin{bmatrix} V_1 \\ \cdot \\ \cdot \\ \cdot \\ V_L \end{bmatrix} \qquad \begin{bmatrix} V_1^m \\ \cdot \\ \cdot \\ \cdot \\ V_L^m \end{bmatrix} \Big\} \begin{matrix} \text{raw material} \\ \text{(prices)} \end{matrix}$$

becomes

$$\begin{bmatrix} V_{L+1} \\ \cdot \\ \cdot \\ \cdot \\ V_M \end{bmatrix} \qquad \begin{bmatrix} V_1^s \\ \cdot \\ \cdot \\ \cdot \\ V_p^s \end{bmatrix} \Big\} \begin{matrix} \text{labor and service} \\ \text{(prices)} \end{matrix}$$

The coefficients $\{a_{ij}\}$, $\{b_{ij}\}$ are similarly partitioned into two groups,

e.g., $\qquad b_{1j} \qquad\qquad\qquad b_{1j}^m$
$$\cdot \qquad\qquad\qquad \cdot$$
$$\cdot \qquad\qquad\qquad \cdot$$
$$\cdot \qquad\qquad\qquad \cdot$$
$$b_{Lj} \qquad\qquad\qquad b_{Lj}^m$$
$$\qquad \text{becomes}$$
$$b_{L+1,j} \qquad\qquad\qquad b_{1j}^s$$
$$\cdot \qquad\qquad\qquad \cdot$$
$$\cdot \qquad\qquad\qquad \cdot$$
$$\cdot \qquad\qquad\qquad \cdot$$
$$b_{Mj} \qquad\qquad\qquad b_{pj}^s$$

These notational changes have no effect whatever on the substance of the model, although the equations become somewhat more cumbersome. The partitioned matrix notation simplifies the restatement of the basic equations. Thus (1b) becomes (9), while (5b) becomes (10).

$$(9) \quad M \left\{ \begin{bmatrix} \vdots \\ r \\ \vdots \end{bmatrix} \equiv \begin{bmatrix} r^m \\ \cdots \\ r^s \end{bmatrix} \begin{matrix} \}L \\ \\ \}P \end{matrix} = M \left\{ \begin{bmatrix} \begin{matrix} \}L & b^m \\ \cdots & \cdots \\ \}P & b^s \end{matrix} \end{bmatrix} \overbrace{}^{N} \begin{bmatrix} Y \end{bmatrix} \right\} N \right.$$

$$(10) \quad \lfloor p_1, \cdots, p_N \rfloor = \lfloor v^m \vdots v^s \rfloor \underbrace{\overbrace{}^{L\ P}}_{M} \begin{bmatrix} b^m \\ \cdots \\ b^s \end{bmatrix} \Big\} M \\ \underbrace{}_{N}$$

$$= [\cdots v^m \cdots] \begin{bmatrix} \cdot & b^m & \cdot \\ \vdots & \cdots & \cdot \end{bmatrix} + [\cdots v^s \cdots] \begin{bmatrix} \cdot & b^s & \cdot \\ \vdots & \cdots & \cdot \end{bmatrix}$$

The equivalent of (5a) is:

$$p_k = \underbrace{\sum_{j=1}^{L} b_{jk}^m v_j^m}_{\substack{\text{prices imputed} \\ \text{to cost of raw} \\ \text{materials}}} + \underbrace{\sum_{j=1}^{P} b_{jk}^s v_j^s}_{\substack{\text{prices imputed} \\ \text{to cost of} \\ \text{services}}}$$

(11)

where $k = 1, \cdots, N$

We wish to focus attention explicitly on the flow of materials through the economy. By definition of the Leontief input coefficients (now related to materials flow), we have:

$C_{kj}X_j$ (physical) quantity transferred

from k to j

$C_{jk}X_k$ quantity transferred from j to k

Hence, material flows *from* the environment to all other sectors are given by:

$$(12) \quad \sum_{k=1}^{N} C_{0k}X_k = \sum_{j=1}^{L} r_j^m = \sum_{j=1}^{L}\sum_{k=1}^{N} a_{jk}^m X_k$$
$$= \sum_{j=1}^{L}\sum_{k=1}^{N} b_{jk}^m Y_k$$

using equation (1), as modified.[18] Obvi-

ously, comparing the first and third terms,

$$(13) \quad \underbrace{C_{0k}}_{\substack{\text{total material} \\ \text{flow (0 to } k)}} = \underbrace{\sum_{j=1}^{L} a_{jk}^m}_{\substack{\text{all raw materials} \\ (0 \text{ to } k)}}$$

Flows into and out of the environmental sector must be in balance:

$$(14) \quad \underbrace{\sum_{k=1}^{N} C_{0k}X_k}_{\substack{\text{sum of all raw} \\ \text{material flows}}} = \underbrace{\sum_{k=1}^{N} C_{k0}X_0 + C_{f0}X_0}_{\substack{\text{sum of all return} \\ \text{(waste) flows}}}$$

Material flows to and from the final sector must also balance:

$$\underbrace{\sum_{k=1}^{N} C_{kf}X_f}_{\substack{\text{sum of all} \\ \text{final goods}}}$$

$$(15) \quad = \underbrace{\sum_{k=1}^{N} C_{fk}X_k}_{\substack{\text{sum of all} \\ \text{materials} \\ \text{recycled}}} + \underbrace{C_{f0}X_0}_{\substack{\text{waste residuals} \\ \text{(plus accumulation}[19])}}$$

---

[18] Ignoring, for convenience, any materials flow from the environment *directly* to the final consumption sector.

[19] For convenience, we can treat accumulation in the final sector as a return flow to the environment. In truth, structures actually *become* part of our environment, although certain disposal costs may be deferred.

Of course, by definition, X_f is the sum of the final demands:

$$(16) \qquad X_f = \sum_{j=1}^{N} Y_j$$

Substituting (16) into the left side of (15) and (2a) into the right side of (15), we obtain an expression for the waste flow in terms of final demands:

$$(17) \quad C_{f0}X_0 = \sum_{j=1}^{N} \sum_{k=1}^{N} (C_{jf} - C_{fj}A_{jk})Y_k$$

The treatment could be simplified slightly if we assumed that there is no recycling per se. Thus, in the context of the model, we could suppose that all residuals return to the environmental sector,[20] where some of them (e.g., waste paper) become "raw materials." They would then be indistinguishable from new raw materials, however, and price differentials between the two would be washed out. In principle, this is an important distinction to retain.

III. *Inclusion of Externalities*

The physical flow of materials between various intermediate (production) sectors and the final (consumption) sector tends to be accompanied by, and correlated with, a (reverse) flow of dollars.[21] However, the physical flow of materials from and back to the environment is only partly reflected by actual dollar flows, namely, land rents and payments for raw materials. There are three classes of physical exchange for which there exist no counterpart economic transactions. These are: (1) private use for production inputs of "common property" resources, notably air, streams, lakes, and the ocean; (2) private use of the assimila-

tive capacity of the environment to "dispose of" or dilute wastes and residuals; (3) inadvertent or unwanted material inputs to productive processes—diluents and pollutants.

All these goods (or "bads") are physically transferred at zero price, not because they are not scarce relative to demand—they often are in developed economies—or because they confer no service or disservice on the user—since they demonstrably do so—but because there exist no social institutions that permit the resources in question to be "owned," and exchanged in the market.

The allocation of resources corresponding to a Pareto optimum cannot be attained without subjecting the above-mentioned nonmarket and involuntary exchanges to the moderation of a market or a surrogate thereof. In principle, the influence of a market might be simulated, to a first approximation, by introducing a set of shadow (or virtual) prices.[22] These may well be zero, where supply truly exceeds demand, or negative (i.e., costs) in some instances; they will be positive in others. The exchanges are, of course, real.

The Walras-Cassel model can be generalized to handle these effects in the following way:

1. One can introduce a set of R common-property resources or services of raw materials $\{r_1^{cp}, \cdots, r_R^{cp}\}$ as a subset of the set $\{r_j\}$; these will have corresponding virtual prices $\{v_j^{cp}\}$, which would constitute an "income" from the environment. Such resources include the atmosphere; streams, lakes, and oceans; landscape; wildlife and biological diversity; and the indispensable assimilative capacity of the environment (its ability to accept and neutralize or recycle residuals).[23]

[20] In calculating actual quantities, we would (by convention) ignore the weight of oxygen taken free from the atmosphere in combustion and return as CO_2. However, such inputs will be treated explicitly later.

[21] To be precise, the dollar flow governs and is governed by a combined flow of materials and services (value added).

[22] A similar concept exists in mechanics where the forces producing "reaction" (to balance action and reaction) are commonly described as "virtual forces."

[23] Economists have previously suggested generalization of the Walras-Cassel model to take account of public goods. One of the earliest appears to be Schles-

2. One can introduce a set of S environmental *disservices* imposed on consumers of material resources, by forcing them to accept unwanted inputs $\{r_1^u, \cdots, r_s^u\}$ (pollutants, contaminants, etc.); these disservices would have negative value, giving rise to *negative* virtual prices $\{u_j\}$.[24]

The matrix coefficients $\{a_{ij}\}$ and $\{b_{ij}\}$ can be further partitioned to take account of this additional refinement, and equations analogous to (9), (10), and (11) can be generalized in the obvious way. Equation (6) carries over unchanged, but (7) must be appropriately generalized to take account of the altered situation. Actually, (7) breaks up into several groups of equations:

$$(18) \qquad \overset{m}{r_k} = G_k^m(p_1, \cdots, p_N)$$
$$k = 1, \cdots, L$$

$$(19) \qquad \overset{s}{r_k} = G_k^s(p_1, \cdots, p_N)$$
$$k = 1, \cdots, P$$

However, as we have noted at the outset, the supplies of common-property resources and environmental services or disservices are *not* regulated directly by market prices of other goods and services. In the case of common-property resources, the supplies are simply constants fixed by nature or otherwise determined by accident or noneconomic factors.

The total value of these services performed by the environment cannot be

calculated but it is suggestive to consider the situation if the natural reservoir of air, water, minerals, etc., were very much smaller, as if the earth were a submarine or "spaceship" (i.e., a vehicle with no assimilative and/or regenerative capacity). In such a case, all material resources would have to be recycled,[25] and the cost of all goods would necessarily reflect the cost of reprocessing residuals and wastes for reuse. In this case, incidentally, the ambient level of unrecovered wastes continuously circulating through the resource inventory of the system (i.e., the spaceship) would in general be nonzero because of the difficulty of 100 percent efficient waste-removal of air and water. However, although the quantity of waste products in constant circulation may fluctuate within limits, it cannot be allowed to increase monotonically with time, which means that as much material must be recycled, on the average, as is discarded. The value of common resources plus the assimilation services performed by the environment, then, is only indirectly a function of the ambient level of untreated residuals per se, or the disutility caused thereby, which depend on the cost efficiency of the available treatment technology. Be this as it may, of course, the bill of goods produced in a spaceship economy would certainly be radically different from that we are familiar with. For this reason, no standard economic comparison between the two situations is meaningful. The measure of worth we are seeking is actually the difference between the total welfare "produced" by a spaceship economy, where 100 percent of all residuals are promptly recycled, vis-à-vis the existing welfare output on earth, where resource inventories are substantial and

---

inger [23]. We are indebted to Otto Eckstein for calling our attention to this key reference.

[24] The notion of introducing the possibility of negative prices in general equilibrium theory has apparently been discussed before, although we are not aware of any systematic development of the idea in the published literature. In this connection, it is worth pointing out the underlying similarity of negative prices and effluent taxes—which have been, and still are being considered as an attractive alternative to subsidies and federal standard-setting as a means of controlling air and water pollution. Such taxes would, of course, be an explicit attempt to rectify an imbalance caused by a market failure.

[25] Any consistent deviation from this 100 per cent rule implies an accumulation of waste products, on the average, which, by definition, is inconsistent with maintaining an equilibrium.

complete recycling need not be contemplated for a very long time to come.

This welfare difference might well be very large, although we possess no methodological tools for quantifying it. In any case, the resource inventory and assimilative capacity of the environment probably contribute very considerably to our standard of living.

If these environmental contributions were paid for, the overall effect on prices would presumably to be push them generally upward. However, the major *differential* effect of undervaluing the environmental contribution is that goods produced by high residual-producing processes, such as papermaking, are substantially underpriced vis-à-vis goods which involve more economical uses of basic resources. This is, however, not socially disadvantageous per se: that is, it causes no misallocation of resources unless, or until, the large resource inventory and/or the assimilative capacity of the environment are used up. When this happens, however, as it now has in most highly industrialized regions, either a market must be allowed to operate or some other form of decision rule must be introduced to permit a rational choice to be made, e.g., between curtailing or controlling the production of residuals or tolerating the effects (disservice) thereof.

It appears that the natural inventory of most common resources used as inputs (e.g., air as an input to combustion and respiratory processes) is still ample,[26] but the assimilative capacity of the environment has already been exceeded in many areas, with important external costs resulting. This suggests a compromise treatment. If an appropriate price could be charged to the producers of the residuals and used to compensate the inadvertent recipients—with the price determined by appropriate Pareto preference criteria—there would be no particular analytic purpose in keeping books on the exchange of the other environmental benefits mentioned, although they are quantitatively massive. We will, therefore, in the remainder of the discussion omit the common-property variables $\{r_j^{cp}\}$ and the corresponding virtual-price variables $\{v_j^{cp}\}$ defined previously, retaining only the terms $\{r_j^u\}$ and $\{u_{jk}\}$. The variable $\{r_j^u\}$ represents a physical quantity of the jth unwanted input. There are S such terms, by assumption, whose magnitudes are proportional to the levels of consumption of basic raw materials, subject to the existing technology. However, residuals production is not immutable: it can be increased or decreased by investment, changes in materials processing technology, raw material substitutions, and so forth.

At first glance it might seem entirely reasonable to assert that the *supplies* of unwanted residuals received will be functions of the (negative) prices (i.e., compensation) paid for them, in analogy with (7). Unfortunately, this assertion immediately introduces a theoretical difficulty, since the assumption of unique coefficients $\{a_{ij}\}$ and $\{C_{ij}\}$[27] is not consistent with the possibility of factor or process substitution or joint-production, as stated earlier. To permit such substitutions, one would have to envision a very large collection of alternative sets of coefficients: one complete set of a's and C's for each specific combination of factors and processes. Maximization of any objective function (such as GNP) would involve solving the entire system of equations as many times as there are combinations of factors and pro-

---

[26] Water is an exception in arid regions; in humid regions, however, water "shortages" are misnomers: they are really consequences of excessive use of watercourses as cheap means of waste disposal. But some ecologists have claimed that oxygen depletion may be a very serious long-run problem; see Cole [10].

[27] Or $\{b_{ij}\}$ and $\{A_{ij}\}$.

(21)
$$[r] = \begin{bmatrix} r^m \\ r^s \\ r^u \end{bmatrix} = M \left\{ \begin{bmatrix} a^m \\ \cdots \\ a^s \\ \cdots \\ a^u \end{bmatrix} X \right\} N = \begin{bmatrix} b^m \\ \cdots \\ b^s \\ \cdots \\ b^u \end{bmatrix} Y$$

$$\underbrace{}_{N}$$

cesses, and picking out that set of solutions which yields the largest value. Alternatively, if the a's and C's are assumed to be continuously variable functions (of each other), the objective function could also, presumably, be parameterized. However, as long as the a's and C's are uniquely given, the supply of the kth unwanted residual is only marginally under the control of the producer, since it will be produced in strict relationship to the composition of the bill of final goods $\{Y_j\}$.

Hence, for the present model it is only correct to assume

(20) $\qquad r_k^u = G_k^u(Y_1, \cdots, Y_N)$

This limitation does not affect the existence of an equilibrium solution for the system of equations; it merely means that the shadow prices $\{u_{jk}\}$ which would emerge from such a solution for given coefficients $\{a_{ij}\}$, $\{b_{ij}\}$, and $\{C_{ij}\}$ might be considerably higher than the real economic optimum, since the latter could only be achieved by introducing factor and process changes.

Of course, the physical inputs are also related to the physical outputs of goods, as in (21).

Written out in full detail (21) is equivalent to:

(22) raw materials $\quad r_k^m = \sum_{j=1}^{N} a_{kj}^m X_j = \sum_{j=1}^{N} b_{kj}^m Y_j$

$$k = 1, \cdots, L$$

(23) labor and technical services $\quad r_k^s = \sum_{j=1}^{N} a_{kj}^s X_j = \sum_{j=1}^{N} b_{kj}^s Y_j$

$$k = 1, \cdots, P$$

(24) unwanted inputs $\quad r_k^u = \sum_{j=1}^{N} a_{kj}^u X_j = \sum_{j=1}^{N} b_{kj}^u Y_j$

$$k = 1, \cdots, S$$

where, of course,

(25) $\qquad L + P + S = M$

The corresponding matrix equation for the prices of goods, in terms of production costs, is

(26)
$$[p_1, \cdots, p_N] = [v^m \vdots v^s \vdots u] \begin{bmatrix} b^m \\ \cdots \\ b^s \\ \cdots \\ b^u \end{bmatrix}$$

Written out in the standard form, we obtain

(27)
$$p_k = \underbrace{\sum_{j=1}^{L} b_{jk}^m v_j^m}_{\substack{\text{cost of raw} \\ \text{materials}}} + \underbrace{\sum_{j=1}^{P} b_{jk}^s v_j^s}_{\substack{\text{cost of labor} \\ \text{and technical} \\ \text{services}}}$$

$$+ \underbrace{\sum_{j=1}^{S} b_{jk}^u v_j^u}_{\substack{\text{cost (compensa-} \\ \text{tion) for pro-} \\ \text{viding environ-} \\ \text{mental disser-} \\ \text{vices}}}$$

$$k = 1, \cdots, N$$

Evidently, the coefficients b_{jk}^u are empirically determined by the structure of the regional economy and its geography. It is assumed that a single overall (negative) price for each residual has meaning, even though each productive sector—and even each consumer—has his own individual utility function. Much the same assumption is conventionally made, and accepted, in the case of positive real prices.

All of the additional variables now fit into the general framework of the original Walras-Cassel analysis. Indeed, we have $2N+2M-1$ variables (r_i, Y_i, p_i, v_i) (allowing an arbitrary normalization factor for the price level) and $2N+2M-1$ independent equations.[28] If solutions exist for the Walras-Cassel system of equations, the arguments presumably continue to hold true for the generalized model. In any case, a discussion of such mathematical questions would carry us too far from our main theme.

IV. *Concluding Comments*

The limited economics literature currently available which is devoted to environmental pollution problems has generally taken a partial equilibrium view of the matter, as well as treated the pollution of particular environmental media, such as air and water, as separate problems.[29] This no doubt reflects the propensity of the theoretical literature to view externalities as exceptional and minor. Clearly, the partial equilibrium approach in particular is very convenient theoretically and empirically for it permits external damage and control cost functions to be defined for each particular case without reference to broader interrelationships and adjustments in the economy.

[28] There is one redundant equation in the system, which expresses the identity between gross product and gross income for the system as a whole (sometimes called "Walras law").

[29] See, for example, the essays in Wolozin [29]

We have argued in this paper that the production of residuals is an inherent and general part of the production and consumption process and, moreover, that there are important trade-offs between the gaseous, liquid, or solid forms that these residuals may take. Further, we have argued that under conditions of intensive economic and population development the environmental media which can receive and assimilate residual wastes are not free goods but natural resources of great value with respect to which voluntary exchange cannot operate because of their common property characteristics. We have also noted, in passing, that the assimilative capacity of environmental media can sometimes be altered and that therefore the problem of achieving Pareto optimality reaches beyond devising appropriate shadow prices and involves the planning and execution of investments with public goods aspects.

We have exhibited a formal mathematical framework for tracing residuals flows in the economy and related it to the general equilibrium model of resources allocation, altered to accommodate recycle and containing unpriced sectors to represent the environment. This formulation, in contrast to the usual partial equilibrium treatments, implies knowledge of all preference and production functions including relations between residuals discharge and external cost and all possible factor and process substitutions. While we feel that it represents reality with greater fidelity than the usual view, it also implies a central planning problem of impossible difficulty, both from the standpoint of data collection and computation.

What, if any, help can the general interdependency approach we have outlined offer in dealing with pollution problems effectively and reasonably efficiently? A minimal contribution is its warning that partial equilibrium approaches, while more tractable, may lead to serious errors. Second, in projecting waste residuals for an economy—a regional economy would usually be the most relevant for planning and control—the inter-industry materials flow model can provide a much more conceptually satisfying and accurate tool for projecting future residuals production than the normal aggregative extrapolations.[30] The latter not only treat gaseous, liquid, and solid wastes separately, but do not take account of input-output relations and the fact that the materials account for the region must balance.

We think that in the next few years it will be possible to make improved regional projections of residuals along the lines sketched above. Undoubtedly, there will also be further progress in empirically estimating external costs associated with residuals discharge and in estimating control costs via various alternative measures. On the basis of this kind of information, a control policy can be devised. However, this approach will still be partial. Interrelations between the regional and national economy must be treated simplistically and to be manageable, the analysis must confine itself to a specific projected bill of goods.

The basic practical question which remains to be answered is whether an iterated series of partial equilibrium treatments—e.g., focusing on one industry or region at a time, *ceteris paribus*—would converge toward the general equilibrium

solution, or not. We know of no theoretical test of convergence which would be applicable in this case but, in the absence of such a criterion, would be willing to admit the possible relevance of an empirical sensitivity test more or less along the following lines: take a major residuals-producing industry (such as electric power) and parametrize its cost structure in terms of emission control levels, allowing all technically feasible permutations of factor (fuel) inputs and processes. It would be a straightforward, but complicated, operations research problem to determine the minimum cost solution as a function of the assumed (negative) price of the residuals produced. If possible industry patterns—factor and process combinations—exist which would permit a high level of emission control at only a small increase in power production cost, then it might be possible to conclude that for a significant range of (negative) residuals prices the effect on power prices —and therefore on the rest of the economy —would not be great. Such a conclusion would support the convergence hypothesis. If, on the other hand, electric power prices are very sensitive to residuals prices, then one would at least have to undertake a deeper study of consumer preference functions to try to determine what residuals prices would actually be if a market mechanism existed. If people prove to have a strong antipathy to soot and sulfur dioxide, for instance, resulting in a high (negative) price for these unwanted inputs, then one would be forced to suspect that the partial equilibrium approach is probably not convergent to the general equilibrium solution and that much more elaborate forms of analysis will be required.

[30] Some efforts to implement these concepts are already underway. Walter Isard and his associates have prepared an input-output table for Philadelphia which includes coefficients representing waterborne wastes (unpublished). The recent study of waste management in the New York Metropolitan region by the Regional Plan Association took a relatively broad view of the waste residuals problem [31]. Relevant data on several industries are being gathered. Richard Frankel's not yet published study of thermal power in which the range of technical options for controlling residuals, and their costs, is being explored is notable in this regard. His and other salient studies are described in Ayres and Kneese [2].

REFERENCES

1. R. U. AYRES, "Stability of Biosystems in Sea Water," Tech. Rept. No. 142, Hudson Laboratories, Columbia University, New York 1967.

2. —— AND A. V. KNEESE, "Environmental Pollution," in U.S. Congress, Joint Economic Committee, *Federal Programs for the Development of Human Resources*, Vol. 2, Washington 1968.

3. W. J. BAUMOL, *Welfare Economics and the Theory of the State*. Cambridge 1967.

4. K. E. BOULDING, "The Economics of the Coming Spaceship Earth," in H. Jarrett, ed., *Environmental Quality in a Growing Economy*, Baltimore 1966, pp. 3–14.

5. J. H. BOYD, "Collective Facilities in Water Quality Management," Appendix to Kneese and Bower [17].

6. J. W. BUCHANAN AND WM. C. STUBBLE-BINE, "Externality," *Economica*, Nov. 1962, *29*, 371–84.

7. —— AND G. TULLOCK, "Public and Private Interaction under Reciprocal Externality," in J. Margolis, ed., *The Public Economy of Urban Communities*, Baltimore 1965, pp. 52–73.

8. G. CASSEL, *The Theory of Social Economy*. New York 1932.

9. R. H. COASE, "The Problem of Social Cost," *Jour. Law and Econ.*, Oct. 1960, *3*, 1–44.

10. L. COLE, "Can the World be Saved?" Paper presented at the 134th Meeting of the American Association for the Advancement of Science, December 27, 1967.

11. R. K. DAVIS, *The Range of Choice in Water Management*. Baltimore 1968.

12. O. A. DAVIS AND A. WHINSTON, "Externalities, Welfare, and the Theory of Games," *Jour. Pol. Econ.*, June 1962, *70*, 241–62.

13. R. DORFMAN, P. SAMUELSON AND R. M. SOLOW, *Linear Programming and Economic Analysis*. New York 1958.

14. I. FISHER, *Nature of Capital and Income*. New York 1906.

15. A. C. HARBERGER, "Monopoly and Resources Allocation," *Am. Econ. Rev.*, Proc., May 1954, *44*, 77–87.

16. K. W. KAPP, *The Social Costs of Private Enterprise*. Cambridge 1950.

17. A. V. KNEESE AND B. T. BOWER, *Managing Water Quality: Economics, Technology, Institutions*. Baltimore 1968.

18. F. H. KNIGHT, *Risk, Uncertainty, and Profit*. Boston and New York 1921.

19. H. LEIBENSTEIN, "Allocative Efficiency vs. 'X-Efficiency,'" *Am. Econ. Rev.*, June 1966, *56*, 392–415.

20. W. P. LOWRY, "The Climate of Cities," *Sci. Am.*, Aug. 1967, *217*, 15–23.

21. E. J. MISHAN, "Reflections on Recent Developments in the Concept of External Effects," *Canadian Jour. Econ. Pol. Sci.*, Feb. 1965, *31*, 1–34.

22. A. C. PIGOU, *Economics of Welfare*. London 1952.

23. K. SCHLESINGER, "Über die Produktionsgleichungen der ökonomischen Wertlehre," *Ergebnisse eines mathematischen Kolloquiums*, No. 6. Vienna, F. Denticke, 1933.

24. D. SCHWARTZMAN, "The Burden of Monopoly," *Jour. Pol. Econ.*, Dec. 1960, *68*, 627–30.

25. T. SCITOVSKY, "Two Concepts of External Economies," *Jour. Pol. Econ.*, Apr. 1954, *62*, 143–51.

26. F. SMITH, *The Economic Theory of Industrial Waste Production and Disposal*, draft of a doctoral dissertation, Northwestern Univ. 1967.

27. R. TURVEY, "On Divergencies between Social Cost and Private Cost," *Economica*, Nov. 1962, *30*, 309–13.

28. L. WALRAS, *Elements d'economie politique pure*, Jaffé translation. London 1954.

29. H. WOLOZIN, ed., *The Economics of Air Pollution*. New York 1966.

30. CONSERVATION FOUNDATION, *Implications of Rising Carbon Dioxide Content of the Atmosphere*. New York 1963.

31. REGIONAL PLAN ASSOCIATION, *Waste Management*, a Report of the Second Regional Plan. New York 1968.

Bibliography

The articles contained in this volume were selected to illustrate particular concepts or mathematical techniques that are relevant to environmental analysis and management. During the course of the compilation of these articles, we found many other papers that were of interest but that were not included for reasons not necessarily related to the quality of the material. This bibliography is a listing of additional papers, books, and journals that may be useful to the reader. The references are grouped under the same six subject headings used in the book to facilitate an in-depth study of any particular area covered by the text.

Ecosystem Analysis

Articles

Dale, M. B., "Systems Analysis and Ecology," *Ecology,* **51**, No. 1, 2–16 (1970). Article includes extensive bibliography.

Lieth, H., "Mathematical Modelling for Ecosystems Analysis," *Productivity of Forest Ecosystems,* Proc. Brussels Symposium, 1969, UNESCO, pp. 567–575 (1971). Compares precomputer modeling and the new postcomputer modeling. Describes construction of a computer model for study of an ecosystem and mentions both digital and analog computation. Good list of references.

Odum, E. P., "The Strategy of Ecosystem Development," *Science,* **164**, No. 3877, 262–270 (1969).

Books

Ehrlich, P. R., and A. E. Ehrlich, *Population, Resources, Environment,* 2nd ed. W. H. Freeman, San Francisco, 1972.

Mooz, W. E., and C. C. Mow, *California's Electricity Quandary: 1. Estimating Future Demand,* Rand Corp. Report R-1084-NSF/CSRA, Santa Monica, Calif., Sept., 1972. Environmental management invariably requires a projection of future supplies and demands. This paper illustrates methods for estimating future demands for electrical energy in California.

Odum, H. T., *Environment, Power and Society,* Wiley, New York, 1970.

Patten, B. C., ed., *Systems Analysis and Simulation in Ecology,* Academic Press, New York, 1972.

Bibliography

Van Dyne, G. M., *Grasslands Management, Research and Training Viewed in a Systems Concept*, Range Science Dept., Science Series No. 3, Colorado State University, 1969. A good introduction to ecosystem modeling.
Watt, K. E. F., ed., *Systems Analysis in Ecology*, Academic Press, New York, 1966.
Watt, K. E. F., *Ecology and Resource Management: A Quantitative Approach*, McGraw-Hill, New York, 1967.

Journals and Periodicals

Environment. Published monthly, except bimonthly Jan.–Feb. and July–Aug. Articles thorough enough to move one into the "advanced layman" to "near expert" category.
Experimental Periodicals: Indexed Article Titles. Published 8 times per year (plus annual index). Begun in 1972. A bibliography featuring serial publications that reflect current awareness in the area of environmental studies.
Environmental Science and Technology. Published monthly. Includes current events, shorts, business and government activities relating to the environment, feature articles at the advanced-layman level, and research articles at the expert–specialist level.

Management of Ecological Systems

Articles

Kaufman, H. G., "Attaining Environmental Quality: The Role of the Technical Decision-Maker," *J. Environ. Sys.*, **2**, No. 3, 191–206 (1972). Discusses the attitudes of engineers and scientists toward attaining environmental quality and their role in decisions affecting the environment.
Mann, S. H., and R. Hobson, "Toward the Development of a Quality of Life Index," presented at the 41st National Meeting of the Operations Research Society of America, New Orleans, Apr. 26–28, 1972.
Swartzman, G. L., and G. M. Van Dyne, "A Simulation–Optimization Approach to Dynamic Ecosystem Management," presented at the 41st National Meeting of the Operations Research Society of America, New Orleans, Apr. 26–28, 1972

Books

Watt, K. E. F., *Ecology and Resource Management: A Quantitative Approach*, McGraw-Hill, New York, 1967.

Journals and Periodicals

Ecology. Published quarterly. Generally for the mature to expert reader with a background in biology.
Journal of Environmental Sciences. Published bimonthly.
Science. Published weekly. Contains a wide variety of articles that range from introductory to advanced.

Air Quality

Articles

Herzog, H. W., Jr., "The Air Diffusion Model as an Urban Planning Tool," *Socio-Econ. Plan. Sci.*, **3**, 329–349 (1969).

Kleinman, F. E., "The Regional Approach to Air Pollution Control," *J. Air Pollution Control Assn.*, **21**, 71–75 (1971).

Lave, L. B., and E. P. Seskin, "Air Pollution and Human Health," *Science*, **169**, 723–733 (1970). Examines the benefits of air pollution abatement in economic terms.

Martin, D. O., "An Urban Diffusion Model for Estimating Long Term Average Values of Air Quality," *J. Air Pollution Control Assn.*, **21**, No. 1, 16–19 (1971). Describes a diffusion model used with a computer and makes a comparison of the model output with St. Louis data.

Panofsky, H. A., "Air Pollution Meteorology," *Amer. Scientist*, **57**, 269–285 (1969). A good introductory article.

Sklarew, R. C., A. J. Fabrick, J. E. Prager, "Mathematical Modeling of Photochemical Smog Using the PICK Method," *J. Air Pollution Control Assn.*, **22**, No. 11, 865–869 (1972). A spatial grid model for the Los Angeles basin is used to calculate pollutant concentrations.

Slade, D. H., "Modeling Air Pollution in the Washington, D. C., to Boston Megalopolis," *Science*, **157**, 1304–1307 (1967).

Smith, D. R., N. G. Edmisten, D. J. de Roeck, "System for Implementing Comprehensive Air Pollution Control Programs," *J. Air Pollution Control Assn.*, **22**, No. 12, 943–949 (1972). An introductory article that describes a systems approach to air quality management.

Wayne, L., R. Danchick, M. Weisburd, A. Kokin, A. Stein, "Modeling Photochemical Smog on a Computer for Decision Making," *J. Air Pollution Control Assn.*, **21**, No. 6, 334–340 (1971). An introductory article on air quality modeling.

Books

Strauss, W., ed., *Air Pollution Control, Part 1,* Wiley-Interscience, New York, 1971.

Journals and Periodicals

Journal of the Air Pollution Control Association. Published monthly. Articles on air quality analysis and management that range from introductory to restricted.

Socio-Economic Planning Sciences. Published bimonthly. Papers are generally introductory to advanced and frequently use a multi-disciplinary approach.

Water Management

Articles

Bedrosian, A. J., W. O. Bennett, J. E. Berry, R. B. Ditton, J. W. Kolka, T. W. Thompson, "Cooperative Community–University Water Resource Planning: An Inter-Disciplinary Approach," *Water Resources Bull.*, **8**, No. 5, 887–899 (1972). Management of a Wisconsin lake used for recreational purposes.

Espey, W. H., Jr., and G. H. Ward, Jr., "Estuarine Water Quality Models," *Water Res.*, **6**, No. 10, 1117–1131 (1972). A review and appraisal of estuarine modeling.

Foster, E. T., Jr., T. C. Chen, J. P. Newton, E. O. Isu, "Improved River Basin Utilization Through Systems Analysis," *Water Resources Bull.*, **8**, No. 5, 863–870 (1972). Linear programming is used to optimize water resource allocation.

Freeze, R. A., "Subsurface Hydrology at Waste Disposal Sites," *IBM J. Res. Develop.*, **16**, No. 2, 117–129 (1972). Some background discussion. Mathematical modeling (partial differential equations). Several references.

Marske, D. M., and L. B. Polkowski, "Evaluation of Methods for Estimating Biochemical Oxygen Demand Parameters," *J. Water Pollution Control Federation*, **44**, No. 10, 1987–2000 (1972). A specialized paper that considers the development of a model from the available data.

Books

Jenkins, S. H., ed., *Advances in Water Pollution Research*, Vols. I and II, Pergamon Press, Oxford, 1971 (International Association on Water Pollution Research). Volumes include articles that may be of interest to specialists in the treatment of waste water, water storage, and related water problems.

Kneese, A. V., and B. T. Bower, *Managing Water Quality: Economics, Technology, Institutions*, Johns Hopkins Press, Baltimore, 1968.

Thomann, R. V., *Systems Analysis and Water Quality Management*, Environmental Research and Applications, Inc., New York, 1972.

International Symposium on Modelling Techniques in Water Resource Systems, Ottawa, 1972. Held May 9–12, 1972. Proceedings of the symposium are presented in three volumes and contain forty-five papers and discussions. They are generally concerned with ecology, hydrology, economics, and water quality.

Journals and Periodicals

Journal of the Water Pollution Control Federation. Published monthly. Started in 1928. Medium to advanced papers. Primarily for the specialist.

Water Research (The Journal of the International Association of Water Pollution Research). Published monthly. Papers are for the most part of moderate complexity.

Water Resources Bulletin. Published bimonthly. Papers of moderate complexity; broad coverage of the water resources field.

Water Resources Research. Published bimonthly. Papers generally on an advanced level.

Waste Management

Articles

Pavoni, J. L., D. J. Hagerty, R. E. Lee, "Environmental Impact Evaluation of Hazardous Waste Disposal in Land," *Water Resources Bull.*, **8**, No. 6, 1091–1107 (1972). Introductory article that considers the hazards of waste disposal and relates these hazards to site considerations.

Spofford, W. O., Jr., "Closing the Gap in Waste Management," *Environ. Sci. Technol.*, **4**, No. 12, 1108–1114 (1970).

Wenger, R. B., and C. R. Rhyner, "Evaluation of Alternatives for Solid Waste Systems," *J. Environ. Sys.*, **2**, No. 2, 89–108 (1972). Describes procedure for using multidimensional criteria in the selection of a solid waste disposal system.

Journals and Periodicals

Environmental Science and Technology. Published monthly. Includes current events, shorts, business and government activities relating to the environment, feature articles at the advanced-layman level, and research articles at the expert–specialist level.

Journal of Environmental Systems. Published quarterly. Primarily general articles covering a wide range of environmental problems.

Coupled-Systems Analysis

Articles

Ackerman, B., and J. Sawyer, "The Uncertain Search for Environmental Policy: Scientific Fact-Finding and Rational Decision-Making Along the Delaware River," *Univ. Penn. Law Rev.*, **120**, No. 3, 419–503 (1972).

Freeman, A. M., III, and R. H. Haveman, "Residuals Charges for Pollution Control: A Policy Evaluation," *Science*, **177**, 322–329 (1972). Good general background discussion (nonquantitative) of the topic described by the title. Many references.

Leontief, W. W., "Input–Output Economics," *Sci. Amer.*, **185**, No. 4, 15–21 (1951). One of the first papers on input–output analysis.

McKusick, R. B., and J. H. Snyder, "A Regional Approach to Project Evaluation," *Water Resources Bull.*, **8**, No. 3, 431–445 (1972). Economic analysis of water resource management.

Solow, R. M., "The Economist's Approach to Pollution and Its Control," *Science*, **173**, 498–503 (1971). Discussion of externalities, use of effluent charges. Cites Paper 26.

Wilson, R. D., and D. W. Minnotte, "A Cost/Benefit Approach to Air Pollution Control," *J. Air Pollution Control Assn.*, **19**, 303–308 (1969).

Books

Kneese, A. V., and B. T. Bower, eds., *Environmental Quality Analysis,* Johns Hopkins Press, Baltimore, 1972.

Journals and Periodicals

IEEE Transactions on Systems, Man and Cybernetics. Published five times per year. Articles are usually mathematical and analysis oriented.

Journal of Environmental Systems. Published quarterly. Primarily general articles covering a wide range of environmental problems.

Science. Published weekly. Journal publishes a wide variety of articles that range from introductory to advanced.

Socio-Economic Planning Sciences. Published bimonthly. Papers are generally introductory to advanced and frequently use a multidisciplinary approach.

Author Citation Index

Federal Water Pollution Control
    Administration, 218, 231
Felton, P. N., 205
Ferguson, G. E., 205
Fischer, H. B., 296
Fisher, A. C., 357
Fisher, D. W., 114
Fisher, I., 385
Ford, J. H., 218
Ford, L. R., 278
Foster, E. T., Jr., 390
Franks, D. E., 88
Frayer, W. E., 101
Fredrickson, H. G., 357
Freeman, A. M., III, 391
Freeze, R. A., 51, 390
Freidlander, S. K., 182
Friend, A. G., 218
Frye, C. H., 278
Fulkerson, D. R., 278

Gardner, W. R., 51
Gardner, W. S., 218
Gass, S. I., 278
Gelpi, M. J., 270
Gerhardt, P. H., 357
Gessel, S. P., 61
Geyer, J. C., 218
Geyer, J. G., 219
Gibbens, R. P., 87
Gifford, F. A., Jr., 164
Ginsburg, A. L., 357
Goebel, J. B., 218
Goetz, C., 357
Goldman, M., 357
Golley, F. B., 61
Gordon, D. T., 88
Gough, W. C., 249
Gould, R. H., 218
Grad, F. B., 357
Gramm, W. P., 357
Green, J. A., 278
Green, L. R., 87, 88
Green, R. E., 51
Greenland, D. J., 61
Grissinger, E. H., 218
Grosch, D. S., 68
Guy, H. P., 205

Hadley, G., 114
Hagerty, D. J., 390
Hagevik, G., 357
Hags, J. C., 357
Haig, I. T., 88

Hall, A. D., 35, 235
Halpern, P. K., 164
Hannum, J. R., 218
Harberger, A. C., 385
Harmeson, R. H., 218
Harris, E. E., 205
Harrison, H. L., 68
Hartesveldt, R., 88
Harvey, H. T., 88
Haushild, W. L., 218
Haveman, R. H., 114, 391
Hawthorne, J. C., 218
Heitkamp, D., 62
Heller, A. N., 155
Helms, B. P., 278
Hembree, C. H., 218
Henderson, C. R., 218
Henning, J., 358
Herzog, H. W., Jr., 182, 389
Heyward, F., 61
Hickey, J. J., 68
Hill, N. J., 51
Hobson, R., 388
Hodgson, A., 87
Hole, F. D., 52
Holzworth, G. C., 155
Horton, R. E., 205
Houston, D. B., 88
Howard, C. S., 219
Howells, G. P., 218
Hu, T. C., 278
Huang, W., 124
Huff, D. D., 68
Huizinga, J., 277
Hunter, W. G., 51
Hynes, H. B. N., 218

Ingram, W., 164
Iorns, W. V., 218
Isaacson, P. A., 68
Isu, E. O., 390
Iverson, K. E., 278

Jaffe, L. L., 328
James, L. D., 205
Jameson, D. A., 100
Jaske, R. T., 218
Jefferys, H., 18
Jenkins, S. H., 390
Jennrich, R. I., 68
Jenny, H., 35, 61
Jensen, L. D., 218
Johnson, W. B., 164

Subject Index

ENVIRONMENTAL MODELING:
Analysis and Management

**Edited by DOUGLAS DAETZ
and RICHARD HARRIS PANTELL
Stanford University**

In recent years, man has become increasingly aware that he can no longer afford to regard air, water, land, and ecosystems as inexhaustible resources. The past decade has witnessed the burgeoning of entire new fields of scientific endeavor focused on environmental planning. Through original writings and authoritative introductory materials, this timely volume illustrates the various systems techniques which have been applied to the identification and management of environmental problems. For the first time, a single book encompasses the economic and social, as well as the physical, factors involved in diverse areas of concern. Works by a distinguished array of authors—including M. K. Hubbert, H. E. Koenig, L. B. Leopold, G. M. Van Dyne, W. Leontief, J. S. Olson, and M. G. Wolman—represent the very foundations and most stimulating highlights of environmental systems analysis.

Six topically organized sections cover ecosystem analysis, ecosystem management, air quality, water management, waste management, and interdisciplinary issues. The general